T0378251

Lecture Notes in Mechanical Engineering

Series Editors

Francisco Cavas-Martínez, Departamento de Estructuras, Universidad Politécnica de Cartagena, Cartagena, Murcia, Spain

Fakher Chaari, National School of Engineers, University of Sfax, Sfax, Tunisia

Francesco Gherardini, Dipartimento di Ingegneria, Università di Modena e Reggio Emilia, Modena, Italy

Mohamed Haddar, National School of Engineers of Sfax (ENIS), Sfax, Tunisia

Vitalii Ivanov, Department of Manufacturing Engineering Machine and Tools, Sumy State University, Sumy, Ukraine

Young W. Kwon, Department of Manufacturing Engineering and Aerospace Engineering, Graduate School of Engineering and Applied Science, Monterey, CA, USA

Justyna Trojanowska, Poznan University of Technology, Poznan, Poland

Lecture Notes in Mechanical Engineering (LNME) publishes the latest developments in Mechanical Engineering—quickly, informally and with high quality. Original research reported in proceedings and post-proceedings represents the core of LNME. Volumes published in LNME embrace all aspects, subfields and new challenges of mechanical engineering. Topics in the series include:

- Engineering Design
- Machinery and Machine Elements
- Mechanical Structures and Stress Analysis
- Automotive Engineering
- Engine Technology
- Aerospace Technology and Astronautics
- Nanotechnology and Microengineering
- Control, Robotics, Mechatronics
- MEMS
- Theoretical and Applied Mechanics
- Dynamical Systems, Control
- Fluid Mechanics
- Engineering Thermodynamics, Heat and Mass Transfer
- Manufacturing
- Precision Engineering, Instrumentation, Measurement
- Materials Engineering
- Tribology and Surface Technology

To submit a proposal or request further information, please contact the Springer Editor of your location:

China: Dr. Mengchu Huang at mengchu.huang@springer.com
India: Priya Vyas at priya.vyas@springer.com
Rest of Asia, Australia, New Zealand: Swati Meherishi at swati.meherishi@springer.com
All other countries: Dr. Leontina Di Cecco at Leontina.dicecco@springer.com

To submit a proposal for a monograph, please check our Springer Tracts in Mechanical Engineering at http://www.springer.com/series/11693 or contact Leontina.dicecco@springer.com

Indexed by SCOPUS. All books published in the series are submitted for consideration in Web of Science.

More information about this series at http://www.springer.com/series/11236

Ranganath M. Singari · Kaliyan Mathiyazhagan · Harish Kumar
Editors

Advances in Manufacturing and Industrial Engineering

Select Proceedings of ICAPIE 2019

Volume 2

Springer

Editors
Ranganath M. Singari
Department of Mechanical Engineering
Delhi Technological University
New Delhi, India

Kaliyan Mathiyazhagan
Department of Mechanical Engineering
Amity School of Engineering and Technology
Noida, India

Harish Kumar
Department of Mechanical Engineering
National Institute of Technology Delhi
New Delhi, India

ISSN 2195-4356 ISSN 2195-4364 (electronic)
Lecture Notes in Mechanical Engineering
ISBN 978-981-15-8541-8 ISBN 978-981-15-8542-5 (eBook)
https://doi.org/10.1007/978-981-15-8542-5

© The Editor(s) (if applicable) and The Author(s), under exclusive license to Springer Nature Singapore Pte Ltd. 2021, corrected publication 2021
This work is subject to copyright. All rights are solely and exclusively licensed by the Publisher, whether the whole or part of the material is concerned, specifically the rights of translation, reprinting, reuse of illustrations, recitation, broadcasting, reproduction on microfilms or in any other physical way, and transmission or information storage and retrieval, electronic adaptation, computer software, or by similar or dissimilar methodology now known or hereafter developed.
The use of general descriptive names, registered names, trademarks, service marks, etc. in this publication does not imply, even in the absence of a specific statement, that such names are exempt from the relevant protective laws and regulations and therefore free for general use.
The publisher, the authors and the editors are safe to assume that the advice and information in this book are believed to be true and accurate at the date of publication. Neither the publisher nor the authors or the editors give a warranty, expressed or implied, with respect to the material contained herein or for any errors or omissions that may have been made. The publisher remains neutral with regard to jurisdictional claims in published maps and institutional affiliations.

This Springer imprint is published by the registered company Springer Nature Singapore Pte Ltd.
The registered company address is: 152 Beach Road, #21-01/04 Gateway East, Singapore 189721, Singapore

Contents

About the Editors

Dr. Ranganath M. Singari is a Professor in the Department of Mechanical, Production & Industrial Engineering and heads the Department of Design, Delhi Technological University, India. He is a graduate in Industrial Production Engineering from Karnataka University. He completed his M.Tech in Computer Technology & Applications and Ph.D. from the Department of Production Engineering from University of Delhi, India. He has more than 60 international publications in conference and reputed journals. He is also a reviewer for reputed journals. Dr. Singari has organised several international conferences, seminars/ workshops, industry-institute interactions and 6 FDP/SDP/STTP. He also serves as Chairman, Production Engineering, Skill India Programme, DTTE, Delhi. He is an expert member of several selection committees for technical, teaching and administrative positions. His research interest is materials, manufacturing, industrial management, production management, CAD/CAM, supply chain management, multi-criteria decision making and sustainable lean manufacturing. He has 25 years of research and teaching experience.

Dr. Kaliyan Mathiyazhagan is currently working as an Associate Professor in the Department of Mechanical Engineering, Amity University, India. He pursued his Ph.D. from the Department of Production Engineering, National Institute of Technology, Tiruchirappalli, Tamil Nadu. He was also a visiting research fellow at the University of Southern Denmark. He has more than 60 international publications in reputed journals and one of his papers received the best paper award in NCAME 2019, NIT Delhi. Dr. K. Mathiyazhagan is an associate editor of Environment, Development and Sustainability. He is also an editorial member of more than 5 international journals. He has served as a guest editor for several special issues in international journals and is an active reviewer of more than thirty reputed international journals. His research interest is green supply chain management, sustainable supply chain management, multi-criteria decision making, third party logistic provider, sustainable lean manufacturing, public distribution system, and Lean Six Sigma. He has more than 10 years of research and teaching experience.

Dr. Harish Kumar is currently working as an Assistant Professor at the National Institute of Technology, Delhi. He has more than 15 years of research and academic experience and has served as a scientist at different grade in CSIR - National Physical Laboratory, India (NPLI). He has been an active researcher in the area of mechanical measurement and metrology. He has worked as a guest researcher at the National Institute of Standards and Technology, USA in 2016. He has been instrumental in the ongoing redefinition of the kilogram in India. He has authored more than 70 publications in peer reviewed journals and conferences. He is an active reviewer of many reputed journals related to measurement, metrology and related areas. He has served as a guest editor of different peer reviewed journals.

Blockchain Technology as a Tool to Manage Digital Identity: A Conceptual Study

Ruchika Singh Malyan and Ashok Kumar Madan

Abstract In recent times, one of the most noticeable developments is of digital identity. Despite its origin in finance, several applications have found digital identification helpful in different ways. Digital identity has been growing quickly. In the near future, individuals around the world shall be concerned with managing their digital identities. This study shows how blockchain technology plays a significant role in digital identity management systems. The shortcomings and weaknesses of different literature are reviewed, and the benefits that blockchain provides for achieving an effective identity management system. This paper will discuss the importance of data privacy in various applications and identify the scope of blockchain technology in digital identity management. Blockchains eliminate the role of middlemen along with reducing the wait lines required for authentication, attestation, and authorization. People are allowed to choose whether to reveal or not to reveal.

Keywords Identity management · Blockchain · Smart contracts · Bitcoin

1 Introduction

Blockchain has received extensive attentions recently. It works as an immutable ledger which allows transactions to happen in a decentralized way. Applications based on blockchain are pouncing up, which cover many fields like financial services, reputation system, and Internet of Things (IoT), etc. Though, there are many challenges like scalability and security-related problems that are yet to be resolved. In this digital age, the consumers are increasingly embracing the Internet to perform online transactions across multiple platforms. Therefore, the concept of digital marketing comes into picture.

R. S. Malyan (✉)
Directorate of Training and Technical Education, Meera Bai Institute of Technology, New Delhi 110065, India
e-mail: rsmalyan04@gmail.com

A. K. Madan
Department of Mechanical Engineering, Delhi Technological University, New Delhi 110042, India

© The Author(s), under exclusive license to Springer Nature Singapore Pte Ltd. 2021

R. M. Singari et al. (eds.), *Advances in Manufacturing and Industrial Engineering*,
Lecture Notes in Mechanical Engineering,
https://doi.org/10.1007/978-981-15-8542-5_55

This came into picture because conventional identity has got several issues. For example, a paper-based birth certificate sitting inactively in a dusty cellar could be lost or stolen. A digital identity decreases the degree of service and increases the pace of procedures within associations by taking into account more noticeable interoperability between departments and various organizations. Moreover, this transforms into a honeypot for hackers if this computerized personality is put away on an integrated database. Since 2017 alone, over 600 million individual subtleties have been hacked or split from connections—such as positions or mastercard numbers. The present identity of a vast majority of the board frameworks is weak and redundant. There is a confirmation process at every stage we need to show something about our character—whether our name, email, or card number. A testing material claims to be true or false the data we say about ourselves. This is usually done by testing our distinctive records. Such proof of identity and authentication methods gives rise to security concerns. A zero-knowledge proof is a verification technique which, using cryptography, allows one substance to prove to another component that it knows a specific data or fulfills a particular requirement without disclosing any actual data that supports that evidence. In this case, the material that governs the verification has "zero-knowledge" about the data that supports the proof is still "persuaded" of its validity. This is particularly valuable when and where the prover component does not believe the testing substance but at the same time needs to show them that it knows a specific data. It allows an individual to demonstrate an identity to the executives in a blockchain situation that their subtleties fulfill those preconditions without uncovering the real subtleties.

Digital identity management systems currently face many risks. These are honeypots for hackers who concentrate on knowledge that is closer to home. Also, there are code violations and releases that put more than 600 million personal data at risk. Nevertheless, with current developments, the technology of blockchain is transforming the board's world of advanced personality. Innovation in the blockchain is another philosophy that strengthens the security of data and counteracts retail fraud. Blockchain technology is effective and commonly used cross-sectionally across different businesses. This guarantees that no single database is compromised and that records do not fall into unauthorized hands. This is why multiple companies that promote the use of blockchain technology require data well-being, bureaucratic reputation, and commitment among customers.

It is essential to ensure that the information is not compromised when working with an online database account. Passwords and IT security, however, are very sensitive and can escape even from the hands of the most confident employees. Non-overseer logins for blockchain technology are not a focal company material. It has a single qualified caretaker and substitutes for an account such as private and free chains. The Internet provides simple and direct connections across the globe between different people. This is achieved by direct messages, the exchange of documents and other phases of process administration. Today, efficient interchanges can be enjoyed, and the price of going out miles to complete those errands can be minimized. Therefore, if you shop or send relevant messages, substantial information is shared online. Blockchain technology ensures the security of interconnected devices updated later.

This ensures that the data made by the Internet of Things (IoT) is increasingly checked or firmly bolted, unregulated, and unapproved. Blockchain technology takes the character of the software to a higher level. It is distributed to provide a cheaper, better quality, and safer computerized customer base.

1.1 Core Components of Blockchain Technology

The blockchain contains the following components:

1. Asymmetric cryptography: It creates secured digital wallet and generates private key that generates digital signatures.
2. Transactions: A group of transactions form a block. They represent the current state of blockchain because conservation of value is a must.
3. Consensus mechanism: There is a common agreement protocol which validates blocks. It helps update ledger, and there are incentives for block miners.
4. Secured distributed ledger, i.e., the blockchain.

1.2 Blockchain Classification

There are four types of blockchains which are as follows:

1. Based on access to blockchain: One is permission—less, where, anyone can join, and the other is permissioned, where, only approved users can join.
2. Based on access to blockchain data: One is public, where, all who access can modify, and the other is private, where, only specific users can write or modify.

Private and permissions are used reciprocally and so are public and permissionless. Depending on the use case, one needs to select an appropriate architecture from above. There are different blockchain-based system configurations against multiple parameters such as performance, cost efficiency, and flexibility. Different dimensions of a blockchain system such as blockchain configuration, storage, computation, a degree of decentralization are considered in coming up with the classification.

Public/Permissionless blockchains: Public blockchains are open for all. Anyone can join them. The participation in mining and consensus process of adding a new block of transaction is open to all. These blockchains usually use proof of work or proof of stake for consensus mechanism. The attack can be avoided by having more number of participants working well for this model. As per Buterin [1] public blockchains enhance the trust factor and also protect the applications from the developers. There is usually sufficient incentive and significant saving when compared to third party dependent systems, for example, there are minimal transaction costs to opt for public blockchain to record transactions.

Private/Permissioned blockchain: As per Buterin [1], permission blockchains are usually built usually by organizations for their specific business need. Blockchains are likely to have interfaces with all the current applications of the organization. Organizations may opt for consortium blockchains. In this case, only few trusted members sign off a transaction. In fully private blockchains, the right permission over the blockchain is given to a central organization. And hence, people prefer private blockchains as it offers more flexibility by increasing control over the rules of transaction, which can be altered by a majority consensus. This becomes easier in a private or consortium blockchain than a public one. There is also increased accountability as all the nodes are named.

1.3 Research Objectives

1. To study the significance of data privacy in several applications.
2. To discuss the scope of blockchain technology in digital identity management.

2 Literature Review

Review of the literature includes the following domains:

2.1 Blockchain Technology

An automated character is relevant in a wide range of circumstances involving individuals and elements to show what their identity is: there is no "one-size-fits-all" all-inclusive solution. This is reflected in the decent variety of methodologies used in various up-to-date frameworks for identification (ID) and access. Continuous discussion and coordinated inter-partner interaction through groups, companies, and local constant dialog and organized inter-partner collaboration across teams, businesses, and locations will cultivate a shared understanding of problems and systems and drive growth worldwide. Although a few steps have been taken to strengthen the business character, by joining the blockchain, space has been upset. As a result, fostering a shared understanding of challenges and frameworks is presented. The ability to collect, verify and negotiate online exchanges, and access services with customer characters is an increasingly important segment of the advanced economy. Current methodologies for providing employee information and data to organizations or computerized stages contribute to safety gaps due to breaches or incorrect executive details. Blockchain makes it easier for people and businesses to manage their secret data.

Moreover, processes along the life cycle of the character collection, verification, and board can be simplified through both more reliable authentication and more transparent information sharing. Before they can continue to access services, many online exchanges require people to share specific individual data. For example, before financial transactions can take place at stages such as Amazon Pay, PayPal, and Google Wallet, consumers are always required to enter their subtleties—finance-related subtleties or other individual details.

Consequently, each time a person discovers this information; it is put on different web servers. They are settling into being advanced copies of one and a related entity through these different stages. It shows many security problems as well. Blockchain can be used to construct a stage that protects people's identities from theft and diminishes fake exercises significantly. The technology can also encourage organizations to produce solid blockchains that address the issues of verification and compromise in a few projects.

Aggarwal et al. [2] monitor the use of blockchain technology for smart networks, focusing on the critical segments of blockchain applications. The researchers are also researching the different procedures and models used to perform secure exchanges. In particular, they present a point-by-point scientific description of the applications, the system models used, and the development of the communication structure needed for different applications.

A distributed method is provided for blockchain technology where the development of the blockchain is described along with integrating into the conveyed frameworks of physical network framework. A four-layered architecture is developed by analyzing the business theoretically and developing the physical Internet and blockchain. This is incorporated into case form of software application, helping the researchers as well as other recruitment professionals to carry out further investigation.

Hebert and De Carbo [3] propose their blockchain advance process for protected appropriation in the big business improvement program. This depends on the likelihood of demonstrating the risk of achieving safety through a planned and sound approach to program security. The methodology guides programming engineers in recognizing whether or not to allow theft of blockchain and on the off chance of considering the different varieties. By integrating an extension to an outstanding and innovative design survey system, the Microsoft-created hazard displays method STRIDE, into our proposal; the authors reflect on the safety implications of blockchain fraud. The authors do as such by proposing a list of the blockchain's specific security risks. They point along these lines to improve the overall suitability of our strategy, talking about the broad, wide scope of STRIDE.

Singh et al. [4] study the IoT blockchain and machine learning union with technology to support the development of a new distributed, autonomous, and permanent IoT network. This paper shows a significant degree of empirical categorization of IoT blockchain and artificial intelligence (AI) with the latest best-in-class methods and implementations. The authors propose BlockIoTIntelligence, a blockchain-enabled intelligent IoT architecture with AI that combines four-level cloud experience, mist awareness, edge information and knowledge of gadgets with blockchain, and machine learning reception at all levels to achieve the goal of adaptable and stable IoT.

They also test the model proposed for standard measures such as reliability, centralization, safety and protection, and dormancy. The proposed design is further analyzed in terms of standard metrics, such as precision.

Lu et al. [5] present a blockchain hybrid as a support stage (uBaaS) to help build and manage blockchain-based applications. The service of uBaaS combines organization as aid, system development as aid, and management support. It is doubtful to send as support in uBaaS, which can keep away from lock-in to clear cloud levels, while the development of applications as assistance applies board configuration designs for data and brilliant agreement configuration to fix the adaptability and security problems of blockchain.

Cheng et al. [6] propose a new polynomial-based blockchain architecture. Fragments of information are sorted by a Lagrange interjection method in each block. To maintain block power, polynomial capabilities are used. The polynomial-based blockchain architecture accomplishes the conversion point as well as the differential control backup technique for adjustment.

Tönnissen and Teuteberg [7] analyze existing conditions with actual cases of use and find answers to our research questions through and based on a specific context study. The findings of the paper integrate bits of insight into the collaborative company's impact of blockchain technology and blockchain innovation and the effect of blockchain innovation on action plans.

Gordon and Catalini [8] identify the interoperability of data well-being and the transition from organizational-driven interoperability to an understanding of interoperability. The authors are buoyant on potential ways that blockchain can promote this advancement and achieve interoperability.

Maesa et al. [9] propose the exploitation of blockchain technology to identify access control systems to ensure that access control techniques are appropriate. The proposal identifies trait-based access control strategies as smart agreements and expresses them on a blockchain, turning the arrangement evaluation cycle into a fully distributed and thorough contract execution. The authors present a reference usage misuse of XACML strategies and solidity composed of intense agreements expressed on the Ethereum's blockchain to illustrate the value of our method.

2.2 *Fundamentals of Digital Identity Management*

Sherif [10] explains that blockchain can be used to track ownership and source of goods. It can also be used to store the identity of people. Imagine that your passport is stored on a blockchain, and the visas you get and your entry and departure from countries are recorded as blockchain transactions. This means that they are fixed, society verified, and decentralized. By adding smart contracts to the system, it may also be possible to encode rules for denying entry to certain people (sanctions against countries of origin, security reasons or any other reason) and have them automatically implemented on the blockchain. The rules would be visible to all and automated which would reduce the possibility of human error entering into the process.

Lin et al. [11] propose a blockchain-based identity management system by developing authentication cryptographic membership scheme. Such a system aims to link the digital identity object to its actual object. The authors individually present another transitively closed undirected graph authentication (TCUGA) that requires node signatures. The project's trapdoor hash function helps the signer to update the statements skillfully without having to re-sign the nodes. The proposal will verify skillfully when there is no edge between two vertices along these lines consisting of the existing intransitive signature on the issue of immovability (the essential tool for checking information on the chart). The authors ultimately explain the safety of their proposed TCUGA in the standard model and test its presentation to illustrate its BIMS feasibility. They affect the inward relationship between open names of vertices to eliminate edge signatures. Modified signature schemes are provided so that for vertices refreshed in the equality group, the signer needs to recalculate the randomness using trapdoor hash functions. In the system of the author, if they are not in a different proportionality class, an administrator can, in any case, legitimately respond to their relationship with the endorsements of the vertices being challenged.

Zhang and Chen [12] propose blockchain (DSSCB)-based data security sharing and storage network used to ensure unwavering accuracy and uprightness when transmitting data to a node, this digital signature, which relies on the principle of bilinear matching of elliptical curves. The consortium of blockchain technology provides a distributed, stable, and reliable database that is managed throughout the processing center. In DSSCB, smart contracts are used to restrict the activation of pre-selected nodes when data coins are transmitted and distributed to vehicles involved in the com. Lee [13] focused on using blockchain technology to provide managers with an additional ID as an advanced personality service (IDaaS). The proposed blockchain-based ID as a service (BIDaaS) is explained with a practical template showing how the proposed BIDaaS fills a character and verification board structure of a scalable data distribution organization. A blockchain-based service ID (BIDaaS) is introduced to a scalable media distribution company with one portable customer model. The proposal contains three elements: user (e.g., phone server), BIDaaS supplier (e.g., media distribution organization), and account.

Lin et al. [14] build another direct homomorphic classification mark conspired by the shortcomings in the use of clear statements of vitality. Under the arbitrary prophet model, on adaptively selected message and ID attack, the plan is proven safe against existential fraud. The ID-based straight homomorphic signature plans can be applied in E-business and distributed computing. In the irregular prophet model, the new program on adaptively selected message and ID attack is proven secure against existential falsification, and it can combine cryptosystem natures based on directly homomorphic mark and character.

2.3 *Overview of Data Privacy Algorithms Suggested by Other Researchers*

Tang et al. [15] build another EHR system that can help manage the integrated cloud-based EHR problem. Their response is to use blockchain technology advancement for EHRs (indicated as EHRs based blockchain accommodation). In the consortium blockchain environment, the application template for blockchain-based EHRs is officially defined. Furthermore, the topic of verification is critical for EHRs. The idea is to conspire a personality-based product with numerous specialists who can counter N − 1 agreement attack by authorities. They divide EHR clients into three groups based on blockchain. The main level, which means level 0, is the EHR's database. For example, the subsequent tier, as indicated by Tier 1, includes hospitals, restorative insurance agencies, foundations for logical study, pharmacy organizations, and so on. The third level, meaning level 2 connected to level 1 client members includes experts, researchers, patients, security operators, and so on. Shi et al. [16] propose a powerful cooperative node game strategy to predict retaliatory actions of near predominance. The status of the mysterious node is assessed by describing the network's organizational reputation trust; the high-trust reference document is used in the general report and node combining to resolve the heaviness of the malignant node and finally reaching the Bayesian harmony. This work incorporates the IoT and the blockchain to propose a secure data storage system for the field of the IoT device to enable the sharing of operations and resources.

Macrinici et al. [17] seek to contribute to the training mix of informed contracts within the technology of blockchain. In the light of a careful mapping analysis, the authors are offering an expansive view of their issues and different arrangements, presenting the slants of exploration within the region, and assembling the distinguished 64 papers gathered from top sources of output, networks, strategies, and methodologies. The mapping analysis empowers us to identify issues relevant to blockchain-based keen agreements and to identify potential answers to locate the problems recognized.

3 Comparison Analysis

This section presents the research gaps that need to be focused in this field from the review of data privacy algorithms. Additionally, the research gaps presented in applications such as identity management systems, e-health records, vehicular ad hoc networks, smart contracts, bank transactions, digital rights management, enterprise operation systems, industrial services, agri-food value chain management, AI, and data access control systems will be presented (Table 1).

Table 1 Author's own compilation

Reference number	Algorithm	Applications	Research gaps
[1]	Cryptographic membership authentication scheme	Blockchain-based applications	Signing with private keys seems to be impractical for the signer
[18]	Blockchain-based electronic health records paradigm on consortium of the blockchain	E-health records in hospitals	It is compulsory to check the authenticity of some doctors' records
[19]	Blockchain-based ID as a service (BIDaaS)	Mobile telecommunication industries	Protection must be given to the partner's user data For mobile users, TEE operation needs to be developed
[13]	A new ID-based linear homomorphic signature scheme	E-business and cloud computing	In the event of returning an incorrect algorithm result, it cannot be detected by the receiver
[20]	A digital evidence framework using blockchain (Block-DEF) with a loose coupling structure	Blockchain applications	Except for Block-DEF, no one knows the relationship between a participant's random public key and traceable public key
[21]	Blockchain technology enabling the physical Internet	Blockchain technology in networking	Except for Block-DEF, no one knows the relationship between a participant's random public key and traceable public key
[22]	Blockchain technology	Agri-food value chain management	Storage capacity is a significant problem involving high cost and different consensus algorithm

(continued)

Table 1 (continued)

Reference number	Algorithm	Applications	Research gaps
[23]	Boundary conditions for traceability using blockchain technology	Food chain management	Organizational improvements are necessary before BCT is used effectively in chain management
[2]	Blockchain-enabled intelligent IoT architecture with AI	IoT applications in companies	Problems such as improving scalability, resource management, lack of norm, reliability of data flow, heterogeneity, power constraints, expense, energy efficiency, and traffic monitoring
[24]	Polynomial-based modifiable blockchain structure	Bank transactions	It is necessary to improve the modification complexity along with fixing the probability of finding a false coordinate in a modifiable blockchain framework

4 Conclusion

Blockchain has shown its potential for transforming traditional industry with its key characteristics: decentralization, persistency, anonymity, and auditability. Cryptographic involvement revealed the signer's difficulty in signing using private keys. Authentication is one of the most significant issues for EHRs provided that we need to guarantee the validity of therapeutic documents in the blockchain consortium. In future research, during information sharing, the safety and ongoing verification and message test procedures should be strengthened, and the effectiveness of our proposed arrangement further enhanced. Improvements to the safety and implementation of signed contracts are required to manage down-to-earth and oriented distributed applications. A number of different security methodologies have been accommodated, defining the agreement's brilliant vulnerabilities and evaluating embedded devices and their use. In addition, different methods are familiar with improving the presentation of the agreement's keen implementation and running the brilliant contract exchanges at the same time. Execution investigation systems are also included to measure the output of various blockchain levels. The organization's work in the future to support Ethereum-based coin for digital rights and trade that supports the latest motivating vision: The right material serves the right customers in the right way for the right value. To further improve the reasonableness of the proposed system, the emphasis will be on high simultaneous preparation and security protection of blockchain information.

Future work includes additional blockchain stages as alternatives to the agreement stage and self-sovereign lifestyle planning as an aid in DBaaS. Checking the credibility of the e-well-being information of some doctors is important. Customer information should be provided by the accomplice in BIDaaS. TEE operation must be generated for the versatile client. Nevertheless, Block-DEF does not know the relationship between an unusual open key and the discernible free key of a member. In blockchain-enabled intelligent IoT architecture with artificial intelligence, issues such as increased usability, the board of resources, lack of standard, data flow uprightness can be found. Block chain can also be used in big data analysis. It could be well combined with big data in two types of data management and data analytics. In data management, blockchain can be used to store important data as it is distributed and secure. It could also ensure that the data is original.

References

1. Buterin V (2015) On public and private blockchains. (Online) https://blog.ethereum.org/2015/08/07/on-public-and-private-blockchains/
2. Aggarwal S, Chaudhary R, Aujla GS, Kumar N, Choo KKR, Zomaya AY (2019) Blockchain for smart communities: applications, challenges and opportunities. J Netw Comput Appl
3. Hebert C, Di Cerbo F (2019) Secure blockchain in the enterprise: a methodology. Pervasive Mob Comput 59:101038

4. Singh SK, Rathore S, Park JH (2019) BlockIoTIntelligence: a blockchain-enabled Intelligent IoT architecture with Artificial Intelligence. Futur Gener Comput Syst 110:721–743
5. Lu Q, Xu X, Liu Y, Weber I, Zhu L, Zhang W (2019) uBaaS: a unified blockchain as a service platform. Futur Gener Comput Syst 101:564–575
6. Cheng L, Liu J, Su C, Liang K, Xu G, Wang W (2019) Polynomial-based modifiable blockchain structure for removing fraud transactions. Futur Gener Comput Syst 99:154–163
7. Tönnissen S, Teuteberg F (2019) Analysing the impact of blockchain-technology for operations and supply chain management: an explanatory model drawn from multiple case studies. Int J Inf Manage 52:101953
8. Gordon WJ, Catalini C (2018) Blockchain technology for healthcare: facilitating the transition to patient-driven interoperability. Comput Struct Biotechnol J 16:224–230
9. Maesa DDF, Mori P, Ricci L (2019) A blockchain-based approach for the definition of auditable Access Control systems. Comput Secur 84:93–119
10. Sherif F (2018) Blockchain and its uses. Available: www.sheriffadelfahmy.org/wp-ontent/uploads/2018/01/doc-1.pdf
11. Lin Q, Yan H, Huang Z, Chen W, Shen J, Tang Y (2018) An ID-based linearly homomorphic signature scheme and its application in blockchain. IEEE Access 6:20632–20640
12. Zhang X, Chen X (2019) Data security sharing and storage based on a consortium blockchain in a vehicular ad hoc network. IEEE Access 7:58241–58254
13. Lee JH (2017) BIDaaS: blockchain-based ID as a service. IEEE Access 6:2274–2278
14. Lin C, He D, Huang X, Khan MK, Choo KKR (2018) A new transitively closed undirected graph authentication scheme for blockchain-based identity management systems. IEEE Access 6:28203–28212
15. Tang F, Ma S, Xiang Y, Lin C (2019) An efficient authentication scheme for blockchain-based electronic health records. IEEE Access 7:41678–41689
16. Si H, Sun C, Li Y, Qiao H, Shi L (2019) IoT information sharing security mechanism based on blockchain technology. Futur Gener Comput Syst 101:1028–1040
17. Macrinici D, Cartofeanu C, Gao S (2018) Smart contract applications within blockchain technology: a systematic mapping study. Telemat Inform 35:2337–2354
18. Wang H, Chen K, Xu D (2016) A maturity model for blockchain adoption. Financ Innov. https://doi.org/10.1186/s40854-016-0031-z
19. Supriya T, Vrushali K (2017) Blockchain and its applications—a detailed survey. Available https://www.ijcaonline.org/archives/volume180/number3/aras-2017-ijca-915994.pdf
20. Liu Y, Liu X, Tang C, Wang J, Zhang L (2018) Unlinkable coin mixing scheme for transaction privacy enhancement of bitcoin. IEEE Access 6:23261–23270
21. Ma Z, Jiang M, Gao H, Wang Z (2018) Blockchain for digital rights management. Futur Gen Comp Sys 89:746–764
22. Rouhani S, Deters R (2019) Security, performance, and applications of smart contracts: a systematic survey. IEEE Access 7:50759–50779
23. Tian Z, Li M, Qiu M, Sun Y, Su S (2019) Block-DEF: a secure digital evidence framework using blockchain. Inf Sci 491:151–165
24. Biswas B, Gupta R (2019) Analysis of barriers to implement blockchain in industry and service sectors. Comput Indu Eng 136:225–241
25. Iuon-Chang L, Tzu-Chun L (2017) A survey of blockchain security issues and challenges. Available https://ijns.jalaxy.com.tw/contents/ijns-v19-n5/ijns-2017-v19-n5-p653-659.pdf
26. Meyer T, Kuhn M, Hartmann E (2019) Blockchain technology enabling the Physical Internet: a synergetic application framework. Comput Ind Eng 136:5–17
27. Zhao G, Liu S, Lopez C, Lu H, Elgueta S, Chen H, Boshkoska BM (2019) Blockchain technology in agri-food value chain management: a synthesis of applications, challenges and future research directions. Comput Ind 109:83–99

28. Behnke K, Janssen MFWHA (2019) Boundary conditions for traceability in food supply chains using blockchain technology. Int J Inf Manage 52:101969
29. Angelis J, da Silva ER (2019) Blockchain adoption: a value driver perspective. Bus Horiz 62(3):307–314

Evolution in Micro-friction Stir Welding

Nadeem Fayaz Lone, Md Masroor Alam, Arbaz Ashraf, Amanullah Mahmood, Nabeel Ali, Dhruv Bajaj, and Soumyashri Basu

Abstract Friction stir welding (FSW) has been a revolutionary development in the joining technology, as it has made possible the welding of metals, similar as well as dissimilar, that could not be joined by conventional methods. Friction stir welding also has the potential to join composites, plastics, and even, the welding of wood has been reported by researchers. Micro-friction stir welding is the downscale version of friction stir welding in which the joining of ultra-thin sheets, usually with thickness less than 1 mm. This downscaling has led to its application in electrical, electronics, and micromechanical assemblies. Although μ-friction stir welding is a recent application, significant work has been reported in this field. This paper aims to highlight the development, challenges, and other related considerations that have occurred due to the downscaling of this process.

Keywords Friction stir welding · Micro-friction stir welding · Mechanical properties · Thin sheets

1 Introduction

Since the development of friction stir welding (FSW) by Wayne Thomas and his team in 1991 at The Welding Institute (TWI), it has become the most efficient method of joining aluminum alloys. FSW is a solid-state welding process, uses a non-consumable tool having a pin bulging out of the shoulder and is rotated at a very high speed. The work surface of the material is constrained by the shoulder of the tool, while the pin is plunged in the joining line, and dwell time is given for material softening before traversing. During the traverse movement of the tool, the material on

N. F. Lone (✉) · M. M. Alam · A. Ashraf · A. Mahmood · N. Ali · D. Bajaj
Department of Mechanical Engineering, Jamia Millia Islamia (A Central University), New Delhi 110025, India
e-mail: nadeemfayaz06@gmail.com

S. Basu
Department of Mechanical Engineering, Netaji Subhas University of Technology, New Delhi, India

© The Author(s), under exclusive license to Springer Nature Singapore Pte Ltd. 2021
R. M. Singari et al. (eds.), *Advances in Manufacturing and Industrial Engineering*, Lecture Notes in Mechanical Engineering,
https://doi.org/10.1007/978-981-15-8542-5_56

the advancing side (AS) undergoes extrusion, and the extruded material goes around the tool and is deposited on the backside. It has been reported that about 86% of the total heat generated is frictional heat between the surface of the base metal and the shoulder of the tool. The Remaining heat is collectively obtained through friction on the pin-workpiece interface and plastic strain [1].

FSW was initially adopted for the welding of aluminum and its alloys which have a high coefficient of thermal expansion and a larger rate of heat transfer, greater electrical conductivity, and high specific heat. The conventional method of fusion welding was not suitable for the welding of aluminum due to its high thermal expansion as it led to the deformation of material [2]. FSW joint is obtained without melting which results in distortion-free welds. FSW is very efficient in joining all kinds of aluminum alloys and thus finds wide applications in the automobile industry and aerospace engineering. The process has also been applied to the joining of other materials in both similar as well as dissimilar combinations such as copper, titanium, brass, and steels.

Micro-friction stir welding (μFSW) is an application or an extended form of FSW which is employed for the joining of ultra-thin sheets with thickness less than 1 mm (1000 μm). Research has been done related to the various aspects accompanying μFSW comprising the defects involved, tools used, and its applications. In 2003, a program was initiated by TWI regarding the welding of ultra-thin sheets and in 2005 reported the successful accomplishment of the project [3]. In electrical applications, the joining of copper to aluminum has been employed [4]. Ultrasonic welding and μFSW are used for wood welding [5]. Though the strength of the weld obtained was weak with μFSW but with optimization of process parameters and tool geometry, the results are most likely going to be improved. Welding of metal matrix composite has also been done for application in the aerospace sector [6].

Earlier, much of the work concerning FSW was done on sheets with a thickness greater than 1 mm. For this much thickness (<1 mm), some work has been done on process parameters as well as on tools being used for similar welds and dissimilar welds. Nishihara and Nagasaka [7] in a successful attempt applied the μFSW process to determine the viability of the process on the AZ31 magnesium alloy. Ahmed [8] analyzed the effect of μFSW on butt welds and lap welds with sheets having a thickness of 0.44 mm. With the main focus placed on butt welds, concluded in applications where both lap and butt welds can be used with sheets having a thickness less than 0.5 mm, lap joints produced better results. Consequently, a greater amount of stirred material leads to a stronger joint. However, hardness in both cases was comparable.

During FSW, heat generation and mechanical work due to the tool lead to the formation of different welding zones. These were first reported by Threadgill [2]. The base metal zone remains unaffected, as there is no change in microstructure or mechanical properties though it may be subjected to light thermal cycles. Heat affected zone (HAZ) undergoes thermal cycles which result in some change in microstructure, leading to change in precipitate morphology, but the temperature is insufficient to cause plastic deformation. Thermo-mechanically affected zone (TMAZ) is the region that undergoes tempted deformation; however, the heat only permits partial recrystallization in some cases. This zone is directly under the tool shoulder. Stir zone (SZ) is the zone where severe plastic deformation prevails and the temperature is high enough to cause recrystallization of the grains.

Scialpi [9] conducted mechanical analysis on micro-friction stir welds of thin sheets of AA2024-T3 and AA6082-T6 alloys having a thickness of 0.8 mm. Unlike the friction stir welds of similar alloys that have four distinct zones, viz. base metal (BM), HAZ, TMAZ, and the SZ, however, in dissimilar weld 8 distinct zones, BM, HAZ, TMAZ, and SZ for AA2024-T3 alloy and SZ, TMAZ, HAZ and BM for AA6082-T6 alloy were investigated. In similar joints for both the alloys, grain refinement was observed, and minimum hardness occurred in the TMAZ. In the case of AA2024 alloy, 1 mm from the weld joining line minimum hardness was observed, while in 6082 Al alloy, this observation occurred at a distance of 3.5 mm from the abutting surfaces. For similar joints, during tensile testing, the fracture arose from the middle of the weld due to thinning. In the case of dissimilar welds, failure did not occur from the weld SZ but from the AA6082 alloy side.

Unlike other solid-state welding processes like ultrasonic welding, which is used for joining ultra-thin sheets, μFSW produces a seam-weld using hefty penetration [10]. Papaefthymiou [11] analyzed the weldability of thin sheets of wrought zinc alloy having 0.7 mm thickness. Klobčar [4] studied the μFSW of copper electrical plates which provided tremendous results in comparison with the fusion welding. Fusion welding of dissimilar metals in electrical connections such as in the case of copper to aluminum resulted in the formation of intermetallics of Cu-Al which consequently increased the resistance and decreased strength of the connection. μFSW minimizes the formation of these inter-metallics.

Galvão used μFSW for aluminum, copper, zinc alloys, and copper-zinc, and his results proved the application of μFSW as feasible and desirable [12]. Myśliwiec and Śliwa used μFSW for the joining of magnesium and aluminum sheets of thickness less than 1 mm, and results showed defect-free welds in both cases [13]. Wang et al. performed μFSW on multilayer aluminum alloy sheets. In this research, the Taguchi optimization method was used to reduce the number of experiments [10]. The high rotational speed of about 20000 rpm was employed to enhance the heat generation for these alloys as they have poor weldability due to their low surface friction coefficient. They found four types of surface morphologies: (a) surface flash (b) discontinuous semi-circular hooks (c) welds with tunneling defect and (d) smooth surfaces. The process parameters and tool geometry were analyzed to find optimum conditions for achieving defect-free welds.

With the development of micro-friction stir spot welding, application for electrical connections in addition to mechanical assembly has emerged. Wang considered mechanical characterization of friction stir spot welding for aluminum sheets of thickness about 300 μm as a replacement of conventional soldering utilized in the joining ultra-thin metals sheets for electronics, microdevices, and medicines. As micro-FSW limits the excessive heat damage [14], it is expected to be a suitable alternative to conventional soldering.

The downscaling of FSW to μFSW faces few challenges as the thickness of plates is very small. Usually, these challenges do not arise with macro FSW. Many researchers have analyzed these problems and have tried to find optimum conditions to overcome these challenges.

2 Challenges in Micro-FSW

The foremost problem in μFSW is insufficient heat generation for the welding process, which is caused due to the high heat dissipation from thin sheets. This problem is neither related to the small diameter of the tool shoulder nor incomplete connection amongst the tool shoulder and the work surface [2]. This problem can be overcome by using the non-conducting backing plates such as acrylic sheets which reduces the heat dissipation into the fixture. This heat loss can also be compensated by using the tool at high rotational speed and smaller traversing speed.

Thinning of the weld zone is another common problem with μFSW which acts as the cause of failure during tensile testing. This reduction is brought to the effect of the forging action of the tool shoulder since it diminishes the mechanical confrontation of the joint. The micro-defects which may be acceptable with thick plates could compromise the joint in terms of its ultimate tensile strength while performing μFSW. In some investigations, the yield strength of the weld zone is about 90% of base metal [15]. With such a low thickness of sheets, the testing equipment should not deform the specimen. Optimum tool geometries should be used for higher strain in the base metal. The length of the probe should be optimized to avoid the over plunging which damages the backing plate. Whereas, under plunging results in the incomplete connection amongst the base metal and tool shoulder. Further, clamping often leads to the deformation of thin sheets.

3 Discussion

3.1 Tool

The μFSW requires precision tools with optimum geometries. The tools generally used for conventional FSW may pose some serious problems with μFSW. Many researchers have investigated the tool geometry, including the tool pin profile, pin length and shoulder diameter. To obtain the defect-free joints in μFSW, the use of a pinless tool has been suggested since enough heat can be generated by the shoulder only [16, 17].

The shape of the tool should be carefully designed to produce mechanically sound welds. Tool geometry requires the design of the shape and diameter of the shoulder pin to the highest of precision. Tilt angle is also an important parameter which is related to the tool geometry. To examine the reduction in thickness of the weld zone, different tilt angles have been investigated.

They reported about the importance of frictional heat produced by the shoulder for the formation of defect-free welds in μ-FSW and difficulties in the manufacturing of miniature tools with complex geometries [3]. Elangovan and Balasubramanian [18] examined the effect of tool pin profile on aluminum alloys. Five profiles, viz. cylindrical, straight, cylindrical threaded, cylindrical tapered, square and triangular were used, and it was inferred that the square tool profile led to sound welds. Parida showed that the complex tool profile wears out easily [19]. Salari studied tool geometry and its implications on the mechanical characteristics of AA5456 alloy and concluded that stepped conical threaded pin improves the joints' surface integrity by improving the material flow [20]. Rodrigues surveyed that tool shoulder has a larger effect than tool pin on mechanical characteristics of the weld [21].

Wang et al. investigated μ-FSW of multilayer aluminum alloy sheets with three types of tool profiles namely straight cylindrical, stepped and cylindrical tapered pins and concluded that the maximum temperature was produced with stepped pin profile because the steps act as additional heat sources [10].

3.2 *Fixture*

The design of the fixture is an important parameter in μFSW as it prevents the sheet separation, deflection and movement during the welding. It should also prevent high heat dissipation from the sheets to ensure that enough heat is available for the softening of materials. The fixture consists of two parts: Backing plates and top clamps. Bakelite sheets can be used as a backing plate with heat resistant top clamps [8]. Sheet separation can be avoided by using clamps, and the upward movement of sheets can be arrested by using a roller in front of the tool [2]. The fixture should be non-conducting to prevent the heat dissipation and should firmly hold the sheets together.

3.3 *Process Parameters*

The selection of suitable parameters for μFSW is of great importance in generating sound welds. The important parameters in μFSW are the rotational speed of the tool and traverse feed. Vijayan has established that rotational speed is a more important factor using multi-objective optimization [22]. Peel [23] Hirata [24] and Laxminarayanan [25] suggested that there exists a set of optimum parameters yielding the most superior weld. These parameters influence the material flow, joint integrity, proper mixing, strength, hardness and ductility of the weld. Buffa examined the role of welding structures on the metallurgical and thermo-mechanical properties of FS welded lap joints of Al alloy [26]. Temperature is affected to a higher degree by tool rotational speed as compared to the traverse speed [27]. Tool rotation speed can be used to increase heat input. However, it cannot be increased beyond some limit in

Table 1 Literature survey employed process parameters with respect to the base material and sheet thickness

S. No	Material	Thickness (mm)	Process parameters (rpm, mm/min)	Citation
1	AA5083 alloy	0.8	1115, 32	[5]
2	Al-Mg-Si (6XXX series) (butt weld)	0.44	1650, 25	[1]
3	Al-Mg-Si (6XXX series) (lap weld)	0.44	1700, 25	[1]
4	AA2024 alloy	0.8	1810, 460	[2]
5	AA6082 alloy	0.8	2085, 762	[2]
6	AA1100 alloy (four layers)	0.254 (one layer)	20,000, 180	[4]

case of thin sheets as it increases the risk of tearing of sheets. Table 1 illustrates the process parameters for various aluminum alloy sheets of different thicknesses employed by the researchers.

3.4 Properties

In the study of lap and butt joints of aluminum alloys, it was observed that lap joints have better strength both in transverse and longitudinal tension tests, and improvement in ductility was observed due to low hardness values [8]. Scialpi examined the welds of ultra-thin sheets of Al alloy with similar and dissimilar materials and observed that 6 AA2024 alloy similar welds showed about 91% strength of the base metal [9]. The AA6082-AA2024 dissimilar and AA6082 alloy similar joints showed the strength of about 69% of the AA6082 base alloy. In the study conducted by Galvão on aluminum, brass, copper and zinc alloys, it was observed that the SZ possesses higher hardness due to the grain refinement. For all the welds, inferior elongation was noted, and the yield strength was low compared to parent metal except for zinc [11]. Initially, it was observed that the maximum tensile load increases when rotation rate increases but a decrease in load with further increase in rotation rate due to extrusion of the softened and stirred material was later observed [10]. For 0.8 mm thick aluminum AA5083 alloy sheets, yield strength of weld was found to be 95% of parent alloy yield strength, and maximum hardness was noted in SZ in comparison with other zones. The region of minimum hardness was found to be at the boundary of TMAZ and HAZ [2].

4 Conclusion

The process of μFSW has been effective in joining the sheets which further enhanced its application in micro-fabrications. The tool and process parameter combination vary for different materials, and hence, there exists room for the investigation to optimize the welding parameters. μFSW has been successful in generating defect-free welds in plates of thickness less than 1000 μm. This process requires careful designing of tools, fixtures and proper selection of their materials to obstruct the heat dissipation from the weld. Micro-friction stir welding in lap joints has been underexplored and thus invites greater room for research work. The mechanical characteristics of the weld can be improved to a greater extent by employing post-weld treatment through a micro-friction stir processing technique.

References

1. Schmidt H, Hattel J, Wert J (2003) An analytical model for the heat generation in friction stir welding. Modell Simul Mater Sci Eng 12(1):143
2. Sattari S, Bisadi H, Sajed M (2012) Mechanical properties and temperature distributions of thin friction stir welded sheets of AA5083. Int J Mech Appl 2(1):1–6
3. Teh N, Goddin H, Whitaker A (2011) Developments in micro applications of friction stir welding. TWI Dense, Cambridge
4. Klobčar D, Tušek J, Bizjak M, Lešer V (2014) Micro friction stir welding of copper electrical contacts. Metalurgija 53(4):509–512
5. Tondi G, Andrews S, Pizzi A, Leban JM (2007) Comparative potential of alternative wood welding systems, ultrasonic and microfriction stir welding. J Adhes Sci Technol 21(16):1633–1643
6. Prater T (2014) Friction stir welding of metal matrix composites for use in aerospace structures. Acta Astronaut 93:366–373
7. Nishihara T, Nagasaka Y (2004) Development of micro-FSW. In: Proceeding of the 5th international symposium on friction stir welding, pp 14–16 (2004)
8. Ahmed S, Shubhrant A, Deep A, Saha P (2015) Development and analysis of butt and lap welds in micro-friction stir welding (μFSW). In: Advances in material forming and joining. Springer, New Delhi, pp 295–306
9. Scialpi A, De Filippis LAC, Cuomo P, Di Summa P (2008) Micro friction stir welding of 2024–6082 aluminium alloys. Weld Int 22(1):16–22
10. Wang K, Khan HA, Li Z, Lyu S, Li J (2018) Micro friction stir welding of multilayer aluminum alloy sheets. J Mater Process Technol 260:137–145
11. Papaefthymiou S, Goulas C, Gavalas E (2015) Micro-friction stir welding of titan zinc sheets. J Mater Process Technol 216:133–139
12. Galvão I, Leitão C, Loureiro A, Rodrigues D (2012) Friction stir welding of very thin plates. Soldagem & Inspeção 17(1):02–10
13. Myśliwiec P, Śliwa RE (2016) Linear FSW technology for joining thin sheets of aluminium and magnesium alloys. In: International scientific conference PRO-TECH-MA
14. Wang DA, Chao CW, Lin PC, Uan JY (2010) Mechanical characterization of friction stir spot microwelds. J Mater Process Technol 210(14):1942–1948
15. Scialpi A, De Giorgi M, De Filippis LAC, Nobile R, Panella FW (2008) Mechanical analysis of ultra-thin friction stir welding joined sheets with dissimilar and similar materials. Mater Des 29(5):928–936

16. Kim KH, Bang HS, Kaplan AFH (2017) Joint properties of ultra thin 430M2 ferritic stainless steel sheets by friction stir welding using pinless tool. J Mater Process Technol 243:381–386
17. Zhang L, Ji S, Luan G, Dong C, Fu L (2011) Friction stir welding of Al alloy thin plate by rotational tool without pin. J Mater Sci Technol 27(7):647–652
18. Elangovan K, Balasubramanian V (2007) Influences of pin profile and rotational speed of the tool on the formation of friction stir processing zone in AA2219 aluminium alloy. Mater Sci Eng, A 459(1–2):7–18
19. Parida B, Mohapatra MM, Biswas P (2014) Effect of tool geometry on mechanical and microstructural properties of friction stir welding of Al-alloy. Int J Curr Eng Technol, Special 2:88–92
20. Salari E, Jahazi M, Khodabandeh A, Ghasemi-Nanesa H (2014) Influence of tool geometry and rotational speed on mechanical properties and defect formation in friction stir lap welded 5456 aluminum alloy sheets. Mater Des 58:381–389
21. Rodrigues DM, Loureiro A, Leitao C, Leal RM, Chaparro BM, Vilaça P (2009) Influence of friction stir welding parameters on the microstructural and mechanical properties of AA 6016-T4 thin welds. Mater Des 30(6):1913–1921
22. Vijayan S, Raju R, Rao SK (2010) Multiobjective optimization of friction stir welding process parameters on aluminum alloy AA 5083 using Taguchi-based grey relation analysis. Mater Manuf Processes 25(11):1206–1212
23. Peel M, Steuwer A, Preuss M, Withers PJ (2003) Microstructure, mechanical properties and residual stresses as a function of welding speed in aluminium AA5083 friction stir welds. Acta Mater 51(16):4791–4801
24. Hirata T, Oguri T, Hagino H, Tanaka T, Chung SW, Takigawa Y, Higashi K (2007) Influence of friction stir welding parameters on grain size and formability in 5083 aluminum alloy. Mater Sci Eng, A 456(1–2):344–349
25. Lakshminarayanan AK, Balasubramanian V (2008) Process parameters optimization for friction stir welding of RDE-40 aluminium alloy using Taguchi technique. Trans Nonferrous Metals Soc China 18(3):548–554
26. Buffa G, Campanile G, Fratini L, Prisco A (2009) Friction stir welding of lap joints: influence of process parameters on the metallurgical and mechanical properties. Mater Sci Eng, A 519(1–2):19–26
27. TWI (2005) Group sponsored project 15718 (Ex: PR8485): Friction stir welding of thin aluminium sheets—phase II. http://www.twi-global.com/EasysiteWeb/getresource.axd?AssetID=55437. Last accessed 2015/07/21

Traffic Noise Modelling Considering Traffic Compositions at Roundabouts

Anupam Thakur and Ramakant Rana

Abstract As the traffic volume is increasing in urban areas, their major effect can be seen in terms of noise near roundabouts. An effort has been made in this work to evaluate the noise level near a roundabout in terms of certain noise-level descriptors such as L_{eq}, L_{10}, L_{50} and L_{90} to estimate the maximum and pervasive noise climate existing at the roundabouts. Multiple regression analysis technique is used to form a model for each of the sound-level descriptors for a roundabout and factors are found that are significant for a particular sound descriptor. Three roundabouts are considered for analysis of varying traffic volume and geometry of roundabout; these were Y.P.S. Roundabout, Thikkari Roundabout, and Fountain Chowk Roundabout, Patiala. The R^2 values were found to be 0.70 for L_{eq}, 0.79 for L_{10}, 0.78 for L_{50}, 0.84 for L_{90} indicating a good correlation. The percentage error was found to be varying between ±3% for all the sound descriptors. "2t-test" was carried out for 25 samples and no significant difference between predicted and measured sound levels was found. A 1/1 octave analysis was done to determine the frequencies that were dominating. Peak noise level at roundabouts was obtained to be at 500 Hz., 1 kHz., 2 kHz., and 4 kHz. These correspond to maximum annoyance range and therefore there is a need for mitigation measures.

Keywords Traffic noise models · Roundabout · Multiple regression · 2t-test · 1/1 octave band

---

A. Thakur · R. Rana
Maharaja Agrasen Institute of Technology, New Delhi, Delhi, India

R. Rana (✉)
Delhi Technological University, New Delhi, Delhi, India
e-mail: 7ramakant@gmail.com

© The Author(s), under exclusive license to Springer Nature Singapore Pte Ltd. 2021
R. M. Singari et al. (eds.), *Advances in Manufacturing and Industrial Engineering*, Lecture Notes in Mechanical Engineering,
https://doi.org/10.1007/978-981-15-8542-5_57

1 Introduction

In India, commercial areas are developing around a large number at intersection regions such as traffic roundabouts. To ease traffic congestion, road intersections are framed into roundabouts. Traffic noise near roundabouts creates great annoyance to the surrounding areas. The prediction model for traffic noise is of great importance for a roundabout design. Makarewicz et al. [1] showed the noise reduction when an intersection is being replaced by a roundabout. Covaciu et al. [2] did a study and found a significant effect seen when an intersection is changed from signalized to a roundabout intersection. Lewis et al. [3] found that the noise on the accelerating lane was found to be within ±1% of the free flow noise. Lewis et al. [4] developed and expressed traffic noise in terms of distance of a vehicle from the roundabout. Chevallier et al. [5] proposed a traffic simulation tool that was developed specially for roundabouts, having integrated noise emission and propagation laws. The American FHWA Traffic Noise Model [6] or the RLS-90 Model [7] roughly considers traffic flows. Qudais et al. [8] proposed some models that can improve noise assessment at Signalized Traffic Intersections. Certu [9] developed certain analytical models based on average noise for each vehicle class considering specific vehicle class speed. Certain studies have been carried out in India that show the effect of various noise parameters such as vehicle flow rate, percentage of heavy vehicles, speed of vehicles, etc. Parida et al. [10] carried out a study in New Delhi and found out that noise levels were found out to be above C.P.C.B. standards, so a barrier design for corridors was suggested that could be used to reduce noise. Anyogita [11] carried out a study in New Delhi of 1/1 octave band analysis for various vehicle categories and stated that Mini Buses (R.T.V.) produced maximum noise levels followed by bus, autos, and taxi. Sooriyaarachchi et al. [12] developed a model considering vehicle class, vehicle speed, and distance from the traffic lane using regression analysis. Banerjee et al. [13] carried out the study considering various parameters such as vehicle speed, percentage of heavy vehicles, road width, open space, and built up areas, residential areas. The hourly traffic volume was found to be influencing the Leq levels in Asansol, India. Rajakumara et al. [14] developed a model for interrupted traffic flow for roads in Bangalore, India. A stop and go model has been formed using regression analysis that could be applied to an acceleration lane and a deceleration lane. The R^2 values were found to be 0.82 for acceleration lane and 0.73 for deceleration lane. Can et al. [15] did a study showing the effect of speed and acceleration on the noise produced by traffic. Li et al. [16] carried out a study showing the effect of entrance and exit lane during a high traffic saturation condition. Tripathi [17] developed a model for Dehradun Roorkee Highway(NH-58) India, and the R^2 values were found to be 0.85. Subramani et al. [18] developed a model for Coimbatore city, Tamil Nadu, India, and considered various parameters such as traffic flow, vehicle speed, atmospheric temperature, and relative humidity. A negative relation was found for atmospheric temperature and positive for relative humidity.

It is found that different countries have developed different traffic noise models to meet their requirements of government regulations and designers. All the above models developed could not be implemented on Indian scenario due to high traffic volume, certain driving conditions like excessive acceleration and deceleration due to a large number of two wheelers, etc. The traffic noise models are being used widely in initial stages of road design. Roundabouts also need equal consideration while designing the dimensions so that there is adequate traffic flow with least traffic noise annoyance to the surrounding commercial and residential areas. In India, not much work has been done on the roundabouts traffic noise modeling. A study is thus presented considering statistical descriptors such as L_{eq}, L_{10}, L_{50}, and L_{90}. From studies, it has been found that the traffic composition, their certain dynamic parameters such as acceleration and deceleration and road characteristics are not being considered and should be introduced in the model for Indian road scenario. Also factors that could affect the individual descriptor are studied; thus, helping to control the traffic noise near roundabouts [20–22].

2 Site Selection and Parameters

Three roundabouts were selected for collecting data with varying traffic flow conditions. The three roundabouts selected are (a) Fountain Roundabout (b) Thikkari Roundabout, and (c) Y.P.S. Roundabout, Patiala City, India. The sites represent predominantly commercial land use pattern. The data was collected on different locations around the roundabouts; the points are marked using "Δ" as shown in Fig. 1. The points selected for the roundabouts lied at certain distance from the roundabout to study the distance attenuation of noise on a roundabout. To study the affect of acceleration and deceleration, points around a roundabout are selected that are just entering or emerging from a roundabout. During site study, various parameters like hourly flow rate for each class of vehicles (Cars S.U.V., Heavy Vehicles,

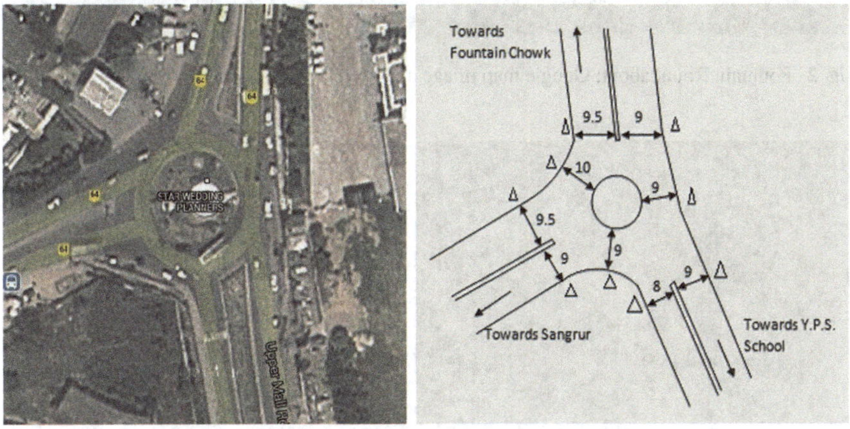

Fig. 1 Thikkari Roundabout; Google map image and sketch

Three wheelers, and Two Wheelers), acceleration (a), deceleration (d), and distance from the road or road width (l) were considered. The instrument was calibrated. Data was collected for a total of 23 days for three roundabouts with each roundabout data collection for a week from 10:00 A.M.to 1 P.M. and from 2 P.M. to 5 P.M. Sound levels such as L_{eq}, L_{10}, L_{50}, and L_{90} were recorded to get an idea of equivalent, maximum, average, and pervasive noise climate near a roundabout. The noise levels were recorded in dB (A) using noise-level meter (CESVA SC-310). The instrument was calibrated before readings were recorded. Wind screen was used to avoid wind effect during recordings on the microphone. Meter was kept at a height of 1.2 above ground level using meter [21].

The data was recorded for one hour with 1-minute integration time. The spot vehicle acceleration and deceleration were recorded using radar gun and using equations of motion. A number of speed data was recorded for assessing an accurate estimate for traffic speeds to calculate acceleration and deceleration [22]. A Google map image and sketch of the three roundabouts showing all dimensions in meter are shown in Figs. 1, 2 and 3. Information regarding the roundabout dimension is shown in Table 1. Each of the point selected around the roundabout has been studied and traffic volume for each of the vehicle class recorded using 5 counter meters. The

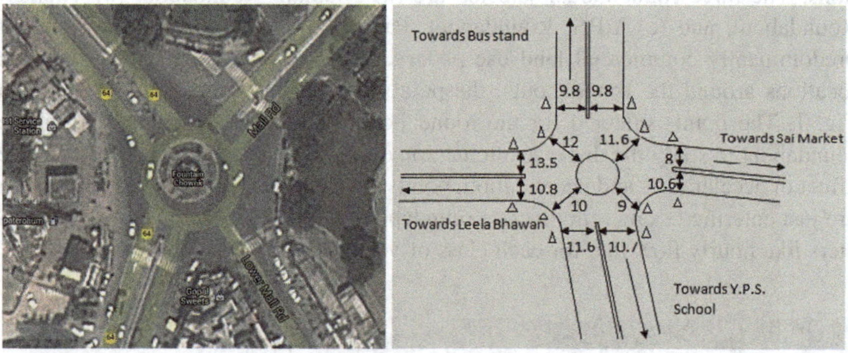

Fig. 2 Fountain Roundabout; Google map image and sketch

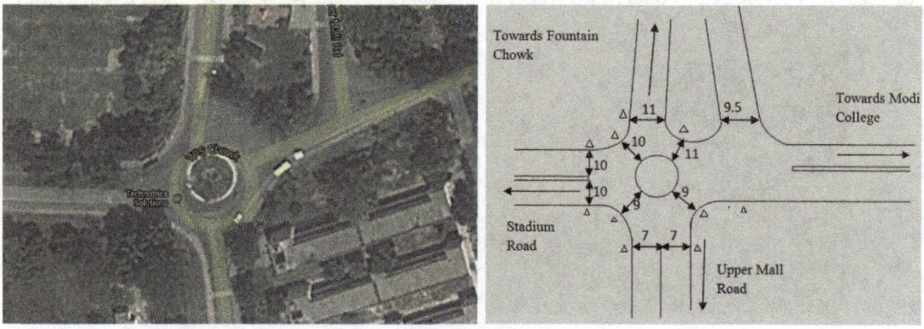

Fig. 3 Y.P.S. Roundabout; Google map image and sketch

Table 1 Roundabout diameter

Site	Diameter of roundabout (m)
(A) Thikkari Roundabout	≅ 13
(B) Fountain Roundabout	≅ 9
(C) Y.P.S. Roundabout	≅ 8

Table 2 List of parameters used and symbol

Parameter	Symbol
Car	C
Heavy vehicle	H.V.
S.U.V.	S
Three wheeler	T1
Two wheeler	T2
Acceleration	*a*
Deceleration	*d*
Road width	*l*

distance of each measuring point from the road has been measured and abbreviated by "*l*". The acceleration and deceleration were measured using equation of motion. A distance of 60 m ahead the measuring point was used for measuring initial speed/final speed and time interval was measured using stopwatch. The Meter was kept at height of 1.2 m above the ground level.

$$S = ut + 1/2(\text{at}2) \tag{2.1}$$

Certain symbols have been used to designate the parameters selected. The list is shown in Table 2. The data thus is collected for each of the roundabout separately to do analysis for each of the roundabout and compare their noise levels.

3 Mathematical Modelling for Roundabout Traffic Noise

Multiple regression analysis has been done by splitting the data for each data roundabout separately and then an equation is formed for each statistical descriptor for the respective roundabout.

3.1 Equivalent Sound-Level Equation (L_{eq})

This noise-level descriptor is measured to get an equivalent noise condition near a roundabout. The equation thus obtained from each independent roundabout data regression analysis is shown below.

For Fountain Roundabout:

$$L_{eq} = 89.26 - 5.59\mathrm{Log(C)} + 1.23\mathrm{Log(S.U.V.)} + 3.73\mathrm{Log(H.V.)} + 0.41\mathrm{Log(T1)} + 1.35\mathrm{Log(T2)} + 0.06a_2 - 0.27d_2 - 0.75l \quad (3.1)$$

For Thikkari Roundabout:

$$L_{eq} = 79.45 + 0.91\mathrm{Log(C)} - 0.98\mathrm{Log(S.U.V.)} + 4.3\mathrm{Log(H.V.)} + 1.62\mathrm{Log(T1)} + 1.79\mathrm{Log(T2)} + 0.33a_2 + 0.04d_2 - 1.87l \quad (3.2)$$

For Y.P.S. Roundabout:

$$L_{eq} = 37.41 + 4.17\mathrm{Log(C)} + 2.71\mathrm{Log(S.U.V.)} + 1.25\mathrm{Log(H.V.)} + 1.16\mathrm{Log(T1)} + 1.58\mathrm{Log(T2)} + 0.03a_2 - 0.02(d_2 + l) \quad (3.3)$$

3.2 10-Percentile Sound-Level Equation (L_{10})

This noise level is captured to get an idea of the maximum or high noise levels that create annoyance near a roundabout. The equation thus obtained from each independent roundabout data regression analysis is shown below.

For Fountain Roundabout

$$L_{10} = 111.67 - 3.31\mathrm{Log(C)} - 0.56\mathrm{Log(S.U.V.)} + 2.04\mathrm{Log(H.V.)} - 0.30\mathrm{Log(T1)} - 4.59\mathrm{Log(T2)} - 0.18a_2 - 0.11d_2 - 0.68l \quad (3.4)$$

For Thikkari Roundabout

$$L_{10} = 76.72 + 1.8\mathrm{Log(C)} - 2.3\mathrm{Log(S.U.V.)} + 1.34\mathrm{Log(H.V.)} + 3.32\mathrm{Log(T1)} - 0.72\mathrm{Log(T2)} + 0.15a_2 + 0.10d_2 - 0.74l \quad (3.5)$$

For Y.P.S. Roundabout

$$L_{10} = 72.45 + 1.78\text{Log(C)} + 0.75\text{Log(S.U.V.)} + 0.63\text{Log(H.V.)} + 0.17\text{Log(T1)} + 0.09(\text{Log(T2)}{-}d_2) + 0.08a_2 - 0.51l \quad (3.6)$$

3.3 50-Percentile Sound-Level Equation (L_{50})

This noise level is captured to get an idea of the average/median noise levels near a roundabout. The equation thus obtained from each independent roundabout data regression analysis is shown below.

For Fountain Roundabout

$$L_{50} = 105.38 - 2.1\text{Log(C)} - 0.66\text{Log(S.U.V.)} + 2.24\text{Log(H.V.)} - 1.43\text{Log(T1)} - 4.61\text{Log(T2)} - 0.13(a_2 + d_2) - 0.75l \quad (3.7)$$

For Thikkari Roundabout

$$L_{50} = 63.42 - 0.24\text{Log(C)} - 1.04\text{Log(S.U.V.)} + 2.07\text{Log(H.V.)} + 0.85\text{Log(T1)} + 1.11\text{Log(T2)} + 0.05a_2 + 0.02d_2 + 0.34l \quad (3.8)$$

For Y.P.S. Roundabout

$$L_{50} = 61.52 + 2.9\text{Log(C)} + 1.18\text{Log(S.U.V.)} + 0.95\text{Log(H.V.)} + 0.26\text{Log(T1)} - 1.05\text{Log(T2)} + 0.05a_2 - 0.03d_2 - 0.23l \quad (3.9)$$

3.4 90-Percentile Sound-Level Equation (L_{90})

See Table 3.

This noise level is captured to get an idea of the pervasive noise levels that create noise near a roundabout. The equation thus obtained from each independent roundabout data regression analysis is shown below.

For Fountain Roundabout

$$L_{90} = 85.16 + 0.36\text{Log(C)} - 0.29\text{Log(S.U.V.)} + 1.34\text{Log(H.V.)} - 0.90\text{Log(T1)}$$

$$- 2.98\text{Log}(\text{T2}) + 0.27a_2 - 0.09d_2 - 0.46l \quad (3.10)$$

For Thikkari Roundabout

$$\begin{aligned} L_{90} = 55.04 &+ 0.01\text{Log}(\text{C}) - 2.36\text{Log}(\text{S.U.V.}) \\ &+ 0.98\text{Log}(\text{H.V.}) + 1.61\text{Log}(\text{T1}) \\ &+ 1.21\text{Log}(\text{T2}) + 0.35a_2 - 1.03l \end{aligned} \quad (3.11)$$

For Y.P.S. Roundabout

$$\begin{aligned} L_{90} = 37.41 &+ 4.17\text{Log}(\text{C}) + 2.71\text{Log}(\text{S.U.V.}) \\ &+ 1.25\text{Log}(\text{HV}) + 1.16\text{Log}(\text{T1}) \\ &+ 1.58\text{Log}(\text{T2}) + 0.03a_2 - 0.02(d_2 + l) \end{aligned} \quad (3.12)$$

Table 3 Summary for three roundabouts individual data regression

Round about name	Statistical noise levels	Percentage error	Multiple R	R^2	Adj. R^2	F-value	P-value	Std. error
Fountain Chowk	L_{eq}	−4.06 to 4.41%	0.68	0.46	0.32	3.24	0.008	1.82
	L_{10}	−3.80 to 3.93%	0.76	0.58	0.47	5.24	0.001	1.33
	L_{50}	−2.99 to 2.98%	0.82	0.68	0.59	8.06	0.000	1.03
	L_{90}	−3.28 to 2.93%	0.77	0.60	0.50	5.73	0.000	1.13
Thikkari Chowk	L_{eq}	−4.27 to 2.83%	0.80	0.63	0.54	7.08	0.000	1.64
	L_{10}	−2.68 to 4.07%	0.71	0.51	0.38	4.18	0.001	1.43
	L_{50}	−2.93 to 2.88%	0.63	0.40	0.25	2.74	0.020	1.24
	L_{90}	−4.68 to 3.58%	0.54	0.29	0.11	1.65	0.150	1.48
Thikkari Chowk	L_{eq}	−3.76 to 4.70%	0.74	0.56	0.48	7.36	0.000	1.42
	L_{10}	−2.68 to 2.09%	0.82	0.67	0.61	11.98	0.000	0.88
	L_{50}	−2.59 to 2.03%	0.69	0.48	0.39	5.38	0.000	0.86
	L_{90}	−2.70 to 3.66%	0.77	0.61	0.53	8.78	0.000	0.94

From the above data analysis, it is found that the error varied within ±5% range. The difference between the R^2 and adjusted R^2 values was found to be less than 0.2, but the F-value was too low (These should be >10) except for L_{10} equation for Y.P.S. roundabout. As the values of R^2 are low thus there is not good correlation found between noise levels and factors [23]. The reason for low value of R^2 could be that the range of the factors considered for a single roundabout was not varying in large range and thus the model could not develop a good correlation. The p-values for all the models formed were good except L_{90} for Thikkari Roundabout, as it is required to be less than 0.05. The models individually showed good prediction for respective roundabout, but were not applicable to other roundabout as the whole scenario changed due to varying traffic volume and changing acceleration and deceleration conditions. Thus, a multiple regression analysis is done with all the three roundabouts data combined. By combining the data, the R^2 values for each descriptor model improved and showed good correlation measured and predicted sound level. The equations thus developed were as under

Equation for L_{eq}:

$$\begin{aligned} L_{eq} &= 59.44 - 1.57\text{Log(C)} + 2.88\text{Log(S)} \\ &+ 2.39\text{Log(H.V.)} + 1.91\text{Log(T1)} \\ &+ 3.39\text{Log(T2)} + 0.11a_2 - 0.04d_2 - 0.45l \end{aligned} \tag{3.13}$$

Equation for L_{10}:

$$\begin{aligned} L_{10} &= 62.07 + 0.17\text{Log(C)} + 0.66\text{Log(S)} + 1.23\text{Log(H.V.)} \\ &+ 4.23\text{Log(T1)} + 1.49\text{Log(T2)} \\ &+ 0.03a_2 - 0.02d_2 - 0.24l \end{aligned} \tag{3.14}$$

Equation for L_{50}:

$$\begin{aligned} L_{50} &= 51.59 + 1.12\text{Log(C)} + 0.84\text{Log(S)} + 0.86\text{Log(H.V.)} \\ &+ 4.02\text{Log(T1)} + 1.47\text{Log(T2)} \\ &- 0.01(a_2 + d_2) - 0.05l \end{aligned} \tag{3.15}$$

Equation for L_{90}:

$$\begin{aligned} L_{90} &= 34.53 + 3.06\text{Log(C)} + 1.18\text{Log(S)} + 0.65\,\text{Log(H.V.)} \\ &+ 3.47\text{Log(T1)} + 3.41\text{Log(T2)} \\ &- 0.01a_2 - 0.02d_2 + 0.13l \end{aligned} \tag{3.16}$$

From the analysis, it was found that L_{eq} was affected significantly by all the flow rate of heavy vehicle, S.U.V., two wheelers, three wheelers, average acceleration (a)

Table 4 Regression output for noise level

Statistical noise levels	%age Error	Multiple R	R^2	Adj. R^2	F-value	P-value	Std. error
L_{eq}	−3.5 to 3.1	0.84	0.70	0.69	38.37	0.000	1.79
L_{10}	−3.6 to 3.2	0.89	0.79	0.78	41.07	0.000	1.62
L_{50}	−4.1 to 3.3	0.85	0.78	0.77	41.59	0.000	1.58
L_{90}	−3.3 to 3.2	0.87	0.84	0.83	52.35	0.000	1.71

of the vehicles and road width l. L_{10} noise level was found to be significantly affected by the flow rate of heavy vehicle, three wheelers, and road width l. L_{50} noise level was found to be significantly affected by the flow rate of three wheelers. L_{90} noise level was found to be significantly affected by the flow rate of car, three wheeler, and two wheelers. Thus, it could be said that higher noise levels were due to heavy vehicles and three wheelers and pervasive noise climate is because of cars and two wheelers near a roundabout. The significance of the factors is evaluated by their p-value less than 0.05 for 95% confidence level. The p-values for the model formed were all less than 0.05. The F-value was found to be 38.37 for L_{eq}, 41.07 for L_{10}, 41.59 for L_{50}, 52.35 for L_{90}. The error between predicted and measured noise levels was found to be within ±4%. The results of analysis can be seen in Table 4.

Out of the total 135 h reading, 25 samples were chosen and a *2t-test* was carried to see if there was any significant difference between measured and predicted values of each noise level. The t-critical value was found to lie within the t-critical range. The results of the analysis are shown in Table 5. There was no significant difference found between the predicted mean noise levels and the actual measured noise levels from the data collected. The scatter plots for predicted versus measured noise levels are shown in Figs. 4, 5, 6 and 7.

Table 5 Two t-test output for noise descriptors

Number of samples	Noise descriptor	Measured mean	Predicted mean	t-statistical	t-critical	$P(T \leq t)$
25	L_{eq}	76.48	77.13	−0.34	2.069	0.74
25	L_{10}	80	80.02	0.133	2.068	0.90
25	L_{50}	73.38	73.45	−0.26	2.068	0.80
25	L_{90}	68.81	68.73	−0.275	2.064	0.79

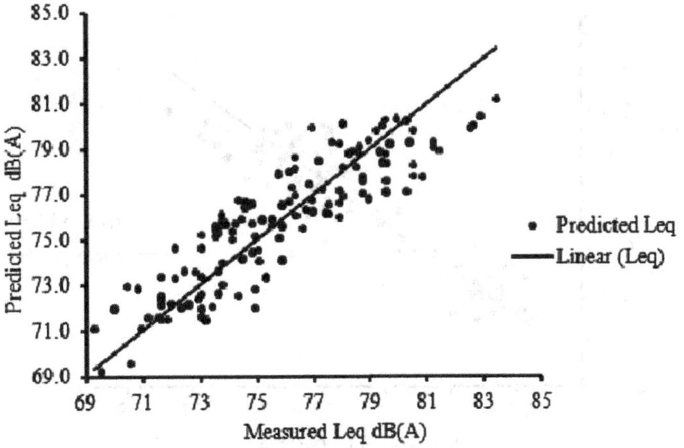

Fig. 4 Predicted versus measured L_{eq}

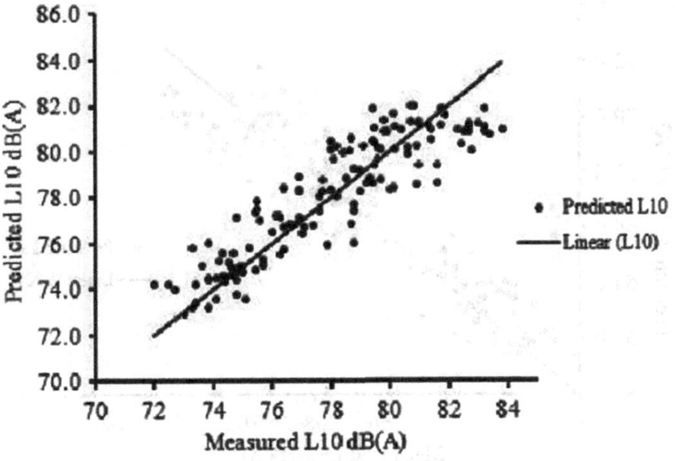

Fig. 5 Predicted Versus measured L_{10}

4 Comparison of the Noise Climate on the Three Roundabouts

See Fig. 8.

Noise measurements for various levels have been done on the three roundabouts and plotted against the days. Thus, a noise pattern is obtained that clearly describes the noise levels on the three roundabouts. It was found that the maximum noise was found to be on the "Thikkari Roundabout" and minimum was found to be on the "Y.P.S. Roundabout". Table 6 clearly depicts the weekly average Leq maximum of

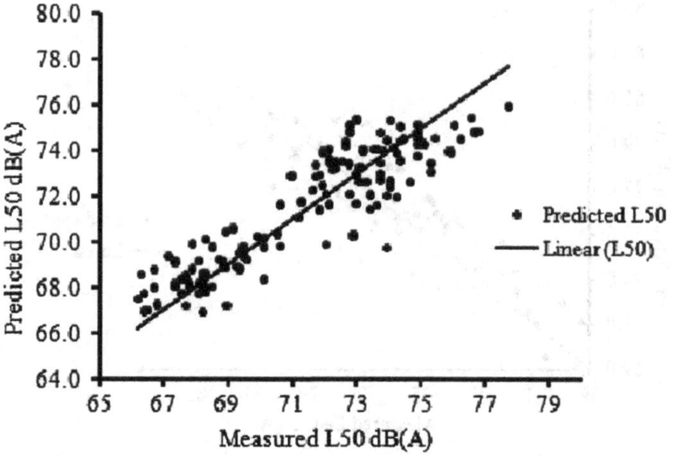

Fig. 6 Predicted versus measured L_{50}

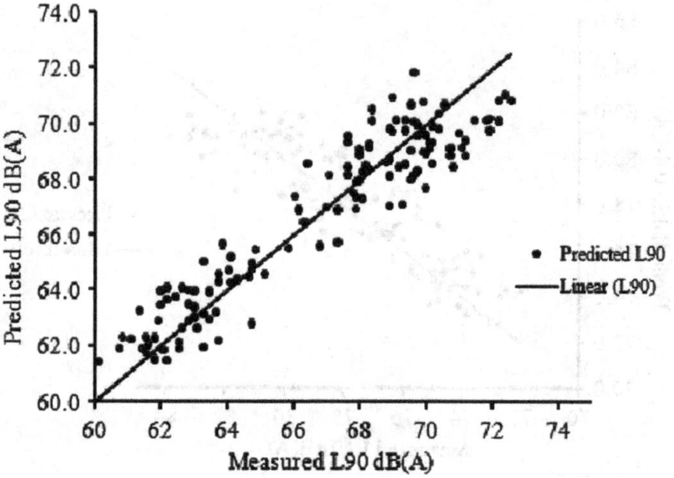

Fig. 7 Predicted versus measured L_{90}

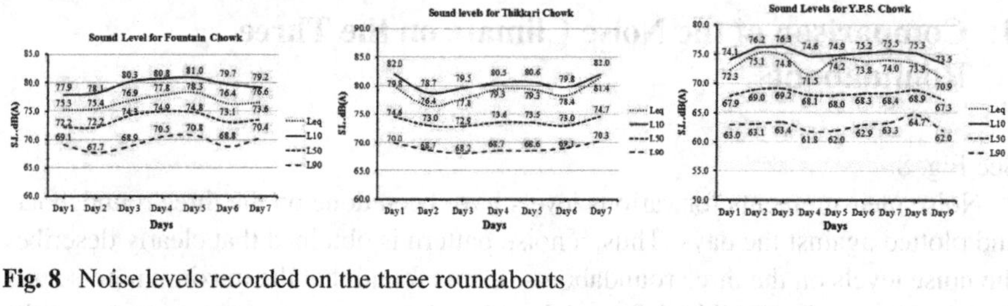

Fig. 8 Noise levels recorded on the three roundabouts

Table 6 Weekly average sound levels for roundabouts

Site	L_{eq}	L_{10}	L_{50}	L_{90}
Fountain Roundabout	76.7	79.6	73.6	69.4
Thikkari Roundabout	78.8	80.4	73.6	69.1
Y.P.S. Roundabout	73.3	75.0	68.3	62.9

Table 7 C.P.C.B Ambient noise standards (India) [24]

S. No.	Area	L_{eq} dB (A)	
		Day time	Night time
1	Industrial area	75	70
2	Commercial areas	**65**	55
3	Residential area	55	45
4	Silence zone	50	40

by 78.8 dB (A) and minimum of 73.3 dB (A) for the same. A band of 15 dB (A) for Fountain Roundabout, 14 dB (A) for Thikkari Roundabout, and 14 dB (A) for Y.P.S. Roundabout is obtained. Thus, it was found that the maximum annoyance due to traffic is found to be Thikkari Roundabout. It was found that the noise levels recorded at all the three roundabout sites were found to be above Indian C.P.C.B. standards, i.e., 65 dB (A). Therefore, measures to mitigate traffic noise levels should be made or at least consideration for barrier design around roundabout should be done. A 1/1 octave band analysis has been done to find out the dominating frequencies in traffic noise near a roundabout (Table 7).

5 1/1 Octave Band Analysis

Frequency analysis has been done for 1/1 octave band for roundabout traffic noise throughout a day. The data collected for each frequency is then averaged and A-weight to each frequency band is applied. This results in a frequency spectrum that would be useful for traffic noise analysis for barrier design to reduce annoyance effect in surrounding areas, and in material selection that could be used in barrier to mitigate the traffic noise level. From Fig. 9, it is clearly visible that the frequencies that are dominating in a traffic noise are found to be 1, 2 and 4 kHz. These are the same frequencies that are known to be the most annoying frequencies in octave band, i.e., 1, 2, and 4 kHz [25] for human hearing. The maximum peak of 96 dB (A), 93.2 dB (A), and 96.4 dB (A) is obtained for L_{10} corresponding to 1, 2, and 4 kHz. Also their dominancy is maintained in other noise levels (L_{eq}, L_{50}, L_{90}). Thus, a measure to reduce their dominancy in traffic noise should be made. This information

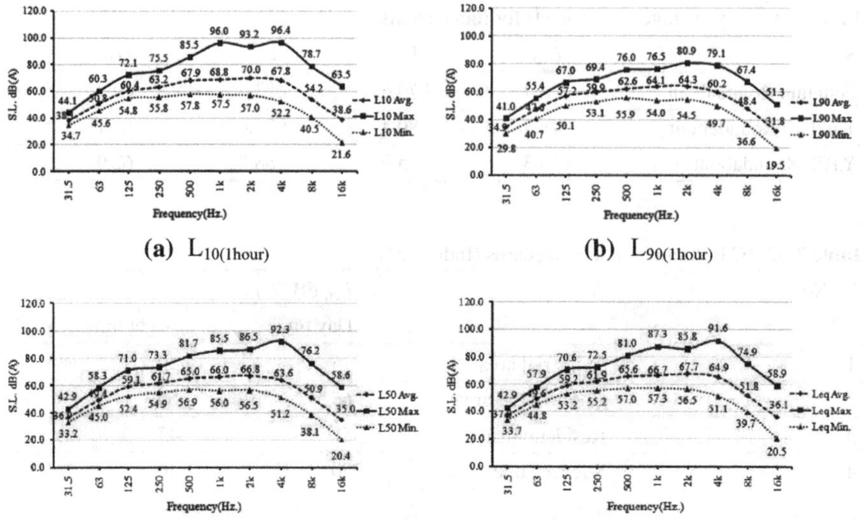

(a) $L_{10(1hour)}$ (b) $L_{90(1hour)}$

(c) $L_{50(1hour)}$ (d) $L_{eq(1hour)}$

Fig. 9 Sound levels versus frequency 1/1 octave band

is quiet useful for the barrier design and for the material selection that could be used in construction of it.

6 Conclusion

A statistical traffic noise model was developed for roundabouts considering various traffic compositions, their average acceleration and deceleration, and road widths. The model developed for each noise descriptor L_{eq}, L_{10}, L_{50}, and L_{90} had R^2 values of 0.70, 0.79, 0.78, and 0.84, respectively. The error between the predicted and the measured noise levels was found to lie between ±4%. *2t-test* was conducted for each noise descriptor for checking any significant difference between measured and predicted noise levels, and the value for "*t-statistical*" was found to lie within "*t-critical*" range and showed no difference for mean values. Comparison of the three roundabouts was also done showing maximum noise climate at the "*Thikkari Roundabout*" and minimum at "*Y.P.S Roundabout*". The noise levels were found to be above the Indian C.P.C.B. (Central Pollution Control Board) with L_{eq} level up to 78.8 dB (A), 76.7 dB (A) and 73.3 dB (A) at the three roundabouts. Measures need to be taken to control the noise levels near the roundabouts. 1/1 octave band analysis was done signifying that 1, 2, and 4 kHz. frequencies were dominant in the traffic noise.

References

1. Makarewicz R, Golebiewski R (2007) Modeling of the roundabout noise impact. Acoust Soc Am, PACS number: 43(50):860–868
2. Covaciu D, Florea D, Timar J (2015) Estimation of the noise level produced by road traffic in roundabouts. Appl Acoust 98:43–51
3. Lewis PT, James A (1978) Noise levels in the vicinity of traffic roundabouts. J Sound Vib 58(2):293–299
4. Lewis PT, James A (1980) On the noise emitted by single vehicles at roundabouts. J Sound Vib 72(1):51–69
5. Chevallier E, Leclercq L (2007) A macroscopic theory for unsignalized intersections. Transp Res Part B Methodol 41:1139–1150
6. Fleming GG, Rapoza AS, Lee CSY(1995) Development of national reference energymean emission levels for the FHWA traffic noise model, version 1.0. Publication No. DOT-VNTSC-FHWA-96-2, Office of Engineering and Highway. Operations Research and Development, p 452
7. RLS (Richtlinien fur den Larmschutz an Strassen) (1990) Model. BM fur Verkehr, Bonn, Germany
8. Qudais A-S, Alhiary A (2007) Statistical models for traffic noise at signalized intersections. Build Environ 42:2939–2948
9. Certu (1980) French guide for transportation noise (Guide du bruit des transports terrestres – Prevision des niveaux sonores). CETUR, ISBN: 2-11-083290-8, p 317
10. Parida M, Mishra RK, Rangnekar S (2010) Evaluation and analysis of traffic noise along bus rapid transit system corridor. Int J Environ Sci Technol 7(4):737–750
11. Anyogita S, Prakash A, Jain VK (2004) A study of noise in CNG driven modes of transport in Delhi. Appl Acoust 65:195–201
12. Sooriyaarachchi RT, Sonnadara DUJ (2006) Development of a road traffic noise prediction model. Proc Tech Sess 22:17–24
13. Banerjee D, Chakraborty SK, Bhattacharyya S, Gangopadhyay A (2008) Modeling of road traffic noise in the industrial town of Asansol, India. Transp Res Part D 13, 539–541
14. Rajakumara HN, Mahalinge Gowda RM (2009) Road traffic noise prediction model under interrupted traffic flow condition. Environ Model Assess 14:251–257
15. Can A, Aumond P (2018) Estimation of road traffic noise emissions: the influence of speed and acceleration. Transp Res Part D 58:155–171
16. Li F, Lin Y, Cai M, Du C (2017) Dynamic simulation and characteristics analysis of traffic noise at roundabout and signalized intersections. Appl Acoust 121:14–24
17. Tripathi V, Mittal A, Ruwali P (2012) Efficient road traffic noise model for generating noise levels in Indian scenario. Int J Comput Appl 38(4):281–286
18. Subramani T, Kavitha M, Sivaraj KP (2012) Modelling of traffic noise pollution. Int J Eng Res Appl (IJERA) 2(3):3175–3182
19. Rana R, Rajput K, Saini R, Lal R (2014) Optimization of tool wear: a review. Int J Mod Eng Res 4(11):35–42
20. Singh RC, Pandey RK, Lal R, Ranganath MS, Maji S (2016) Tribological performance analysis of textured steel surfaces under lubricating conditions. Surf Topogr: Metrol Prop 4:034005
21. Daljeet Singh SP, Nigam VP, Agrawal, Maneek Kumar (2016) Vehicular traffic noise prediction using soft computing approach. J Environ Manag 183(Part 1):59–66
22. Lata S, Rana R, Hitesh (2018) Investigation of chip-tool interface temperature: effect of machining parameters and tool material on ferrous and non-ferrous metal. Mater Today: Proc 5:4250–4257

23. Lal R, Singh RC (2019) Investigations of tribodynamic characteristics of chrome steel pin against plain and textured surface cast iron discs in lubricated conditions. World J Eng 16(4):560–568
24. Shukla AK (2011) An approach for design of Noise Barriers on flyovers in urban areas. Int J Traffic Transp Eng 1(3):158–167
25. Harris CM Handbook of Acoustical and noise control, 3rd edn

Hydrogen Embrittlement Prevention in High Strength Steels by Application of Various Surface Coatings-A Review

Sandeep Kumar Dwivedi and Manish Vishwakarma

Abstract The applicability of various coatings such as graphene, reduced graphene, al, cu, zn, carbon, nitrogen, oxygen, niobium, boron nitride, and TiAlN as a protective barrier to high strength steel for hydrogen embrittlement was studied. These all coatings were applied by different coating techniques such as(chemical vapor deposition (CVD), electroplating discharge (EPD), electrolysis, gas diffusion, plasma diffusion, high velocity oxygen fuel (HVOF), magnetically enhanced plasma ion plating system, plasma vapor deposition, and ion beam sputter.) on the high strength steel substrate followed by characterization of applied coating and mechanical testing. Reduction in life cycle due to the hydrogen embrittlement was analyzed.

Keywords Hydrogen embrittlement · Graphene coating · Scanning electron microscopy

1 Introduction

The phenomenon of hydrogen embrittlement (HE) will occur when metal surfaces are exposed to hydrogen environment which leads to catastrophic failure. It was found that HE considered as an important problem in several application areas, such as nuclear plant reactor vessel [1, 2], high-pressure gaseous hydrogen storage tanks [3, 4], petroleum and natural gas pipelines [5, 6].

Generally, when metal comes to contact with the hydrogen or working in hydrogen atmosphere, then some common ways by metals failed are hydrogen embrittlement (HE), hydrogen-induced-blistering, hydrogen attack, cracking due to precipitation of internal hydrogen, and hydride-formation cracking [7]. Generally, there are two mechanisms which responsible for the HE in steels. First one is that at cracks or crack tip, hydrogen atoms are accumulated, and decreases the fracture energy while inspiring cleavage-like failure [8, 9]. The second mechanism includes mobility of

S. K. Dwivedi (✉) · M. Vishwakarma
Department of Mechanical Engineering, Maulana Azad National Institute of Technology, Bhopal, M.P. 462003, India
e-mail: Sandeep0183@gmail.com

© The Author(s), under exclusive license to Springer Nature Singapore Pte Ltd. 2021
R. M. Singari et al. (eds.), *Advances in Manufacturing and Industrial Engineering*, Lecture Notes in Mechanical Engineering,
https://doi.org/10.1007/978-981-15-8542-5_58

dislocations increased due to effect of hydrogen at crack tip through shielding effect, reducing the shear strength, and ultimately responsible for local plasticity enhancement [10]. In order to protect the equipment from the hydrogen diffusion, it is necessary to measure how much concentration of hydrogen has been diffused into the material at the time of failure, but thrush hold concentration of hydrogen which is causing the equipment failure is not found clearly. And the basic mechanism which is responsible for the HE is also not found clearly still some discussion is going on this [11].

It was found that hydrogen diffusion causes mechanical properties broadly tensile strength and fatigue strength are reduced with and without changes in yield strength. Temperature, source of hydrogen, and surface condition are responsible for HE [12]. Various methods which were responsible for hydrogen entrance in the material are cathodic reaction or charging, electroplating, during welding, etc. [13]. Literature said that the cathodic charging plays a very significant role for hydrogen diffusion and current density considered as an important factor for hydrogen absorption and propagation in steel. For hydrogen diffusion and embrittlement occurrence in steel, current density must be in range of 0.02–40 mA/cm^2 [14, 15]. It was concluded that at higher current density, hydrogen absorption and propagation are faster than the lower current density, so more hydrogen is diffused at higher current density.

In this paper, the prevention of HE in high strength steels by application of different surface coatings such as graphene, reduced graphene, al, cu, zn, carbon, nitrogen, oxygen, niobium, boron nitride, and TiAlN, etc., is discussed and selection of suitable coatings is given to reduce the HE phenomena.

2 Hydrogen Embrittlement in Steel

High strength steels are mostly subjected to hydrogen embrittlement effect. The embrittlement refers to change from ductile phase to brittle phase. Hydrogen caused the embrittlement is termed as "hydrogen embrittlement." HE can be noticeable in different ways. HE causes reduction in the following

- Ultimate-tensile strength and braking strain or ductility
- Fracture strength and toughness.

Furthermore, there is a case where unstressed components are subjected to failure. However, there is a case in which ductility will decrease with no change in tensile strength [16]. Examples of internal hydrogen embrittlement is solutions used during fabrication like acid cleaning, electroplating, pickling, and providing protective coatings, phosphating, etching, paint stripping, etc. Second way of HE is external hydrogen embrittlement which occurs during the working condition of equipment. The following mechanisms by which hydrogen is entering and tapped in the lattice are

- Hydrogen-enhanced decohesion mechnaism (HEDE),

- Hydrogen-enhanced local plasticity model (HELP), and
- Adsorption-induced dislocation emission (AIDE) [7, 17, 18].

After entering into the metal lattice hydrogen will be tapped in different regions. Examples of hydrogen taps are dislocations, grain boundaries, voids, and phase boundaries, micro cracks, precipitates, interfaces, solutes, surface cracks, surface oxides, and 0D, 1D, 2D lattice defects. In last 50 years, various authors focus on research related to hydrogen diffusion and hydrogen embrittlement (HE) [19–21]. Hydrogen causes slip localization [22–24], softening and hardening [24–32], interaction between hydrogen–dislocation [32–35] and creep [36] have been also reported, and apart from all these effects, HE has severely effect fatigue crack rate behavior as well as tensile properties.

3 Hydrogen Embrittlement Prevention

Complete prevention of HE in high strength steels is laborious but we can prevent some extent so that we can save the life of the equipment. To reduce the hydrogen diffusion from metal surface, sharp and very steady variation, notches must be prevented or avoided and removal of residual stresses are performed before the processing step [37, 38]. Sometimes, baking operation is performed to remove hydrogen which was available in lattice or surface during the processing. In broad view, baking is considered as a heat-treatment-process and processing temperature is depending on the process material [39]. Prevention from HE can also be achieved by adding specific alloying material to the base or parent material [7].

Our main focus on prevention of HE by surface coatings, surface coatings can be done by different ways. Very well-known methods are chemical vapor deposition, physical vapor deposition, electroplating, plasma vapor deposition, gas diffusion, etc.

3.1 HE Prevention by Reduced Graphene Oxide (RGO) Coating

Graphene is most suitable element to reduce the HE from high strength steel materials. It has a thin film of carbon atom which has some unique properties [40–42]. Its chemical inertness properties make him to use at various operating condition. It is also stable in atmosphere condition temperature to 400 °C and reduce the oxidation of the substrate material because of hydrophobic characteristic or property [43], and that is due to its non-polar covalent double bonds, which is responsible for prevention of hydrogen atom bonding with the water [44, 45]. That is the reason, we are selected the graphene material as a coating due to its light weight and halt charge transmit at the metal–electrolyte interface. Thus, most of the researchers have studied and selected

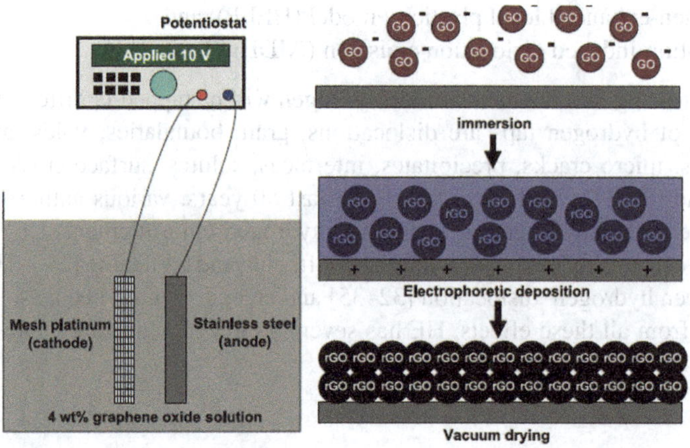

Fig. 1 rGO deposition in the stainless-steel substrate by using EPD technique [47]

the graphene as the coating material [46]. However, graphene coating can diminish corrosion properties of stainless steel as it is coated by CVD. Graphene has been replaced by reduced graphene oxide (rGO) because of its high-volume, low in cost, and processing capabilities at room temperature. One of the economical methods to develop rGO on stainless steel is electrophoretic-deposition method (EPD).

Kim and Kim [47] explained the effect of reduced graphene oxide coating to reduce HE in stainless-steel material. They used the electrophoretic-decomposition technique to coat the rGO above the substrate material. Evaluation of mechanical properties was done by slow strain rate test (SSRT) during hydrogen charging condition [47].

Synthetization of purified graphene to graphene oxide (GO) was done by modified-hammers method [48]. After this synthetization, GO was spreaded out in water and sonicated at room temperature for 4 h in 4 mg/ml concentration solution. After preparation of GO solution, electrophoretic-deposition technique was used at 10 V DC current supply to deposit rGO. For achieving required thickness of rGO above the substrate material time required to be 8 min. After completion of deposition process, samples were taken in vacuum for drying to prevent from oxidation [47].

A below representation showing electrophoretic-deposition process for the establishment of reduced graphene oxide on the stainless-steel substrate material is illustrated in Fig. 1.

3.1.1 Characterization of Sample

The investigation of the rGO in surface and cross-section of substrate material (stainless steel) was done by scanning electron microscopy (SEM).

3.1.2 Electro Chemical Measurements

Electro chemical behavior of rGO on stainless-steel specimen was analyzed using the conventional three electrode cell. Cathodic-polarization experiment or examination and electrochemical-impedance spectroscopy were conducted on rGO. Initially, 0.5 M sulfuric acid solution (H_2SO_4) with 250 mg/L arsenic trioxide (As_2O_3) (pH = 1.0) was prepared and samples were kept in that solution for 2 h. So that hydrogen diffusion and penetration were taken place easily. Nitrogen gas flows through the solution to remove the dissolved oxygen for 2 h and then measurement was done [47].

3.1.3 Slow Strain Rate Tests (SSRT)

SSRT have been widely applied to environmental induced cracking like corrosion cracking and HE. In this case, SSRT testing was performed at strain rate of 1.0 $\times$ 10^{-6}/s to evaluate the effect of rGO deposit or coating to reduce the hydrogen embrittlement effect in stainless-steel material. Here, HC was done at a negative current (-0.5 A/cm^2) for investigation of behavior or effect of HE in coated and bare samples by SSRT testing.

During the testing, it was found that the mechanical properties like the yield, tensile stresses, and strain, of stainless-steel specimen in hydrogen charging condition are reduced. After applying the rGO coating to stainless steel specimen, it was found that the mechanical properties of coated sample return back to initial value of uncharged specimen. Elongation behavior of hydrogen-charged specimen was also affected and reduced from percentage elongation of 40.1% to the value of 23.5% for uncharged to hydrogen-charged specimen. While elongation percentage of rGO-coated hydrogen-charged specimen comes around 35.6%. So, effect of HE in coated sample was reduced and these coating acts a barrier for hydrogen diffusion in stainless steel [47].

3.2 Coating of Zn, Ni, Cu, Al, PVD-Ti-DLC, Carbon, Nitrogen, and Oxygen Diffusion Layers

Electroplating of Zinc (Zn), Nickel (Ni), Copper (Cu), Aluminum (Al), PVD-Ti-DLC, various diffusion layers such as oxygen, nitrogen, and carbon, and electroless NiP coatings are examined and evaluated to reduce the effect of HE in 304 austenitic stainless-steel material. SSRT testing was performed for mechanical properties estimation [17].

Metastable grades like AISI 304 material rigorously experience the HE effects because of austenite to ferrite phase transformation at the time of operation [49–51].

Table 1 Overview of various coatings [62]

Coating material	Coating type	Deposition technique	Thickness μm	Hardness	Adhesion acc [HF]	Remarks
Al	On top	Elecrtroplating	22	15–25 HB	4	Globular structure
Cu	On top	Elecrtroplating	2	50–110 HB	1	Thin coating with pin load
Ni	On top	Elecrtroplating	12	510 HV	1	Ni adhesion coating
Zn	On top	Elecrtroplating	11	92 HV	2	Ductile zinc coating
NiP	On top	Electroless	10	550 HV	2	Amorphous NIP type coating
Ti-DLC	On top	PVD	3	1440 HV	1	Ti adhesion coating
Carbon	Diffusion	Gas diffusion	25	1225 HV	1	Long term gas diffusion
Nitrogen	Diffusion	Gas diffusion	500	300 HV	1	Small term gas diffusion
Oxygen	Diffusion	Plasma diffusion	1	1200 HV	2	Oxidizing at −550 °C

Calculation of HV was done by some formula HB = 0.95 HV

Zn, Ti-DLC, Al, Cu, and the oxygen diffusion layers as well as amorphous Nip coatings were examined because of their unique quality of low hydrogen diffusion coefficient [52–61].

Two AISI 304 stainless steel of internal heat number 40, 4 were used as a substrate material for testing various coatings and diffusion layers. The coatings properties were illustrated by metallo-graphic cross sections, Vickers and Rockwell testing, and X-ray diffraction (XRD) technique was used for internal stresses and phase analysis [62]. A simple coating and some diffusion layer properties are given in Table 1.

Cylindrical specimen is having an outer diameter 8 mm, notch diameter 6 mm, radius on notch is 0.2 mm, and angle of notch is 35^0. Stress concentration factor is 3.4. Fatigue testing was performed in that specimen and operated in atmosphere of hydrogen (10 MPa) until the specimen has broken or failure occurs.

3.3 Niobium Coating on API 5CT P110 Steel

In gas and oil industry, the components are subjected to extreme conditions and aggressive environment. The corrosion phenomena in API 5CT P110 steels is very

Table 2 Niobium coatings spraying parameter [83]

Parameters used	
Flow rate of oxygen (L/min)	164
Flow of propane (L/min)	144
Compressed air flow (L/min)	181
Nitrogen flow—carrier gas (L/min)	290
Gun to substrate distance (mm)	300
Power feed rate (g/min)	41
Speed of Gun transverse (m/s)	0.2
Number of passes	5

fast [63] and HE occurs in that steel due to working in aggressive condition [64]. By application of coating in P110 steel, they exhibit a good corrosion resistance and also work effectively in hydrogen environment as they show resistance for, HE [65].

Thermal spray coatings have received very good attention for protection of oil and gas industry components against HE [66–69]. These kinds of coatings have growing and expanding application in protection against fatigue operation [70] and oxidation process [71–73] and economical and very efficient choice to meet all requirement [74]. Recently, various authors are working in development of nanostructured coating which have better and superior properties [75, 76]. Other coating techniques such as high velocity oxygen fuel (HVOF) is also very efficient in case of corrosion in API 5CT P110 steel [77].

Niobium has very good corrosion resistant in various media [78, 79], and refractory properties [80, 81]. HVOF coating technique was used to deposit niobium in API 5CT P110 steel as it acts as a protective film over the base material against the corrosion.

The parameters which are responsible for spraying process are depicted in Table 2. The surface roughness of coating substrate was measured by contact profilometer in which by Ry values were obtained. The microhardness of both materials was obtained by microhardness tests and this hardness test was performed at the 3 N load and 15 s holding or processing time, as specified and given in the ASTM standard [82].

4 Conclusions

i. Hydrogen embrittlement resistance of rGO coating in stainless steel was examined and investigated by Kim and Kim. Their results showed that rGO-coated samples have less tendency to absorb the hydrogen than the uncoated sample.
ii. Resistance for hydrogen was increased by rGO coating because of C-H bond formation and increase the diffusion length, during the charging condition. So

it is clear that rGO can act as a barrier for hydrogen penetration and acts as a resistance for, HE.

iii. During hydrogen charging condition in coated samples, some surface irregularities, defects such as pores, coating and surface interface act as a hydrogen accumulator and hydrogen adsorbed sites. So, these sides inhibiting the entry of hydrogen from coated sample APT CT P110 steel. It is concluded that hydrogen-trapping capacity of coated sample was 7.5 time more than the normal steel specimen.

iv. Thermal spraying coating of niobium using HVOF also reduces the susceptibility of hydrogen embrittlement and act as barrier for hydrogen diffusion.

v. Electroplating of Zn, Cu, Ni, Al, PVD-Ti-DLC and electrolysis NiP coatings along with oxygen, nitrogen, and carbon diffusion layers was examined for reducing the susceptibility of HE. It was found that carbon and nitrogen diffusion layers work effectively to reduce hydrogen diffusion.

vi. It was found that very little work is carried out for tensile and ductility improvement when worked on hydrogen atmosphere. The Ni and C diffusion layers reduced crack propagation in hydrogen atmosphere.

References

1. Harries DR, Broomfield GH (1963) Hydrogen embrittlement of steel pressure vessels in pressurized water reactor systems. J Nucl Mater 9:327–338
2. Lucas GE (2010) An evolution of understanding of reactor pressure vessel steel embrittlement. J Nucl Mater 407:59–69
3. Zheng J, Liu X, Xu P, Liu P, Zhao Y, Yang J (2012) Development of high pressure gaseous hydrogen storage technologies. Int J Hydrogen Energy 37:1048–1057
4. Takasawa K, Wada Y, Ishigaki R, Kayano R (2010) Effects of grain size on hydrogen environment embrittlement of high strength low alloy steel in 45 MPa gaseous hydrogen. Mater Trans 51:347–353
5. Capelle J, Gilgert J, Dmytrakh I, Pluvinage G (2008) Sensitivity of pipeline steel API X52 to hydrogen embrittlement. Int J Hydrogen Energy 33:7630–7641
6. Hadianfard MJ (2010) Failure in a high pressure feeding line of an oil refinery due to hydrogen effect. Eng Fail Anal 17:873–881
7. Dwivedi SK, Vishwakarma M (2018) Hydrogen embrittlement in different materials: A review. Int J Hydrogen Energy 43(46):21603–21616
8. Troiano AR (1960) The role of hydrogen and other interstitials in the mechanical behavior of metals. Trans ASM 52:54–80
9. Song J, Curtin WA (2013) Atomic mechanism and prediction of hydrogen embrittlement in iron. Nat Mater 12:145–151
10. Devanathan MAV, Stachurski Z (1962) The absorption and diffusion of electrolytic hydrogen in palladium. Proc R Soc A270:90–102
11. Dwivedi SK, Vishwakarma M, Ahmed S (2018) Experimental investigation of hydrogen embrittlement during coating process and effect on mechanical properties of high strength steel used for fasteners. Mater Today: Proc 5(9):18707–18715
12. Grabke HJ, Gehrmann F, Riecke E (2001) Hydrogen in microalloyed steels. Steel Res Int 72:225–235

13. Birnbaum HK, Sofronis P (1994) Hydrogen-enhanced localized plasticity-mechanism for hydrogen-related fracture. Mater Sci Eng A 176:191–202
14. Robertson IM, Sofronis P, Nagao A, Martin ML, Wang S, Gross DW et al (2015) Hydrogen embrittlement understood. Metall Mater Trans 46(6):2323–2341
15. Vergani L, Colombo C, Gobbi G, Bolzoni FM, Fumagalli G (2014) Hydrogen effect on fatigue behavior of a quenched & tempered steel. Procedia Eng 1(74):468–471
16. Dong CF, Liu ZY, Li XG, Cheng YF (2009) Effects of hydrogen charging on the susceptibility of X100 pipeline steel to hydrogen-induced cracking. Int J Hydrogen Energy 34(24):9879–9884
17. Dwivedi SK, Vishwakarma M (2019) Effect of hydrogen in advanced high strength steel materials. Int J Hydrogen Energy 44(51):28007–28030
18. Pradhan A, Vishwakarma M, Dwivedi SK (2020) A review: The impact of hydrogen embrittlement on the fatigue strength of high strength steel. Materials Today: Proceedings, Mar 12
19. Mertens G, Duprez L, De Cooman BC, Verhaege M (2007) Hydrogen absorption and desorption in steel by electrolytic charging. Adv Mat Res 15:816–821. Trans Tech Publications
20. Danford MD (1987) Hydrogen trapping and the interaction of hydrogen with metals. NASA; Technical Paper 2744
21. A review of hydrogen embrittlement of martensitic advanced high-strength steels
22. Farrell K, Quarrell AG (1964) Hydrogen embrittlement of an ultra-high-tensile steel. J Iron Steel Ins 1002–1011
23. Shih DS, Robertson IM, Birnbaum HK (1988) Hydrogen embrittlement of alpha titanium: in situ TEM studies. Acta Metall 36(1):111–124
24. Brass AM, Chene J (1998) Influence of deformation on the hydrogen behavior in iron and nickel base alloys: a review of experimental data. Mater Sci Engng A 242:210–211
25. Birnbaum HK, Sofronis P (1994) Hydrogen-enhanced localized plasticity: a mechanism for hydrogen-related fracture. Mater Sci Engng A 176:191–202
26. Heller WR (1961) Quantum effects in diffusion: internal friction due to hydrogen and deuterium dissolved in/cap alpha/-iron. Acta Metall 9:600–613
27. Matsui H, Kimura H, Kimura A (1979) The effect of hydrogen on the mechanical properties of high-purity iron. III.—the dependence of softening on specimen size and charging current density. Mater Sci Engng 40(22):227–234
28. Kimura H, Matsui H (1979) Reply to "further discussion on the lattice hardening due to dissolved hydrogen in iron and steel" by Asano and Otsuka. Scripta Metall 13:221–223
29. Hirth JP (1980) Effects of hydrogen on the properties of iron and steel. Metall Mater Trans A 11(6):861–890
30. Dufresne F, Seeger A, Groh P, Moser P (1976) Hydrogen relaxation in a-iron. Phys Stat Sol A 36:579–589
31. Senkov ON, Jonas JJ (1996) Dynamic strain aging and hydrogen-induced softening in alpha titanium. Metall Mater Trans A 27:1877–1887
32. Au JJ, Birnbaum HK (1973) Magnetic relaxation studies of hydrogen in iron: relaxation spectra. Scripta Metall 7:595–604
33. Magnin T, Bosch C, Wolski K, Delafosse D (2001) Cyclic plastic deformation behaviour of Ni single crystals oriented for single slip as a function of hydrogen content. Mater Sci Engng A 314:7–11
34. Clum JA (1975) The role of hydrogen in dislocation generation in iron alloys. Scripta Metall 9:51–58
35. Birnbaum HK, Robertson IM, Sofronis P (2000) Hydrogen effects on plasticity. In: Lepinoux J (ed) Multiscale phenomena in plasticity. Kluwer Academic Publishers, Dordrecht
36. Mignot F, Doquet V, Sarrazin-Baudoux C (2004) Contributions of internal hydrogen and room-temperature creep to the abnormal fatigue cracking of Ti6246 at high Kmax. Mater Sci Engng A 380:308–319
37. _Cwiek J (2010) Prevention methods against hydrogen degradation of steel. J Achiev Mater Manuf Eng 43(1):214–221
38. Timmins PF. Solutions to hydrogen attack in steels

39. Grobin AW Jr (1988) Other ASTM committees and ISO committees involved in hydrogen embrittlement test methods. Hydrogen Embrittlement: Prev Contr 962:46
40. Geim AK, Novoselv KS (2007) The raise of graphene. Nat Mater 6:183–191
41. Zou C, Yang B, Bin D, Wang J, Li S, Yang P et al (2017) Electrochemical synthesis of gold nanoparticles decorated flower-like graphene for high sensitivity detection of nitrite. J Colloid Interface Sci 488:135–141
42. Zhang K, Xiong Z, Li S, Yan B, Wang J, Du Y (2017) Cu3P/RGO promoted Pd catalysts for alcohol electro-oxidation. J Alloys Compd 706:89–96
43. Bunch JS, Verbridge SS, Alden JS, van der Zande AM, Parpia JM, Craighead HG et al (2008) Impermeable atomic membranes from graphene sheets. Nano Lett 8:2458–2462
44. Leenaerts O, Partoens B, Peeters FM (2009) Water on graphene: hydrophobicity and dipole moment using density functional theory. Phys Rev B 79:235–244
45. Chen S, Brown L, Levendorf M, Cai W, Ju YS, Edge-worth J et al (2011) ACS Nano 5:1321–1327
46. Liu Y, Zhang J, Li S, Wang Y, Han Z, Ren L (2014) Fabrication of a super hydrophobic graphene surface with excellent mechanical abrasion and corrosion resistance on aluminium alloy substrate. RSC Adv 4:45389–45396
47. Kim Y-S, Kim J-G (2017) Electroplating of reduced-graphene oxide on austenitic stainless steel to prevent hydrogen embrittlement. Int J Hydrogen Energy 1–10
48. Hummers WS, Offeman RE (1958) Preparation of graphitic oxide. J Am Chem Soc 80:1339–1339
49. Hecker SS, Stout MG, Stauidhammer KP, Smith JL (1982) Metal Trans 13A:619
50. Han G, He S, Fukuyama S, Yokogawa K (1998) Acta Mater 46(13):4559
51. Maksimovich GG, Tret'yak IY, Ivas'kevich LM, Slipchenko TV (1985) Fiz.-Khim. Mekh. Mater 21(4):29
52. Mindyuk AK, Svist EI, Koval VP (1974), Vasilenko II, Babei YuI (1974) Mater Sci 8(1):98
53. Chen CL, Lee PY, Wu JK, Chiou DJ, Chu CY, Lin JY (1993) Corros Prev Control 40(3):71
54. Vainonen E, Likonen J, Ahlgren T, Haussalo P, Wu CH (1997) J Appl Phys 82(8):3791
55. Scully JR, Young GA, Smith SW (2000) Mat Sci Forum 331–337:1583
56. Young GA, Scully JR (1998) Acta Mater 46:6337
57. Hashimoto E, Kino T (1983) J Phys F Met Phys 13:1157
58. Caskey GR, Dexter AH, Holzworth ML, Louthan MR, Derrick RG (1976) Corrosion 32:370
59. Rudd DW, Vose DW, Johnson S (1961) J. Phys. Chem. 65:1018
60. Begeal DB (1978) J Vac Sci Technol 15:1146
61. Luu WC, Kuo HS, Wu JK (1997) Corros Sci 39(6):1051
62. Michler T, Naumann J (2009) Coatings to reduce hydrogen environment embrittlement of 304 austenitic stainless steel. Surface & Coatings Technology 203:1819–1828
63. Zhu SD, Wei JF, Bai ZQ, Zhou GS, Miao J, Cai R (2011) Failure analysis of P110 tubing string in the ultra-deep oil well. Eng Fail Anal 18:950–962
64. Covered T Analysis of high-collapse grade P110 coupling failures—a case study by element materials technology
65. Lin N, Guo J, Xie F, Zou J, Tian W, Yao X, Zhang H, Tang B (2014) Comparison of surface fractal dimensions of chromizing coating and P110 steel for corrosion resistance estimation. Appl Surf Sci 311:330–338. https://doi.org/10.1016/j.apsusc.2014.05.062
66. El Rayes MM, Abdo HS, Khalil KA (2013) Erosion—corrosion of cermet coating. Int J Electrochem Sci 8:1117–1137
67. Pombo Rodriguez RMH, Paredes RSC, Wido SH, Calixto A (2007) Comparison of aluminum coatings deposited by flame spray and by electric arc spray. Surf Coatings Technol 202:172–179. http://dx.doi.org/10.1016/j.surfcoat.2007.05.067
68. Guilemany JM, Miguel JM, Armada S, Vizcaino S, Climent F (2001) Use of scanning white light interferometry in the characterization ofwear mechanisms in thermal-sprayed coatings. Mater Charact 47:307–314. https://doi.org/10.1016/S1044-5803(02)00180-8
69. Motta FP (2011) Propriedades de revestimentos de nióbio obtidos por aspersão térmica a plasma sobre aço API 5L X65. Universidade Federal do Rio Grande do Sul

70. Ibrahim A, Berndt CC (2007) Fatigue and deformation of HVOF sprayed WC–Co coatings and hard chrome plating. Mater Sci Eng A 456:114–119
71. Cha SC, Gudenau HW, Bayer GT (2002) Comparison of corrosion behavior of thermal sprayed and diffusion-coated materials. Mater Corros 53:195–205
72. Gu L, Zou B, Fan X, Zeng S, Chen X, Wang Y, Cao X (2012) Oxidation behavior of plasma sprayed Al @ NiCr with cyclic thermal treatment at different temperatures. Corros Sci 55:164–171. https://doi.org/10.1016/j.corsci.2011.10.017
73. Matthews S, James B, Hyland M (2013) High temperature erosion—oxidation of Cr_3C_2—NiCr thermal spray coatings under simulated turbine conditions. Corros Sci 70:203–211
74. Tan JC, Looney L, Hashmi MSJ (1999) Component repair using HVOF thermal spraying. Mater Process Technol 93:203–208
75. Wang Y, Bai Y, Liu K, Wang JW, Kang YX, Li JR, Chen HY, Li BQ (2015) Microstructural Evolution of Plasma Sprayed Submicron-/nano-zirconia-based thermal barrier coatings
76. Xu K, Wang A, Wang Y, Dong X, Zhang X, Huang Z (2009) Surface nano crystallization mechanism of a rare earth magnesium alloy induced by HVOF supersonic micro particles bombarding. Appl Surf Sci 256:619–626. https://doi.org/10.1016/j.apsusc.2009.06.098
77. de Brandolt CS, Ortega-Vega MR, Menezes TL, Schroeder RM, de Malfatti CF (2016) Corrosion behavior of nickel and cobalt coatings obtained by high-velocity oxyfuel (HVOF) thermal spraying on API 5CT P110 steel. Mater Corros 67:368–377
78. Kouřil M, Christensen E, Eriksen S, Gillesberg B (2012) Corrosion rate of construction materials in hot phosphoric acid with the contribution of anodic polarization. Mater Corros 63:310–316
79. Wang W, Mohammadi F, Alfantazi A (2012) Corrosion behaviour of niobium in phosphate buffered saline solutions with different concentrations of bovine serum albumin. Corros Sci 57:11–21
80. Tommaselli MAG, Mariano NA, Pallone EMJA, Kuri SE (2004) Oxidation of niobium particles embedded in a sintered ceramic matrix. Mater Corros 55:531–535
81. Galetz MC, Rammer B, Schutze M (2015) Refractory metals and nickel in high temperature chlorine containing environments—thermodynamic prediction of volatile corrosion products and surface reaction mechanisms: a review. Mater Corros 66:1206–1214
82. Mathieu S, Knittel S, Berthod P, Mathieu S, Vilasi M (2012) On the oxidation mechanism of niobium-base in situ composites. Corros Sci 60:181–192
83. de Souza Brandolt C, Noronha LC, Hidalgo GEN, Takimi AS et al (2017) Niobium coating applied by HVOF as protection against hydrogen embrittlement of API 5CT P110 steel. Surf Coat Technol

Commencement of Green Supply Chain Management Barriers: A Case of Rubber Industry

Somesh Agarwal, Mohit Tyagi, and R. K. Garg

Abstract Green supply chain management (GSCM) is one of the recent revolutions to improve supply chain management services. Due to the presence of many obstacles, green business practices are not easy to adopt and implement. Therefore, industries must implement strategies to reduce the impact of their products and services on the environment. Most SMEs (small and medium enterprises) that play a crucial role in the Indian economy face great challenges in the implementation of green initiatives. The purpose of this research is to reduce the effect of the barriers that prevent the Indian rubber industry from implementing the GSCM. The aim of this research is to reduce the effect of barriers preventing the adoption of GSCM by Indian RUBBER industries situated in north India. For that reason, contextual relationships among the identified barriers were identified using Interpretive Structural Modeling (ISM) methodology. Literature survey of various articles and expert suggestions from industries and academia has been done and twenty-five most prominent barriers have been found out. Data was gathered through a survey based on the questionnaire, and the results were gathered. Data collected has been analyzed using ISM. However, a structural model barrier design for implementing the GSCM has also been proposed for the case of Indian rubber industries. A strong understanding of these barriers allows businesses to prioritize more efficiently and effectively managing their resources. In addition, strategies for overcoming barriers in the adoption of GSCM are proposed.

Keywords Green supply chain management · Barriers · Rubber industries · ISM · MICMAC analysis

---

S. Agarwal · M. Tyagi (✉) · R. K. Garg
Department of Industrial and Production Engineering, Dr. B R Ambedkar National Institute of Technology, Jalandhar, Punjab, India
e-mail: mohitmied@gmail.com

© The Author(s), under exclusive license to Springer Nature Singapore Pte Ltd. 2021

R. M. Singari et al. (eds.), *Advances in Manufacturing and Industrial Engineering*, Lecture Notes in Mechanical Engineering,
https://doi.org/10.1007/978-981-15-8542-5_59

1 Introduction and Background

Degradation of the atmosphere is a growing global issue. Increasing production and consumption has resulted in increased use of raw material and energy, contributing to natural resource depletion. In addition, there has also been a significant increase in waste production and pollution [2]. Therefore, there is a great need for improvement in the manufacturing of products in industrial processes. Thus, companies are faced with the dual challenge of responding consequently to competitive and environmental demands [3]. GSCM is viewed as a method of incorporating environmental concerns, principles, and philosophy into the supply chain. For environmental factors, GSCM has attracted the attention of most multinationals and small and medium-sized enterprises (SME's) over the past two decades. The main aim of GSCM is to minimize or reduce waste in the form of energy, emission, dangerous solid and waste chemicals. The addition of the 'green' component to SCM includes addressing the impact and connection between SCM and the natural environment [22, 24]. According to Gilbert [8], the necessity for GSCM has increased due to public knowledge, cultural, environmental or legislative reasons. That's not an easy task, though. The adoption and application of sustainable business practices was hampered by many obstacles [4].

The objective of this research is to identify the various barriers to the implementation of GSCM in the Indian rubber industry and identifying the contextual relationship between the identified barriers. Classification of barriers was achieved using MICMAC analysis based on dependency and driving power. In addition, a functional analysis of obstacles to the introduction of GSCM in the Indian rubber industry has also been developed using the Interpretive Functional Modeling (ISM) methodology. Various types of rubber industries situated in North India and East India are taken into account. Mostly, rubbers are used in belts, braces, sleepers, dampers, tubing, sports equipment, water insulation.

2 Identification of GSCM Barriers

GSCM is an excellent practice to deal with the environmental issues. Many researchers contribute to this topic, and they had found barriers related to different kinds of industries, namely plastic, manufacturing, automobile, etc. After reviewing research articles related to GSCM barriers, a total of the most prominent 25 barriers have been identified which hinders in implementation of GSCM in the Indian rubber industry.

Literature was reviewed to identify barriers to implement GSCM, and findings are listed in Table 1. They are arranged according to the mean values of the responses obtained through questionnaire survey.

Table 1 Listing of the barriers and their description

S. N.	Barrier	Description	Author(s)
1	Inflexible organizational systems and processes	The structure of the organization must be planned so that any future changes can rearranged	Chen et al. [5], Roarty [19]
2	An unwillingness to share risks and rewards	Risk management is a common success feature. If someone does not take risks, then possibility to improve decreases and even fail to progress	Sharma [21]
3	Market competition and uncertainty	Due to global competition and consumer demands, market volatility is very high	Hosseini [10], Mudgal et al. [15]
4	Poor quality of human resources	A business that has higher human resources efficiency such as better training or education will help in implementing Green Supply Chain Management	Yu Lin [26], Hsu et al. [11]
5	Lack of training for new mindsets and skills	Continuous improvement is a compulsion. It is done through training. Skill and attitude training is crucial to survival and survival	Chien et al. [6]
6	Material used is not easily biodegradable	For material requires more complexity in the recycling process creates the problem	AlKhidir et al. [1]
7	Restrictive company policies toward product/process stewardship	Product management is the management code to ensure the safe handling and use of products over their life cycle, including from development, design, construction, marketing, distribution, use. Product management is the management code	Roarty [19], Jose [12]
8	Lack of government support systems	No compensation to industries from the government for GSCM implementation	Yu Lin et al. [26], Tyagi et al. [25]
9	Difficult to carry out the reverse logistics	It is not easy to carry out the reverse logistics of the product	Yu Lin et al. [26]

(continued)

Table 1 (continued)

S. N.	Barrier	Description	Author(s)
10	Unawareness of customers	Customer demands become the most important type of external pressure as they have to be strictly followed	Lamming et al. [13], Mudgal et al. [16]
11	Lack of support and guidance from regulatory authorities	Support and guidance from regulatory bodies in the areas of waste management, waste reduction and recycling regulations, green energy income	Luthra et al. [14], Tyagi et al. [23]
12	Fright of product failure	Fear that the product will fail in the market	Zhu and Sarkis [27]
13	Limited forecasting and planning	Accurate return forecasts are hardly available. This is a direct barrier for both strategic and reverse chain due to the degree of diversity of goods and flows	Lamming et al. [13], Holland [9]
14	Lack of top management support	Support of top management can modify the existing system from all points of consideration. Lacking which will have a negative impact on the system	Luthra et al. [14], Tyagi et al. [23]
15	Inability or unwillingness to share information	A very likely problem is Communication lag. If the flow of information is disrupted by ignorance or reluctance, the system flow is greatly disturbed and the possibility of further system transformation is reduced	Chen et al. [5], Roarty [19]
16	Lack of corporate social responsibility	Corporate social responsibility indicated that businesses are prepared to go beyond enforcement and take the public implications of organizational steps	Sharma [21]
17	Time demand is more for its outcome	GSCM requires a lot of time for its proper implementation and to find its results outcomes	Freeman et al. [7]
18	Non-aligned strategic and operating philosophies	All stakeholders in the system should have a clear system definition that the company adopts	Freeman et al. [7]

(continued)

Table 1 (continued)

S. N.	Barrier	Description	Author(s)
19	Recycling cost is too high	Very costly to recycle the product	Chien et al. [6]
20	Incapable of achieving the exact design standard	They are incapable of achieving the same design as that obtained by a simple process	Hosseini [10], AlKhidir et al. [1]
21	Lack of trust among supply chain members	The psychological barrier is a lack of trust. It is extremely harmful because a single individualistic attitude is involved. Missing trust slows the progress	Jose [12]
22	Communication gap between management and shop floor workers	There must not be proper interaction and proper understanding between management and workers. Also, it is due to under-educated workers	Ravi and Shankar [18]
23	Small market demand	The market demand for the product is quite less	Sharma [21]
24	Benefits are not clear	Benefits related to implementation related to GSCM are not being cleared to the industry	Ravi and Shankar [18]
25	Not in priority list	For most of the industry, their own goals are so much that they cannot able to focus on greening	Tyagi et al. [24]

3 Methodology

Firstly, a questionnaire is sent to the relevant persons of the industries. All the replies of the survey are collected. Then, these replies are converted into numeric values, and the barriers are ranked according to their mean values. After that, the ISM-based model is applied to the prioritized barriers. For that firstly, SSIM matrix has been formed by giving a relation of one barrier with all other. After that, the SSIM matrix is converted into binary form to form an initial reachability matrix. After that final reachability matrix has been formed by using the concept of transitivity. On iteration levels of barriers are found out. MICMAC analysis is done to classify the barriers in four categories.

3.1 Interpretive Structural Modeling (ISM)

Interpretive Structural Modeling (ISM) was developed in the 1970 s and used as a technique to describe the relationship between the particular elements that characterize an issue or problem [20]. ISM is an interactive learning mechanism in which a number of different variables are organized into a full structural model, direct and indirectly linked to the person [15, 17].

3.2 ISM Approach

In the beginning, SSIM has been developed, the following four symbols have been used to denote the direction of the relationship between two barriers *i* and *j* [24].

V—Barrier i will lead to barrier j;
A—Barrier j will lead to barrier i;
X—Barrier i and j will lead to each other;
O—Barrier i and j are unrelated.

Based on the contextual relationships, the SSIM has been developed (Table 2).

After that, this matrix is used to classify the barriers into reachability, intersection and antecedent sets (Table 3). Further level partition iteration has been done in Table 3, and barrier with equivalent accessibility is shown in the ISM hierarchy or Level 1 at the intersection set as a top-level barrier. It is then discarded to find additional levels after reaching level 1. The iterative process is continued until the barrier level of each barrier is formed. In our study, 12 levels were identified.

Further, on the basis of level partition, the ISM-based model or framework has been formed in Fig. 1 by converting. According to the Fig. 1 Lack of Government, support systems is on the top which shows the effectiveness of GSCM activities depends has appeared on top of the hierarchy. In Fig. 1, various levels have been obtained ranging from 1 to 12.

3.3 MICMAC Analysis

This was subsequently carried out, based on the study of the barriers dependency power and driver power, categorized into four sectors [14] and shown in Fig. 2. There are low driving power and dependency on autonomous variables (first cluster), and one can disconnect these variables from the system. In our research, no major barrier lies in this scope. The second cluster is called dependent variables which have weak driving power but strong dependence power. In our research, there is only one barrier named lack of trust among supply chain members is falling in this range. The third cluster named linkage variables having strong driving power and also strong

Table 2 Structured self-intersection matrix (SSIM) for barriers

	2	3	4	5	6	7	8	9	10	11	12	13	14	15	16	17	18	19	20	21	22	23	24	25
1	V	O	X	V	O	A	O	O	O	V	O	V	V	V	V	O	V	O	O	O	V	O	A	V
2		O	O	O	O	O	O	O	O	O	O	O	A	O	O	O	A	O	O	X	V	O	O	O
3			O	O	O	O	O	O	O	O	V	V	O	O	O	V	O	O	O	O	X	X	X	O
4				A	O	A	O	O	O	A	O	O	O	O	O	O	O	O	A	O	X	O	O	O
5					O	A	O	V	O	A	O	O	O	O	O	O	O	O	A	O	O	O	O	O
6						A	O	V	O	O	O	O	O	O	O	V	O	O	X	O	O	O	O	O
7							A	O	A	A	A	A	V	V	V	O	V	X	O	X	O	O	O	V
8								O	V	V	O	X	V	O	O	X	O	O	X	O	A	A	V	V
9									O	A	X	O	A	O	O	O	O	X	A	O	O	A	A	O
10										O	O	O	A	O	O	A	O	A	A	V	V	O	A	O
11											A	O	A	V	V	O	O	O	V	O	O	A	O	V
12												O	A	X	O	O	O	A	A	V	A	A	O	O
13													A	A	X	O	A	O	O	A	V	O	O	A
14														V	O	O	V	O	A	O	A	O	V	O
15															O	O	O	O	O	O	O	X	V	O
16																O	V	O	V	X	O	X	A	V
17																	O	O	A	O	O	A	V	X
18																		O	O	O	O	V	A	V
19																			O	O	O	O	V	V
20																				O	O	O	O	V
21																					O	O	O	O

(continued)

Table 2 (continued)

	2	3	4	5	6	7	8	9	10	11	12	13	14	15	16	17	18	19	20	21	22	23	24	25
22																						O	O	O
23																							O	V
24																								V

Based on SSIM matrix, initial reachability matrix has been formed, and then, final reachability has been developed using the concept of transitivity Sharma [21]
Here D. P. represents dependence power, and Dr. P. denotes driving power

Table 3 Classification of reachability and antecedent and iteration for level partition

S. N.	Barriers	Reachability set	Antecedent set	Intersection set	Level
1.	8	1 2 3 4 8 9 10 11 12 13 15 16 17 18 20 21 22 23 24 25	8 15	8 15	I
2.	21	1 2 3 5 6 7 9 10 11 12 15 16 17 18 19 20 21 22 23 24 25	8 9 10 21	9 10 21	II
3.	13	1 2 4 5 9 10 11 12 13 14 15 16 17 18 19 20 22 23 24 25	1 8 13 17 18	1 8 13 17 18	III
4.	6	1 2 4 6 7 9 10 11 12 14 15 16 18 19 20 22 23 24 25	3 4 5 6 10 12 16 20 21 22 23 24 25	4 6 10 12 16 20 22 23 24 25	IV
5.	3	1 2 3 4 5 7 8 11 12 14 15 17 18 20 21 22 23 24 25	1 2 3 5 7 8 11 12 14 15 17 18 20 21 22 23 24	1 2 3 4 5 7 8 9 10 11 12 13 14 15 16 17 18 20 21 22 23 24	V
6.	22	1 2 3 4 5 6 7 9 10 11 12 15 17 18 19 23 24 25	1 2 3 5 6 7 8 9 10 11 12 13 15 16 17 19 21 23 24 25	1 2 3 4 5 6 7 9 10 11 12 15 17 19 23 24 25	VI
7.	7	1 2 3 4 5 6 7 9 10 11 12 13 14 15 16 17 18 19 20 21 22 23 24 25	1 2 4 6 7 9 10 11 12 14 15 16 18 19 20 21 23 24 25	1 2 4 6 7 9 10 11 14 15 18 19 20 22 23 24 25	VII
8.	9	1 2 3 5 7 8 9 10 12 13 14 16 17 18 19 21 22 24 25	2 4 5 6 7 9 10 11 12 14 15 16 17 18 19 20 21 22 23 24 25	2 5 7 9 10 11 12 14 16 17 18 19 20 21 22 24 25	VIII
9.	14	1 2 3 5 7 9 10 11 14 15 16 17 18 19 23 25	1 2 3 5 6 7 9 10 11 12 13 14 16 17 18 19 20 22 23 25	1 2 3 5 7 9 10 11 14 16 17 18 19 23 25	VIII
10.	16	1 4 5 6 7 9 10 11 12 14 15 16 17 18 19 20 22 24 25	1 2 3 5 6 8 9 10 11 13 14 15 16 17 19 20 21 22 23 25	1 5 6 9 10 11 14 15 16 17 19 20 22 25	VIII
11.	12	1 2 3 4 5 6 7 9 10 11 12 14 15 17 18 19 20 22 23 24 25	1 2 4 5 6 8 9 10 11 12 13 15 16 17 18 19 20 21 22 23 24 25	1 2 4 5 6 9 10 11 12 15 17 18 19 20 22 23 24 25	IX

(continued)

Table 3 (continued)

S. N.	Barriers	Reachability set	Antecedent set	Intersection set	Level
12.	17	1 2 3 4 5 9 10 11 12 13 14 16 17 18 19 20 22 23 24 25	1 2 3 4 5 7 8 9 10 11 12 13 14 15 16 17 18 19 21 22 23 24	1 2 3 4 5 9 10 11 12 13 14 16 17 18 19 22 23 24	IX
13.	24	1 2 3 4 5 6 7 9 10 11 12 15 17 18 19 22 23 24 25	1 2 3 5 6 7 8 9 10 11 12 13 15 16 17 18 19 20 21 23 24 25	1 2 3 5 6 7 9 10 11 12 15 17 18 19 23 24 25	IX
14.	25	1 2 4 5 6 7 9 10 11 12 14 15 16 18 19 20 22 23 24 25	1 2 4 5 6 8 9 10 11 12 13 14 15 16 17 18 20 21 22 23 24 25	1 2 4 5 6 9 10 11 12 14 15 16 18 20 22 23 24 25	IX
15.	2	1 2 3 4 5 7 9 10 11 12 14 15 16 17 18 19 20 23 24 25	1 2 3 5 6 7 8 9 10 11 12 13 14 15 17 18 19 20 21 2.2 23 24 25	1 2 3 5 6 7 9 10 11 12 13 14 15 17 18 19 20 23 24 25	X
16.	4	1 2 3 4 5 6 7 8 9 10 11 12 13 15 16 17 18 19 21 22 23 24 25	1 4 5 6 7 10 11 12 15 17 18 19 20 22 23 25	1 4 5 6 7 10 11 12 15 17 18 19 20 22 23 25	X
17.	5	1 2 3 4 5 6 9 10 11 12 14 15 16 17 18 19 20 23 24 25	1 2 3 4 5 7 9 10 11 12 13 14 15 16 17 18 19 20 21 22 23 24 25	1 2 3 4 5 6 7 8 9 10 11 12 14 15 16 17 18 19 20 23 24 25	X
18.	20	1 2 3 5 6 7 9 10 11 12 14 15 16 18 19 20 23 24 25	1 2 3 4 5 6 7 8 9 10 11 12 13 15 16 17 18 19 20 21 22 23 25	1 2 3 5 6 7 9 10 11 12 15 16 18 19 20 23 25	X
19.	1	1 2 3 4 5 6 7 9 10 11 12 13 14 15 16 17 18 19 20 23 24 25	1 2 3 4 5 6 7 8 10 11 12 13 14 15 16 17 18 19 20 21 22 23 24 25	1 2 3 4 5 6 7 10 11 12 13 14 15 16 17 18 19 20 21 22 23 24 25	XI
20.	15	1 2 3 4 5 7 8 10 11 12 15 16 17 18 19 20 22 23 24 25	1 2 3 4 5 6 7 8 9 10 11 12 13 14 15 16 18 19 20 21 22 23 24 25	1 2 3 4 5 7 8 10 11 12 15 16 18 19 20 22 23 24 25	XI

(continued)

Table 3 (continued)

S. N.	Barriers	Reachability set	Antecedent set	Intersection set	Level
21.	19	1 2 4 5 7 9 10 11 12 14 15 16 17 18 19 20 22 23 24	1 2 3 4 5 6 7 9 10 11 12 13 14 15 16 17 18 19 20 21 22 23 24 25	1 2 4 5 7 9 10 11 12 14 15 16 17 18 19 20 22 23 24	XI
22.	23	1 2 3 4 5 6 7 10 11 12 14 15 16 17 18 19 20 22 23 24 25	1 2 3 4 5 6 7 8 9 10 11 12 13 14 15 17 18 19 20 21 22 23 24 25	1 2 3 4 5 6 7 10 11 12 14 15 17 18 19 20 22 23 24 25	XI
23.	10	1 2 4 5 6 7 9 10 11 12 14 15 16 17 18 19 20 21 23 24 25	1 2 3 4 5 6 7 8 9 10 11 12 13 14 15 16 17 18 19 20 21 22 23 24 25	1 2 4 5 6 7 9 10 11 12 14 15 16 17 18 19 20 21 23 24 25	XII
24.	11	1 2 3 4 5 7 10 11 12 14 15 16 17 18 19 20 22 23 24 25	1 2 3 4 5 6 7 8 9 10 11 12 13 14 15 16 17 18 19 20 21 22 23 24 25	1 2 3 4 5 7 10 11 12 14 15 16 17 18 19 20 22 23 24 25	XII
25.	18	1 2 3 4 5 7 9 10 11 12 13 14 15 17 18 19 20 23 24 25	1 2 3 4 5 6 7 8 9 10 11 12 13 14 15 16 17 18 19 20 21 22 23 24 25	1 2 3 4 5 7 9 10 11 12 13 14 15 17 18 19 20 23 24 25	XII

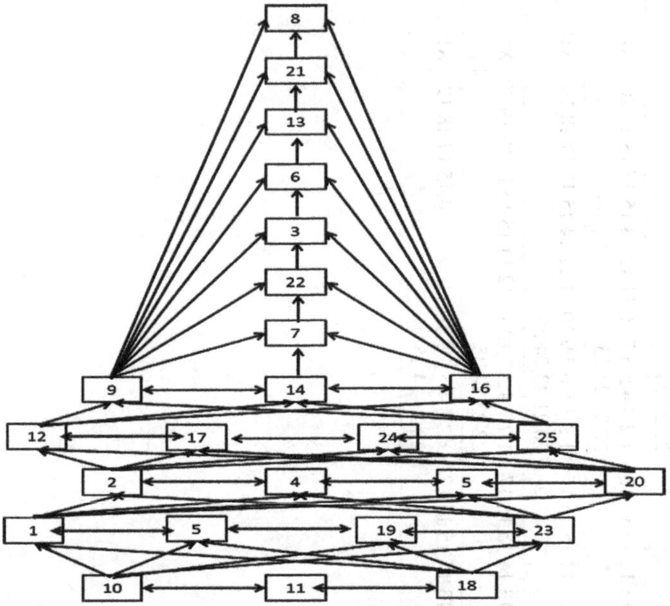

Fig. 1 ISM framework

dependence power. In our study, 1, 2, 3, 4, 5, 6, 7, 9, 10, 11, 12, 14, 15, 16, 17, 18, 19, 20, 21, 22, 24 and 25 lie in this range. The fourth cluster named independent variables has high driving power but weak dependence power. In our study, two barriers named lack of government support systems and limited forecasting and planning (8 & 13) lie in this range. Figure 2 shows the graph between dependence power and driving power for the barriers.

Higher dependency values for a variable mean a large number of barriers to be resolved before eliminating it, and high barrier driving value means a large number of barriers can be overcome after removing it.

4 Results and Discussion

An ISM-based model was developed in this research to analyze the interactions between various GSCM barriers. For success in GSCM programs, these barriers need to be overcome. The driver dependence model offers few valuable insights into the relative importance of obstacles and their interdependencies. A few of the major implications of this research are as follows:

- There is no autonomous barrier in the driver dependence matrix (Fig. 2). The unavailability of autonomous barriers in this study draws attention to the fact that

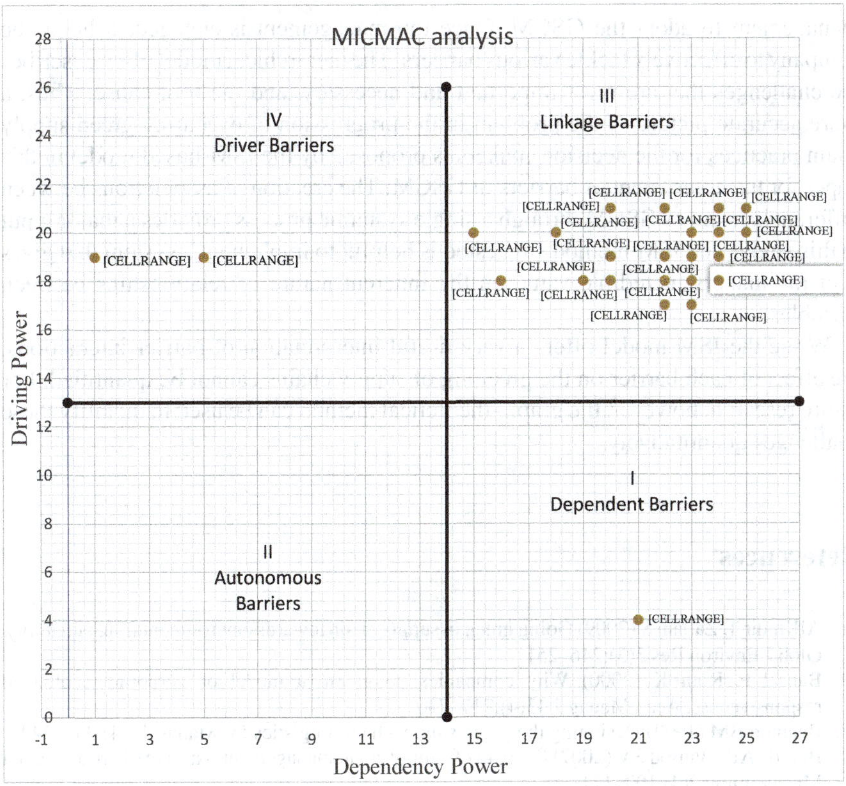

Fig. 2 Driver and dependence power

all the barriers considered affecting the practices of GSCM. 21 is found to be dependent barrier and are weak drivers but strongly depend on other barriers.

- Some barriers such as 1, 2, 3, 4, 5, 6, 7, 10, 11, 12, 14, 15, 16, 17, 18, 19, 20, 22, 23, 24 and 25 are barriers to linkage. They have powerful driving power as well as strong dependence. These will be the ones influenced by lower-level barriers and, in turn, have an impact on the model's other barriers.
- However, barriers like 8 and 13 are individual barriers, i.e., they have high driving power and weak dependence over other barriers. They can be treated as the root cause of the remaining barriers. A detailed strategic plan of GSCM should be implemented to overcome these barriers in order to achieve sustainability.

5 Conclusion and Future Work

This study can help an enterprise to address the barriers of GSCM in strategic and tactical ways. The key strategic decision is based on the commitment of the top management to adopt the GSCM. Once top management is engaged, it helps the company to effectively tackle various barriers. The hierarchical model often describes the challenges that are most important and necessary, and the root cause. Thus, a more accurate picture of the problem in the progression of executing green supply chain practices for the decision-makers is proposed by the ISM-based model in this paper for the assessment of barriers of GSCM. The creation of associations between different barriers of GSCM through a single systemic process provides a major input to this study. The ISM method suggested is helpful to implement a system that gives decision-makers incredible value on the inherent nature of relationships between variables.

While the ISM model offers a very useful interpretation of barrier interactions, the effect of each barrier on the greening of supply chains cannot be quantified. For future research, however, the graph–theoretical method can be used to quantify these challenges quantitatively.

References

1. AlKhidir T, Zailani S (2009) Going green in supply chain towards environmental sustainability. Glob J Environ Res 3(3):246–251
2. Bansal P, Roth K (2000) Why companies go green: a model of corporate ecological responsiveness. Acad Manag J 43(4):717–736
3. Beamon BM (1999) Designing the green supply chain. Logistics Inf Manag 12(4):332–342
4. Borade AB, Bansod SV (2007) Domain of supply chain management—a state of art. J Technol Manag Innov 2(4):109–121
5. Chen Y-S, Lai S-B, Wen C-T (2006) The influence of green innovation performance on corporate advantage in Taiwan. J Bus Ethics 67(4):331–339
6. Chien MK, Shih LH (2007) An empirical study of the implementation of green supply chain management practices in the electrical and electronics industries and their relation to organizational behavior. Int J Sci Technol 4(3):383–394
7. Freeman H, Harten T, Springer J, Randall P, Curran MA, Stone K (1992) Industrial pollution prevention: a critical review. J Air Waste Manag Assoc 42(5):617–656
8. Gilbert S (2000) Greening supply chain: enhancing competitiveness through green productivity. Report of the top forum on enhancing competitiveness through green productivity held in the Republic of China, 25–27 May (2000)
9. Holland I, Gibbon J (1997) SMEs in the metal manufacturing, construction and contracting service sectors: environmental awareness and actions. Eco-Manag Audit 4:7–14
10. Hosseini A (2007) Identification of green management of system's factors—a conceptualized model. Int J Manag Sci Eng Manag 2(3):221–228
11. Hsu CW, Hu AH (2008) Green supply chain management in the electronic industry. Int J Sci Technol 5(2):205–216
12. Jose PD (2008) Getting serious about green. Real CIO World 3(8):26–28
13. Lamming R, Hamapson J (1996) The environmental as a supply chain management issue. Br J Manag 7(March Special Issue):45–62
14. Luthra S, Manju, Kumar S., Haleem A suggested implementation of the green supply chain management in automobile industry of India: a review. In: Proceedings of national conference on 'advancements and futuristic trends of mechanical and industrial engineering, GITM, Bilaspur (INDIA), 12–13 Nov, (2010)
15. Mudgal RK, Shankar R, Talib P (2010) Modeling of green supply chain management barriers; Indian perspective. Int J Logistics Syst Manag 7(1):81–107

16. Mudgal RK, Shankar R, Talib P, Raj T (2009) Greening the supply chain practices: an Indian perspective. Int J Logistics Res Appl 8(4):513–520
17. Pun KF (2006) Determinants of environmentally responsible operations: a review. Int J Qual Reliab Manag 23(3):279–297
18. Ravi V, Shankar R (2005) Analysis of interactions among the barriers of reverse logistics. Technol Forecast Soc Change 72:1011–1029
19. Roarty M (1997) Greening business in a market economy. Eur Bus Rev 97(5):244–254
20. Sarkar A, Mohapatra PK (2006) Evaluation of supplier capability and performance: a method for supply base reduction. J Purchas Supply Manag 12(3):148–163
21. Sharma S (2000) Managerial interpretations and organizational context as predictors of corporate choice of environmental strategy. Acad Manag J 43(4):681–697
22. Srivastava SK (2007) Green supply-chain management: a state-of-the-art literature review. Int J Manag Rev 9(1):53–80
23. Tyagi M, Kumar P, Kumar D (2015) Analysis of interaction among the drivers of green supply chain management. Int J Bus Perform Supply Chain Model 7(1):92–108
24. Tyagi M, Kumar P, Kumar D (2017) Modeling and analysis of barriers for supply chain performance measurement system. Int J Oper Res 28(3):392–414
25. Tyagi M, Kumar D, Kumar P (2018) Assessment of CSR based supply chain performance system using an integrated fuzzy AHP-TOPSIS approach. Int J Logistics Res Appl 21(4):378–406
26. Yu L, Hills P, Welford R (2008) Extended producer responsibility and eco-design changes: perspectives from China. Corp Soc Responsib Environ Manag 15:111–124
27. Zhu Q, Sarkis J (2008) Relationships between operational practices and performance among early adopters of green supply chain management practices in Chinese manufacturing enterprises. J Oper Manag 22(3):265–289

Estimation of Critical Key Performance Factors of Food Cold Supply Chain Using Fuzzy AHP Approach

Neeraj Kumar, Mohit Tyagi, and Anish Sachdeva

Abstract The success of any food cold supply chain (FCSC) depends upon the degree of extent to which a range of sustainable temperature can be maintained which must be achieved in order to maintain the self-life, quality and potency of the food from production to end consumption stage. Fluctuation in permissible sustainable temperature range starts microbiological reactions in food products which results into loss of quality, taste and texture and finally into risk of food poisoning and food spoilage. The failure of cold chain increases the risk for human lives and environmental changes. Therefore, in order to reduce the food spoilage, energy consumption, global warming and maintaining the quality, taste and texture of food products, it becomes necessary to identify the key performance factors (KPF) of FCSC. This research work identifies the most critical key performance factors of FCSC and provides priority weights for these factors which represent interrelationship amongst the factors and thus helps the decision makers to take decisions so that the performance of the same can be improved. In order to achieve the desired objective of the research, fuzzy AHP approach has been used. The objective of the research is not only to prioritize the key performance factors but also to make aware about these factors and enable the decision makers to take decisions so that the performance of FCSC can be improved. Results of the research work reveal that the factors: climatic impact of FCSC, high volume of food waste and high amount of energy consumption are most critical and must be taken at first priority by the decision-makers while establishing the benchmark for performance improvement of FCSC.

Keywords Food cold supply chain · Food waste · Carbon footprint · Energy consumption · Fuzzy- AHP · GHG emission · Global warming

N. Kumar · M. Tyagi (✉) · A. Sachdeva
Department of Industrial and Production Engineering, B. R. Ambedkar National Institute of Technology, Jalandhar, India
e-mail: mohitmied@gmail.com

© The Author(s), under exclusive license to Springer Nature Singapore Pte Ltd. 2021
R. M. Singari et al. (eds.), *Advances in Manufacturing and Industrial Engineering*, Lecture Notes in Mechanical Engineering,
https://doi.org/10.1007/978-981-15-8542-5_60

1 Introduction and Background

From the past few decades, the trends of online purchasing of perishable food products have been raised very abruptly. Perishable food products defined as the food items which are likely to lose their quality and potency and turn out to be insecure to consume if they are not preserved at their sustainable temperature range. In high temperature and humid countries where the frequent variations in temperature are observed, it becomes very difficult to maintain the quality, integrity, taste and texture of perishable and temperature sensitive food products. Therefore, in order to maintain the quality, potency, integrity, taste and texture of the perishable food products from production to last mile delivery, it becomes necessary to integrate the terms quality, safety, refrigeration and green concept in food supply chain. A food cold supply chain is defined as the grid of refrigeration units, cold stores, refrigerated wagons and trucks, freezers and cold boxes organised in such way that a right range of temperature can be maintained so that the perishable food items remain fresh and intoxicated throughout the supply chain, i.e., from production to end consumption. The refrigeration process in cold chain is responsible for slow down the biological decay processes and deliver safe and high-quality foods to consumers. Though the refrigeration unit becomes as a soul measure for preservation and transportation of perishable foods, it faces several challenges to achieve its intended objective. The structure of a food cold supply chain comprises several elements such as suppliers, transport units, refrigerated storage, warehouses and end consumers. Therefore, the performance of a food cold supply chain depends on all the factors which directly or indirectly involved in operating a food cold chain. Mercier et al. [1] stated that any variation in the permissible sustainable temperature range of the perishable food products results into escalations of the microorganism's growth which under a time span increases the food spoilage and render the product inedible.

In a case study on U.S. food chain, Scharff et al. [2] estimated 3000 deaths and more than $50 billion annual illness cost due the hygienic food-borne diseases. Insufficient cold storage facility, results into high spoilage of the temperature sensitive and perishable food products. A research report published by U.S. Department of Commerce (2016) on top markets of cold supply chain estimated that nearly 50% of the global agricultural food produce losses due to lack of sufficient cold storage facilities which cost nearly $750 billion annually. India is the second largest populated country, third largest producer of agricultural produce and amongst the highest consumer of food in world, the food loss accounts for 40% of total agricultural produce which cost nearly $14 billion annually. James et al. [3] presented that an increase in surrounding temperature increases the risk of food poisoning and substantial increase in food loss and climate change. Similar effect of climate change on food cold chain was observed by Brander et al. [4]; Gregory et al. [5, 6]; Patterson and Lima [7]. The reverse effect of food loss also impacts the performance of food cold supply chain. As the food waste increases, the amount of greenhouse gas (GHG) emission from the food cold supply chain increases.

Inadequate temperature monitoring of the surrounding and food items in FCSC also affects the performance of cold supply chain [8]. Since the food items are biological products and a very little fluctuation in permissible sustainable range starts biochemical and microbial spoilage of food products. Therefore, it becomes necessary to monitor the temperature conditions throughout the FCSC. In addition to temperature monitoring, it is also required to monitor the other parameters such as humidity throughout the food cold chain (in warehouse as well as in transit). Similar results of inadequate temperature monitoring on FCSC also observed by Mercier et al. [1] and Sainathan et al., [9]. Sainathan et al. [9] presented that by tracking the temperature and surrounding conditions, the real-time temperature monitoring in transit as well as in warehouse reduces the risk of food waste and thus improves the performance of FCSC.

Many researchers have identified energy consumption as one of the important criteria in performance measure of FCSC. To maintain the sustainable temperature of perishable food items, cold supply chain (CSC) consumes a very huge amount of energy consumption. Around 40% of the food products requires its transportation in controlled temperature, and this required refrigeration of food items at a certain temperature range and consumes around 15% of the total electricity produced worldwide [10]. This huge amount of energy consumption accounts for 2 to 2.4% of total GHG emission worldwide [11]. It increases the global temperature which intern increases the rate of food spoilage and reduces the performance of food cold supply chain [12].

In a FCSC, flow of information of data plays a very critical role in the performance measure, and it is estimated that around 75% of the total food waste arises during production and distribution only because of lack of communication between supervisor and the other parties involved in FCSC. Liu et al. [13] presented that to make the FCCS success and effective, there should be uninterrupted flow of information between every member of the cold chain which must comprise environmental and processing data records from multiple sources. Many researchers [9, 14, 15] have presented that lack of traceability of food product in the FCSC plays a major role in reducing the food waste in FCSC. The temperature and surrounding conditions of product changed in every step the CSC (during supply, storage, packaging, transportation and dispatching); therefore, it is necessary to trace the product to check the functionality and surrounding status of product. The product traceability becomes as major challenge during primary, secondary and end consumer processing states. Traceability which is defined as the ability to track the food items throughout the FCSC is used to locate the food item and trace the geographical conditions.

2 Model Development

From a deep study of previous researches, identified literature gaps, discussion with experts and motivation, the objective of the present research work has been defined as:

1. Identify the most critical key performance factors of food cold supply chain; and
2. Find out the priority weights for the most critical key performance factor.

On the basis of existing literature review and discussion with field experts, eight most critical key factors which are responsible for performance measure of FCSC have been identified. Majority of the field experts belong to food processing industries and different academia located near Jalandhar district in Punjab state of India. The identified critical factors (criteria) are namely climatic impact of FCSC (F_1), energy consumption (F_2), high volume of food waste (F_3), inadequate temperature monitoring (F_4), insufficient cold storage facility (F_5), lack of product traceability (F_6), improper packaging and handling (F_7) and lack of efficient information flow network (F_8).

3 Research Methodology

In the research methodology section, it is aimed to give a brief introduction about the methodology which has been used to prioritize the KPF. To prioritize the identified most KPF, fuzzy AHP approach has been used. Fuzzy theory is integrated with other MCDM techniques by various authors and has been extensively used to assess the performance of intended system. Tyagi et al. [16–18] integrate fuzzy with DEMATEL approach to measure the performance of flexible manufacturing system. Tyagi et al. [16–18] used fuzzy AHP in integration with AHP approach to prioritize the alternatives of a supply chain performance system of automobile industry. To make a deep understanding about the implementation of fuzzy AHP, step-by-step algorithm of fuzzy AHP methodology has been discussed in the following section.

While implementing fuzzy decision-making approaches, the linguistic terms are converted into fuzzy numbers (FNs) which are available in various shapes (such as triangular FN, trapezoidal FN and Gaussian FN). Triangular FN and trapezoidal FN are most commonly used fuzzy numbers [19].

3.1 *Fuzzy Analytic Hierarchy Process (F-AHP)*

The Analytic Hierarchy Process (AHP) which is used as a multicriteria decision-making tool was first introduced by Thomas Saaty in [20]. Due to its simplicity to use and robustness, nowadays, it is extensively used as the standard tool to solve multicriteria decision-making (MCDM) problem. In another study preformed on an organization [21] defined AHP as a tool for solving a complex decision-making problem having a set of criteria/attributes by decomposing it into subproblem and assigning the definite scores based on views and opinions of decision-makers. Though the traditional AHP approach is used as a standard tool in decision-making problem, it faces several weaknesses such as it does not consider uncertainty and vagueness

Table 1 Saaty's scale for intensity of relative importance

Definition	Intensity of relative importance
Equally important/preferred	1
Weakly important/preferred	3
Strongly important/preferred	5
Very strongly more important/preferred	7
Absolutely more important/preferred	9
Intermediate importance between two adjacent judgements	2, 4, 6, 8

factors. To encounter these uncertainty and vagueness, the concept of fuzziness has been introduced into the traditional AHP method. Integration of fuzzy set theory into traditional AHP was first made by Laarhoven et al. in [22]. Tyagi et al. [23] proposed fuzzy AHP tool to measure the performance of a supply chain which is based upon corporate social responsibility (CSR). In the current research work, geometric mean method of fuzzy AHP which was proposed by Buckley in [24] has been used to calculate the priority weights of the considered criteria of FCSC. The step-by-step procedure to calculate the priority weights is given as follows:

Step 1. Construction of pairwise comparison matrix. It is constructed with the help of Saaty's scale of relative importance for pairwise comparison as given in Table 1.

The pairwise comparison matrix for criteria or attributes should satisfy the following criteria

$$a_{ij} = 1, \text{ when } i = j; \text{ and } a_{ji} = \frac{1}{a_{ij}} \tag{1}$$

where i indicates *i*th row and *j* for *j*th column

Step 2. This step is known as fuzzification of step one. In this, linguistic terms are converted into membership function, i.e., scale of crisp numeric value is converted into fuzzy number (i.e., TFN) using five point conversion scale as given in Table 2.

Any reciprocal number can be converted into fuzzy number by using following equation-

In (*j*, *k*, *l*) denotes a TFN and Ã denotes any crisp numeric value, then

$$\tilde{A}^{-1} = (j, k, l)^{-1} = \left(\frac{1}{l}, \frac{1}{k}, \frac{1}{j}\right) \tag{2}$$

Step 3. After converting all the crisp numeric value into TFN, next fuzzy geometric mean $(\tilde{r}_i)$ for all the rows has to be calculated as given in Eq. (3)

$$(\tilde{r}_i) = \left(\sqrt[n]{\prod_1^n l_i}, \sqrt[n]{\prod_1^n j_i}, \sqrt[n]{\prod_1^n k_i}\right) \tag{3}$$

Table 2 Linguistic scale for alternatives and their corresponding TFNs

Definition	Crisp numeric value	Corresponding TFN
Equally important/preferred	1	1, 1, 1
Weakly important/preferred	3	2, 3, 4
Strongly important/preferred	5	4, 5, 6
Very strongly more important/preferred	7	6, 7, 8
Absolutely more important/preferred	9	9, 9, 9
Intermediate importance between two adjacent judgements	2,4,6,8	(1, 2, 3), (3, 4, 5), (5, 6, 7), (7, 8, 9) for 2, 4, 6, 8 respectively

Step 4. After calculating fuzzy geometric mean, next step calculate the fuzzy priority weights for each criteria as given in Eq. (4).

$$\tilde{w}_i = \tilde{r}_i \times \left(\sum_1^n \tilde{r}_i\right)^{-1} \tag{4}$$

Step 5. Calculate the crisp priority weights for each criteria as given in the following Eq. (5).

$$\text{CWi} = \frac{\sum_1^n \tilde{w}_i}{n} \tag{5}$$

Step 6. Consistency check for pairwise comparison between criteria or attributes.

4 Numerical Illustration

To analyse the criticality and to find out the priority weights of the identified eight most critical criteria for the performance improvement of FCSC, at first a questionnaire has been structured and dispatched to field experts, academic experts and food industries by email on Google doc and obtain the opinions and rating about the severity of each criteria. Industries which have been selected for data collection are situated in Punjab region and the field experts belong to different food industries and academies located in Punjab, India region. The rating of relative severity of identified criteria is obtained on Saaty's 1–9 scale of relative importance as given in Table 1. To find out the priority weights and to select most critical factors among them, fuzzy AHP approach has been used as (Table 3).

After preparing the pairwise comparison matrix, in the next step crisp numeric values are converted into TFN using Table 2 and is given in Table 4.

Table 3 Pairwise comparison matrix for criteria

Criteria	F1	F2	F3	F4	F5	F6	F7	F8
F1	1	2	1	5	3	6	5	6
F2	1/2	1	1/2	3	4	5	2	3
F3	1	2	1	4	2	5	4	3
F4	1/5	1/3	1/4	1	1/3	1/2	2	2
F5	1/3	1/4	1/2	3	1	4	6	3
F6	1/6	1/5	1/5	2	1/4	1	3	1/2
F7	1/5	1/2	1/4	1/2	1/6	1/3	1	3
F8	1/6	1/3	1/3	1/2	1/3	2	1/3	1

Table 4 Conversion matrix for pairwise comparison matrix into TFN

F_i	F1	F2	F3	F4	F5	F6	F7	F8
F1	(1, 1, 1)	(1, 2, 3)	(1, 1, 1)	(4, 5, 6)	(2, 3, 4)	(5, 6, 7)	(4, 5, 6)	(5, 6, 7)
F2	(1/3, 1/2, 1)	(1, 1, 1)	(1/3,1/2,1)	(2, 3, 4)	(3, 4, 5)	(4, 5, 6)	(1, 2, 3)	(2, 3, 4)
F3	(1, 1, 1)	(1, 2, 3)	(1, 1, 1)	(3, 4, 5)	(1, 2, 3)	(4, 5, 6)	(3, 4, 5)	(2, 3, 4)
F4	(1/6, 1/5, 1/4)	(1/4, 1/3, 1/2)	(1/5, 1/4, 1/3)	(1, 1, 1)	(1/4, 1/3, 1/2)	(1/3, 1/2, 1)	(1, 2, 3)	(1, 2, 3)
F5	(1/4, 1/3, 1/2)	(1/5, 1/4, 1/3)	(1/3, 1/2, 1)	(2, 3, 4)	(1, 1, 1)	(3, 4, 5)	(5, 6, 7)	(2, 3, 4)
F6	(1/7, 1/6, 1/5)	(1/6, 1/5, 1/4)	(1/6, 1/5, 1/4)	(1, 2, 3)	(1/5, 1/4, 1/3)	(1, 1, 1)	(2, 3, 4)	(1/3, 1/2, 1)
F7	(1/6, 1/5, 1/4)	(1/3, 1/2, 1)	(1/5, 1/4, 1/3)	(1/3, 1/2, 1)	(1/7, 1/6, 1/5)	(1/4, 1/3, 1/2)	(1, 1, 1)	(2, 3, 4)
F8	(1/7, 1/6, 1/5)	(1/4, 1/3, 1/2)	(1/4, 1/3, 1/2)	(1/3, 1/2, 1)	(1/4, 1/3, 1/2)	(1, 2, 3)	(1/4, 1/3, 1/2)	(1, 1, 1)

After converting all the crisp numeric value into TFN, next fuzzy geometric mean ($\tilde{r}_i$) for all the rows has been calculated using Eq. 3 (Table 5).

After calculating fuzzy geometric mean values for each row, next fuzzy weight for each criterion (factor) has been calculate for which Eq. 4 has been used. For example, the fuzzy weight for factor one can be given as;

Table 5 Calculated fuzzy geometric mean ($\tilde{r}_i$) for rows of Table 4

Factor	Fuzzy geometric mean value, ($\tilde{r}_i$)
F1	(2.306, 2.928, 3.473)
F2	(1.233, 1.755, 2.482)
F3	(1.707, 2.359, 2.928)
F4	(0.403, 0.569, 0.811)
F5	(1.000, 1.316, 1.763)
F6	(0.389, 0.516, 0.688)
F7	(0.357, 0.462, 0.654)
F8	(0.342, 0.529, 0.663)
$\sum_{i=1}^{8} \tilde{r}_i$	(7.737, 10.434, 13.462)
$\left(\sum_{i=1}^{8} \tilde{r}_i\right)^{-1}$	$\left(\frac{1}{13.462}, \frac{1}{10.434}, \frac{1}{70737}\right)$

$$\begin{aligned} \text{WF1} =& (2.306, 2.928, 3.473) \times \left(\frac{1}{13.462}, \frac{1}{10.434}, \frac{1}{70737}\right) \\ &= (0.1713, 0.2806, 0.4490) \end{aligned}$$

While calculating the priority weights for KPF, if the summation of calculated priority weights is more than one, then in order to make the sum of priority weights equal to one, normalization should be done. The summary of fuzzy priority weights and their normalized weights for each criterion is given in Table 6.

Next to check the consistency of the calculated priority weights, consistency test has been performed. From consistency check, the observed values are;

Table 6 Summary of fuzzy priority weights, for each criteria

Factor	Fuzzy priority weights for criteria	Aggregate/combined weight (Wi) $= \sum \frac{\tilde{w}_i}{n}$	Normalized weight for criteria $= \left(\frac{W_i}{\sum w_i}\right)$
F1	(0.1713, 0.2806, 0.4490)	0.3003	0.2717
F2	(0.0916, 0.1682, 0.3208)	0.1935	0.1751
F3	(0.1268, 0.2261, 0.3784)	0.2438	0.2206
F4	(0.0299, 0.0545, 0.1048)	0.0631	0.0571
F5	(0.0743, 0.1261, 0.2279)	0.1428	0.1292
F6	(0.0289, 0.0495, 0.0889)	0.0558	0.0505
F7	(0.0265, 0.0443, 0.0845)	0.0518	0.0468
F8	(0.0254, 0.0507, 0.0857)	0.0839	0.0487
$\sum w_i$		1.105	0.9997

Random Index, RI = 1.41; for $N = 8$ (from Saaty' RI table); Consistency Index, CI = 0.13796 and consistency ratio, CR = 0.097845

Since the calculated CR value for pairwise comparison between factors (Table 3) is less than 0.1 which is Saaty's upper bound limit for consistency to exist. Therefore, the pairwise comparison between identified factors is consistent, and their priority weights are acceptable.

5 Results and Discussion

The main objective of the current research is to identify and prioritize the most critical key performance factors of FCSC using fuzzy AHP approach. On the basis of existing literature review, identified literature gaps and discussion with experts, eight most critical key performance factors which are responsible for performance measure of FCSC have been identified. The identified key performance factors are namely climatic impact of FCSC (F_1), energy consumption (F_2), high volume of food waste (F_3), inadequate temperature monitoring (F_4), insufficient cold storage facility (F_5), lack of product traceability (F_6), improper packaging and handling (F_7) and lack of efficient information flow network (F_8). To analyse the criticality of the identified key performance factors and to find out the priority weights, fuzzy AHP approach has been used in this research work. The results of the research work show that the factor F1 (climatic impact of FCSC) having priority weight 0.2717 is most critical factor and F7 (improper packaging and handling) forms the least priority weight (0.0468). The reason behind the high priority weight for factor F1 is that in FCSC, a huge amount of energy consumption takes place. Larger the energy consumption, higher is the GHG emission which leads to increase in global temperature and the phenomena of global warming. Due to the lack of sufficient cold storage facility, FCSC faces a huge amount of food waste which became as the third largest emitter of GHG after USA and China if it is considered as a country. Therefore, while stablish benchmark for FCSC, decision-maker must aim to reduce the GHG emission from their designed FCSC and should consider F1 at their first priority so that performance of the same can be improved. Figure 1 shows the calculated priority weights for identified key

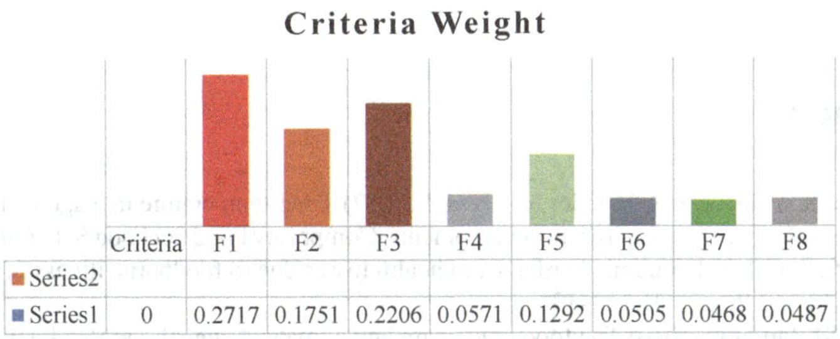

Fig. 1 Bar diagram for priority weights of identified key performance factors of FCSC

performance factors of the FCSC.

The priority weights for other factors are: 0.1751, 0.2206, 0.0571, 0.1292, 0.0505, and 0.0487 for F2, F3, F4, F5, F6 and F8, respectively. The sequence of priority for different key performance factors can be considered as: F1 (0.2717) > F3 (0.2206) > F2 (0.1751) > F5 (0.1292) > F4 (0.0571) > F6 (0.0505) > F8 (0.0487) > F7 (0.0468).

6 Conclusion

In the current research work, on the basis of existing literature review, identified literature gaps and discussion with experts, eight most critical factors responsible for performance measure of FCSC, have been identified. To find out the priority weights of the factors, fuzzy AHP approach has been proposed. The results of the research work show that the factors F1 (climatic impact of FCSC), F_3 (high volume of food waste), F_2 (energy consumption) are three most critical and must be taken at first priority by the decision-makers while stablishing the benchmark for performance improvement of FCSC. The research work also enlights the mutual relationship among the various identified key performance factors numerically. The results of the research work can be applied to all perishable and temperature-sensitive food processing supply networks such as meat, poultry, fish, dairy products, fruits, vegetables and all cooked food items. In this research work, a fuzzy AHP approach has been used and thus tackles all the uncertainty and vagueness about the collected data and results. Thus, the findings of current investigation may offer a cosiness to the mangers in improving the decision-making policies for improving their food cold supply chain performance system.

Though the current research work provides incredible aids to the decision-makers to improve the performance of the FCSC, it has several limitations. In the current research, it is aimed to find out the key performance factors of FCSC, interested researcher may extend this work by considering cofactors of the identified factors and their alternatives to measure the performance. This work may also be extended by using other MCDM techniques such as ANP, Interpretive Structural Modelling (ISM), technique for order preference by similarity to ideal solution and extend analysis method of fuzzy AHP.

References

1. Mercier S, Villeneuve S, Mondor M, Uysal I (2017) Time-temperature management along the food cold chain: a review of recent developments. Compr Rev Food Sci Food Saf 16(4):647–667
2. Scharff RL (2012) Economic burden from health losses due to foodborne illness in the United States. J Food Prot 75(1):123–131
3. James SJ, James C (2010) The food cold chain and climate change. Food Res Int 43(7):1944–1956
4. Brander K (2010) Impacts of climate change on fisheries. J Mar Syst 79(3–4):389–402

5. Gregory NG (2010) How climatic changes could affect meat quality. Food Res Int 43(7):1866–1873
6. Miraglia M, Marvin HJP, Kleter GA, Battilani P, Brera C, Coni E et al (2009) Climate change and food safety: an emerging issue with special focus on Europe. Food Chem Toxicol 47(2009):1009–1021
7. Paterson RRM, Lima N (2010) How will climate change affect mycotoxins in food? Food Res Int 43(7):1902–1914
8. Estrada- Flores S (2010) Achieving temperature control and energy efficiency in the cold chain. In: 1st IIR international cold chain conference
9. Sainathan P, Time R (2018) 10 potential risks in cold chain management. Retrieved from: https://blog.roambee.com/supply-chain-technology/10-potential-risks-in-cold-chain-management
10. Mattarolo et al (1990) Refrigeration and food processing to ensure the nutrition of the growing world population. In: Progress in the science and technology of refrigeration in food engineering, proceeding of meetings of commissions B2, C2, D1, D2-D3, September 24-28, Dresden, Germany, Paris, France. Institute international du Froid, pp 43–54
11. Garnett T (2007) Food refrigeration: what is the contribution to Greenhouse gas emissions and how might emissions be reduced. Food Clim Res Network
12. Schmidhuber J, Tubiello FN (2007) Global food security under climate change. Proc Natl Acad Sci USA (PNAS) 104(50):19703–19708
13. Liu L, Liu X, Guangchen Liu G (2018) The risk management of perishable supply chain based on coloured Petri Net modelling. Inf Proc Agricul 5(1):47–59
14. Gardas BB, Raut RD, Narkhede BE (2018) Evaluating critical causal factors for post-harvest losses (PHL) in the fruit and vegetables supply chain in India using the DEMATEL approach. J Clean Prod 199:47–61
15. Joshi R, Banwet DK, Shankar R (2011) A Delphi-AHP-TOPSIS based benchmarking framework for performance improvement of a cold chain. Expert Sys Appl 38:10170–10182
16. Tyagi M, Kumar P, Kumar D (2015) Assessment of critical enablers for flexible supply chain performance measurement system using fuzzy DEMATEL approach. Glob J Flex Syst Manage 16(2):115–132
17. Tyagi M, Kumar P, Kumar D (2015) Analyzing CSR issues for supply chain performance system using preference rating approach. J Manufact Technol Manage 26(6):830–852
18. Tyagi M, Kumar P, Kumar D (2015) Permutation of fuzzy AHP and AHP methods to prioritizing the alternatives of supply chain performance system. Int J Ind Eng Res Appl 21(2015):729–752
19. Nadaban S, Dzitac S, Dzitac I (2016) Fuzzy topsis: a general view. Proc Comput Sci 91(2016):823–831
20. Saaty TL (1980) The analytic hierarchy process. McGraw-Hill, New York
21. Tyagi M, Kumar P, Kumar D (2014) A hybrid approach using AHP-TOPSIS for analysing e-SCM performance. Proc Eng 97(2014):2195–2203
22. Laarhoven PJMV, Pedrycz W (1983) A fuzzy extension of Saaty's priority theory. Fuzzy Sets Syst 11(1–3):199–227
23. Tyagi M, Kumar P, Kumar D (2018) Assessment of CSR based supply chain performance system using an integrated fuzzy AHP-TOPSIS approach. Int J Logistics: Res Appl 21(4):378–406
24. Buckley JJ (1985) Fuzzy hierarchical analysis. Fuzzy Sets System 17(1):233–247

A Short Review on Machining with Ultrasonic MQL Method

Rahul Katna, M. Suhaib, Narayan Agrawal, and S. Maji

Abstract Cutting fluids are required in huge amount in the modern machining methods. Cutting fluids are needed for providing lubrication as well as cooling the workpiece during machining. Cutting fluids are made of mineral oils and have many drawbacks including many health hazards and environment impact. Various other methods of cutting fluid delivery are being tested in order to reduce the effects of cutting fluids in machining. One such method is minimum quantity lubrication. Many researchers have shown the effectiveness of MQL over the conventional flood method. However, the MQL method has drawbacks in terms of heat carrying capacity. In order to increase the effectiveness of the MQL method, many advanced methods are being tested for improving the cooling efficiency of the MQL method. One of the methods is ultrasonic method. This method is currently in nascent stages and is being researched upon as a viable alternative to the conventional MQL method. This paper describes this hybrid delivery method in machining.

Keywords Green · Manufacturing · Minimum · Quantity · Lubrication · Ultrasonic · Nanoparticle

1 Introduction

1.1 MQL

MQL stands for minimum quantity lubrication. It is being used as an alternate to the conventional flood method of machining in the manufacturing industries. The

---

R. Katna (✉) · M. Suhaib
Department of Mechanical Engineering, Jamia Millia Islamia (a Central University), New Delhi 110025, India
e-mail: katnarahul@gmail.com

N. Agrawal · S. Maji
Delhi Institute of Tool Engineering, Okhla, New Delhi 110020, India

© The Author(s), under exclusive license to Springer Nature Singapore Pte Ltd. 2021
R. M. Singari et al. (eds.), *Advances in Manufacturing and Industrial Engineering*, Lecture Notes in Mechanical Engineering,
https://doi.org/10.1007/978-981-15-8542-5_61

benefits it draws are use of very little quantity of lubricant, thereby saving costs associated with using huge quantity of lubricant and associated hazards.

MQL or minimum quantity lubrication is slowly gaining attention from researchers worldwide due to its better lubrication properties and considerably low quantity of cutting fluid [1–5]. MQL has many advantages like it increases tool life and surface finish [6–13]. A typical MQL involves use of high-pressure liquid from 2-bar pressure to 5-bar pressure and lubricant in atomized form as shown in Fig. 1. The atomized lubricant is in the range of microns and in combination with the high pressure air is able to penetrate the cutting zone effectively [3, 14, 15]. Machining using MQL method thus provides better lubricity than the conventional flood cooling method. MQL typically uses 50–200 ml/h lubricant. An MQL setup consists of a spray nozzle in which the atomized lubricant is mixed either internally or externally depending upon the type of configuration used [16, 17]. Researchers have tested MQL method in different machining methods and have found it to be better than the conventional flood delivery method of cutting fluid delivery. MQL is delivered into the cutting zones as shown in Fig. 2.

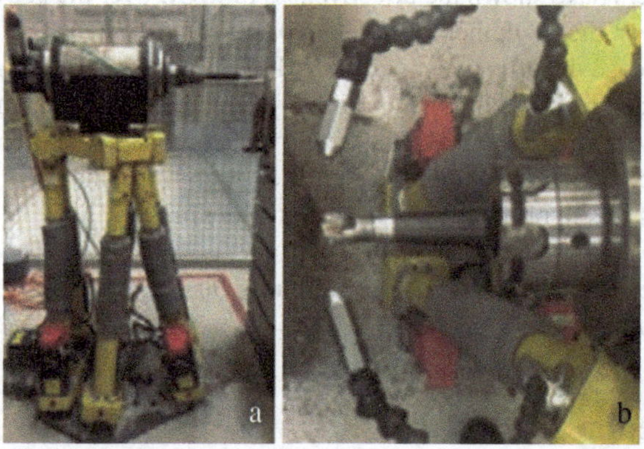

Fig. 1 Typical MQL setup [18]

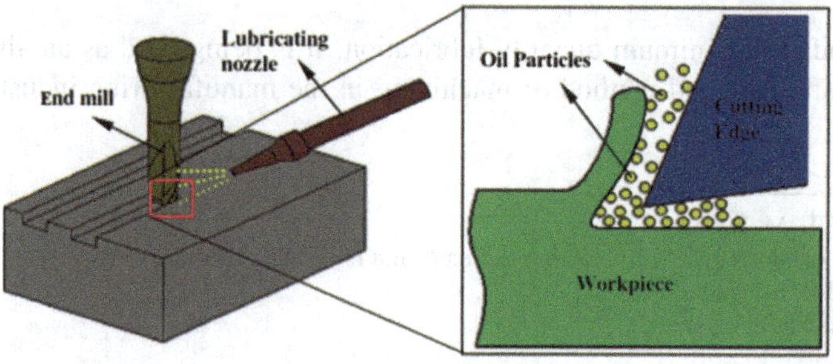

Fig. 2 MQL delivery into cutting zone [19]

2 Ultrasonic MQL

Although MQL has proved to be better than the conventional method, researchers have found that combining the MQL method with some other methods like cold air MQL, cryogenic gas, electrically charging the lubricant has performed better than the conventional MQL method itself [20–26]. One such emerging and promising research is ultrasonic MQL or UMQL. The ultrasonic MQL method actually combines three methods—first is the MQL itself, i.e., atomized lubricant, second is the nanoparticles mixed in the lubricant, and third is the ultrasonic mixing mechanism. The result of these triple parts is an efficient lubricity during machining with a constant homogenous delivery of cutting fluid in to the cutting zone. It has been shown that addition of nanoparticles increase the lubricity of a lubricant. However, it is difficult to keep the nanoparticles dispersed in the lubricant [27–29].

2.1 Application of UMQL

The nanoparticles have a tendency to agglomerate and form lumps within the lubricant bulk. Thus, without proper dispersion of the nanoparticles the whole process will get rendered useless. Thus, it is important to keep the nanoparticles dispersed in the lubricant for an effective fluid delivery process [30–33]. If the nanoparticles agglomerate together, there will be time patches where the lubricant is delivered without nanoparticles and there will be instances where lumps are delivered into the machining zone which will eventually be unable to reach the cutting zone thus failing the whole process. Emulsifiers are used for this purpose [34]. An emulsifier is a chemical compound which makes the dispersion of the nanoparticles in the lubricant easy and for a long time. However, addition of emulsifiers in the lubricant has shown to affect the properties of the nanoparticles in the lubricant. Another way of tackling this problem is employing a mechanism that can keep the nanoparticles dispersed in the lubricant. Ultrasonic dispersion is one such method in which the nanoparticles [35–39]. This paper considers the latest developments in the ultrasonic method with nanoparticles. Ultrasonic MQL has been used in grinding operation mostly. For efficiently utilizing in other operations, it has to be properly setup. Secondly, nanoparticles are costly affair and are lost during machining, so proper care has to be taken for the toxicity effects of nanoparticles on human.

In an attempt to increase the thermal efficiency of the MQL method, Rabiei et al. [40] used water-based nanoparticles in ultrasonic. They employed six different nanoparticles like oxides of titanium, silicon, aluminum, copper, nickel, and multi-walled carbon nanotube in grinding. From the experimental results, it was observed that there was a reduction of 20 and 24.6% in the grinding force when compared to grinding done without any lubricant or coolant. Also, a superior surface finish was achieved without any visible surface defect and damage. Also, no plastic deformation or side flow defect was observed which indicate good lubrication during machining.

Interestingly, the type of chip obtained with MQL ultrasonic grinding was similar to that obtained with the flood method of lubrication which points to the fact that both these methods are similar in terms of machining quality.

Huang et al. [41] tried to improvise the conventional machining process by using nanoparticles in the cutting fluid and also amalgamating MQL with ultrasonic dispersion. Nanoparticles being solid powders have e tendency to agglomerate in water and cause lumping problems which decreases their performance. An effective way is to use an emulsifier to keep the nanoparticles suspended in the solution. However, use of emulsifier increases the cost and can also affect the performance of the nanoparticles. Hence, ultrasonic dispersion method was utilized to keep the nanoparticles suspended for a homogenous solution of cutting fluid. In machining mold steel with this method, the experimental results showed that agglomeration of the nanoparticles was reduced to a very large extent. Also, low grinding forces were achieved with this method in comparison to MQL method alone. This hybrid method also showed lower temperature rise and excellent surface finish than the conventional MQL method without nanoparticles and conventional MQL method with nanoparticles.

Ni et al. [42] used ultrasonic vibration assisted with the conventional MQL machining method in machining titanium alloy. Introduction of the lubricant during the ultrasonic friction between mating surfaces of the tool and workpiece caused increased lubrication. From the experimental results, it was found that the combined method of ultrasonic vibration and MQL method caused reduction in tool wear to a very large extent. On further analyzing the surface of the tool, it was found that the major cause of tool failure was fracture of the surface upon impact which was visible on the surface. From the results, it can be concluded that the combined method of ultrasonic with MQL is a better way of machining titanium alloy and can also enhance the tool life considerably.

Helmy et al. [43] used ultrasonic method of machining combined with MQL method in machining of composite laminates at different machining parameters. The parameters such as cutting speed, feed, and depth of cut were varied in machining the composite with a diamond tool, and the performance was measure in terms of cutting forces and surface roughness. Experimental results however indicated that flood method of machining produced better results than the MQL method, but the MQL method performed nearly equal to the flood method in machining of the composite laminate. This indicates that ultrasonic MQL method is comparable to the flood method and can be improved by changing the machining parameters and optimizing the input parameters.

In another study using oil-based nanoparticles, Molaie et al. used molybdenum disulfide nanoparticles in oil assisted with ultrasonic method in grinding operation [44]. The experimental results clearly indicate that the combined method of ultrasonic grinding clubbed with the conventional MQL method increased the effectiveness of the conventional MQL method and resulted in lower grinding forces. It was also seen from the experimental results that the combined method resulted in lower surface roughness in comparison to the conventionally MQL method which clearly indicates the better lubricity achieved with this hybrid method of fluid delivery.

Madarkar et al. [45] used ultrasonic-assisted MQL method in machining titanium alloy. An indigenous horn was fabricated in order to produce ultrasonic frequency. The authors used sunflower oil with nanoparticles in 1, 5, and 10% concentration and ultrasonic method of fluid delivery. The combined method of fluid delivery performed better than the conventional MQL method as it yielded lower grinding forces than the conventional MQL method. However, the conventional method yielded lower surface roughness than the ultrasonic-assisted grinding process which was due to enhancement in the capacity to retain the sharpness over long period of time than that observed in the conventional MQL method of fluid delivery. Experimental results, however, prove the enhancement in the grindability of titanium alloy with the hybrid method involving ultrasonic and nanoparticles.

Rasidi et al. [46] employed a piezoelectric transducer in micro-machining with MQL method. They compared dry method with MQL method combined with the ultrasonic method in each case and at two different flow rate of the MQL fluid. The experimental results obtained show a slight improvement in the surface finish with the MQL assisted with ultrasonic method. However, the hybrid method improved the tool wear significantly. Similarly, Li et al. [47] reported significant improvement in the tool life of the cutting tool with the ultrasonic-assisted machining with MQL.

Alemayehu et al. [48] evaluated the machining performance in turning Inconel 718 with a new hybrid method. The authors employed ultrasonic vibration with MQL method in machining the alloy. The authors reported that with the unique method of hybrid delivery lower cutting forces were achieved. Thus, this process saves not only input energy but also produces better quality machined surface.

Isobe et al. [49] used carbide drill for drilling with carbide drill vibrating ultrasonically. The authors drilled 302 holes by ultrasonic drilling combined with MQL process. Experimental results showed that the micro-drilling method combined with MQL method was able to reduce the deflections occurring in the drill bit. Not only this, the authors achieved higher tool life with tool wear reducing by almost half.

3 Conclusion

As seen ultrasonic MQL is a relatively newer field and very little literature is published. There is a huge research scope in MQL machining combined with MQL method in machining. It was seen that most of the studies are done in grinding operation only. However, MQL method combined with the ultrasonic machining can be tested in milling and turning operations also. It is recommended to test the effectiveness of MQL in ultrasonic machining on super alloys also.

References

1. Benedicto E, Carou D, Rubio EM (2017) Technical, economic and environmental review of the lubrication/cooling systems used in machining processes. Procedia Eng 184:99–116
2. "Micro/Nanofluids in Sustainable Machining | IntechOpen." [Online]. Available: https://www.intechopen.com/books/microfluidics-and-nanofluidics/micro-nanofluids-in-sustainable-machining. Accessed: 23 Nov 2019.
3. Dixit US, Sarma DK, Davim JP (2012) Machining with minimal cutting fluid. In: Dixit DKS, Davim JP (eds) Environmentally friendly machining, U. S. Springer US, Boston, MA, pp 9–17.
4. Paul S, Sarkar S (2019) Integration of cryogenic machining technologies in advance manufacturing systems
5. Mannekote JK, Kailas SV, Venkatesh K, Kathyayini N (2018) Environmentally friendly functional fluids from renewable and sustainable sources-a review. Renew Sustain Energy Rev 81:1787–1801
6. Dhar NR, Kamruzzaman M, Ahmed M (2006) Effect of minimum quantity lubrication (MQL) on tool wear and surface roughness in turning AISI-4340 steel. J Mater Process Technol 172(2):299–304
7. Khan MMA, Mithu MAH, Dhar NR (2009) Effects of minimum quantity lubrication on turning AISI 9310 alloy steel using vegetable oil-based cutting fluid. J Mater Process Technol 209(15):5573–5583
8. Dhar N, Islam S, Kamruzzaman M (2010) Effect of minimum quantity lubrication (MQL) on tool wear, surface roughness and dimensional deviation in turning AISI-4340 steel. Gazi Univ J Sci 20(2):23–32
9. Kamata Y, Obikawa T (2007) High speed MQL finish-turning of Inconel 718 with different coated tools. J Mater Process Technol 192–193:281–286
10. Davim JP, Sreejith PS, Silva J (2007) Turning of brasses using minimum quantity of lubricant (MQL) and flooded lubricant conditions. Mater Manuf Process 22(1):45–50
11. Attanasio A, Gelfi M, Giardini C, Remino C (2006) Minimal quantity lubrication in turning: effect on tool wear. Wear 260(3):333–338
12. Hamdan A, Sarhan AAD, Hamdi M (2012) An optimization method of the machining parameters in high-speed machining of stainless steel using coated carbide tool for best surface finish. Int J Adv Manuf Technol 58(1):81–91
13. Tasdelen B, Wikblom T, Ekered S (2008) Studies on minimum quantity lubrication (MQL) and air cooling at drilling. J Mater Process Technol 200(1):339–346
14. Ezugwu EO (2005) Key improvements in the machining of difficult-to-cut aerospace superalloys. Int. J Mach Tools Manuf 45(12):1353–1367
15. Sadeghi MH, Haddad MJ, Tawakoli T, Emami M (2009) Minimal quantity lubrication-MQL in grinding of Ti–6Al–4V titanium alloy. Int J Adv Manuf Technol 44(5):487–500
16. Zaman PB, Dhar NR (2019) Design and evaluation of an embedded double jet nozzle for MQL delivery intending machinability improvement in turning operation. J Manuf Process 44:179–196
17. Rahim EA, Dorairaju H (2018) Evaluation of mist flow characteristic and performance in Minimum Quantity Lubrication (MQL) machining. Measurement 123:213–225
18. Tunc LT, Gu Y, Burke MG (2016) Effects of minimal quantity lubrication (MQL) on surface integrity in robotic milling of austenitic stainless steel. Procedia CIRP 45:215–218
19. Khatri A, Jahan MP (2018) Investigating tool wear mechanisms in machining of Ti-6Al-4V in flood coolant, dry and MQL conditions. Procedia Manuf 26:434–445
20. Sanchez JA et al (2010) Machining evaluation of a hybrid MQL-CO_2 grinding technology. J Clean Prod 18(18):1840–1849
21. Pereira O et al (2015) The use of hybrid CO_2+MQL in machining operations. Procedia Eng 132:492–499
22. Stachurski W, Sawicki J, Wójcik R, Nadolny K (2018) Influence of application of hybrid MQL-CCA method of applying coolant during hob cutter sharpening on cutting blade surface condition. J Clean Prod 171:892–910

23. Wakabayashi T et al (2015) Near-dry machining of titanium alloy with MQL and hybrid mist supply. Key Eng Mater 2015. [Online]. Available: https://www.scientific.net/KEM.656-657.341. Accessed: 23 Nov 2019
24. Neugebauer R, Drossel W, Wertheim R, Hochmuth C, Dix M (2012) Resource and energy efficiency in machining using high-performance and hybrid processes. Procedia CIRP 1:3–16
25. Kannan C, Ramanujam R, Balan ASS (2018) Machinability studies on Al 7075/BN/Al_2O_3 squeeze cast hybrid nanocomposite under different machining environments. Mater Manuf Process 33(5):587–595
26. Xu J, Mkaddem A, El Mansori M (2016) Recent advances in drilling hybrid FRP/Ti composite: a state-of-the-art review. Compos Struct 135:316–338
27. Ilie F, Covaliu C (2016) Tribological properties of the lubricant containing titanium dioxide nanoparticles as an additive. Lubricants 4(2):12
28. Kolodziejczyk L, Martínez-Martínez D, Rojas TC, Fernández A, Sánchez-López JC (2007) Surface-modified Pd nanoparticles as a superior additive for lubrication. J Nanoparticle Res 9(4):639–645
29. Wu YY, Tsui WC, Liu TC (2007) Experimental analysis of tribological properties of lubricating oils with nanoparticle additives. Wear 262(7):819–825
30. Lee P-H, Nam JS, Li C, Lee SW (2012) An experimental study on micro-grinding process with nanofluid minimum quantity lubrication (MQL). Int J Precis Eng Manuf 13(3):331–338
31. Kalita P, Malshe AP, Arun Kumar S, Yoganath VG, Gurumurthy T (2012) Study of specific energy and friction coefficient in minimum quantity lubrication grinding using oil-based nanolubricants. J Manuf Process 14(2):160–166
32. Wang X, Xu X, Choi SUS (1999) Thermal conductivity of nanoparticle—fluid mixture. J Thermophys Heat Transf 13(4):474–480
33. Mao C, Tang X, Zou H, Huang X, Zhou Z (2012) Investigation of grinding characteristic using nanofluid minimum quantity lubrication. Int J Precis Eng Manuf 13(10):1745–1752
34. Amrita M, Shariq SA, Manoj, Gopal C (2014) Experimental investigation on application of emulsifier oil based nano cutting fluids in metal cutting process. Procedia Eng 97:115–124
35. Taurozzi JS, Hackley VA, Wiesner MR (2011) Ultrasonic dispersion of nanoparticles for environmental, health and safety assessment—issues and recommendations. Nanotoxicology 5(4):711–729
36. Chung SJ et al (2009) Characterization of ZnO nanoparticle suspension in water: effectiveness of ultrasonic dispersion. Powder Technol 194(1):75–80
37. Sato K, Li J-G, Kamiya H, Ishigaki T (2008) Ultrasonic dispersion of TiO_2 nanoparticles in aqueous suspension. J Am Ceram Soc 91(8):2481–2487
38. Sauter C, Emin MA, Schuchmann HP, Tavman S (2008) Influence of hydrostatic pressure and sound amplitude on the ultrasound induced dispersion and de-agglomeration of nanoparticles. Ultrason Sonochem 15(4):517–523
39. Xiaochun L, Yang Y, Yang Y, Weiss D, Weiss D (2008) Theoretical and experimental study on ultrasonic dispersion of nanoparticles for strengthening cast Aluminum Alloy A356. Metall Sci Tecnol 26:2
40. Rabiei F, Rahimi AR, Hadad MJ, Saberi A (2017) Experimental evaluation of coolant-lubricant properties of nanofluids in ultrasonic assistant MQL grinding. Int J Adv Manuf Technol 93(9–12):3935–3953
41. Huang WT, Liu WS, Wu DH (2016) Investigations into lubrication in grinding processes using MWCNTs nanofluids with ultrasonic-assisted dispersion. J Clean Prod 137:1553–1559
42. Ni C, Zhu L, Yang Z (2019) Comparative investigation of tool wear mechanism and corresponding machined surface characterization in feed-direction ultrasonic vibration assisted milling of Ti–6Al–4V from dynamic view. Wear 436–437:203006
43. Helmy MO, El-Hofy MH, El-Hofy H (2018) Effect of cutting fluid delivery method on ultrasonic assisted edge trimming of multidirectional CFRP composites at different machining conditions. Procedia CIRP 68:450–455
44. Molaie MM, Akbari J, Movahhedy MR (2016) Ultrasonic assisted grinding process with minimum quantity lubrication using oil-based nanofluids. J Clean Prod 129:212–222

45. Madarkar R, Agarwal S, Attar P, Ghosh S, Rao PV (2018) Application of ultrasonic vibration assisted MQL in grinding of Ti–6Al–4V. Mater Manuf Process 33(13):1445–1452
46. Rasidi I, Rahim EA, Ibrahim AA, Maskam NA, Ghani SC (2019) The effect on the application of coolant and ultrasonic vibration assisted micro milling on machining performance. Appl Mech Mater [Online]. Available: https://www.scientific.net/AMM.660.65. Accessed: 23 Nov 2019
47. Li K-M, Wang S-L (2014) Effect of tool wear in ultrasonic vibration-assisted micro-milling. Proc Inst Mech Eng Part B J. Eng Manuf 228(6):847–855
48. Alemayehu H, Ghosh S, Rao PV (2020) Evaluation of cutting force and surface roughness of inconel 718 using a hybrid ultrasonic vibration-assisted turning and minimum quantity lubrication (MQL). In: Advances in unconventional machining and composites, Singapore, pp 325–333
49. Isobe H, Hara K (2015) Improvement of drill life for nickel super alloy by ultrasonic vibration machining with minimum quantity lubrication. Key Eng Mater [Online]. Available: https://www.scientific.net/KEM.625.581. Accessed: 23 Nov 2019

Professional Values and Ethics: Challenges, Solutions and Different Dimension

Shalini Sharma

Abstract Values and ethics both are very essential for personal and professional success in the life of any individual. Values refer to those rules which help an individual to take right decision in life, and ethics are those moral principles that basically govern the behavior of a person. The word 'Ethics' has been derived from the Latin word 'Ethos' which means habit or custom. As far as professional values and ethics are concerned, they may be defined as those professional behaviors which are professionally accepted, inculcated by different institution/organization to instruct or guide all employees to behave and work as per consistent and real ethical principle. The major component of professional values is dedication toward work, self-motivation and to motivate others, responsibility, honesty, discipline, positive attitude toward every type of situation, etc. Similarly, the main component of professional ethics is accountability, transparency, adherence to law, etc. Both professional values and ethics help to enhance professionalism at the workplace. They do not only help to grow and develop the business but also create a good impression of the organization in the market and society. But sometimes, several ethical dilemmas popped up in the workplace such as sexual harassment at workplace, gender discrimination, unethical workplace culture, unattainable and unrealistic objective of the company, etc. Though solving theses major ethical issues is a complex process but stepwise recognition, identification, consideration, and implementation may solve this problem and create a better, congenial and healthy workplace for all.

Keywords Values · Ethics · Accountability · Transparency · Ethical dilemmas

1 Introduction

Professional values and ethics are very important for the tremendous success in professional sphere. The way an individual or a group of people interacts with other

S. Sharma (✉)
Department of Mechatronics, Delhi Institute of Tool Engineering (DITE),
Government of NCT of Delhi, New Delhi 110020, India
e-mail: Shalini.dite@gmail.com

© The Author(s), under exclusive license to Springer Nature Singapore Pte Ltd. 2021

R. M. Singari et al. (eds.), *Advances in Manufacturing and Industrial Engineering*,
Lecture Notes in Mechanical Engineering,
https://doi.org/10.1007/978-981-15-8542-5_62

truly reveals his/her real character because action speaks louder than words. The people who have strong value system and ethical standard are easy to be recognized by their actions, and they are always motivated to do right things even when nobody is watching them. The origin of values and ethics are same—family, friends, school, society, spiritual belief, etc. In simple words, professional values and ethics are nothing but what we learn from our surroundings before joining the workforce. So, the beliefs, learnt early in life, are carried forward in professional field and have positive or negative impact on the career growth.

2 Professional Values and Ethics

Professional values exhibit the real business practice of any company. They play the role of a guiding force and mentor which help to take right decision and also help for resolution of the conflicts. Ethics are considered in relation with values because they are basically moral philosophy of one's values. Ethics tell us what is righteous, fair and honorable. Generally, codes of ethics are set forth by different business companies, and it is expected that employees shall follow it. Every employee should familiarize themselves with the ethical codes of the company for the ultimate growth of the individual and company itself.

3 Major Component of Professional Values and Ethics

3.1 Dedication Toward Work

The employee should be completely dedicated toward work. In addition to working hard, it is also important to work smart. It is also important to complete the assigned task in the most efficient way in/before the allotted time while maintaining the positive attitude.

3.2 Responsibility and Dependability

Responsible behavior and attitude show you a trustworthy person. For example, updating the progress of assigned project and keeping your boss abreast with the change of the schedule as you are going late shows that you may be trust upon.

3.3 Positive Attitude

A positive person creates positive atmosphere in the workplace and becomes a positive role model for others as well. He solves the workplace problems with a different and positive way leaving no chance for chaos in the company.

3.4 Self-motivated and Motivate Others

A professional with high standards of value system seeks less/no supervision to complete the assigned task which is done by him/her in professional and timely manner. They also motivate other employees and create a supportive and safe work environment for all.

3.5 Strong Self-confidence

It is the key component that differentiates successful people to unsuccessful one. The self-confident person does what is right, keeps courage to take risks and admits his mistakes, if committed. He knows his strengths as well as his weakness and try to work on the latter.

3.6 Professionalism

Professionalism refers to try to learn every minute detail of the work and give their hundred percent to the assigned work. Professionals are detail-oriented and are optimistic about the company and its future.

3.7 Loyalty, Integrity and Honesty

A real professional is loyal toward his company and job. He keeps confidentiality of official records for the sake of his company. The employers also value those employees who show a great sense of honesty and integrity toward professional commitments.

3.8 Respect and Discipline

Respect refers to esteem, and it is a positive feeling that shows that you are valued upon. It is generally said that respect is earned (by good deeds). Similarly, a disciplined person follows rules and behaves in calm, composed and controlled way.

3.9 Adherence to Rules/Law

Adherence to the rules/laws/guidelines of company or organization is important for the individual and further the growth of the company. Violation of it may lead to punishments as observed in many cases.

4 Importance of Values and Ethics in Professional Life

Maintaining good ethical standard and value system is essential for the development of the individual and for the company. They create an honest image of the person and ultimately of the company and keep us away from those activities which may lead to disrespect to the company. When professional values and ethics are followed, employees respect one another, try to work in a team, bring fruitful results and ultimately create a healthy and happy work environment. Values alignment helps the company to gain its basic mission. On the contrary, when professional values and ethics are not aligned, employees work toward different goals and different intentions. This may damage productivity, creative potential and job relations in the company.

5 Professional Values and Ethics: How Useful They Are

5.1. When a task is performed ethically as per the code of conduct of the company, it is easier for employees to know what to do in certain circumstances.

5.2. The employees feel proud to work in the organization when they come to know that the organization supports them in all possible way to maintain the high standards.

5.3. Professional training boosts friendly and healthy environment, with less room for misunderstanding among employees. It also ensures teamwork to help for the development of the organization.

6 Ethical Dilemmas at Workplace: Challenges and Solution

6.1 Sexual Harassment and Discrimination at Workplace

Law enforces equal growing opportunities to all employees irrespective to their gender, caste, age, etc. It is the duty of organization to employ diverse workforce and better utilize the potential of employees. Unfortunately, when harassment and discrimination (based on gender, caste, race, etc. occurs, the environment of the workplace becomes more and more toxic. Harassment may be sexual, verbal or written. Harassment is any unwelcome behavior by which the other person feels humiliated and offended, man also sometimes feel harassed, but it affects more women in comparison to men.

All types of harassment, irrespective of how small or big it is and who is involved, requires immediate and appropriate response/action from the employer. There are several laws/rules/guidelines to prevent such type of harassment at workplace. Besides these rules and regulations, the first and one of the most important ways to curb it is speaking out about it. Don't keep mum; let others know actually what he did. Silence instigates the harasser, but visibility stops them. Moreover, taking initiative against harassment and undue advances is the right of every person. The victim should report immediately to his/her immediate boss/other concerned functionary/NGO, etc., at the same time; it is the responsibility of the employer to take appropriate remedial action as per rule against this type of behavior.

Awareness among employees is another important tool to curb it. The guidelines related with the rights of women should be notified at all prominent places so that they may know what their rights are and what steps they may take in a particular circumstance.

Discrimination occurs when at the workplace any employee is favored because of his/her gender/religion/age/origin, etc. Though discrimination is not legal, still it is being observed generally at most of the workplaces. It is unfair treatment based on prejudices. Discrimination can be easily noticed through less diverse workforce at the workplace, fixed and certain roles for men and women, pending or almost denied promotions, sore or not-so healthy communication between employer and employees. To stop such type of toxic environment created by discrimination, it is very important to aware/educate all employees about discrimination, encourage every employee to respect other, make a rule/policy to stop it, provide training to people on high posts about how to deal with it in an appropriate manner. Whatever policy is made, make sure that it is properly enforced. Review and amend the policy from time to time.

6.2 Unethical Workplace Culture: Doing/Completing Personal Work During Office Time

It is true that most of the time of the employees is consumed in the office, they find less time to complete their own personal work. In this case, they often feel tempted to do own work during office time, for example, taking appointment of the doctor in the office by using company's phone, making reservations of train/bus, etc. in company's computer, doing side-business while physically present in the office. To curb this tendency, company should make/set policy, but at the same time it should also be remembered that this policy should be based on humanitarian ground.

6.3 Taking Undue Credit for What Is Not Done

In a company, generally teams are made to accomplish any project/work but sometimes what happens; one or two, out of five-six take the full credit for the accomplishment. This type of ethical dilemma creates confusion and resentment in the team. So, it is the duty of team manager to understand, recognize and appreciate well all team members for successful completion of any given task.

6.4 Toxic Workplace

In the company, the people who are holding higher posts are corrupt, for example, they take bribes, pressurizing their subordinates to do/favor in a particular task, bully employees, the environment becomes more and more toxic. It suffocates other and makes them difficult to survive. In this type of situation, either the top management should remove this leadership or make them understand to work in a proper way in the company so that the environment of the company becomes healthy for all.

6.5 Misuse of the Technology of the Company

It has been observed that most of the employees use company telephone and Internet service, etc. for their personal use in company time. It should also be stopped and discouraged in a polite manner.

7 Conclusion

7.1 It Should Always Be Remembered that:

7.1. Good professional values and ethics create good and healthy environment which helps to serve the customer in a better way.
7.2. Choosing right path is not always easy, but ultimately it pays off in long run.
7.3. Don't face dilemmas alone, talk about it to your colleague/manager. Follow your instinct and do what is appropriate according to the situation.

Basically, values and ethics are important in every part of your life—personal or professional. The difference between personal and professional values and ethics should also be considered. The main aim of ethics is knowing and doing what is right and acceptable in society. Professional ethics must be looked upon as a time tested, collective and comprehensive wisdom that is passed to others. It is important for building reputation and image of the company in this competitive world. It is also a very important ingredient for a trustworthy professional. The relationship among employees and clients is more fruitful when they are characterized by certain moral quality. A certain ethical attitude of the company helps to make right decision for the interest of the employees and clients. It is also important to mention here professional ethics and profit go hand in hand always. The company that follows certain professional values and ethics becomes successful in long run. An ethical oriented management may prevent many undesirable issues like pollution, protects and ensures the health and general well-being of the public much before being mandated by rules/regulation/law etc.

At the same time, it also needs to be mentioned here in this regard that dealing and resolving ethical challenges/dilemmas are also equally important for overall growth and development of the company. For this, try to know where the problem/challenge/dilemma is and then develop a plan of attack to resolve it. Take appropriate step and measures to avoid recurrence of that dilemma in future. For example, when you encounter any problem at the workplace, once when you get over with it, consider what you have learnt from that problem and start policy and procedure which stop it happening again.

Design of 3D Printed Fabric for Fashion and Functional Applications

Arpit Singh, Pradeep Kumar Yadav, Kamal Singh, Jitendra Bhaskar, and Anand Kumar

Abstract Fused deposition modeling (FDM) enables the production of prototypes directly from 3D models by laying down consecutive deposits of material to obtain the final geometry. FDM based printing has an advantage over other methods as very complex 3D shapes may be produced in quick time without generation of recyclable waste. In the present work, 3D printing of polymeric material poly lactic acid (PLA) has been performed on cotton-knitted and tulle fabrics to improve the adhesion of polymeric material to fabrics. The process may also be gainfully utilized to improve design patterns and geometry to add style and uniqueness to fashion apparels.

Keywords Fused deposition modeling (FDM) · Adhesion force · 3D printing · Infill pattern · z-offset distance · Printed fabric

1 Introduction

Additive manufacturing is a fast-emerging technology being used for solving problems associated with conventional manufacturing processes. FDM as a part of rapid prototyping technology enables the production of prototypes directly from a three-dimensional model by resting down consecutive layers of material until the final geometry is obtained. These days, different kinds of 3D printers and printing innovations are accessible around the world. First 3D printing innovation was stereolithography (SLA) which was designed by Chuck Hull during the 1980s, yet FDM based 3D printing innovation is modest and most generally utilized when contrasted with other 3D printing advancements. FDM uses a continuous filament of a thermoplastic material which can be fed from a large coil, through a moving heated extruder head. The molten material is forced out of the nozzle and gets deposited on the heated

A. Singh (✉) · P. K. Yadav · K. Singh · J. Bhaskar · A. Kumar
Department of Mechanical Engineering, Harcourt Butler Technical University, Kanpur, Uttar Pradesh, India
e-mail: arpit1496singh@gmail.com

A. Kumar
e-mail: kranandhbti@gmail.com

© The Author(s), under exclusive license to Springer Nature Singapore Pte Ltd. 2021

R. M. Singari et al. (eds.), *Advances in Manufacturing and Industrial Engineering*, Lecture Notes in Mechanical Engineering,
https://doi.org/10.1007/978-981-15-8542-5_63

bed layer by layer one over other, thus forming the required 3D shape. As compared to other 3D printing methods, FDM is a relatively slow process. In FDM based 3D printer, there are two types of approaches: first direct type and second is indirect type. In the case of direct approach, the extruder is typically mounted directly on top of the nozzle's hot end, and therefore, the filament is holed tightly by a wheel and gear. In the case of indirect type approach, the recent end is separated physically from the extruder. Usually, the extruder is mounted anywhere on the inside of the 3D printer. FDM based 3D printing is becoming popular in the fashion industry.

Conventional printing techniques on fabric include dye-based printing and inkjet printing. In dye-based printing, we use engraved design on dye to make impressions on fabric material. In inkjet printing, we use deposition tool for precision droplets of ink on the surface of fabric material. Problem associated with conventional printing technique is that it produces lot of chemical waste which usually gets discarded into the water bodies. Using FDM based 3D printing can overcome this issue as well as provide more precise printing of designs on fabric materials. That is why 3D printed designs on clothes are becoming a point of attraction in fashion shows. Besides making fabric attractive, 3D printing on fabric can also be used to impart functional characteristics such as wear-resistance and stiffness. Due to variety of filaments available for FDM based 3D printing, we can custom print our designs as per requirement, e.g., conductive filament are available, which may print conductive paths on fabric for wearable technologies. Material selection for printing designs on fabrics is also very crucial. Flexible materials like soft PLA, NinjaFlex and PolyFlex filaments are used to print 3D designs [1]. The effect of infill pattern design on the adhesion quality of the first layer has been investigated in the present work. PLA plus filament of 1.75 mm procured from 3DXTECH is used for printing designs. Printing speed, nozzle temperature, printing bed temperature, and first layer height are some key parameters that affect the quality of the print.

2 State of the Art

The conventional printing technique for fabric uses a dye-based method or inkjet printer for printing 2D designs. These methods produce a lot of liquid-based waste which contains harmful chemicals. FDM based printing on fabric has an upper edge over the conventional methods. It enables us to print very complex 3D designs directly on the fabrics. It produces less harmful and recyclable waste which is good for our environment. FDM can be used to 3D print the whole dress or directly print design patterns on the fabric.

Many tests are linked with interactions of textile surface properties on the adhesion strength of printing various polymers [2, 3]. The interactions show that FDM prints on cotton, polyester, wool and viscose may result in great adhesion properties [4, 5]. Further, the z-offset distance between nozzle and printing bed has a significant effect on the measured adhesion force [6]. Various applications of 3D printing on textile fabric are considered for smart and functional textiles which are not possible with the conventional printing processes [7].

Pei et al. examined the adhesion properties of PLA, ABS and nylon filaments printed using a commercial FDM printer onto various woven and knitted fabrics. They printed various 3D structures such as parallel strips, circular rings, circles, and triangular shapes of 1 mm height, Braille characters, articulate parts, functional hooks and latches. PLA show the best results with low warping and higher bonding, print quality and flexural strength. Among the fabrics woven cotton, woven polyester–wool and knit soy, woven polyester–wool had good adhesion properties than other two fabrics [8, 9].

Grimmelsmann et al. extended the study of printing on textile fabrics to examine the impact of the z-offset distance on the adhesion properties of the materials [10]. They reported that physical 'locking' between fabrics and printing polymers (PLA and ABS) was the major factor that caused adhesion, not the chemical bonding. Prisca Aude et al. investigated the stress–strain and deformation of PLA material on polyethylene terephthalate. They concluded that platform temperature has a limited role in improving adhesion properties, whereas fiber direction in fabric does contribute to better adhesion characteristics.

First layer of the print is very crucial for adhesion properties. No reference is available in existing literature related to role of infill pattern design. Present work is aimed at determining the effect of varying infill percentage on adhesion property. It also studies the extent of bonding of the first layer on the fabric.

3 3D Printing of Design Pattern on Fabrics

Thin wire of 1.75 mm of PLA is used for 3D printing. A CAD model of circular disc-shape is provided with a hook designed in Creo 2.0 and converted into stereolithographical (.stl) format. G-codes for these STL files are generated by using Slic3r as shown in Fig. 1.

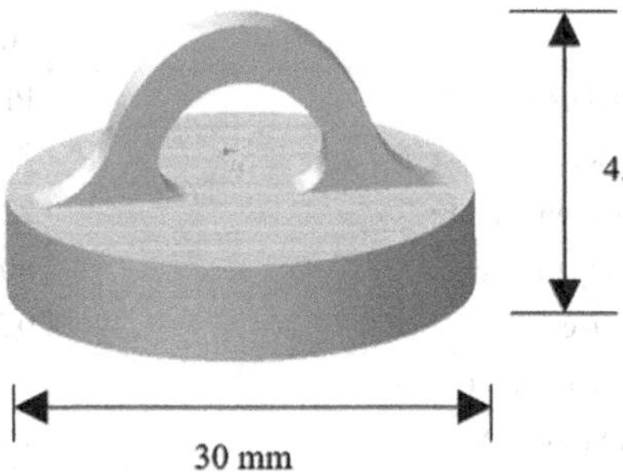

Fig. 1 Disc shape model

4 Experimental Investigation

Cotton-knitted fabric and tulle net fabric (nylon) have been used in the present investigation. Netted fabrics are selected for better deposition of the fused plastic material deep inside the fabric net as shown in Fig. 2. Process parameters are selected with the help of available literature prior to the investigation for ensuring a better quality of print as given in Table 1. Fabric is placed on the printing bed by means of a heating tape as shown in Fig. 3. Optimum z-offset distances (distance between nozzle and fabric) are crucial to ensure better deposition of molten plastic deep inside the fabric voids. Samples of a circular disc of diameter 30 mm and 3 mm thickness are printed on cotton-knitted fabric and tulle fabric by changing infill patterns of the first layer as shown in Fig. 4.

Infill patterns are varied for the first layer of the print. First layer is crucial for good adhesion between the printed object and the fabric. Slic3r is provided with different types of infill patterns. One sample has been printed on each fabric (tulle and cotton) with each infill pattern as shown in Fig. 5.

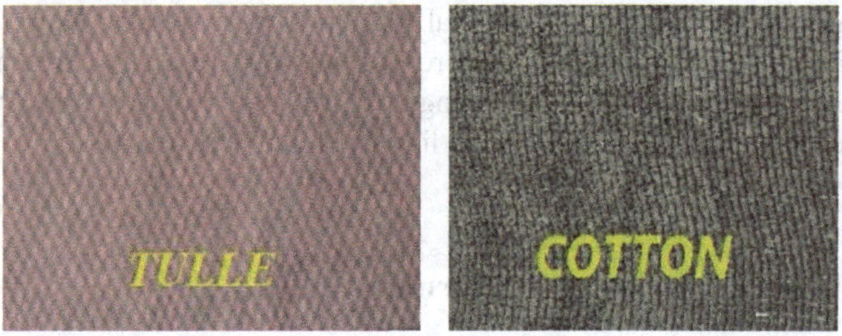

Fig. 2 Cotton-knitted and tulle fabric

Table 1 Process parameters used in the present investigation

Parameters	Description
Nozzle diameter	0.20 mm
Infill	20, 60 and 100%
Material used	PLA (1.75 mm white)
Print speed	22.5 mm/s
Nozzle temperature	220 °C
Bed temperature	60 °C
Layer size	0.20 mm
First layer height	0.30 mm
z-offset (cotton)	1.30 mm
z-offset (tulle)	0.06 mm

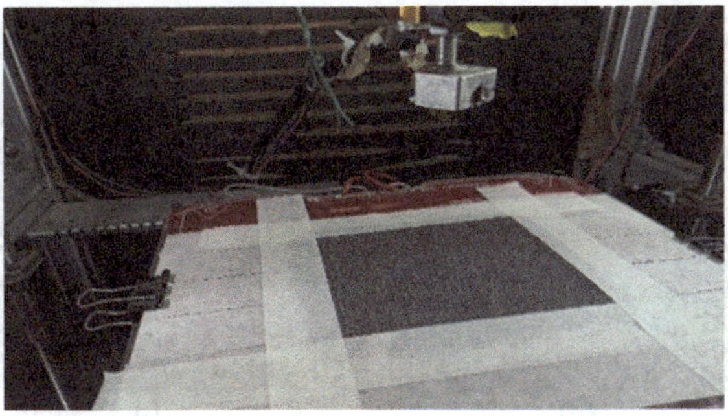

Fig. 3 Fabric printing setup

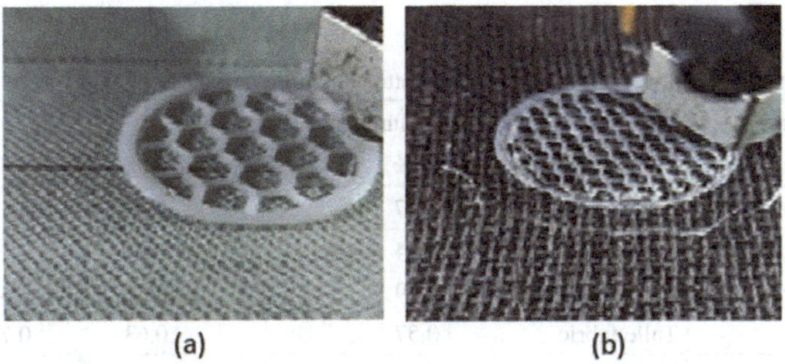

Fig. 4 Honeycomb infill pattern design on **a** tulle fabric and **b** cotton fabric

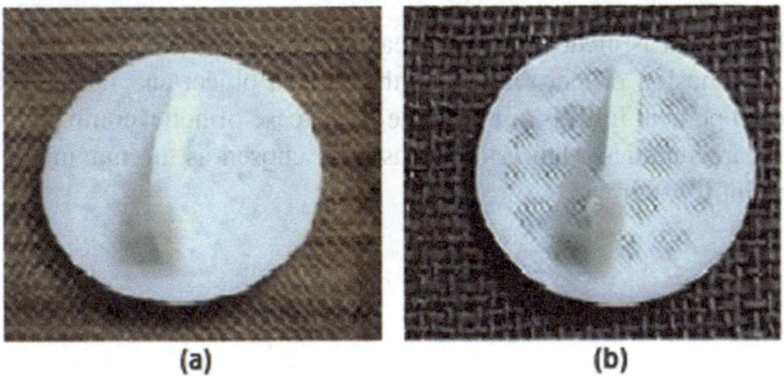

Fig. 5 Printed sample on **a** tulle fabric and **b** cotton fabric

5 Results and Discussion

Failure load for each sample until the complete separation of the object from the fabric is investigated by changing the weights on the hook. Failure load for each printed sample (Fig. 6) with different infill design, infill percentage, and fabric is

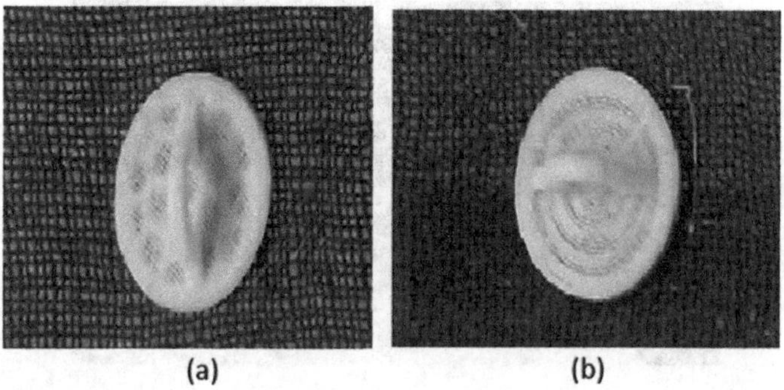

Fig. 6. 3D printed sample with 20% infill on cotton fabric with **a** honeycomb and **b** concentric infill styles

Table 2 Failure load for different samples of cotton and tulle fabrics

Infill pattern	Fabric	Failure loads (kg)		
		20%	60%	100%
Honeycomb	Cotton knitted	0.37	0.49	–
	Tulle fabric	0.14	0.27	–
Concentric	Cotton knitted	0.29	0.31	0.44
	Tulle fabric	0.57	0.63	0.78
Rectilinear	Cotton knitted	0.51	0.69	0.44
	Tulle fabric	0.20	0.67	0.19

given in Table 2. It is clear that an increase in infill percentage will significantly increase the adhesion property between the printed object and fabric (refer Table 2). A hundred percent infill is not possible in the case of honeycomb pattern due to its design characteristics. Hence, 60% has been chosen as the maximum infill for honeycomb infill patterns.

6 Conclusions

Followings are the major conclusions of the present investigation:

- The design of the first layer is crucial for better adhesion characteristics of a direct 3D printed object on the fabric material.
- An increase in infill percentage leads to enhanced adhesive bonding between printed object and the fabric.
- Rectilinear infill of 60% in case of cotton-knitted fabric gives the best adhesion.
- Concentric infill of 100% gives the best adhesion for tulle fabric.

Due to greater porosity, the adhesion quality is better in tulle fabric in comparison with the cotton-knitted fabric as infill can deposit deep inside the fabric. Fused deposition modeling-based process may be gainfully utilized to design fashionable patterns and geometry to the apparels to add new style and uniqueness.

References

1. Tadesse MG, Dumitrescu D, Loghin C, Chen Y, Wang L, Nierstrasz V (2018) 3D printing of NinjaFlex filament onto PEDOT: PSS-coated textile fabrics for electroluminescence applications. J Electron Mater 47(3)
2. Neuß J, Kreuziger M, Grimmelsmann N, Korger M, Ehrmann A (2016) Interaction between 3D deformation of textile fabrics and imprinted lamellae. In: Proceedings of Aachen-Dresden-Denkendorf, International Textile Conference
3. Riviera ML, Moukperian M, Ashbrook D, Mankoff J, Hudson SE (2017) Stretching the bounds of 3D printing with embedded textiles. In: Conference on Human Factors in Computing Systems, Denver, Colorado, USA
4. Korger M, Bergschneider J, Lutz M, Mahltig B, Finsterbusch K, Rabe M (2016) Possible applications of 3D printing technology on textile substrates. In: IOP Conference Series: Materials Science and Engineering, vol 141, no 012011
5. Sabantina L, Kinzel F, Ehrmann A, Finsterbusch K (2015) Combining 3D printed forms with textile structures—Mechanical and geometrical properties of multi-material systems. In: IOP Conference Series: Materials Science and Engineering, vol 87, no 012005
6. Döpke C, Grimmelsmann N, Ehrmann A (2016) 3D printing on knitted fabrics. In: Proceedings of 48th IFKT Congress, Mönchengladbach, Germany
7. Santgar RH, Christine C, Nierstrasz V (2017) Investigation of the Adhesion properties of Direct 3D printing of Polymers and Nanocomposites on textiles: Effect of FDM Printing Process Parameters. Appl Surf Sci 403:551–563
8. Spahiu T, Grimmelsmann N, Ehrmann A, Shehi E, Piperi E (2017) Effect of 3D printing on textile fabric. In: 1st International Conference Engineering and Entrepreneurship Proceedings (ICEE-2017)
9. Pei E, Shen J, Watling J (2015) Direct 3D printing of polymers onto textiles: Experimental studies and applications. Rapid Prototyp J 21:556
10. Grimmelsmann N, Kreuziger M, Korger M, Meissner H, Ehrmann A (2017) Adhesion of 3D printed material on textile substrates. Rapid Prototyp J 24:166

Fabrication and Characterization of PVA-Based Films Cross-Linked with Citric Acid

Naman Jain, Gaurang Deep, Ashok Kumar Madan, Madhur Dubey, Nomendra Tomar, and Manik Gupta

Abstract Recently, the development of biodegradable polymer-based film/composites has acknowledged more attention due to their environment-friendly properties. Polyvinyl alcohol (PVA) is a biodegradable, non-toxic and biocompatible polymer with major limitation of water solubility in water. The present work is an attempt to decrease the solubility limitation with improvement in mechanical properties. In the present work, citric acid was used as cross-linking agent to reduce the water solubility of PVA. Concentration of citric acid and curing time was considered as the important parameter, and their effect on fabricated films was evaluated. Further, water absorption test was performed to test the water solubility. Mechanical characterization of films was done by tensile test which determines the ultimate tensile strength, % elongation and Young's modulus. It was observed that cross-linking reaction was optimum when the curing time for preparing films was kept more than 48 h. It was also observed that maximum UTS was found for 30 wt% CA when curing time was kept 96 h. The maximum UTS was increased up to 300% as compared to neat PVA. ANOVA result states that curing time was the major critical factor which affects the mechanical properties of cross-linked films.

Keywords Cross-linking · Bio-composites · Polyvinyl alcohol

N. Jain (✉) · G. Deep · M. Dubey · N. Tomar · M. Gupta
Meerut Institute of Engineering and Technology, Affiliated To AKTU, Lucknow, Meerut 250005, India
e-mail: namanjainyati@gmail.com

G. Deep · A. K. Madan
Delhi Technological University, New Delhi, Delhi 110078, India

© The Author(s), under exclusive license to Springer Nature Singapore Pte Ltd. 2021
R. M. Singari et al. (eds.), *Advances in Manufacturing and Industrial Engineering*, Lecture Notes in Mechanical Engineering,
https://doi.org/10.1007/978-981-15-8542-5_64

1 Introduction

Polyvinyl alcohol (PVA) is a biodegradable, synthetic, thermoplastic and non-toxic polymer that is soluble in water. It is creamy or whitish, biocompatible, thermostable, non-toxic, colourless, odourless and semi-crystalline polymer. Moreover due to its resistant to grease, oils and solvents researcher gain interest in PVA for many application. For commercially purpose, as the raw material PVA is either in granular or powdered form. Polymerization technique is not employed for fabrication of PVA unlike the other vinyl polymers. Hydrolysis is employed to obtain PVA by hydrolyzing the polyvinyl acetate by removal of an acetate group, (replaced with hydroxide group –OH) which is prepared through the polymerization of vinyl acetate. Thus, key raw material to create PVA is the vinyl acetate monomer. Physical, thermal and mechanical properties of PVA depend upon the degree of hydrolysis [1]. Important parameter which affects the hydrolysis process is the time period in the reaction.

Semi-crystalline nature of PVA occurs due to the formation of hydrogen bonding between the PVA chains due to the presence of hydroxyl group. Hydroxyl group also plays an important role in mechanical, optical and electrical properties of PVA-based films/composites [2]. Properties of PVA had been modified by different researcher in past few years. González-Guisasola and Ribes-Greus [3] prepared PVA composite membrane by using different proportion of graphene oxide (GO) as a reinforcement material. During preparation of composite, it was cross-linked with sulfosuccinic acid (SSA) to reduce the water solubility and has application in proton exchange membrane fuel cells (PEMFCs). GO dispersion was confirmed by TEM image, and cross-linking was confirmed by FTIR spectroscopy. Moreover, thermal stability of membrane is also increased by reinforcing GO which is desirable for fuel cell application. Sonkar et al. [4] fabricated PVA-based cross-linked composites using suberic and terephthalic acid as cross-linking agent. Both the acid are dicarboxylic acids contain two carboxyl group (molecular formula: HO2C–R–CO2H, where R can be aliphatic or aromatic). Tensile and thermal properties of the cross-linked PVA were found. It was noted that suberic acid gives better properties than terephthalic acid. Prepared sample was also kept for biodegradation which shows desirable results. Sonker et al. [5] prepared PVA-based composite film using bacterial cellulose nanowhiskers [BCNW] as a reinforcement material. In 2017, they have used citric acid, whereas in 2018 tartaric acid [6] was as cross-linking agent to reduce water uptake. Due to the covalent bond formation, cross-linked PVA composite film showed better mechanical properties and improved thermal stability. With citric acid, composite film shows 10 times percentage reduction in water swelling test. It was observed that as the concentration of citric acid and cross-linking time (in hour) increases, there is significant reduction in swelling percentage. With tartaric acid, the composite film was prepared by two methods, viz. microwave (MW)-assisted rapid synthesis and by conventional hot air oven heating (CH). Cross-linked composite prepared by MW method took 1/8th time as compared to CH method. Cross-linked PVA with citric acid was also used as coating purpose on to the empty fruit bunch paper by Shakir et al. [7]. Cross-linked PVA solution was applied to the dried hand sheet paper using a spray gun. Swelling test and dimensional changes of paper were observed. It was concluded that with the increase in cross-linked PVA coating reduces volume of water absorbed and slows down the speed of water absorption in the paper. Aparicio et al. [8] fabricated PVA/TiO_2-based nano-composite membranes which are cross-linked by glutaraldehyde solution (GA) as an agent. Impedance spectroscopy was used to measure ionic conductivity. With increase in the reaction time, the proton conductivity also increases significantly. The prepared membrane was suggested to be used as proton exchange membrane for fuel cells.

2 Material and Method

2.1 *Material*

PVA in powdered form having molecular weight 2000–96,000, pH value ranging from 5.0 to 8, having viscosity (4% aqueous solution at 20 °C) 35–50 cP and 87% degree of hydrolysis and was purchased from Thermo Fisher Scientific India Pvt Ltd., India. Citric acid ($C_6H_8O_7$) was also purchased from Thermo Fisher Scientific India Pvt Ltd. It was received in fine granular form. It was anhydrous, 99.9% pure, having molecular weight 192.12 g/mol.

2.2 *Method*

All samples were prepared by solution casting method. In this method, aqueous solution PVA, citric acid and distilled water were prepared on magnetic stirrer and then casted into polypropylene moulds. The mould is then dried in an oven at 70 °C for different time. To prepare neat PVA film, 5 g PVA is poured into 100-ml-distilled water and the solution is left for approximately 1 h 30 min on magnetic stirrer at 70 °C and 250 RPM. To prepare different cross-linked PVA films, 1 g, 1.5 g and 2 g citric acid are mixed in above solution for another 1 h on magnetic stirrer. Each casted film was dried in oven for 24, 48, 72 and 96 h detail shown in Table 1. After completion of respective time in oven, films were peeled off from the mould. To avoid moisture absorption, all prepared films were kept in a close glass box containing calcium chloride at the bottom without touching the films. Nomenclature of fabricated films was presented in Table 1.

Table 1 Sample description

Nomenclature	Description
Neat	5 g PVA + 100 ml distilled water (dried for 24 h)
PVACA20-I	5 g PVA + 100 ml distilled water + 20 wt% citric acid (dried for 24 h)
PVACA20-II	5 g PVA + 100 ml distilled water + 20 wt% citric acid (dried for 48 h)
PVACA20-III	5 g PVA + 100 ml distilled water + 20 wt% citric acid (dried for 72 h)
PVACA20-IV	5 g PVA + 100 ml distilled water + 20 wt% citric acid (dried for 96 h)
PVACA30-I	5 g PVA + 100 ml distilled water + 30 wt% citric acid (dried for 24 h)
PVACA30-II	5 g PVA + 100 ml distilled water + 30 wt% citric acid (dried for 48 h)
PVACA30-III	5 g PVA + 100 ml distilled water + 30 wt% citric acid (dried for 72 h)
PVACA30-IV	5 g PVA + 100 ml distilled water + 30 wt% citric acid (dried for 96 h)
PVACA40-I	5 g PVA + 100 ml distilled water + 40 wt% citric acid (dried for 24 h)
PVACA40-II	5 g PVA + 100 ml distilled water + 40 wt% citric acid (dried for 48 h)
PVACA40-III	5 g PVA + 100 ml distilled water + 40 wt% citric acid (dried for 72 h)
PVACA40-IV	5 g PVA + 100 ml distilled water + 40 wt% citric acid (dried for 96 h)

2.3 Water Absorption Test

In the present work, cross-linking of PVA film was done with citric acid to overcome the major drawback of PVA, i.e. solubility in water and some organic solvent. To verify the cross-linking reaction, water absorption test was performed on cross-linked PVA-based film. For performing the test, 40 × 10 mm^2 sample film was cut from each prepared cross-linked PVA-based film. Weight of each cut sample (W_1) was measured by an electronic weighting machine (having least count 0.001). Each sample was dipped into 100 ml distilled water and left to absorbed water for 24 h at room temperature. Afterwards, all samples were withdrawn from water. Tissue paper was used to wipe out extra amount of water present on the surface of the film. After that, weight of water absorbed sample was measured (W_2) and compared with initial weight of cut samples. Water absorption percentage is calculated by the following expression [9, 10].

$$\text{WA\%} = \frac{W_2 - W_1}{W_1} \times 100 \tag{1}$$

where
W_1 = initial weight of samples and
W_2 = specimen weight after 24 h of water absorption.

2.4 Tensile Test

Tensile testing is considered as an essential step to determine the suitable application in different field. In the present work, mechanical properties such as ultimate tensile strength, Young's modulus and percentage of elongation had been studied by tensile test. The tensile tests were carried out on the 5 kN servo hydraulic universal testing machine (model AMT-SC). All of the specimens for tensile testing were prepared according to the ASTM D638 [11–13] type IV standard having rectangular sample of size 60 mm × 10 mm with gauge length of 40 mm and thickness less than 1 mm approximately. Thickness of films was measured at three points from span of gauge length for calculating cross-sectional area. Test was conducted at room temperature with speed of 2 mm/min.

3 Result and Discussions

3.1 Water Absorption Test

Results of water absorption percentage were presented in Fig. 1 and Table 2. Water absorption test was performed to analyse the effect of cross-linking agent CA on to solubility of PVA in water medium. Results show that fabricated samples which were dried for 24 h (in the oven at 70 °C) found completely soluble in water [14]. This concluded that no cross-linking reaction had been done between PVA chain and CA irrespective of the concentration of CA. With increase in cross-linking curing time water absorption percentage of PVA-based samples were restricted. Samples which were dried for 48 h (in the oven at 70 °C) at 10 wt% of CA was completely soluble, 20 wt% had large amount of water absorption and 30 wt% had 193%. As water absorption percentage is very high, it can be analysed that degree of cross-linking was very low even after 48 h curing time. Further, it was found that the degree of cross-linking further increases with increase in dried time from 48 to 72 h and all films were insoluble in water. Above result shows that time has major effect on the cross-linking reaction as compared to concentration of CA. With further increase in the dried time from 72 to 96 h, water absorption percentage almost become same for 20 and 30 wt% cross-lined composites. Restriction in solubility of PVA/CA films conformed the cross-linked reaction, i.e. CA react to PVA to remove hydroxyl group (–OH).

3.2 Tensile Test

Values of ultimate tensile strength (UTS), percentage elongation and Young's modulus are evaluated from tensile tests for PVA and CA cross-linked PVA film

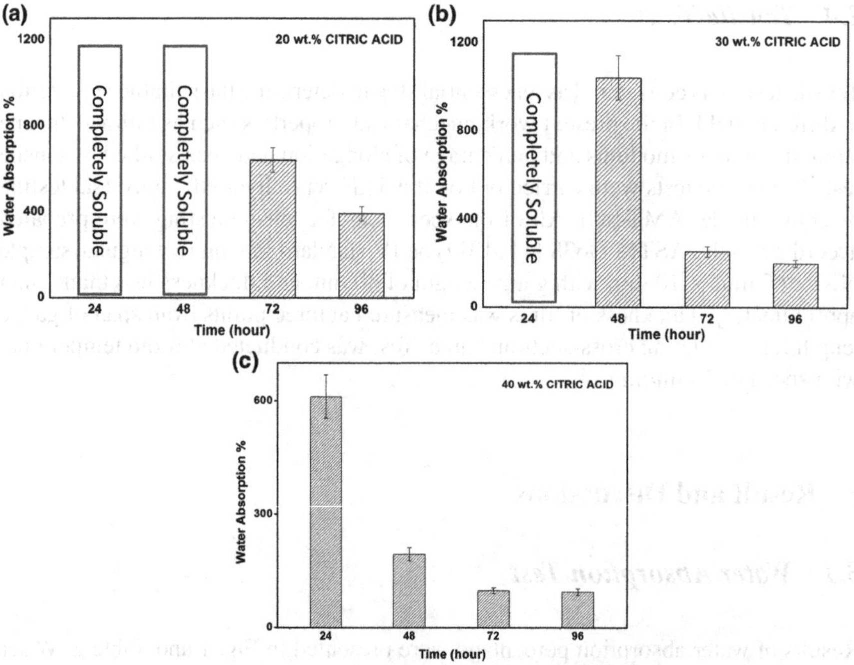

Fig. 1 Result of water absorption test. **a** 20 wt% citric acid, **b** 10 wt% citric acid and **c** 30 wt% citric acid

Table 2 Water absorption test

Curing time	wt% of CA		
	20	30	40
24	Completely soluble	Completely soluble	610.7 ± 58.1
48	Completely soluble	1041.6 ± 101.2	193.3 ± 17.2
72	640.2 ± 57.5	253.2 ± 23.3	96.8 ± 8.6
96	392.7 ± 33.4	199.5 ± 17.7	93.5 ± 9.3

and shown in Table 3. Behaviour of material under tensile loading is represented in Fig. 2. Neat PVA has ultimate tensile strength of about 30 MPa with percentage elongation of about 10%. From Table 3, it can be observed that maximum strength among cross-linked PVA films is 90.1 MPa for PVA30CA-IV, which is about 300% more that neat PVA. Both curing time and concentration of CA act as the major factors which affect the mechanical properties of PVA-based films. It was observed that UTS, % elongation and Young's modulus of fabricated cross-linked films varies with respect to curing time (film dried time in oven) and weight percentage of citric acid. It is observed from the results that with respect to curing time, UTS of PVA/CA cross-linked film increases, whereas % elongation decreases. On the other end with

Table 3 Ultimate tensile strength (UTS) of cross-linked PVA sample films

Sample film	UTS (MPa)	% Elongation	Young's modulus (MPa)
Neat PVA	30.09 ± 2.5	10.7 ± 1.2	1310 ± 86
PVA20CA-I	36.48 ± 2.1	165 ± 11.5	1077 ± 66
PVA20CA-II	41.79 ± 3.3	48.6 ± 3.2	1796 ± 105
PVA20CA-III	66.33 ± 5.2	24.2 ± 1.9	3278 ± 196
PVA20CA-IV	66.16 ± 4.5	9 ± 1.1	3026 ± 208
PVA30CA-I	40 ± 2.9	233 ± 19.6	545 ± 39
PVA30CA-II	44.91 ± 3.2	117 ± 8.90	1085 ± 74
PVA30CA-III	68.16 ± 4.23	65.4 ± 4.80	2931 ± 21
PVA30CA-IV	90.1 ± 5.8	25.1 ± 1.75	5793 ± 376
PVA40CA-I	19.5 ± 2.1	322 ± 12.8	21 ± 2
PVA40CA-II	34.29 ± 2.2	182 ± 10.1	903 ± 65
PVA40CA-III	41.35 ± 3.1	110 ± 9.4	1831 ± 132
PVA40CA-IV	58.06 ± 4.12	21 ± 1.3	881 ± 51

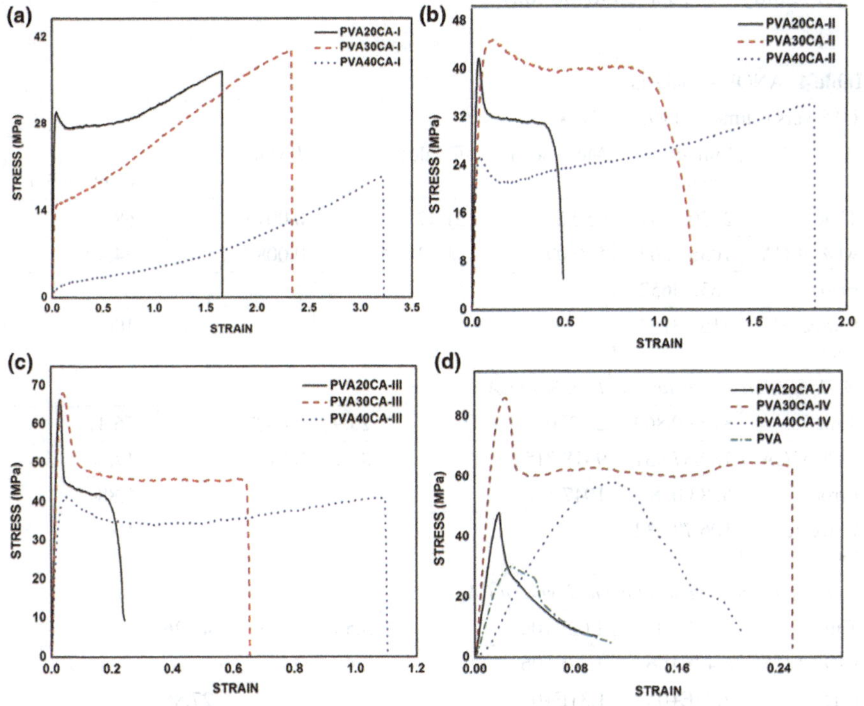

Fig. 2 Stress–strain curve for cross-linked PVA sample films. **a** Curing time is 24 h, **b** curing time is 48 h, **c** curing time is 72 h and **d** curing time is 96 h

increase in wt% of CA, % elongation increases in each case. Whereas UTS first increases up to 30 wt% CA, then decreases for 40 wt% CA. In 10 wt% CA films UTS and YM almost become constant after 72 h curing time. It was observed that maximum UTS and Young's modulus were achieved for 30 wt% of CA and 96 h curing time.

From the water absorption test, this can also be analysed that cross-linking with CA was achieved after 48 h of curing time. Water absorption percentage significantly reduces for the 72 h curing time which shows successful cross-linking reaction. Further, UTS decreases for 40% concentration of CA which shows incomplete cross-linking reaction. This shows that CA molecules got stuck between PVA chains which hamper the cross-linking reaction partially and cause reduction in UTS. This fact can also be confirmed by observing UTS value of PVA40CA-I sample (19.5 MPa), which shows no cross-linking reaction. From the ANNOVA analysis Table 4, it is observed that curing time has major influence on the obtained values of UTS (68.61%), percentage elongation (74.47%) and YM (49.26%) as compared to wt% CA. So curing time is the most important control parameter which affects tensile properties of all sample films. Also in case of UTS value, wt% of citric acid affects significantly (24.99% contribution).

Table 4 ANOVA analysis

UTS versus time and wt. % of CA					
	Sum of squares	Mean square	*F* value	*P* value	% Contribution
Time	2850.5997	950.2	21.47	0.0013	68.62
wt% of CA	1038.1802	519.09	11.73	0.008	24.99
Error	265.53652				6.39
Corrected total	4154.3164				100
% elongation versus time and wt. % of CA					
Time	81,660.869	27,220.28	25.9	0.0007	76.47
wt% of CA	18,837.431	9418.715	8.99	0.01	17.64
Error	6283.028	1047.17			5.884
Corrected total	106,781.32				100
Young's modulus versus time and wt% of CA					
Time	1.39E+07	4.62E+06	3.53	0.008	49.26
wt% of CA	6.43E+06	3.22E+06	2.46	0.16	22.8
Error	7.84E+06	1.31E+06			27.86
Corrected total	2.81E+07				100

4 Conclusions

The effect of curing time and concentration of cross-linking agent CA on the water absorption and tensile properties of sample films was studied. In water absorption test, it was observed that curing time has major impact. Water uptake percentage decreases with respect to increase in curing time as degree of cross-linking also increases; similar observation is also seen for concentration of citric acid. The sample which was dried from 72 to 96 h in oven was found insoluble in water with minimum water absorption percentage. In tensile test, it was observed with increasing curing time, UTS was increased and % elongation was reduced while with increase in concentration of CA, UTS increases only up to 30 wt% of CA. It shows that degree of cross-linking reduces when CA was added beyond 40 wt% which conclude that more curing time is required for cross-linking reaction between PVA and CA. But on the other hand, it is undesirable on the basic of manufacturing point of view. Maximum UTS and maximum YM were obtained for 96 h curing time and at 30 wt% of CA which is 300% higher than the neat PVA.

Acknowledgements The authors would like to thank the Department of Mechanical Engineering, MIET, Meerut, for proving the necessary facilities for fabricating the cross-linked films. Authors also express gratitude to AKTU for providing necessary found under TEQIP-III for procurement of raw material and testing/analysing the fabricated films.

References

1. Aslam M, Kalyar MA, Raza ZA (2018) Polyvinyl alcohol: a review of research status and use of polyvinyl alcohol based nanocomposites. Polym Eng Sci. https://doi.org/10.1002/pen.24855
2. Reddy N, Yang Y (2009) Citric acid cross-linking of starch films. Food Chem. https://doi.org/10.1016/j.foodchem.2009.05.050
3. González-Guisasola C, Ribes-Greus A. Dielectric relaxations and conductivity of cross-linked PVA/SSA/GO composite membranes for fuel cells. Polym Test. https://doi.org/10.1016/j.polymertesting
4. Sonker AK, Rathore K, Nagarale RK, Verma V (2017) Crosslinking of polyvinyl alcohol (PVA) and effect of crosslinker shape (aliphatic and aromatic) thereof. J Polym Environ. https://doi.org/10.1007/s10924-017-1077-3
5. Sonker AK, Teotia AK, Rajaram AK, Nagarale K, Verma V (2017) Development of polyvinyl alcohol based high strength biocompatible composite films. Macromol Chem Phys J. https://doi.org/10.1002/macp.201700130
6. Sonker AK, Rathore K, Teotia AK, Kumar A, Verma V (2018) Rapid synthesis of high strength cellulose–poly(vinyl alcohol) (PVA)biocompatible composite films via microwave crosslinking. J Appl Polym Sci. https://doi.org/10.1002/app.47393
7. Shakir MA, Yhaya MF, Ahmad MI (2017) The effect of crosslinking fibers with polyvinyl alcohol using citric acid. Imperial J Interdisc Res (IJIR) 3(4)
8. Aparicioa GM, Vargasb RA, Buenoc PR (2019) Protonic conductivity and thermal properties of cross-linked PVA/TiO_2 nanocomposite polymer membranes. J Non-Cryst Solids. https://doi.org/10.1016/j.jnoncrysol.2019.119520

9. Jain N, Ali S, Singh VK, Kumar N, Chohan S (2019) Creep and dynamic mechanical behaviour of cross-linked polyvinyl alcohol reinforced with cotton fiber laminate composites. J Polym Eng. https://doi.org/10.1515/polyeng-2018-0286
10. Jain N, Verma A, Singh VK. Dynamic and creep-recovery analysis of polyvinyl alcohol based cross-linked composite reinforced with basalt fiber. Mater Res Express. https://doi.org/10.1088/2053-1591/ab4332
11. Verma A, Jain N, Singh VK (2019) Fabrication and characterization of chitosan coating sisal fiber reinforced phytagel modified soy protein based green composite. J Compos Mater. https://doi.org/10.1177/0021998319831748
12. Jain N, Singh VK, Chauhan S (2018) Dynamic and creep analysis of polyvinyl alcohol based films blended with starch and protein. J Polym Eng 39(1):26–35. https://doi.org/10.1515/polyeng-2018-0032
13. Deepmala K, Jain N, Singh VK, Chauhan S (2018) Fabrication and characterization of chitosan coated human hair reinforced phytagel modified soy protein-based green composite. J Mech Behav Mater. https://doi.org/10.1515/jmbm-2018-0007
14. Jain N, Singh VK, Chauhan S (2017). A review on mechanical and water absorption properties of polyvinyl alcohol based composites/films. J Mech Behav Mater. https://doi.org/10.1515/jmbm-2017-0027

Characterization of Bael Shell (*Aegle marmelos*) Pyrolytic Biochar

Monoj Bardalai and D. K. Mahanta

Abstract Biochar is found to be an important tool in improving the soil quality for agriculture. In this work, biochar was derived from *Aegle marmelos* (commonly known as bael) shell (AMS) by pyrolysis for evaluation and analysis of different properties. In the experiment of pyrolysis, a fixed bed reactor was used for the production of biochar from AMS. The range of particle size of the biomass feedstock was 0.5–1.0 mm, while pyrolysis temperature was fixed at 450 °C with the rate of heating as 15 °C/min. The heating value of the AMS biochar (i.e., 24.91 MJ/kg) is found to be higher when compared with the raw AMS (i.e., 18.11 MJ/kg). Thermogravimetric (TGA) and derivative thermogravimetric (DTG) analyses show that biochar decomposes at higher temperature due to the significant contamination of lignin. The surface morphology of AMS biochar reveals few small pores (e.g., 0.88–1.4 μm), and the surface area of the biochar according to Brunauer, Emmett, and Teller (BET) was measured to be low, i.e., 3.9 m^2/g. The AMS biochar shows the pH value to be quite high as 9.3. The spectrum of Fourier transform infrared (FTIR) of the biochar indicates the presence of some compounds of aromatic functional groups with C=C stretching. The alkaline biochar of AMS becomes useful for the improvement of soil of acidic nature. Further, aromatic functional groups of AMS biochar enhance the soil stability.

Keywords *Aegle marmelos* · Biochar · Pyrolysis · Morphology · Soil

---

M. Bardalai (✉)
Department of Mechanical Engineering, Tezpur University, Tezpur 784028, India
e-mail: monojtezu@gmail.com

D. K. Mahanta
Department of Mechanical Engineering, AEC, Gauhati University, Guwahati 781013, India

© The Author(s), under exclusive license to Springer Nature Singapore Pte Ltd. 2021
R. M. Singari et al. (eds.), *Advances in Manufacturing and Industrial Engineering*,
Lecture Notes in Mechanical Engineering,
https://doi.org/10.1007/978-981-15-8542-5_65

1 Introduction

Biomasses are found almost every part of the world and considered as wastes. A major part of the biomass can be used as the substituting agent for soil fertilization. However, conversion of biomass into liquid and gaseous fuel and chemicals are the major concerns of the recent research. Although many conversion techniques are available, the pyrolysis and gasification are the most common and easy to perform [1]. In the process of pyrolysis, the feedstock is generally put inside a reactor in the temperature range of 400–700 °C in an environment of limited oxygen. During pyrolysis, the biomass undergoes many chemical reactions leading to convert the biomass into liquid (tar or pyrolytic oil), solid (biochar), and gas (incondensable gas). It is found that the properties of pyrolysis products such as biochar show better quality in comparison with the raw biomass. Therefore, biochar can be utilized as an agent for the improvement of soil properties [2, 3]. Although pyrolysis is mostly carried out to produce bio-oil, a significant amount of biomass (approximately 15–20%) is found to be solid product, i.e., biochar [4]. The biochar obtained by pyrolysis contains the residual of carbonaceous substance and can be the cause of natural fire with negligible amount of oxygen [5]. Biochar can also be used for the purpose of fertilization of soil, plant growth, decontamination of adulterant such as acaricides, heavy metallic substance, and compounds of hydrocarbon [6, 7].

In the past years, the use of biochar has been found as a substance for improvement and reclusion of C in soil [8]. Biochar can be considered as a by-product of pyrolysis in solid state, which is available in the form of black carbon with the presence of elemental carbon or graphite to aromatic carbon [9]. In general, biochar has much smaller specific area and micropore volume than the carbon which is activated commercially. However, the capacity of adsorption relative to organic pollutants and heavy metals are equivalent to and sometimes higher in comparison with the activated carbon [9]. Moreover, the detail study of the previous publications reveals that the biochar is useful for upgradation of soil properties, improving the productivity of crop, fixation of CO_2, and absorption of unnecessary ingredients [10]. Biochar is relatively stable in comparison with some other organic substances and sustainable for a long period of time in soil [11–13]. Further, the biochar has the ability of reduction of the release of major greenhouse gases such as CH_4 and N_2O, particularly from the soil of rice paddy field [14–16].

The variation of biochar in terms of physiochemical properties is influenced by the nature of biomass and conditions of pyrolysis experiment. In particular, the temperature used in pyrolysis process has the significant effects on the yield of biochar and its properties. It was observed that by increasing temperature in pyrolysis process, the yield of biochar is found to be decreased and at higher temperature, the biochar losses carbon, some important functional groups and development of useful microstructure [17, 18]. The pyrolysis temperature also affects the properties such as thermal stability, pH value, morphology of surface, and chemical composition [19].

Gottipatti and Mishra [20] used *Aegle marmelos* fruit shell in order to prepare microporous activated carbon (MAC) by KOH activation. This study was concentrated mainly on the porous characteristics of the activated carbon, such as pore volume, surface area, and pore size distribution. The same biomass was taken for the production of microporous activated carbon by activating $ZnCl_2$, and this MAC was applied for the removal of Cr (VI) from the aqueous solution [21]. Another study performed by Ahmed and Kumar [22] showed the possible application of AMS carbon as an absorbent to eliminate the congo red dye from aqueous solution. Roy et al. [23] performed the research on characterization and application of bael shell biochar in the removal of Patent Blue dye solution. Recently, Palniandy et al. [24] have studied the application of bio-char derived from different biomasses such as rubberwood and rice husk as fuel source in direct carbon fuel cells (DCFC) for power generation. However, the detail investigation of different properties of the pyrolyzed biochar of AMS dust is not much available in the previous publications. Although the bael tree is found in different parts of the world, it is abundantly available in the northeastern region of India. The inner part of bael fruit is very useful as food for various medicinal purposes. On the other hand, the outer part, i.e., the shell of the fruit is treated as waste. The biochar has many properties which are helpful in the improvement of soil quality for agricultural purposes such as plant nutrition. In order to reduce the wastes and to explore the properties of biochar, the bael shell was taken in the present study as the raw material for biochar production. The present work is concentrated on the production of AMS biochar by the process of pyrolysis and to characterize all the physiochemical properties with a detailed comparative study.

2 Materials and Methods

The AMS biochar was obtained as a solid product during the production of pyrolysis oil from AMS dust as explained by Bardalai and Mahanta [25]. Pyrolysis was carried out with a heating rate of 15 °C/min in the range of temperature 450–600 °C. However, the biochar sample produced at 450 °C was considered for various analyses at which the yield of biochar was about 45 wt%.

The proximate analysis, elemental analysis, TGA, FTIR, and pH measurement were carried out according to the methodology used in the previous publication [25]. For pH measurement, a mixture of biochar and deionized water was prepared in the ratio 1:20 (w/v) in order to obtain a homogeneous suspension to evaluate the pH value within 1.5 h. An analytical instrument was used to perform the crystallographic study known as XRD diffractometer (RIGAKU Mini flex, Japan). The operating voltage of XRD diffractometer was 30 kV while current density was 15 mA. The range of scanning was set at $2\theta = 10°–70°$, and the scanning speed was 0.05°/s.

The morphology study of the biochar surface was conducted with the help of scanning electron microscope—SEM (JEOL, JSM 6390 LV). BET (QUANTACHROME, NOVA 1000E) was used to measure the specific surface area (SSA) of biochar at

350.35 °C using N_2 sorption data. The timing of outgassing analysis was 6 h with the temperature of 100 °C.

For characterization of all physical and chemical properties of AMS biochar, the experiments were carried out for three times (standard deviation 0.5–1.5) and the average values of these results are tabulated in this paper.

3 Results and Discussion

Different analyses for AMS biochar were carried out and the results are presented in Table 1 and discussed in the following sections.

3.1 Proximate and Ultimate Analysis

The moisture content in AMS biochar is reasonably lower (refer Table 1) than the biomass as it was removed during pyrolysis process and found to be consistent with EFB biochar [26]. However, AMS biochar contains relatively higher moisture content when compared with the biochar obtained from coconut shell (CSB) and mesquite wood (MWB) as reported in Table 1 [27]. It can be seen that fixed carbon has increased by an amount of 60.87 wt% when AMS was converted into biochar indicating that the biochar is more carbonaceous than biomass and thus suitable for soil amendment. The fixed carbon in AMS biochar is found to be quite comparable with other biochar as seen in Table 1. High value of fixed carbon content helps the biochar in improving soil properties. During pyrolysis, a series of thermochemical reactions take place

Table 1 Proximate and ultimate analysis of AMS and AMS biochar

Property	AMS	AMS biochar	Other biochar
Moisture (wt%)	10.88	5.42	0.35–5.15 [26, 29, 32]
Volatile matter (dry basis, wt%)	90.56	29.82	7.2–17.62 [26, 29]
Fixed carbon (dry basis, wt%)	5.46	66.33	65.4–72.94 [26, 29]
Ash (dry basis, wt%)	3.98	3.00	7.9–67 [8, 26, 28, 29]
Calorific value (MJ/kg)	18.11	24.91	26.6–28.8 [33]
C (wt%)	41.80	70.87	22.5–74.19 [8, 26, 28, 29]
H (wt%)	5.80	3.99	1.6–3.48 [26, 28, 33]
N (wt%)	2.10	1.14	6.83 [26]
O (calculated by difference, wt%)	50.30	24	23.9–24.6 [26, 28]
H/C	1.67	0.68	0.62 [28]
O/C	0.90	0.25	0.27 [28]
pH	–	9.3	7.2–10.9 [8, 27–29, 34, 35]

leading to produce carbonaceous hydrocarbons from the oxygenated compounds and thus content of oxygen is found to be decreased in biochar. The amount of ash content in AMS biochar is significantly lower in comparison with many biochar which was obtained from the biomasses such as bamboo shoot shell (BSS) [28], palm empty fruit bunch (EFB) [26], and palm kernel shell (PKS) [29] as listed in Table 1. The biochar which contains low ash is preferred for the purpose of soil improvement, since the possibility of the presence of contaminants of heavy metallic compounds in ash is high and leads to pollute the soil [30]. Moreover, the biochar containing low ash content carries high calorific value, because the high ash content is responsible for diluting the energy value of the biochar [31]. In this study, it is seen that percentage of ash in AMS biochar is slightly more when compared with CSB and MWB as they ranged within 1.29–1.46 wt% [32].

Due to high carbon content, low ash, and oxygen content, the heating value of AMS biochar is relatively higher when compared with biomass (refer Table 1). The amount of carbon content in the present biochar is higher in comparison with the biochar obtained from sewage sludge (SS), municipal waste (MW), cattle digestate (CD), poultry litter (PL), BSS, and EFB [8, 26, 28]. However, the carbon content in AMS biochar is relatively lower than the biochar of PKS and cashew nutshell (CNS) produced at 450 °C [29, 33]. The calorific value of AMS biochar is relatively lower in comparison with CNS biochar due to lower carbon content [33].

The pH value of AMS biochar (see Table 1) can be compared with many biochar which are available in a significant number of publications. It is found that, pH of AMS biochar is higher when compared with the biochar of SS (pH = 7.2), MW (pH = 7.4), CSB (pH = 8.66), and MWB (pH = 8.73) [8, 32]. However, in AMS biochar, the pH value is slightly lower relative to the biochar produced from rice straw (RS), BSS, PKS, and argan shells (AS) [27–29, 34, 35]. The highly alkaline behavior of the biochar is found to be favorable for improving acidic soil.

The atomic ratios, i.e., O/C and H/C of AMS biochar, are significantly lower in comparison with raw AMS (Table 1). The reason of lower atomic ratios is the liberation of H and O at the time of pyrolysis relative to carbon according to the explanation of previous publications [27, 34–36]. The biochar with low value of H/C is highly aromatic. The highly aromatic biochar resists the decomposition and remains intractable, which is helpful in sequestrating the carbon in soil [17, 37]. However, the atomic ratio, O/C of AMS biochar is higher in comparison with the biochar of wheat residue and activated carbon (e.g., 0.06–0.09). This reveals that biochar of AMS is highly hydrophilic due to the presence of more polar groups [9]. Further, in AMS biochar, H/C ratio is slightly higher and O/C ratio is lower when compared with BSS biochar indicating the release of higher amount of hydrogen and lower amount of oxygen relative to carbon [31]. These differences of atomic ratios are supposed to be due to polymerization of dehydrogenative radicals and dehydrating polycondensation at the time of pyrolysis [38].

3.2 Thermal Degradation Analysis

TGA profile of AMS dust (Fig. 1a) indicates that about 70 wt% of biomasses have decomposed in temperature range of 450–500 °C. The highest decomposition rate of AMS is observed at 259 °C as shown by DTG profile (Fig. 1a). This thermal behavior of AMS biomass is consistent with other biomasses, such as in CNS, where about 60 wt% degradation was observed in nitrogen atmosphere in the temperatures between 450 and 500 °C, while the decomposition rate was found to be highest at 300 °C [33]. Further, about 70% weight loss was observed upto 450 °C when EFB was thermally degraded and the highest rate of decomposition was at 300 °C [26]. Thus, the decomposition of AMS within the temperature of 450–500 °C is found to be quite similar to the other biomasses, and hence, the pyrolytic temperature for AMS was fixed at 450 °C. As the pyrolysis temperature increases, the lignin components start decomposing leading to the formation of tar which enhances in partially blocking the pores on biochar surface [39].

The TGA curve of AMS biochar (Fig. 1b) shows that about 3% till 100 °C followed by about 2% upto 300 °C mass loss has occurred. This is due to the release of physisorbed water. According to the TGA curve (Fig. 1b), significant degradation of the biochar starts from 300 °C, which reveals the degradation of lignin and residual cellulose available in the biochar. The DTG plot of AMS continues to rise upto about 450 °C, which represents the degradation of residual cellulose and lignin content [40]. The peak at around 448 °C on the DTG curve (Fig. 1b) indicates the highest rate of lignin and cellulose decomposition present in the biochar. Thus, the biochar derived from AMS with lignin content is useful for the promotion of Fusarium oxysporum cucumerinum (*f.o.c.*) survival in soil by growing the peculiar fungal flora [41].

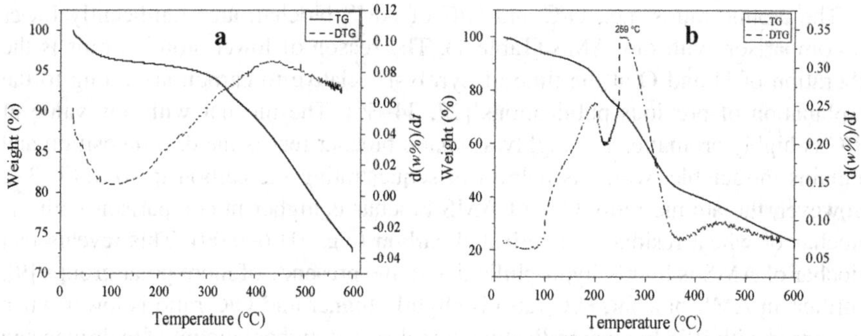

Fig. 1 Profiles of TGA and DTG for **a** AMS dust and **b** AMS biochar

3.3 FTIR Analysis

The peaks observed in the FTIR spectra of AMS and AMS biochar (Fig. 2) have been used in order to identify the functional groups of various compounds. A broadband which centered near 3414 cm^{-1} is assigned to –OH functional group in AMS. However, intensity of this peak is less when compared with the spectra of AMS biochar indicating the deterioration of hydroxyl compounds during conversion of biomass to biochar. The peaks at near to 2927 cm^{-1} are due to the presence of alkane group with C–H stretching vibration which are found on both biomass and biochar. Similar type of functional group was identified in the FTIR spectrum of MWB biochar [32]. Few peaks in the band of 1600 and 1800 cm^{-1} in the spectrum as seen in Fig. 2 indicate the existence of functional groups of the compounds of alkanes, carbonyl (C=O), water, and oxygenated hydrocarbon with plane bending –OH, which disappear in AMS biochar [40]. This is the indication of reduction of oxygenated compounds from biomass to biochar. The intense peak near 1592 cm^{-1} with stretching vibration represents the ring bonding of aromatic and alkynes (C≡C) characteristic in the AMS biochar similar to MWB [32], which is not found in the raw AMS. Pituello et al. [8] also found the aromatic functional groups in the silage digestate biochar spectrum. The carbon–carbon double-bonded stretching aromatic peak reveals the presence of the ring of aromatic compounds such as benzene in the AMS biochar, which helps in stabilizing the soil. The strong peak of AMS spectrum at 1031 cm^{-1} indicates the stretching bond of C–C–O or C–O–C, which are not significant in the spectrum of AMS biochar. Further, the peaks found within the band 500–700 cm^{-1} in AMS spectrum (centered at 611 cm^{-1}) is supposed to be hydroxyl group in the mode of out plane bending [40]. In FTIR spectrum of AMS biochar, the appearance of peak nearby 827 cm^{-1} is the agreement of the presence of aromatic functional group. This can be attributed to aromatic C–H group which is to

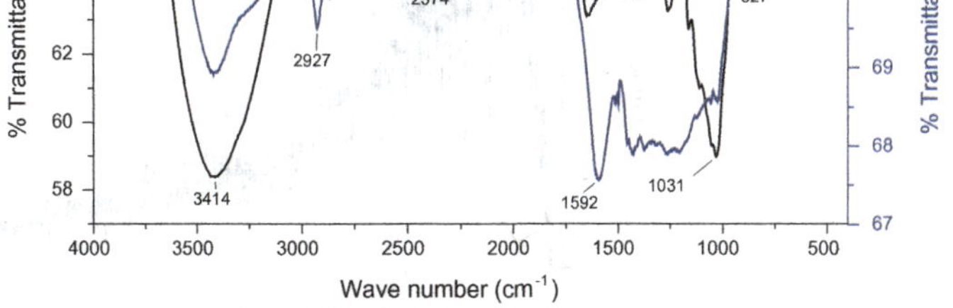

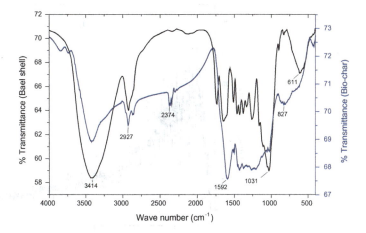

Fig. 2 FTIR spectra of AMS and AMS biochar

be out of plane deformation [32, 42, 43]. Thus, the FTIR analysis indicates that AMS consists of compound of oxygenated hydrocarbons, while aromaticity dominates the characteristics of AMS biochar.

3.4 Crystallographic Analysis

The XRD diffractograms of AMS and the AMS biochar are shown in Fig. 3a and b, respectively, where both the spectra clearly show the amorphous behavior. However, the peak spacings observed at 5.8 Å, 5.52 Å, 5.3 Å, 4.42 Å, 3.88 Å, 2.57 Å in Fig. 3a

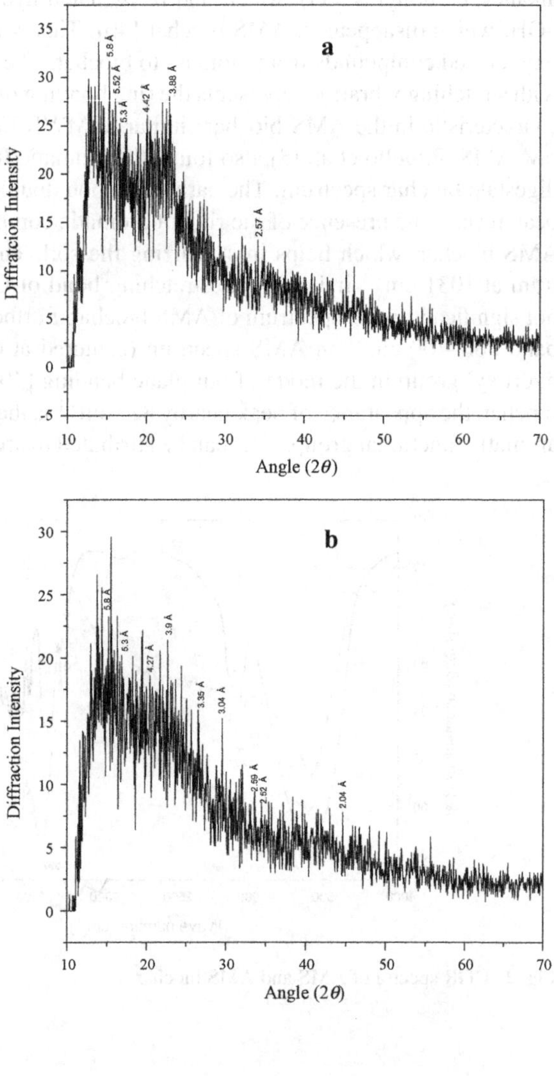

Fig. 3 X-ray diffractograms of **a** AMS and **b** AMS biochar

indicate the crystal structure of cellulose in AMS [4, 19, 44]. This is due to the fact that cellulose in biomass may consist of a wide range of segments, which contain both amorphous and crystalline portions [45]. The peaks observed at 5.8 Å, 5.3 Å, 3.9 Å, 2.59 Å, and 2.52 Å in the diffractogram (Fig. 3b) of AMS biochar represent the crystalline structure of cellulose [19, 42]. The presence of SiO_2 in AMS biochar can be expected by the peaks at 4.27 Å and 3.35 Å (Fig. 3b) which is consistent with RS biochar [34]. Further, the possibility of calcite ($CaCO_3$) contamination in AMS biochar is detected by the peak identified at 3.04 Å [34]. The calcite content in biochar improves the soil of acidic nature and makes it useful for agricultural purposes. The presence of graphitic structure in the biochar of AMS can be explored by the peaks at the spacings of 3.35 Å and 2.04 Å [46].

3.5 *Morphology Study*

The SEM images of AMS biochar are shown with the magnifications 500×, 2500×, and 5500× in Fig. 4a–d, respectively. At low magnifications (Fig. 4a, b), the micrographs show no defined morphology which are similar to CNS biochar [33]. However, a few numbers of pores with diameter in the range of 0.51–1.40 μm are observed on the biochar surface (Fig. 4c, d). Therefore, the SSA of the biochar recorded by BET is found to be very low (3.9 m^2/g). The SSA of AMS biochar is quite comparable with the results published by Shariff et al. [26], i.e., 0.1301–7.9890 m^2/g. As the pyrolysis was carried out at low temperature, the surface area of AMS biochar was found to be relatively low [46]. The ash content and its composition are responsible for the occurrence of pores on the biochar surface. Inorganic compounds present in biochar such as ash and leads to plug the pores caused by pyrolysis [31, 47]. The less porosity on the biochar surface is also the result of the production of tar in pyrolysis which mixes with the biochar and tends to plug the pores. Few fibrillar structures can be seen in Fig. 4b, c, which is the indication of the presence of cellulose on AMS biochar and consistent with the results obtained in thermal and crystallographic analyses [8, 40].

The isotherms obtained in adsorption and desorption of nitrogen in AMS biochar reveal the growth of micropores on the biochar (Fig. 5). The volume of adsorption and desorption is not much significant at low relative pressure (e.g., 0.9), but becomes very high beyond that. At low relative pressure, desorption is higher than adsorption, while at higher relative pressure, both adsorption and desorption curves coincide and good agreement of the study carried out by some previous researchers with the analysis gas CO_2 and N_2 [40, 46]. This study reveals the partial desorption of nitrogen in the AMS biochar.

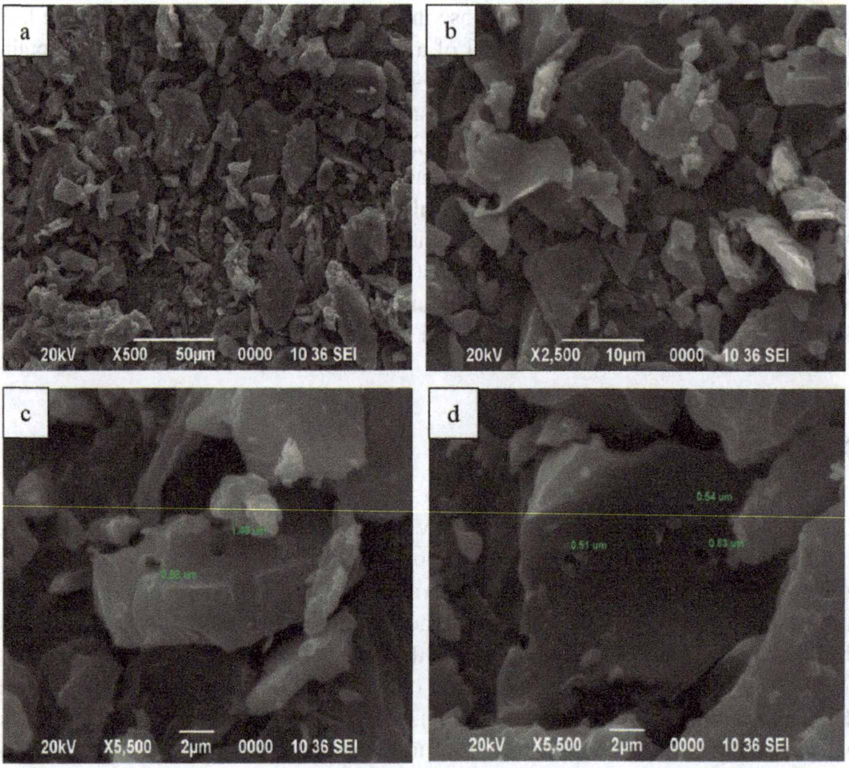

Fig. 4 SEM images of AMS biochar at various magnifications

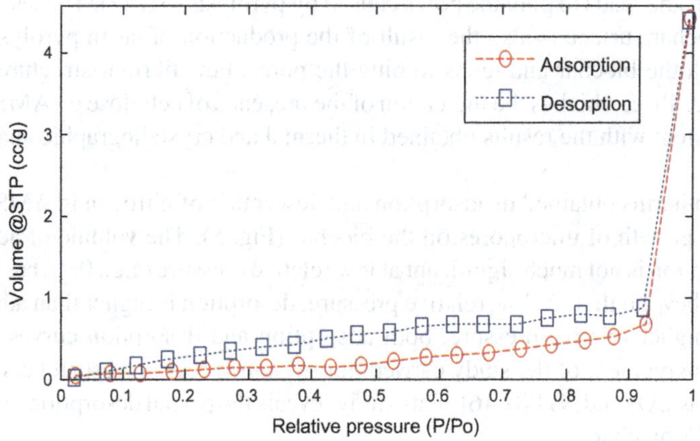

Fig. 5 Adsorption and desorption of AMS biochar

4 Conclusions

The physical and chemical properties of the biochar produced by pyrolysis process from AMS were evaluated, and a comparative study was performed with different biochar. The decomposition of AMS biochar begins at about 300 °C, which indicates that it consists of lignin along with some undecomposed amount of cellulose. The biochar contains higher heating value relative to biomass. Due to the high pH (9.3), the AMS biochar is alkaline in nature and thus it can be used for soil improvement. The biochar of AMS is composed of the compounds which are less oxygenated but more carbonaceous with low O/C atomic ratio when compared with the biomass. Again, the AMS biochar is highly aromatic due to low H/C ratio and helpful for the removal of carbon and resisting the decomposition in soil. The FTIR analysis also shows the functional groups of aromatic carbon due to the peaks with carbon stretching vibration in AMS biochar. The diffractogram of XRD of AMS biochar shows the presence of cellulose with few local crystalline structures. Moreover, the study also reveals the possible existence of SiO_2, $CaCO_3$, and graphitic carbon in AMS biochar. The surface of AMS biochar is mostly nonporous except few micropores (0.54–1.40 μm). The results of BET analysis indicate the small SSA on AMS biochar which is similar to some other biochar. As further study, the biochar can be produced from the AMS at different temperatures and the detailed study for the various properties can be performed.

Acknowledgements The authors like to acknowledge the Tezpur University for supporting financially in the development of setup for pyrolysis experiments and utilizing different instruments available in various laboratories.

References

1. Yanik J, Kommayer C, Saglam M, Yuksel M (2007) Fast pyrolysis of agricultural waste: characterization of pyrolysis products. Fuel Process Technol 88:942–947. https://doi.org/10.1016/j.fuproc.2007.05.002
2. Yin R, Liu R, Mei Y, Fei W, Sun X (2013) Characterization of bio-oil and bio-char obtained from sweet sorghum bagasse fast pyrolysis with fractional condensers. Fuel 112:96–104. https://doi.org/10.1016/j.fuel.2013.04.090
3. Mukherjee A, Lal R (2013) Biochar impacts on soil physical properties and greenhouse gas emissions. Agronomy 3:313–339. https://doi.org/10.3390/agronomy3020313
4. Kim KH, Kim JY, Cho TS, Choi JW (2012) Influence of pyrolysis temperature on physiochemical properties of biochar obtained from the fast pyrolysis of pitch pine (*Pinus rigida*). Bioresour Technol 118:158–162. https://doi.org/10.1016/j.biortech.2012.04.094
5. Brown R (2009) In: Lehman J, Joseph S (eds) Biochar for environmental management: science and technology, 2nd edn. Earthscan, London, pp 127–146
6. Beesley L, Moreno-Jiménez E, Gomez-Eyles JL, Harris E, Robinson B, Sizmur T (2011) A review of biochars' potential role in the remediation, revegetation and restoration of contaminated soils. Environ Pollut 159:3269–3282. https://doi.org/10.1016/j.envpol.2011.07.023

7. Cabrera A, Cox L, Spokas KA, Celis R, Hermosín MC, Cornejo J, Koskinen WC (2011) Comparative sorption and leaching study of the herbicides fluometuron and 4- chloro-2 methylphenoxyacetic acid (MCPA) in a soil amended with biochars and other sorbents. J Agric Food Chem 59:12550–12560. https://doi.org/10.1021/jf202713q
8. Pituello C, Francioso O, Simonetti G, Pisi A, Toreggiani A, Berti A, Morari F (2015) Characterization of chemical-physical, structural and morphological properties of biochars from biowastes produced at different temperatures. J Soil Sediment 15:792–804. https://doi.org/10.1007/s11368-014-0964-7
9. Chun Y, Sheng GY, Chiou CT, Xing BS (2004) Compositions and sorptive properties of crop residue-derived chars. Environ Sci Technol 38:4649–4655. https://doi.org/10.1021/es035034w
10. Tan C, Yaxin Z, Hongtao W, Wenjing L, Zeyu Z, Yuancheng Z, Lulu R (2014) Influence of pyrolysis temperature on characteristics and heavy metal adsorptive performance of biochar derived from municipal sewage sludge. Bioresour Technol 164:47–54. https://doi.org/10.1016/j.biortech.2014.04.048
11. Lehmann J (2007) Bio-energy in the black. Front Ecol Environ 5:381–387. https://doi.org/10.1890/1540-9295(2007)5[381:BITB]2.0.CO;2
12. Masiello CA, Druffel ERM (1998) Black carbon in deep-sea sediments. Science 280:1911–1913. https://doi.org/10.1126/science.280.5371.1911
13. Pessenda LCR, Gouveia SEM, Aravena R (2001) Radiocarbon dating of total soil organic matter and humin fraction and its comparison with ^{14}C ages of fossil charcoal. Radiocarbon 43:595–601. https://doi.org/10.1017/S0033822200041242
14. Knoblauch C, Maarifat AA, Pfeiffer EM, Haefele SM (2011) Degradability of black carbon and its impact on trace gas fluxes and carbon turnover in paddy soils. Soil Bio Biochem 43:1768–1778. https://doi.org/10.1016/j.soilbio.2010.07.012
15. Liu Y, Yang M, Wu Y, Wang H, Chen Y, Wu W (2011) Reducing CH_4 and CO_2 emissions from waterlogged paddy soil with biochar. J Soils Sediments 11:930–939. https://doi.org/10.1007/s11368-011-0376-x
16. Zhang A, Cui L, Pan G, Li L, Hussain Q, Zhang X, Zheng J, Crowley D (2010) Effect of biochar amendment on yield and methane and nitrous oxide emissions from a rice paddy from Tai Lake plain, China. Agric Ecosyst Environ 139:469–475. https://doi.org/10.1016/j.agee.2010.09.003
17. Yuan H, Lu T, Zhao D, Huang H, Noriyuki K, Chen Y (2013) Influence of temperature on product distribution and biochar properties by municipal sludge pyrolysis. J Mater Cycles Waste Manage 15:357–361. https://doi.org/10.1007/s10163-013-0126-9
18. Krull ES, Baldock JA, Skjemstad JD, Smernik RS (2009) Characteristics of biochar: organo-chemical properties. In: Lehman J, Joseph S (eds) Biochar for environmental management: science and technology. Earthscan, London, pp 53–66
19. Al-Wabel MI, Al-Omran A, El-Nagar AH, Nadeem M, Usman ARA (2013) Pyrolysis temperature induced changes in characteristics and chemical composition of biochar produced from conocarpus wastes. Bioresour Technol 131:374–379. https://doi.org/10.1016/j.biortech.2012.12.165
20. Gottipatti R, Mishra S (2013) Preparation of microporous activated carbon from *Aegle marmelos* fruit shell by KOH activation. Can J Chem Eng 91:1215–1222. https://doi.org/10.1002/cjce.21786
21. Gottipatti R, Mishra S (2016) Preparation of microporous activated carbon from *Aegle marmelos* fruit shell and its application in removal of chromium (VI) from aqueous phase. J Ind Eng Chem 36:355–363. https://doi.org/10.1016/j.jiec.2016.03.005
22. Ahmad R, Kumar R (2010) Adsorptive removal of congo red dye from aqueous solution using bael shell carbon. Appl Surf Sci 257:1628–1633. https://doi.org/10.1016/j.apsusc.2010.08.111
23. Roy K, Verma KM, Vikrant K, Goswami M, Sonwani RK, Rai BN, Vellingiri K, Kim K-H, Giri BS, Singh RH (2018) Removal of patent blue (v) dye using indian bael shell biochar: characterization, application and kinetic studies. Sustainability 10:2669. https://doi.org/10.3390/su10082669
24. Palniandy LK, Yoon LW, Wong WY, Yong S-T, Pang MM (2019) Application of biochar derived from different types of biomass and treatment methods as a fuel source for direct carbon fuel cells. Energies 12:2477

25. Bardalai M, Mahanta DK (2016) Characterization of pyrolysis oil derived from bael shell (*Aegle mamellos*). Environ Eng Res 21:180–187. https://doi.org/10.4491/eer.2015.142
26. Shariff A, Aziz NSM, Abdullah N (2014) Slow pyrolysis of oil palm empty fruit bunches for biochar production and characterisation. J Phys Sci 25:97–112
27. Park J, Lee Y, Ryu C, Park YK (2014) Slow pyrolysis of rice straw: analysis of products properties, carbon and energy yields. Bioresour Technol 155:63–70. https://doi.org/10.1016/j.biortech.2013.12.084
28. Ye L, Zhang J, Zhao J, Luo Z, Tu S, Yin Y (2015) Properties of biochar obtained from pyrolysis of bamboo shoot shell. J Anal Appl Pyrol 114:172–178. https://doi.org/10.1016/j.jaap.2015.05.016
29. Kong SH, Loh SK, Bachmann RT, Choo YM, Salimon J, Rahim SA (2013) Production and physico-chemical characterization of biochar from palm kernel shell. AIP Conf Proc 1571:749–752. https://doi.org/10.1063/1.4858744
30. Brewer CE, Schmidt-Rohr K, Satrio JA, Brown RC (2009) Characterization of biochar from fast pyrolysis gasification systems. Environ Prog Sustain Energy 28:386–396. https://doi.org/10.1002/ep.10378
31. Ronsse F, Hecke SV, Dickinson D, Prins W (2013) Production and characterization of slow pyrolysis biochar: influence of feedstock type and pyrolysis conditions. GCB Bioenergy 5:104–115. https://doi.org/10.1111/gcbb.12018
32. Angalaeeswari K, Kamaludeen SPB (2017) Production and characterization of coconut shell and mesquite wood biochar. Int J Chem Stud 5:442–446
33. Moreira R, dos Reis Orsini R, Vaz JM, Penteado JC, Spinace EV (2017) Production of biochar, bio-oil and synthesis gas from cashew nut shell by slow pyrolysis. Waste Biomass Valor 8:217–224. https://doi.org/10.1007/s12649-016-9569-2
34. Wu W, Yang M, Feng Q, McGrouther K, Wang H, Lu H, Chen Y (2012) Chemical characterisation of rice straw derived biochar for soil amendment. Biomass Bioenergy 47:268–276. https://doi.org/10.1016/j.biombioe.2012.09.034
35. Bouqbis L, Daoud S, Koyro H-W, Kammann CI, Ainlhout LFZ, Harrouni MC (2016) Biochar from argan shells: production and characterization. Int J Recycl Org Waste Agric 5:361–365. https://doi.org/10.1007/s40093-016-0146-2
36. Xiao R, Chen X, Wang F, Yu G (2010) Pyrolysis pretreatment of biomass for entrained-flow gasification. Appl Energy 87:149–155. https://doi.org/10.1016/j.apenergy.2009.06.025
37. Choudhury ND, Chutia RS, Bhaskar T, Kataki R (2014) Pyrolysis of jute dust: effect of reaction parameters and analysis of products. J Mater Cycles Waste Manage 16:449–459. https://doi.org/10.1007/s10163-014-0268-4
38. Filippis PD, Palma LD, Petrucci E, Scarsella M, Verdone N (2013) Production and characterization of adsorbent materials from sewage sludge by pyrolysis. Chem Eng Trans 32:205–210. https://doi.org/10.3303/CET1332035
39. Bourke J, Manely-Harris M, Fushimi C, Dowaki K, Nunoura T, Antal MJ Jr (2007) Do all carbonized charcoals have the same chemical structure? 2. A model of the chemical structure of carbonized charcoal. Ind Eng Chem Res 46:5954–5967 (2007). https://doi.org/10.1021/ie070415u
40. Ghani WAWAK, Mohd A, da Silva G, Bachmann RT, Taufiq-Yap YH, Rashid U, Al-Muhtaseb AH (2013) Biochar production from waste rubber-wood-sawdust and its potential use in C-sequestration: chemical and physical characterisation. Ind Crop Prod 44:18–24. https://doi.org/10.1016/j.indcrop.2012.10.017
41. Adachi K, Kobayashi M, Takahashi E (1987) Effect of the application of lignin and/or chitin to soil inoculated with *Fusarium oxysporum* on the variation of soil microflora and plant growth. Soil Sci Plant Nutr 33:245–259. https://doi.org/10.1080/00380768.1987.10557570
42. Keiluweit M, Nico PS, Johnson MG, Kleber M (2010) Dynamic molecular structure of plant biomass-derived black carbon (biochar). Environ Sci Technol 44:1247–1253. https://doi.org/10.1021/es9031419
43. Özçimen D, Erosy-Meriçbouy A (2010) Characterization of biochar and bio-oil samples obtained from carbonization of various biomass materials. Renew Energy 35:1319–1324. https://doi.org/10.1016/j.renene.2009.11.042

44. Shabaan A, Se SM, Mitan NMM, Fairuz DM (2013) Characterization of biochar derived from rubber wood sawdust through slow pyrolysis on surface porosities and functional groups. Procedia Eng 68:365–371. https://doi.org/10.1016/j.proeng.2013.12.193
45. Xiao LP, Sun ZJ, Shi ZJ, Xu F, Sun RC (2011) Impact of hot compressed water pre-treatment on the structural changes of woody biomass for bioethanol production. BioResources 6:1576–1598
46. Azargohar R, Dalai AK (2006) Biochar as a precursor of activated carbon. Appl Biochem Biotech 131:762–773. https://doi.org/10.1385/ABAB:131:1:762
47. Novak JM, Lima I, Xing B, Gaskin JW, Steiner C, Das KC, Ahmedna N, Rehrah D, Watts DW, Busscher WJ, Schomberg H (2009) Characterization of designer biochar produced at different temperatures and their effects on a loamy sand. Annal Environ Sci 3:195–206

Metal Foam Manufacturing, Mechanical Properties and Its Designing Aspects—A Review

Rahul Pandey, Piyush Singh, Mahima Khanna, and Qasim Murtaza

Abstract The implementation of metal foams (especially aluminium alloy foams) has made an impact in the automobile and aerospace industries where crash energy absorption, vibration and sound damping and weight reduction is necessary. This paper includes a study in the field of metal foams encompassing different aspects such as its purpose, manufacturing methods, primary study about the mechanical and analytical behaviour of these upcoming materials and simulation-based model development for further experimentation. The mechanics focus on different models which are suggested by different researchers and the empirical formulas suggested by them to calculate various mechanical properties of the aluminium metal foam. Using LS-DYNA, the behaviour of foam has also been observed. Observations showed that, as opposed to normal metal grid structures, it was observed that the metal foam reinforced structure showed major improvement in mechanical properties such as yield stress and crushability and is thereby a great alternative to the hollow tubes without compromising on the weight.

Keywords Metal foams · Manufacturing · Mechanics of foams · Analytical models · Simulation

1 Introduction

Metal foams are porous materials that consist of solid metal with gas-filled pores. With the increase in the need for materials having higher crash energy absorption, vibration absorption, sound damping and weight reduction, especially in the automobile and aerospace industries, it has become necessary to explore new areas. These also have applications in orthopaedics alongside thermal applications (in heat exchangers). There are two forms of metal foams: closed-cell and open-cell type.

R. Pandey (✉) · P. Singh · M. Khanna · Q. Murtaza
Delhi Technological University (Formerly Delhi College of Engineering), Government of NCT of Delhi, Shahbad Daulatpur, Bawana Road, New Delhi, Delhi 110042, India
e-mail: rahul.dtu1997@gmail.com

© The Author(s), under exclusive license to Springer Nature Singapore Pte Ltd. 2021
R. M. Singari et al. (eds.), *Advances in Manufacturing and Industrial Engineering*, Lecture Notes in Mechanical Engineering,
https://doi.org/10.1007/978-981-15-8542-5_66

The most commonly used metal foam today is the aluminium alloy foam. Due to its varying properties, it is widely used in the automobile and aerospace industries. However, there are certain other materials such as copper, titanium and magnesium [1] (not used extensively), which do not find wide applications as aluminium foams, are still used as metal foams. Magnesium metal foams even though lighter than aluminium foams (due to lower density) have not been used much due to the high reactivity of Mg and low corrosion resistance.

Closed-cell Al foam has extensive applications due to its exceptional properties such as high specific strength and stiffness, thermal insulation capabilities, energy absorption ability, vibration and sound absorption [2]. Another form of implementation includes tubular structures. The advantage of this form is that of lightness in weight, cost lesser and comparatively easier manufacturing processes [3]. This paper encompasses the various methods of manufacturing metal foams, their types and classification, a basic overview of the mechanics behind foam deformations and the stress–strain curves, and an FEA model inspecting the deformation pattern under compressive loads using LS-DYNA.

2 Review of Literature

Over the past 20 years, the use of heavy solid materials has been down line. Engineers have always been searching for lightweight, high strength materials, excellent vibration damping characteristics, particularly, high energy absorption capacity and recyclability. The authors focused on the design of class materials that can be adapted to the needs of the application. These materials include different alloys, composites of metal and polymer, metal foams. Thanks to its impressive mechanical and physical properties, aluminium foams are the most favoured materials of today. Such foams have gained popularity in various applications due to the advancement in production processes, foaming and thickening agents. In structural as well as practical applications, aluminium foams have boundless applications. These can be used with a small increase in weight to boost energy absorption. These also have a low density, high strength to weight ratio and vibration-reducing properties. A lot of work has been done on enhancing energy absorption using aluminium foams due to excellent energy absorption efficiency. The aluminium foam manages plateau stress, which depends on the structure of pores, is in direct relationship with the strain in the material and densification strain. Using the hollow tubular structures and aluminium foam as the filler in the tubes is preferably the enhancement. Investigations are carried out to adjust the structural parameters of tubular structures for specific loading, modification of pore size and aluminium foam wall thickness. Researches have shown that the ability to absorb energy varies with changes in pore morphology. The ability to absorb energy is a result of buckling of cell walls and collapse of cell walls. Rajak et al. researched the ability of empty tubes (ETs) and foam-filled tubes (FFTs) to absorb energy at various strain levels. We showed that the stress–strain graph obtained from the compression test showed three distinct regions, while the ability to absorb energy increased by 33.45%. At elevated temperatures, Movahedi and colleagues conducted research on ETs and FFTs; the results showed an increase in energy absorption.

3 Manufacturing Methods of Metal Foams

Manufacturing metal foams essentially boils down to the question, how to create voids inside the solid bulk metal. Voids or pores can be generated by either gas expansion or using space-holders inside the metal. The easiest method to create porosity in metal is to inject gas in molten metal by means of a nozzle or using a particular blowing agent which releases gas when heated. These methods can be termed as direct foaming of melt. When gas is injected in molten metal, resulting bubbles quickly rises to the surface due to high buoyant forces of liquid metal. Same is the case when a blowing agent is used, to control this upward rise of the bubbled and stabilise the foam, a stabiliser is used. Stabilisers are solid particles that thoroughly wet the liquid metal. Ceramic powders, alumina, titanium diboride, zirconia and metal powders which form their oxides when mixed with molten metals can be used as a stabiliser [4–10]. Stabiliser particle needs to be in a certain range as too small particle size results in poor mixing and too large particle size means they will settle down in the melt. Experiments indicate that stabiliser particle size can be in the range 0.1–100 μm [11, 12]. Addition of carbides as stabilisers can introduce brittleness in the foam. Additive-free melt has to be made to make foam less brittle, thus, metal has to be melted close to its melting point so that its remains in viscous form and compensates for the lack of stabilisation by additives. In direct gas injection, metal is melted and a gas (e.g. air, nitrogen, or CO_2) from an external source is injected into the melt by either a nozzle or by a propeller [13]. Norsk Hydro, Norway and Cymat, USA make aluminium foam by this method with porosity ranging between 80–98% and density 0.069–0.54 g/cm^3.

By using a blowing agent like titanium hydride or zirconium hydride for in situ gas generation in the melt, we eliminate the use of a gas injection setup. Shinko Wire, Japan makes aluminium foam under trade name ALPORAS by this method. They use calcium metal powder for stabilising, as it increases the viscosity of melt by forming CaO, Al_2O_3 and $CaAl_2O_4$ [14] inside the melt, and titanium hydride as a blowing agent. When heated TiH_2 releases hydrogen gas which expands inside the melt and form voids. To make sure, TiH_2 does not release hydrogen prematurely, it can be heat treated at 400 °C for 24 h and at 500 °C for an hour to create a diffusion barrier layer of titanium oxide. The blowing agent can be mixed directly in the melt in case of ALPORAS and Formgrip process [15] or pre-mixed with metal powders in case of Foaminal process. Expanded foam due to blowing agents and even generated by direct injection needs to compacted to be used in engineering applications. Extrusion or pressing is done by dies, roller or presses to achieve this [16, 17]. Metal sheets can be bonded on either side of the pressed foam to make Foam Sandwiches. Aluminium Foam Sandwiches, or AFS panels made by Pohltec metal foam, GmbH, Germany are a great example of this [18]. A metal slurry with a blowing agent can also be used to make porous metal. A slurry of metal powder, blowing agent, solvent and additives is prepared, mixed and poured into a mould. Under elevated temperatures, this slurry gets expanded by the blowing agent and can then be carefully preserved, dried and sintered to make porous metal foam.

Space-holders are like cores used in the casting process. They create voids in metal without the needs of gas injection and blowing agent. To create porous foam using space-holders, we can use a polymer foam, thermally decomposable material, clay pellets or hollow spheres made out polymer, glass, metal, ceramic or salts [19–25]. These space-holders can be mixed with metal powders, suitable solvents, and an organic binder to make syntactic foams or they can be used as a base skeleton in casting or metal slurries to add layers of metal on it. A polymer foam space-holder is used by ERG, USA and Mayser, Germany [26] to create a mould by dipping the foam in a slurry of refractory material. Upon curing that, mould can be used in a simple casting process to make exact metal foam specimen as that of the polymer foam used as a casting pattern. Same polymer foam can also be dipped in a slurry of metal powder, solvent, binders to create a thick layer of dried slurry. The slurry coated polymer is heated to pyrolyse the polymer and sinter the metallic structure of the dried slurry shell.

A polymer foam can be also be used to deposit metal by vapour deposition or electroplating the metal on it. "Incofoam" [27] is made by this method. Physical vapour deposition is first done to make polymer electrically conductive, after which standard electroplating can be done easily. Chemical vapour deposition of nickel tetracarbonyl can also be used to make polymer electrically conductive. Space-holders can also be used to make "syntactic foams". In this hollow spheres of polymer, glass, ceramic or lightweight metal are mixed with metal powder to make a homogenous mix which can be later sintered. Hollow spheres can also be placed in a die, and molten metal can be then injected to make porous metal foam. Fraunhofer IFAM, Germany makes syntactic foam using glass spheres of type S60HS and diameter of 35 μm. Aluminium foam of density 1.1–1.4 g/cm^3 can be made by this process.

Other manufacturing methods which do not fall in the above two categories are rapid prototyping, gas entrapment, reaction sintering, freeze casting and solid–gas eutectic solidification. Rapid prototyping or more commonly called 3D printing is a novel method to make perfect metal foam down to every pore. SEBM or selective electron beam melting uses an electron laser to melt and fuse metal powder layer by layer to make a metal foam specimen. This requires a CAD model of the foam required and slicing software to generate cross-section of every layer. SLS or selective laser sintering is also good a process where laser does not fully melt the powder. In rapid prototyping shape of every pore or cell can be accurately controlled, and the resulting foam is a perfect model with properties close the theoretical optimum limit. Gas entrapment technique closely resembles powder metallurgy, but at the compaction step, gas is allowed to be trapped inside the powder precursor. Further, the heat treatment in the next step makes the metal expand due to the internal pressure created by entrapped gas [28]. This method is used to make titanium structures by Boeing, USA. Titanium powder filled in a can which is then evacuated and filled with argon at 3–5 bar [29]. In reaction, sintering two or more components which are being sintered undergo a chemical reaction. Metal powder mixtures of titanium and aluminium, iron and aluminium, and titanium and silicon can make porous structures by this method [30–32]. A slurry of metal powder in water or camphene is first prepared. When freezing dendritic growth pattern emerges, leaving interconnected pores after freeze-drying. The frozen carrier fluid is then removed, and the leftover pattern is sintered to form open-celled metal foam [33]. Solid–gas eutectic solidification method is used for metals which form a eutectic system with hydrogen [34]. Melting certain metals in a hydrogen-rich atmosphere at high pressures (50 bar) results in a homogenous mixture of molten metals and dissolved hydrogen gas. The mixture will change into a heterogeneous two-phase mixture when the temperature is lowered. This results in hydrogen being released in the melt, which rises upwards causing directional solidification, which results in porous metal with elongated pores [35].

4 Mechanics of Metal Foams

Metal foams are porous materials that consist of gas bubbles divided by thin metallic walls and exhibit unique properties owing to their constitutional structure. The metal foam consists of a combination of metal and gas-filled pores. Few mechanical properties which can be determined via analytical, FEA and experimental methods are Young's modulus, absorption capacity (area under the stress–strain curve), specific energy absorption and finally the energy absorption efficiency. The properties of these materials obtained analytically are based on theoretical models from various sources, through FEA using LS-DYNA and experimentally mainly from namely: compression tests, bending tests, crashworthiness test, etc. [36].

The predominant mode of failure is that of progressive collapse. The failure mechanism that is seen in uniaxial tension is unlike the behaviour in uniaxial compression. Under compression, the foam is progressively crushed because of the plastic buckling of the cell wall, whereas in tension a ductile crack is formed. Under dynamic load, the foam undergoes initial compression followed by gradual crushing. The increase in the density of the foam due to the crushing stage leads to a rise in the energy absorption capacity [36] (Fig. 1).

Fig. 1 Stress–strain curve of aluminium metal foam (compression)

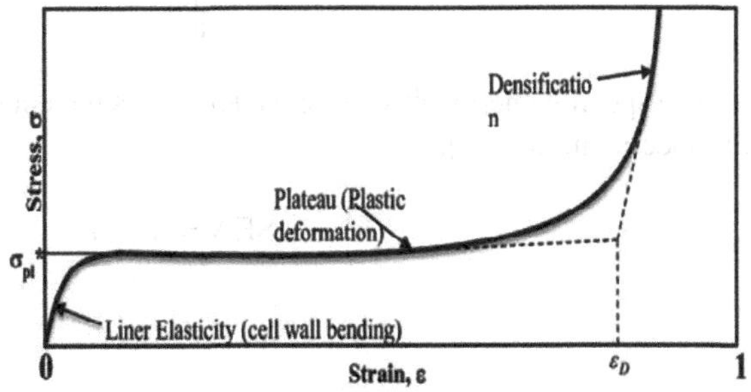

Metal foams also undergo three major deformations processes as shown in the schematic diagram above, namely: elastic deformation, plastic deformation and foam densification. It has been observed that a sudden increase in the compressive force after the foam reaches its densification displacement. Small-sized pores tend to collapse at the elastic stage (plasticity appears at sufficiently high loads).

Few basic formulae and terms:

$$\sigma_n = F/A \tag{1}$$

$$\text{Poisson's ratio} = \nu = \epsilon_{\text{lateral}}/\epsilon_{\text{longitudnal}} \tag{2}$$

$$\text{Porosity of metal foams} = \text{Pore volume}/\text{volume of porous material} \tag{3}$$

$$\text{Relative density} = \text{Density of metal foam}/\text{density of cell wall material} \tag{4}$$

By performing a simple impact or quasi-static experimental test, the load–displacement curve of the material can be obtained. The area under the graph will then represent the strain energy per unit volume or in other words, the amount of energy that can be stored in a material before its failure. This represents the absorption capacity of the foam, and consequently, the energy absorption efficiency can be calculated through the formula [37]:

$$\eta(a) = \left(\frac{1}{P(a)h}\right)\int_{a_y}^{a} P(a)\mathrm{d}a \tag{5}$$

where $P(a)$ is the compressive force, h is the height of the aluminium foam and a_y is the displacement at the yield point.

The energy absorption capacity of the foam can be determined, as mentioned above by integrating the area under the load–displacement curves, given by [37]:

$$E_a = \int_{0}^{a_d} P(a)\mathrm{d}a \tag{6}$$

The specific energy absorption is defined as the energy absorbed per unit mass of the specimen, i.e. [37]:

$$\text{SEA} = E_a/m \tag{7}$$

5 Models Used for Analytical Computations

Theoretical approaches based on the models by Gibson and Ashby are used to determine the Young's modulus and the plateau stress of the foam during compression loading. These values are critical in plotting a stress–strain curve of the foam up to the point of densification initiation. Also, approaches following the Cowper-Symonds model have been used to define the relations between different parameters such as strain rate and dynamic increase factors (for support structures for metal foam specimens such as curved plates used in connectors).

According to the Gibson and Ashby theoretical approach, the plateau stress can be represented as [38]:

$$\frac{\sigma_p}{\sigma_y} = 0.3\left(\varphi\left(\frac{\rho_f}{\rho_s}\right)\right)^{\frac{3}{2}} + 0.4(1-\varphi)\left(\frac{\rho_f}{\rho_s}\right) + (p_o - p_{at})/\sigma_y \tag{8}$$

where σ_p = plateau stress of the foam, σ_y = yield stress of the cell wall, ρ_f = density of the metal foam, ρ_s = density of the cell wall material, p_o = gas pressure inside the cell and p_{at} = atmospheric pressure.

The value of Young's modulus of the metal foam can be determined based on the above-mentioned model by [38]:

$$\frac{E_f}{E_s} = \varphi^2\left(\frac{\rho_f}{\rho_s}\right)^2 + (1-\varphi)\left(\frac{\rho_f}{\rho_s}\right) + p_o(1-2\nu)/E_s\left(1 - \frac{\rho_f}{\rho_s}\right) \tag{9}$$

where E_f = Young's modulus of metal foam, E_s = Young's modulus of the cell wall and ν = Poisson's ratio. Φ = fraction of solid present in the cell edges.

6 Finite Element Modelling of Metal Foam

Metal foams, despite being prevalent in the automobile industry and aerospace industry, are a relatively newer discovered material and hence it is not readily available in popular finite element modelling softwares such as ANSYS. The only readily available models are found in ABAQUS and LS-DYNA. In ABAQUS, the only a Deshpande/Fleck foam model is available, whereas, in LS-DYNA, there are multiple models available such as MAT 5 Soil and Crushable Foam, MAT 26-Anisotropic Honeycomb Model, MAT 26 Honeycomb, MAT 38 Blatz Ko, MAT 57 Low Density Foam, MAT 62 Viscous Foam, MAT 63 Crushable Foam and MAT 126 Modified Honeycomb.

After due considerations to the mechanical properties that are being used in the applicable models, we utilise MAT 26 Honeycomb model as it resembles the actual material properties of metal foam under consideration.

Table 1 Physical properties

Density	Young's modulus	Poisson's ratio	Yield stress	Relative volume
00.5×10^{-7} kg/mm^3	68.94 GPa	0.3	0.18 GPa	0.2

Table 2 Strength values in various directions

Eaau	Ebbu	Eccu	Gabu	Gbcu	Gacu
1.02 GPa	0.34 GPa	0.34 GPa	0.434 GPa	0.214 GPa	0.434 GPa

In the modelling of the material in LS-DYNA, various properties needed to be specified including the density, the Young's modulus and Poisson's ratio. These values were derived experimentally and were used to create an accurate model (Tables 1 and 2).

6.1 Finite Element Model

The below-shown schematic depicts a finite element model created using the above data (Figs. 2 and 3).

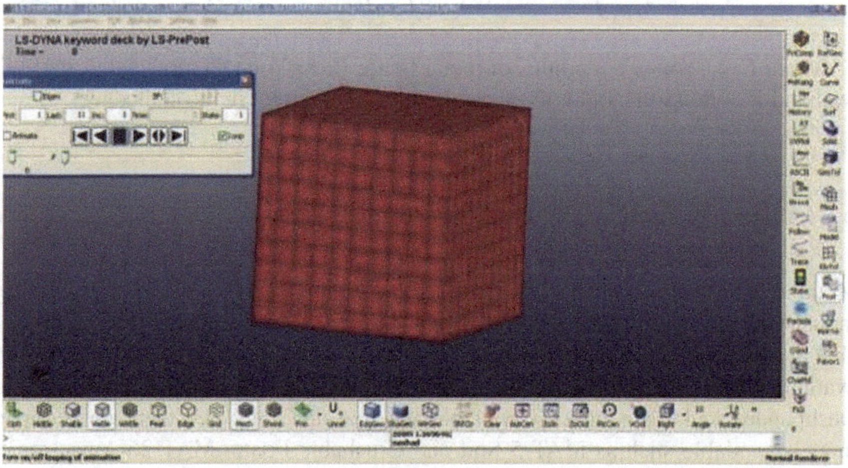

Fig. 2 Before compression

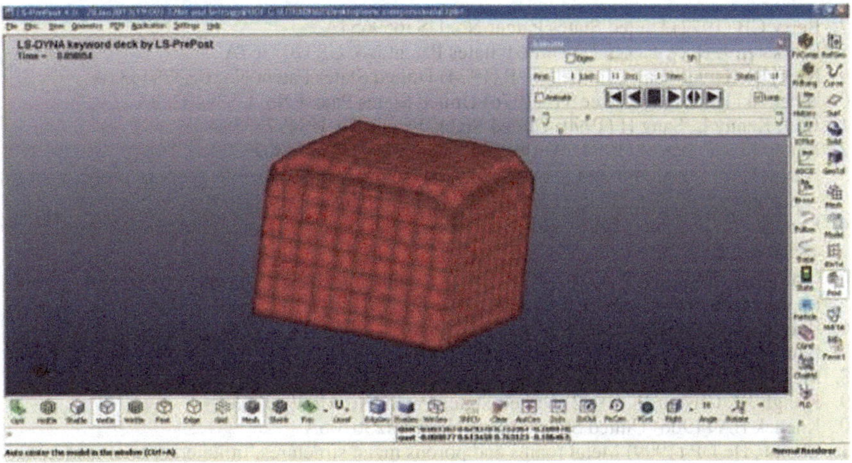

Fig. 3 After compression

7 Conclusion

From all the above-discussed sections, it can be seen that metal foams, especially aluminium metal foams are a good way to improve the mechanical properties of elements with minimal addition to their weight due to the small density of the metal foams. Hence, it is seen that the use of aluminium metal foam in automobile industry, especially in the bumpers of the cars to improve its crushing strength and also as members of the chassis, is growing at an exponential rate. Other metal foams are also known such as titanium metal foams (used in medical purposes) as well as copper metal foams, but their usage is limited due to a lack of availability. Despite the growing use of metal foams, there is a huge scope for more work in this field, and the usage can be extended to other industries in order to optimise the strength without compromising on the weight constraints.

References

1. Neu TR, Mukherjee M, Garcia-Moreno F, Banhart J (2011) Magnesium and magnesium alloy foams. In: 7th international conference on porous metals and metallic foams (MetFoam2011), p 133
2. An Y, Yang S, Zhao E, Wang Z (2017) Characterization of metal grid-structure reinforced aluminium foam under quasi-static bending loads. Compos Struct 178:288–296
3. Rajak DK, Mahajan NN, Linul E (2019) Crashworthiness performance and microstructural characteristics of foam-filled thin-walled tubes under diverse strain rate. J Alloy Compd 775:675–689
4. Elliot JC (1961) United States Patent No. US2983597A
5. Fiedler W (1965) United States Patent No. US3214265A
6. Wilson HP, William PG (1967) United States Patent No. US3300296A

7. Berry CB (1972) United States Patent No. US3669654A
8. Bjorksten J, Rock EJ (1972) United States Patent No. US3707367A
9. Niebylski L, Jarema C, Immethun P (1974) United States Patent No. US3794481A
10. Niebylski L, Jarema C, Lee TE (1976) United States Patent No. US3940262A
11. Jin I, Kenny L, Sang H (1990) United States Patent No. US4973358A
12. Kenny L, Thomas M (1994) Worldwide Patent No. WO1994009931A1
13. Asholt P (1999) Aluminium foam produced by the melt foaming route process, properties, and applications. In: Metal foams and porous metal structures, pp 133–140
14. Simone A, Gibson L (1998) Aluminum foams produced by liquid state process. Acta Mater 46:3109–3123
15. Gergely V, Degischer H, Clyne T (2000) Recycling of MMCs and production of metallic foams. In: Comprehensive composite materials, pp 797–820
16. Baumeister J (1990) Germany Patent No. DE4018360C1
17. Baumeister J, Schrader H (1991) Germany Patent No. DE4101630C2
18. Metalfoam (n.d.) AFS sheets, from https://en.metalfoam.de/afs-sheets/
19. Kreigh JR, Gibson JK (1962) United States Patent No. US3055763A
20. Kuchek HA (1966) United States Patent No. US3236706A
21. Chen F, He DP (1999) Metal foams and porous metal structures. In: International conference. MIT Press, Bremen, Germany, p 163
22. Zwissler M (1997) Germany Patent No. DE19725210C1
23. Grote F, Busse P (1999) A new casting method for open-celled metal foams. GIESSERE I:75
24. Thiele W (1971) Germany Patent No. DE1933321A1
25. Hartmann M, Reindel K, Singer RF (1998) Fabrication and properties of syntactic magnesium foams. Porous Cell Mater Struct Appl 521:211
26. Mayser (n.d.) Metal foam m.pore technical data. Retrieved from mayser.com: https://www.mayser.com/de/download?c=44
27. ElectronicsWeb (n.d.) INCOFOAM for batteries, catalysis and filters. Retrieved from electronicsweb.com: https://www.electronicsweb.com/doc/incofoam-for-batteries-catalysis-and-filters-0001
28. Kearns MW, Blenkinsop PA, Barber AC, Farthing TW (1986) Manufacture of novel porous metal. Int J Powder Metall 59–64
29. Schwartz DS, Shih DS, Lederich RJ, Martin RL, Deuser DA (1998) Development and scale-up of the low-density core process for Ti-64, p 225
30. Kubo Y, Igarashi H (1991) United States Patent No. US4331477A
31. Wang GX, Dahms M (1993) Reaction sintering of cold extruded. Metall Trans A 24(7):1514–1526
32. Krueger BR, Mutz AH, Vreeland T (1992) Shock-induced and self-propagating high-temperature synthesis reactions in two powder mixtures: 5:3 atomic ratio Ti/Si and 1:1 atomic ratio Ni/Si. Metall Trans A 23(1):55–58
33. Yook SW, Yoon BH, Kim HE, Koh YH, Kim YS (2008) Porous titanium (Ti) scaffolds by freezing TiH2/camphene slurries. Mater Lett 62:4506–4508
34. Drenchev L, Sobczak J, Malinov S, Sha W (2006) Gasars: a class of metallic materials with ordered porosity. Mater Sci Technol 22(10):1135–1147
35. Shapovalov VI (1998) Formation of ordered gas-solid structures via solidification in metal-hydrogen systems. MRS online proceedings library archive, 521
36. Roszkos CS, Bocko J, Kula T, Šarloši J (2019) Static and dynamic analyses of aluminium foam geometric models using the homogenization procedure and the FEA. Compos Part B Eng 171:361–374
37. Wang Y, Zhai X, Ying W, Wang W (2018) Dynamic crushing response of an energy absorption connector with a curved plate and aluminium foam as an energy absorber. Int J Impact Eng 121:119–133
38. Papadopoulos DP, Konstantinidis IC, Papanastasiou N, Skolianos S, Lefakis H, Tsipas DN (2004) Mechanical properties of Al metal foams. Mater Lett 58:2574–2578

Emerging Trends in Internet of Things

Yash Agarwal and K. A. Nethravathi

Abstract Recent technological eco-space has observed an exponential increase in the number of devices connected to the Internet. The data generated by these devices has reached astronomical figures. Due to this, there exists a need for managing and processing the big data, at the same time maintaining the reasonable latency. This requirement of today's world in the field of Internet of things has given rise to technologies like fog computing, edge computing, mist computing, etc. This paper focuses on new technologies which are improvements of existing cloud computing. A computational analysis of fog computing is performed using iFogSim and cloud-Analyst simulator to carry out latency and cost comparisons between fog and cloud computing. As simulation results conclude fog has a lower latency of 159 ms, but at the same time, it has a higher total network implementation cost of $2.39, while the data transfer cost remains the same.

Keywords Big data · Fog computing · Edge computing · Applications · IoT (internet of things)

1 Introduction

Internet is expanding like never before. Each and every day, new devices are becoming the part of it. Today, there are about 5 billion active mobile subscribers in the world. The total number of devices connected to the Internet is expected to be 50 billion by the year 2020 from 8.7 billion in 2012 [1]. Such a drastic increase is mainly due to two reasons—firstly, the need to make everything a part of a network to facilitate communication among various devices or entities which may be far apart or not physically accessible—secondly the quest to make every device **SMART** like smart televisions, phones, machines, home appliances, cities, etc. with the motive of improving the general standards of living of the people [2]. All these smart or network-connected devices generate huge amounts of data. This data is expected to

Y. Agarwal (✉) · K. A. Nethravathi
R.V. College of Engineering, Mysore road, Bengaluru 560059, India
e-mail: yashagrawal.ec16@rvce.edu.in

© The Author(s), under exclusive license to Springer Nature Singapore Pte Ltd. 2021
R. M. Singari et al. (eds.), *Advances in Manufacturing and Industrial Engineering*, Lecture Notes in Mechanical Engineering,
https://doi.org/10.1007/978-981-15-8542-5_67

reach 4.4 ZB by 2020 from just 0.1 ZB in 2013. Such an explosion in produced data has given rise to scope of technologies like big data analysis or analytics. This big data can be stored and processed by means of machine learning algorithms and neural networks to make sense out of it and calculate certain statistical parameters known as key performance indicators (KPIs). These KPIs can prove to be extremely beneficial for various firms and institutions to take crucial decisions [3]. In today's world, data is being considered as the most important asset. It is set to become a trillion dollar industry. The annual revenue from IOT sales is expected to hit $1.6 trillion by 2025.

Traditionally, the data gathered from IoT devices is sent to a cloud-based platform where it is processed by using various algorithms to perform the necessary prediction or to make some decision. As the number of IoT devices has increased, the associated data traffic has also increased. Therefore, the cloud platforms should be capable of storing and processing such huge traffic and should have required computational capacity. But this kind of system has some drawbacks. The bandwidth requirement is extremely huge to send the data from large number of devices to the cloud. The computationally extensive cloud platform will lead to higher costs and power consumption. Also there might be a latency problem in case of time sensitive data which needs to be responded with some action quickly but has to travel long distances to reach and get processed by a remotely located cloud data center. This problem can be very well held by fog and edge computing techniques. The term fog computing was first used by CISCO in the year 2014. It focuses on moving the operations of storage, processing and network services between the end user and the cloud data center. Edge computing involves processing of data locally at the edges or as close as possible to the edges [4]. This reduces the latency and improves the response time.

2 Why Fog Computing?

The existence of gigantic number of IoT devices and the huge data produced by them when integrated with the cloud computing can pose several problems. This huge data creates a lot of traffic on the network, and it can become really troublesome to handle it. Also the bandwidth consumption will be large. Apart from this, there is a problem of high latency or response time. The data from these devices can be identified into three broad categories depending upon the response time requirements.

- Time-sensitive data
- Less time-sensitive data
- Non-time-sensitive data

In case of time-sensitive data and applications, the latency problem will be exacerbated. In regular cloud computing systems all the data has to be sent to a centralized cloud data center, which can be located at a large distance from the device generating data. This is not a desirable condition in case of time sensitive data which require quick response and might result in unwanted results. Figure 1 illustrates the latency

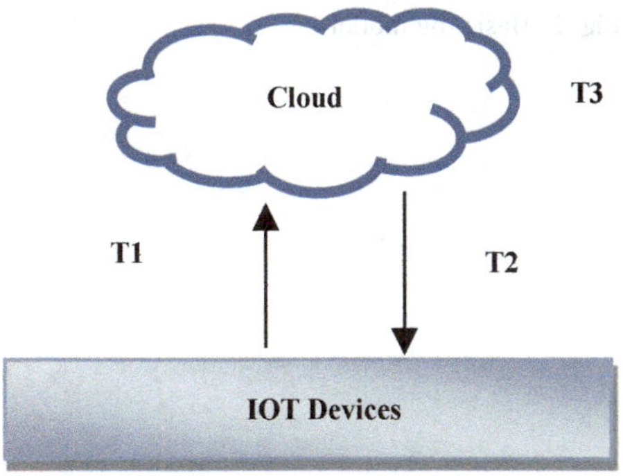

Fig. 1 Latency problem

problem.

T1 Time from devices to cloud.
T2 Time from cloud to device.
T3 Cloud processing time.

Latency = T1 + T2 + T3

In case of huge volumes of data, the processing time T3 will be large which will give longer response time and larger latency. This problem led to the development of modified cloud models like fog and edge computing.

3 Fog Computing

3.1 Basics

The conventional cloud model faces many issues while handling big data and large number of devices. To overcome these issues, fog computing came into picture. The basic idea of fog is to extend the cloud capabilities closer to IoT devices. It acts as an intermediary layer between the cloud data center and the end IoT devices (Fig. 2).

The fog performs processing before sending it to the cloud which improves the response time and latency. The fog layer consists of various fog nodes which can be routers, embedded servers, switches, etc. These nodes are virtual instances of IoT devices and provide enhanced processing, storage and networking. Each node is associated with the aggregate fog node [5]. Unlike the cloud data center, the fog layer provides transient data storage. A fog node also provides the connection with IoT devices, other fog nodes and the centralized cloud.

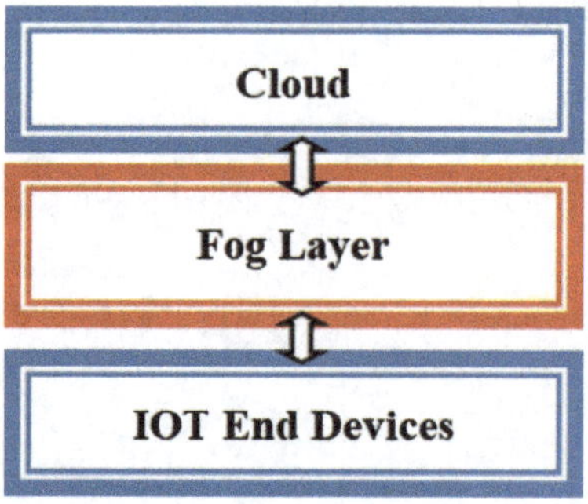

Fig. 2 Basic fog model

3.2 *Fog Architecture*

A fog layer cannot exist independently. It is not a substitute of cloud computing and works in conjunction with it. Figure 3 illustrates the detailed fog architecture.

1. Non-time-sensitive data
2. Summary for historical analysis and storage
3. Ingestion of time sensitive data
4. Immediate response
5. Less time-sensitive data
6. Response to less time-sensitive data

The fog architecture operates on the basis of type of data generated by devices. If the data is time sensitive, it is sent to the nearest fog node which processes the data

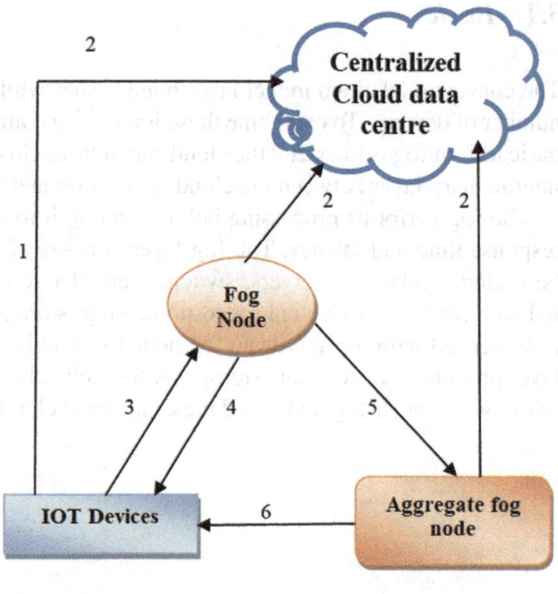

Fig. 3 Fog architecture

and responds with an immediate action. This data is also sent to the cloud by the node as summary for analysis and historical storage. In case of less time-sensitive data, the data is forwarded to the aggregate fog node which performs the required processing and performs the necessary action. From here, the data is forwarded to the cloud as summary. In case of non-time-sensitive data, the data is directly forwarded to cloud for processing and storage [6].

3.3 Fog Computing Versus Edge Computing

Fog computing enables the storage, networking and processing operations nearer to the end IoT devices (Edge). On the other hand, edge computing enables processing of data locally at the edge or as close as possible to the IoT device. It occurs directly on the devices to which sensors are connected by means of Programmable Automation Controllers (PAC's) [7]. It does not involve sending of data to centralized remote data centers. This reduces the distances and time to send data to cloud which improves speed (latency) and performance of data transport. Edge computing in fact is a concept, and fog computing is a standard that defines how it should operate or work. Fog acts as a jumping off point for edge computing [4]. There is another term mist computing which is generally used synonymously with edge computing. The difference between the two is that mist uses lightweight computing objects with microchips and microcontrollers [7].

3.4 Latency Comparisons with Cloud Computing

As mentioned earlier, the latency in a centralized cloud data center is higher than fog architecture. In a system without a fog, data from all the devices has to travel to a centralized cloud data center for processing. This leads to increased latency because of time taken to reach the remotely located cloud and processing time taken at the cloud. Unlike this fog architecture does the processing part near the device and reduces latency.

Figure 4 shows a cloud topology which is having six user bases (consisting of ten devices or users each) and one centralized cloud data center. This topology is created using a Java-based iFogSim simulator and is simulated with cloud analyst tool to obtain Fig. 5 which shows average, maximum and minimum response times for each user base with respect to the centralized data center [8]. The cloud analyst allows us to select a server broker policy, which in this case is closest data center. But it will not have an impact on the response times since there is only one centralized cloud data center. Figures 6 and 7 show the simulation results of the fog architecture with multiple fog nodes apart from a cloud data center.

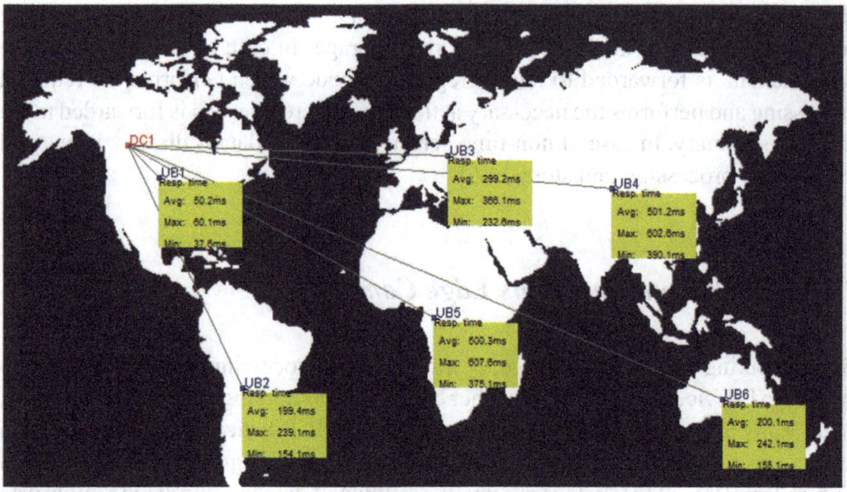

Fig. 4 Cloud topology

Overall Response Time Summary

	Average (ms)	Minimum (ms)	Maximum (ms)
Overall Response Time:	292.10	37.61	607.62
Data Center Processing Time:	0.28	0.02	0.86

Export Results

Response Time By Region

Userbase	Avg (ms)	Min (ms)	Max (ms)
UB1	50.173	37.606	60.107
UB2	199.446	154.112	239.114
UB3	299.227	232.616	366.059
UB4	501.173	390.116	602.614
UB5	500.337	375.116	607.617
UB6	200.115	155.112	242.114

Fig. 5 Overall response time of cloud topology

DC Data center.
UB User base.

Figure 6 shows fog cloud topology with six user bases, one centralized cloud data center (DC1) and three fog nodes (fogcloud1, fogcloud2 and fogcloud3). Figure 7 shows the results of the simulation of this topology in the cloud analyst in terms of response times. As it is clearly evident from the simulation results, the average overall response time of fog cloud topology (159 ms) is lesser than the cloud topology (292.10 ms). The same trend can be observed in response times of individual user bases. This is because of the presence of multiple fog nodes in the fog cloud topology. When we select the server broker policy as closest data center, the data from the user bases is sent to the nearest fog node and not without any decision making to the centralized cloud. This reduces the response time and thus latency. The above simulation results clearly show how fog architecture is an improvement of regular

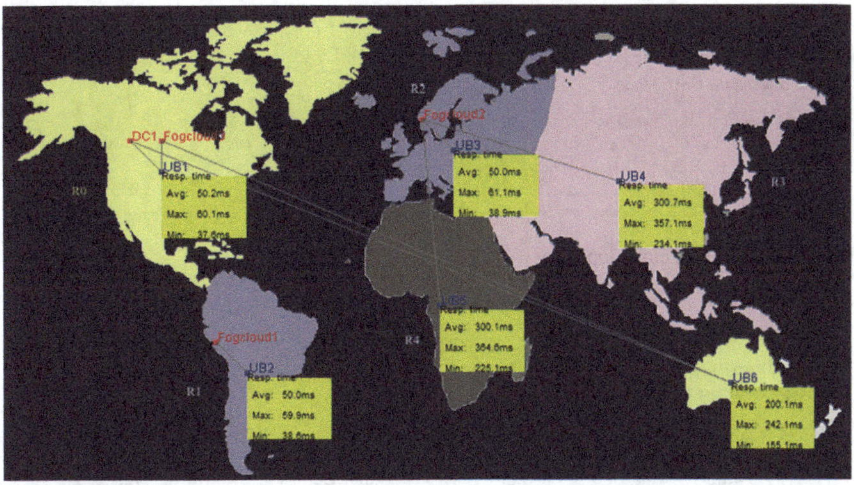

Fig. 6 Fog cloud topology

Overall Response Time Summary

	Average (ms)	Minimum (ms)	Maximum (ms)
Overall Response Time:	159.00	37.61	364.65
Data Center Processing Time:	0.39	0.02	0.89

Export Results

Response Time By Region

Userbase	Avg (ms)	Min (ms)	Max (ms)
UB1	50.207	37.606	60.107
UB2	49.974	38.641	59.893
UB3	49.989	38.881	61.058
UB4	300.726	234.141	357.146
UB5	300.098	225.141	364.648
UB6	200.137	155.137	242.142

Fig. 7 Overall response times of fog cloud topology

cloud computing in terms of latency or response time. But the same cannot be said regarding the cost of the network. The following figures illustrate the simulation of cost analysis of the two topologies.

From Figs. 8 and 9, it is clearly evident that the total cost of implementation of a cloud topology ($0.89) is lesser than that of fog topology ($2.39). This is because fog involves setting up of multiple fog nodes (fogcloud1, fogcloud2 and fogcloud3) rather than a single centralized data center. This accounts for additional costs in fog. However, total data transfer cost is same in both the cases since the same data is transferred, with the difference that the total data transfer cost is distributed among various fog nodes in fog topology instead of being aggregated in case of regular cloud.

Cost

Total Virtual Machine Cost: $0.50

Total Data Transfer Cost: $0.38

Grand Total: $0.89

Data Center	VM Cost	Data Transfer Cost	Total
DC1	0.502	0.385	0.886

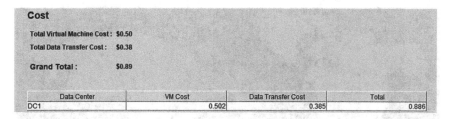

Fig. 8 Cost table for cloud topology

Cost

Total Virtual Machine Cost: $2.01

Total Data Transfer Cost: $0.38

Grand Total: $2.39

Data Center	VM Cost	Data Transfer Cost	Total
DC1	0.502	0.064	0.566
Fogcloud1	0.502	0.065	0.567
Fogcloud2	0.502	0.192	0.693
Fogcloud3	0.502	0.064	0.566

Fig. 9 Cost table for fog topology

3.5 Advantages of Fog Architecture

The fog architecture is a significant improvement over the regular cloud computing. Some of its major advantages are mentioned below.

- It provides better data security, and all the fog nodes can use the same security policy [5].
- Bandwidth consumption and data traffic is significantly reduced, since not all the data is being sent to cloud.
- It provides better privacy as the confidential data of an organization can be stored on the local servers.
- It facilitates quicker decision making, owing to less latency.
- Fog architecture provides a lot of flexibility as the fog nodes can join and leave the network at any time.
- Every industry can analyze their data locally using fog architecture.
- It allows deployment in remote or harsh environmental conditions.
- The operation costs are lesser than cloud computing.

3.6 Challenges Faced by Fog Architecture

Despite its many benefits, fog architecture also faces some challenges which need to be worked on. Some of the major challenges are mentioned below.

- Fog uses multiple fog nodes which increase the power consumption, and in fact, the power consumption of the fog is even higher than cloud.
- Since the data producing nodes are distributed and not centralized, providing authentication and authorization to the whole system of nodes is not an easy task [5].
- Maintenance of data integrity and ensuring availability of millions of nodes is extremely difficult.
- Fog nodes may be mobile, and they can join or leave the network at any time. But many data processing frameworks might be statically configured. Therefore, these frameworks will not be able to provide required scalability and flexibility [5].

4 Mobile Cloud Computing

Nowadays, mobile devices are used for hosting and processing various types of services in entertainment, social media, news, business, games, health, etc. However, the increased demand for these services has led to certain problems with mobile devices like low energy, poor resources and low connectivity. This created the concept of mobile cloud computing, where a resource rich cloud server can be used by a thin mobile client for running a service. It allows the mobile nodes to play the role of resource provider in case of peer-to-peer network. But this type of computing may also face issues faced regular cloud computing like latency, bandwidth and connectivity issues. This led to the concept of cloudlets which later on leads to mobile edge computing. A cloudlet is very much similar to fog architecture. It is connected to a centralized cloud and aims at bringing the cloud services nearer to the end mobile user.

In the cloudlet concept, the mobile device offloads its workload to a resource-rich local cloudlet. These cloudlets situated in common areas like libraries, coffee shops, universities, etc. which allows the mobile devices to connect and function as thin clients to the cloudlet. It can be considered to be the first hop node at the edge of the network.

The cloudlet is considered to be as the central tier of a three-tier cloudlet architecture, which consists of mobile devices, local cloudlet and distant cloud [9].

A cloudlet has four key attributes as follows:

- It has only soft state.
- It should be resource rich and well connected.
- It should have low end to end latency.
- It should have a certain standard for workload offloading like virtual machine migration.

All these attributes together define a local cloudlet in mobile cloud computing.

5 Applications

The multiple advantages of fog computing make it useful in a wide number of applications. It can also be used for existing cloud applications to achieve better performance. Some of the popular applications are:

- Real-time health analysis—It involves real-time monitoring of patients suffering from chronic diseases by monitoring multiple body parameters. Due to reduced latency of fog, the doctors can be intimated quickly in case of emergency situations. The collected historical data can be used for predicting any future dangers to the patients.
- Pipeline optimization—Oil and gas pipelines require the monitoring of parameters like pressure and flow rate. In case of very long pipelines, terabytes of data could be generated. In such situation, fog computing is preferable [5].
- Supply chain and inventory management—Cloud computing is extensively used in warehouses and departmental stores for keeping a track of the inventory using radio frequency identification (RFID) tags. There could be millions of such tags sending data, which could not be efficiently managed by cloud and thus will require fog architecture [10].

6 Potential Applications (Case Study)

Agriculture is a domain which is attracting a lot of attention for research in recent times. Over use of fertilizers, poor yield, deteriorating quality and forever reducing area available for cultivation are some of the main reasons. There has been an idea for growing the crops in artificially created eco-systems with complete control over growth parameters. The soilless farming techniques like aeroponics, aquaponics, hydroponics and techniques prevalent in urban landscapes like vertical faming can be a part of such system. Such kind of a setup will require a large number of sensors, actuators and other data generating devices to monitor and control the growth parameters. Such systems on commercially viable scale will generate big data and a lot of traffic. This makes it an excellent potential application for fog computing. The following is an example of a smart agriculture setup, to which fog computing could be applied—Smart Vertical Farm.

Figure 10 shows an abstract view of a smart agriculture farm. Here multiple sensors and actuators are interfaced to a microcontroller to monitor and control essential growth parameters like soil moisture content, soil pH, light intensity and gas concentration. A similar setup can be used in urban landscapes in the form of vertical farms where such sensor–actuator setups can be implemented in multiple stacks for better area utilization. Such a setting on commercial scale will require a large number of sensors. To give an idea of the number of sensors required, an instance of a vertical farm of 3000 ft^2, with five stacks providing an effective area of 15,000 ft^2, is considered in Table 1.

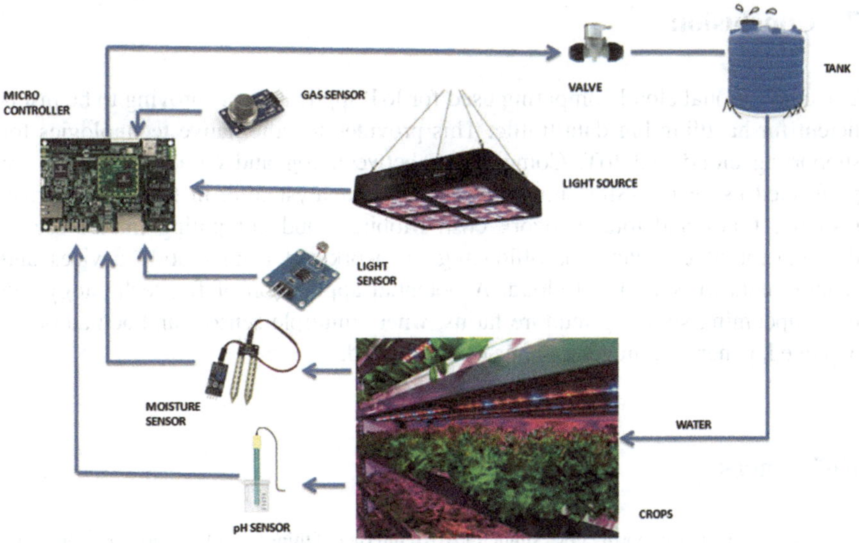

Fig. 10 Abstract view of a small agriculture

Table 1 Sensor requirements

Entity	Number of units
Microcontrollers	200
pH sensors	200
Soil moisture sensor	200
Gas sensor	1
Light intensity sensor	5
Total	606

As evident from the table, a 15,000 ft^2 vertical farm will require approximately 600 sensors. The data generated from them will be huge and a computationally extensive cloud data center will be required to process such data. These sensors will keep sending the growth parameter data to the cloud. But in case if some sort actuation is required to control or change a parameter, it might take a lot time for the necessary action to happen due to large volumes of data being sent and processed by the cloud. There might be a situation where the ultraviolet wavelength light sources used in such systems need to be switched off when there is human interference in otherwise isolated setups for the purpose of biological safety. All these applications require low latency which can be achieved by deploying fog nodes. These nodes will process the data from the sensors depending upon which node is closest to a sensor. This will facilitate quicker decision making and immediate action. The fog nodes can forward this data to the centralized cloud data center for permanent storage and historical analysis.

7 Conclusion

The conventional cloud computing used for IoT applications is proving to be insufficient for handling big data traffic. This provides for alternative technologies for supporting cloud and IoT. Comparisons between fog and cloud computing are performed using iFogSim simulator and cloud analyst tool, in terms of average response times and total network cost. Mobile cloud computing makes use of the concept of cloudlets for offloading the workload from mobile devices and sending data to centralized cloud. A potential application of fog technology can be in upcoming smart agriculture farms, where multiple sensors and actuators are deployed, generating huge amounts of real-time data.

References

1. Bedi G, Venayagamoorthy GK, Singh R (2016) Internet of things (IoT) sensors for smart home electric energy usage management. In: 2016 IEEE international conference on information and automation for sustainability
2. Pa Y, Suzuki S, Yoshioka K, Matsumoto T, Kasama T, Rossow C (2015) IoTPOT: analysing the rise of IoT compromises. USENIX WOOT 2015
3. Qliksense business intelligence resource library. https://www.qlik.com/us/resource-library
4. Technical article by Cisco on "Edge computing versus fog computing". https://www.cisco.com/c/en/us/solutions/enterprisenetworks/edge-computing.html
5. NPTEL lectures on internet of things by Professor Sudeep Misra. https://nptel.ac.in/courses/106105166/
6. Yi S, Hao Z, Qin Z, Li Q (2015) Fog computing: platform and applications. In: 2015 third IEEE workshop on hot topics in web systems and technologies
7. Technical article on "iFogSim: an open source simulator for edge computing, fog computing and IoT". https://opensourceforu.com/2018/12/ifogsim-an-open-source-simulator-for-edge-computing-fog-computing-and-iot/
8. Gupta H, Dastjerdi AV, Ghosh SK, Buyya R (2016) iFogSim: a toolkit for modelling and simulation of resource management techniques in internet of things, edge and fog computing environments, June 2016
9. Neishaboori M, Jaaski AY (2015) Implementation and evaluation of mobile-edge computing cooperative caching. Master's Thesis, Aalto University, Espoo, July 28, 2015
10. Ali MM, Haseebuddin M (2015) Cloud computing for retailing industry: an overview. Int J Emerg Trends Technol Comput Sci 19(1):51–56

Selection of Best Dispatching Rule for Job Sequencing Using Combined Best–Worst and Proximity Index Value Methods

Shafi Ahmad, Ariba Akber, Zahid A. Khan, and Mohammed Ali

Abstract Job sequencing is imperative in shop floor management as it impacts the performance of the shop floor very significantly. In order to select a job for processing on a machine, as soon as the machine becomes free, dispatching rules (DRs) are used. DRs do not necessarily generate the best job sequence; rather, they provide a high-quality job sequence in a very short span of time. There are several standards as well as newly developed DRs that have been proposed in the last two decades. However, these rules perform better for one performance measure but lack to perform for others. In order to select the best DR which will perform optimally for a number of performance measures, multi-criteria decision-making (MCDM) has been employed in this work. Different performance measures used to evaluate the performance of DRs have been identified from the literature. The identified performance measures are prioritized using the widely used best–worst method. On the basis of identified performance measures and their priorities, eleven static and dynamic sequencing rules are ranked. It was highly possible that the results drawn might be affected when the problem is changed. Therefore, ten thousand randomly generated problems are solved and the results so achieved are used to select out the best DR. It has been found in the study that the best DR is processing time and due date total (PDT) followed by weighted processing time and due date total (WPDT) rule.

Keywords Dispatching rules · Best–worst method · Proximity index value method

S. Ahmad (✉) · A. Akber · Z. A. Khan
Department of Mechanical Engineering, Jamia Millia Islamia, New Delhi 110025, India
e-mail: shafiahmad.amu@gmail.com

M. Ali
Department of Mechanical Engineering, ZHCET, Aligarh Muslim University, Aligarh, India

© The Author(s), under exclusive license to Springer Nature Singapore Pte Ltd. 2021

R. M. Singari et al. (eds.), *Advances in Manufacturing and Industrial Engineering*, Lecture Notes in Mechanical Engineering,
https://doi.org/10.1007/978-981-15-8542-5_68

1 Introduction

An important feature of the shop floor management system is job sequencing, which influences the performance of the shop floor. DRs are the set of rules that find out which job will be loaded on the machine for processing as soon as the machine becomes free. For determination of job sequence, DR implements a priority function for ranking of jobs and the job with the finest priority value is processed on machine. Over the last two decades, a significant number of DRs have been proposed. However, based on the implementation of DRs in the system, the derived results can be desirable for a certain performance objective but it might not be the same for another. Although a vast number of DRs have been proposed for job sequencing, researches related to the comparison of the DRs are very diminutive.

As maximizing the performance of multiple performance measures is the basic intention of scheduling decisions, selection of best rule for optimum conditions for conflicting performance measures is a tedious task. This problem can be resolved using multi-criteria decision-making (MCDM) techniques. An MCDM technique is used for ranking different alternatives for optimum performance of multiple conflicting decision criteria. BWM combined with PIV method is used in this work to select the best DR among eleven DRs for optimum performance of multiple performance measures. Ten thousand randomly generated problems are solved, and rank to the considered DRs has been defined. The average ranks to each of the DRs for ten thousand problems have been used to select the best DR. The rest of the paper is organized as follows: Sect. 2 explains the methodology adopted to achieve the objective of this work. BWM and PIV methods have also been explained in this section. Section 3 of the paper presents the results obtained by solving ten thousand problems. Finally, Sect. 4 of the paper presents the conclusion drawn from the results so obtained.

2 Methodology

The methodology used in this work to achieve the research objective has been shown in Fig. 1.

In the first step, an exhaustive literature review has been done to identify performance measures used for mapping the performance of DRs for job sequencing on a single machine (JSSM) environment. The best–worst method (BWM) which is a widely used tool in MCDM to prioritize different performance measures is used to calculate the weights of the identified performance measures. Then, different DRs used for the JSSM environment are identified. PIV method is employed to rank the identified DRs in order to select the best DR. Ten thousand randomly generated problems are solved using the above-said methodology to draw robust conclusion about the rank so obtained.

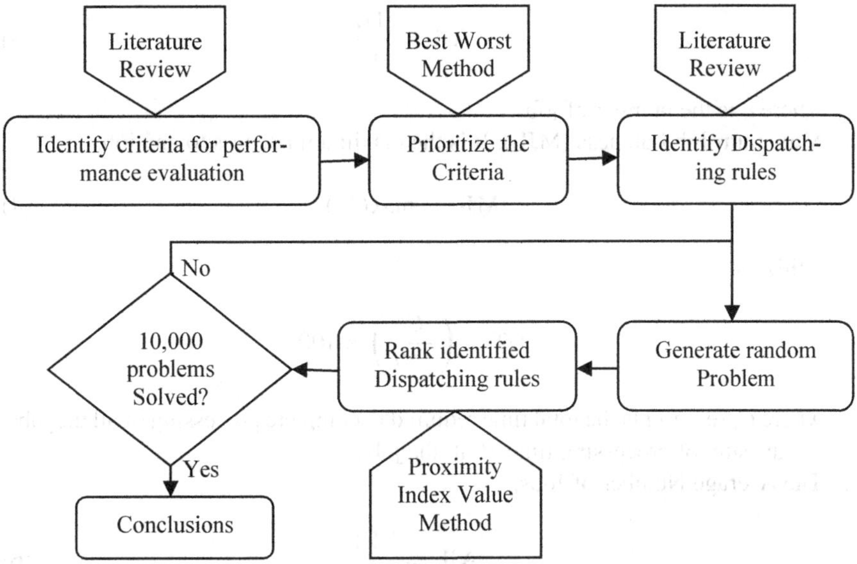

Fig. 1 Research methodology

2.1 *Job Sequencing*

Job sequencing is a process of determining the priority of each job on the basis of which jobs are processed on a machine. The effectiveness of any given sequence obtained using a DR is examined by the performance of the system so obtained. Seven performance measures have been used in this work which are as follows [1–3]:

1. Total Flow Time (TFT): It is the sum of flow times of all the jobs.

$$\mathrm{TFT} = \sum_{i=1}^{n} F_{t,i} \tag{1}$$

   where $F_{t,i}$ represents the time at which job i exits the system after finishing its execution on the machine.
2. Total Lateness (TL): It shows the sum of the lateness of the jobs which are late.

$$\mathrm{TL} = \sum_{i=1}^{n} L_i \tag{2}$$

   where L_i represents the time interval a job spent in the system after its due date.
3. Average Job Lateness (AJL): It denotes the average time a job is considered to be late.

$$\text{AJL} = \frac{\text{TL}}{n} \tag{3}$$

where n is the number of jobs.

4. Maximum Job Lateness (MJL): It is the maximum lateness to any job.

$$\text{MJL} = \max(L_i) \tag{4}$$

5. Utilization:

$$\text{UT} = \left(\frac{C_\text{t}}{\text{TFT}}\right) \times 100 \tag{5}$$

where C_t represents the total time required to complete processing of all the jobs, i.e. the sum of processing time of all the jobs.

6. The Average Number of Jobs:

$$\text{AJL} = \frac{\text{TFT}}{C_\text{t}} \tag{6}$$

7. Average Job Completion Time:

$$\text{AJCT} = \frac{\text{TFT}}{n} \tag{7}$$

DRs are used to sequence these jobs so as to obtain better performance of the defined performance measures. There exist several different DRs, which are based on a certain defined priority function. Eleven different DRs are identified from the literature which is most commonly used for JSSM environment. DRs along with their priority function and reference are shown in Table 1.

2.2 *Best–Worst Method*

For the ranking of dispatching rules, their performance needs to be estimated on the basis of the selected performance measures. The performance measures possibly will have a different importance characterized by their weights. Therefore, it is essential to compute the weight of different performance measures. BWM has been used in this work for the purpose of determining the weights of the performance measures. The BWM makes use of the opinion of the decision-makers/experts.

The step-by-step procedure to implement BWM given by [9] is as follows:

Table 1 Dispatching rules and their priority function

Symbol	Dispatching rule	Priority function	References
FCFS	First come first served	$\text{Min}\{At_i\}$	[2]
LCFS	Last come first served	$\text{Max}\{At_i\}$	[2]
SPT	Shortest processing time	$\text{Min}\{Pt_i\}$	[4]
LPT	Longest processing time	$\text{Max}\{Pt_i\}$	[5]
EDD	Earliest due date	$\text{Min}\{D_i\}$	[2, 5]
SLACK	Minimum slack value	$\text{Min}\{D_i - Pt_i - t\}$	[2]
CR	Critical ratio	$\text{Min}\{(D_i - t)/Pt_i\}$	[2]
SD	Maximum standard deviation	$\text{Max}\{(Pt_i - \text{mean}(Pt_i))\text{\^{}}2\}$	[6]
MDD	Modified due date	$\text{Min}\{\text{Max}(D_i, Pt_i + t\}$	[7]
PDT	Processing time and due date total	$\text{Min}\{Pt_i + D_i\}$	[8]
WPDT	Weighted processing time and due date total	$\text{Min}\{w{*}Pt_i + (1 - w)D_i\}$	[8]

Step 1 Identify performance measures ($P_1, P_2, \ldots, P_n$).

Step 2 Select the best performance measures (P_B) and the worst performance measures (P_W) among the identified performance measures.

Step 3 The P_B is compared with each performance measure using a number from 1 to 9 as defined in Table 2. The comparison results in a vector caller as best to other vector ($\overrightarrow{B^o}$) as shown in Eq. (8).

$$\overrightarrow{B^o} = (B^o{}_{B1}, B^o{}_{B}, \ldots, B^o{}_{Bn}) \tag{8}$$

where $B^o{}_{Bi}$ represents the relative importance of P_B over P_i.

Step 4 Each performance measure is compared with the P_W using the scale defined in Table 2. The comparison results in others to the worst vector ($\overrightarrow{O^W}$) as

Table 2 Number and relative importance

Importance	Number
A is equally important to B	1
A is between equal and moderate important to B	2
A is moderate to B	3
A is between moderate and strong important to B	4
A is strongly important to B	5
A is between strong and very strong important to B	6
A is very strong important to B	7
A is between very strong and extremely important to B	8
A is extremely important to B	9

given in Eq. (9).

$$\overrightarrow{O^{\mathrm{W}}} = (O^{\mathrm{W}}{}_1, O^{\mathrm{W}}{}_2, \ldots, O^{\mathrm{W}}{}_n) \tag{9}$$

where $O^{\mathrm{W}}{}_n$ represents the relative importance of the P_i over P_{B}.

Step 5 Calculate the weights of performance measures (w_i*), such that the absolute maximum difference, i.e. $\left\{\left|w_{\mathrm{B}}^* - B^{\mathrm{o}}{}_{\mathrm{B}1} * w_i^*\right|, \left|w_{\mathrm{W}}^* - O^{\mathrm{W}}{}_i * w_{\mathrm{B}}^*\right|\right\}$, is minimum.

The mathematical form of modified BWM is given in Eq. (10).
Minimize $Z = \xi^{\mathrm{L}}$
Subject to:

$$\begin{aligned} &\left|w_{\mathrm{B}}^* - B^{\mathrm{o}}{}_{\mathrm{B}1} * w_i^*\right| \le \xi^{\mathrm{L}} \\ &\left|w_{\mathrm{W}}^* - O^{\mathrm{W}}{}_i * w_{\mathrm{B}}^*\right| \le \xi^{\mathrm{L}} \\ &\sum_i w_i = 1 \\ &w_i \ge 0 \end{aligned} \tag{10}$$

where ξ^{L} is a consistency ratio which signifies the consistency of the comparisons. w_i is the relative weight of the ith performance measure.

Hence in BWM, it is required to solve the linear programming problem given in Eq. (10) so as to obtain the weights, i.e. w_1*, w_2*, ..., w_n* of the n performance measures and the corresponding consistency ratio ξ^{L}*. The consistency ratio is required to be close to zero to represent consistent comparisons.

2.3 Proximity Index Value Method

The proximity index value (PIV) method is a not long developed method used for ranking different alternatives in an MCDM method [10]. This method prevents the rank reversal occurrence of the TOPSIS method which is why it is widely used for ranking of different alternatives [11, 12]. PIV method utilized the proximity value (P_{V}) for ranking different alternatives. The step-by-step procedure to calculate P_{V} is as follows:

Step 1 Arrange different alternatives in rows and the value for each performance measure in the column which results in a matrix (P_{DM}) known as decision matrix as given in Eq. (11)

$$P_{\mathrm{DM}} = [x_{lm}]_{p \times q} \tag{11}$$

where x_{lm} represents performance value of lth alternative for mth criterion, $l = 1, 2, \ldots, p$ and $m = 1, 2, \ldots, q$, where p represents the number of alternatives and q represents the number of performance measures.

Step 2 Normalize the P_{DM} using Eq. (12)

$$x_{lm}^{n} = \frac{x_{lm}}{\sqrt{\sum_{l=1}^{p} x_{lm}^{2}}} \tag{12}$$

Step 3 Multiply the weight of each performance measure with the corresponding normalized value of all the alternatives using Eq. (14). This results in a normalized decision matrix.

$$w_{lm}^{n} = w_{m} \times x_{lm}^{n} \tag{13}$$

where w_m represents the weight of the mth criterion.

Step 4 Determine the weighted proximity index (A_{D}), using Eq. (14)

$$A_{\mathrm{D},l} = \left\{ \begin{array}{ll} w_{\max}^{n} - w_{l}^{n}; & \text{for beneficial criteria} \\ w_{l}^{n} - w_{\min}^{n}; & \text{for cost criteria} \end{array} \right\} \tag{14}$$

Step 5 Calculate proximity value, P_{V} using Eq. (15)

$$P_{\mathrm{V},m} = \sum_{m=1}^{n} A_{\mathrm{D},m} \tag{15}$$

The P_{V} which represents the deviation from the best alternative is used to rank the alternatives. The alternative with the least value is considered to be very close to the finest alternative and therefore is ranked first; subsequently, other alternatives are ranked in ascending order of P_{V} value.

3 Result and Discussion

Our aim is to rank the available DRs for optimum performance of identified performance measures. As there are a number of performance measures, their priorities need to be defined. In order to determine the priority of the performance measures, a group of four experts was formed: two from industry and two from academic universities. Several interview sessions were performed to take their opinion regarding the importance of performance measures. They were requested to give their opinion regarding the best performance measure (P_{B}) and the worst performance measure (P_{W}). Further, the P_{B} and the P_{W} identified by each expert are used to inquire about the importance of the P_{B} over others and other performance measures over P_{C}. The response given by each expert was used to solve the linear optimization problem

Table 3 Overall weights

Performance measures	Weights
ANJ	0.1860
UT	0.0719
AJL	0.1867
TFT	0.0766
MJL	0.1032
TL	0.1631
ACT	0.2125
ξ^{L*}	0.078281

given in Eq. (10). The solution to the problem resulted in the optimal weights of the performance measures and consistency ratio. As for each expert $\overrightarrow{B^{o}}$ and $\overrightarrow{O^{w}}$ vectors are different, the simple average weight for each performance measure was calculated to obtain a single weight vector. The single weight vector thus obtained is shown in Table 3.

It can be observed from Table 3 that the value ξ^{L*} is close to zero representing the high consistency of the weights so obtained. Further, the most important performance measure among the considered performance measures is found to be average completion time (ACT) followed by ANJ followed by AJL followed by TL followed by MJL followed by TFT followed by UT.

The methodology used in this work helps in ranking the different DRs used for the JSSM environment, which can be used to make decisions about which DR will be suitable for maximizing the output from the system. It is likely that the results may change with change in the parameters of the problem, i.e. processing time and due date of jobs. Hence, robust conclusion cannot be drawn from the results obtained from solution of the single problem. In order to draw robust conclusions, ten thousand randomly generated problems are solved. The basic parameters of jobs are held constant; i.e. processing time and due date is generated randomly in the same interval. For each of the problems, the ranking of DRs has been done using PIV method. Table 4 shows the mean rank of DRs, standard deviation in rank and the final ranking on the basis of all the solved problems.

The results obtained show that the mean rank of PDT rule found to be lowest among all the DRs. Hence, PDT rule found to perform better than other DRs. Also, the deviation of this rule from mean rule is found to be lowest. It represents that for all the solved problems the rank of PDT varies between 1 and 3. The ranking of the dispatching rules from the obtained results can be regarded as PDT followed by WPDT followed by SPT followed by MDD followed by EDD followed by SL followed by CR followed by SD followed by LCFS followed by FCFS followed by LPT.

Table 4 Mean rank of DRs over ten thousand problems

Symbol	Mean rank	Standard deviation	Final rank
FCFS	7.97	1.37	10
LCFS	7.97	1.42	9
SPT	2.66	1.66	3
LPT	10.11	0.88	11
EDD	4.22	1.28	5
SL	6.02	1.54	6
CR	6.26	1.09	7
SD	7.22	2.63	8
MDD	2.82	1.06	4
PDT	1.70	0.78	1
WPDT	2.42	1.06	2

4 Conclusion

The main concern in shop floor activities is the efficiency of the system which relies on how the jobs are processed on the machine. The sequence of job processing on the shop floor is determined using DRs. DRs help in organizing the sequence of jobs in a very short period of time for better performance of the system. For operation managers, it seems difficult to select a DR which results in healthier performance of multiple performance measures. This work aims to identify the finest possible DR which can be used for JSSM environment for optimum performance of multiple performance measures, viz. ANJ, UT, AJL, TFT, MJL, TL and ACT. This has been achieved by ranking the identified DRs used for JSSM environment using combined BWM and PIV methods. BWM is used to provide a priority weight to the performance measures, and PIV method is used to rank DRs. This is bound to happen that the rank so obtained may be different for a different set of problem. Hence, a set of ten thousand randomly generated problems are solved and the ranks so obtained are used to draw conclusions. The conclusions from the results obtained are as follows:

- The priority weight obtained using BWM shows that among the considered performance measures, total lateness has maximum weight. This shows that TL is the most important performance measure for the evaluation of DR in JSSM environment.
- The best rule for the JSSM environment is found to be PDT as the mean rank is lowest for PDT followed by WPDT which is a modified form of PDT rule.
- The standard deviation in the rank of the PDT and WPDT shows that the variation in the rank of PDT and WPDT is very less. It can be concluded that the rank of PDT rule varies between 1 and 3 and for WPDT rule it varies from 2 to 4 for all the solved problems.

References

1. Tyagi N, Tripathi RP, Chandramouli AB (2016) Single machine scheduling model with total tardiness problem. Indian J Sci Technol 9(37):1–14
2. Kumar KK, Nagaraju D, Gayathri S, Narayanan S (2017) Evaluation and selection of best priority sequencing rule in job shop scheduling using hybrid MCDM technique. IOP Conf Ser Mater Sci Eng 197:012059
3. Đurasević M, Jakobović D (2018) A survey of dispatching rules for the dynamic unrelated machines environment. Expert Syst Appl 113:555–569
4. Maheswaran M, Ali S, Siegel HJ, Hensgen D, Freund RF (1999) Dynamic mapping of a class of independent tasks onto heterogeneous computing systems. J Parallel Distrib Comput 59:107–131
5. Pinedo ML (2012) Scheduling: theory, algorithms, and systems, 4th edn. Springer, New York
6. Munir EU, Li J, Shi S, Zou Z, Yang D (2008) MaxStd: a task scheduling heuristic for heterogeneous computing environment. Inf Technol J 7:679–683
7. Baker KR, Kanet JJ (1983) Job shop scheduling with modified due dates. J Oper Manage 4(1):11–22
8. Hamidi M (2016) Two new sequencing rules for the non-preemptive single machine scheduling problem. J Bus Inq 15(2):116–127
9. Rezaei J (2015) Best-worst multi-criteria decision-making method. Omega 53:49–57
10. Mufazzal S, Muzakkir SM (2018) A new multi-criterion decision making (MCDM) method based on proximity indexed value for minimizing rank reversals. Comput Ind Eng 119:427–438
11. Khan NZ, Ansari TSA, Siddiquee AN, Khan ZA (2019) Selection of E-learning websites using a novel proximity indexed value (PIV) MCDM method. J Comput Educ 6:241–256
12. Singh S, Yuvaraj N, Uz K, Khan Z, Siddiquee AN, Khan ZA (2018) Multi-response optimization of friction stir welding process parameters using standard deviation (SD) based preference indexed value (PIV) method. Int J Adv Prod Ind Eng 3(4):44–48

Thermal Performance Investigation of a Single Pass Solar Air Heater

Ovais Gulzar, Adnan Qayoum, and Rajat Gupta

Abstract One of the most promising energy sources being utilized in much direct and indirect processes of energy conversion is solar energy, of which solar air heaters are the foremost application of direct use of solar energy. Solar air heaters use air as a medium for thermal conduction; however, air has low thermal conductivity resulting in lower heat transfer coefficient absorber plate and the medium of air. Artificial roughness induced by ribs reduces thermal resistance by breaking the viscous sublayer while promoting turbulence to increase the coefficient of heat transfer. In the present work, evaluation of the performance of conventional solar air heater with an absorber plate containing V-shaped ribs has been done. V-shaped ribs have been used on absorber plate at the relative roughness pitch of 10 and an optimum value of relative roughness height of 0.02, with 45° as angle of attack. The values for solar insolation and Reynolds number are varied from 600 to 1200 W/m^2 and 5000 to 20,000, respectively. The results show an enhancement in Nusselt number of ribbed duct from 1.9 to 2.25 times as compared to a smooth duct. The use of ribs on absorber plate has significantly increased the efficiency by about 7.75–18.33%. However, the variation in outlet temperature with increasing Reynolds number is less significant for lower values in contrast to the higher values of solar insolation. The variation of overall loss coefficient, thermal efficiency and useful heat gain is also studied for the range of operational scenarios.

Keywords Solar air heaters · Reynolds number · Solar flux · Useful heat gain · Heat transfer coefficients

O. Gulzar (✉)
Department of Mechanical Engineering, Islamic University of Science and Technology, Awantipora, Jammu and Kashmir, India
e-mail: bhatovais@gmail.com

A. Qayoum
Department of Mechanical Engineering, National Institute of Technology Srinagar, Srinagar, India

R. Gupta
Department of Mechanical Engineering, National Institute of Technology Mizoram, Aizawl, India

© The Author(s), under exclusive license to Springer Nature Singapore Pte Ltd. 2021
R. M. Singari et al. (eds.), *Advances in Manufacturing and Industrial Engineering*, Lecture Notes in Mechanical Engineering,
https://doi.org/10.1007/978-981-15-8542-5_69

1 Introduction

The incident solar radiation at the ground surface can be thermally harvested for heat or naturally converted to biomass by photosynthesis or converted to electricity using photovoltaic materials. Wind and ocean energy are direct consequence of convention currents caused due to the solar heating. Solar energy causes the water to evaporate which subsequently produces rainfall in higher elevations leading to the potential for hydropower. With climate change happening, it is more imperative to shift toward solar-based technologies especially in developing countries [1]. Also, with the growing energy demands and the overexploitation of the nonrenewable sources of energy, the need for sustainable engineering materials and systems is increasing [2, 3]. In order to convert solar energy to thermal energy, solar air heaters are primarily and most widely systems used [4]. These find many engineering applications including heating systems, agri-drying, air preheaters, etc. A conventional solar air heater comprises transparent cover, rear bottom plate and an absorber plate. The air flows between the bottom and absorber plate. These heaters offer design simplicity with minimal maintenance. Since air possesses low thermal conductivity of about 0.0273 W/m^2 [4], it results in lower heat transfer between the air and the absorber plate [5].Thus, air heaters are subjected to many configuration changes for improving the coefficient of heat transfer. In certain cases, the intermediary fluid is changed to nanofluids which also offers a promising alternative to increase the overall thermal efficiency [6, 7]. In the present study, V-shaped ribs are used to improve surface roughness to break the viscous sublayer on the absorber plate, thereby increasing the coefficient of heat transfer between plates and air. The single pass solar air heater (SPSAH) is investigated theoretically for various operational scenarios. In doing so, the air heater is considered for two different configurations such as absorber plate without ribs and absorber plate with V-shaped ribs. The solar flux incident on the collector area is varied with constant Re and vice versa. The efficiency curves, outlet temperature, etc., are plotted using MATLAB and Origin software. The performance parameters such as instantaneous efficiency, useful heat gain, outlet temperature are investigated with the variation in Re and solar insolation. There are many other applications of solar air heaters, particularly in developing countries such as India. There are many states in India that receive very high amount of solar radiations, and these heaters can be used to tap this solar energy for useful purposes. These heaters can find special application in the states like Jammu and Kashmir, which face harsh winters but at the same time, receive enough solar energy to be used for thermal energy applications. The notion of using these heaters in remote areas and far flung areas where direct electricity cannot be delivered gains more popularity and possibility of wider applications.

2 Theoretical Analysis

The theoretical analysis of SPSAH is based on the energy balance equations and solving them analytically to investigate the behavior under certain operational scenarios. The energy balance provides an insight about the utilization of incident solar radiant energy into useful form of energy and losses that occur during the process of conversion. It quantifies the amount of losses which further can be minimized by means of various techniques. The energy balance formulations for various elements are based on below assumptions.

The energy balance formulations for various elements are based on below assumptions [8, 9]:

1. For air, the bulk mean temperature increases from T_f to $T_f + dT_f$.
2. The convective heat transfer coefficient between absorber plate and air (h_{fp}) and that between bottom rear plate and air is (h_{fb}) is same.
3. Loss coefficient U_b for bottom is significantly smaller than loss coefficient of the top and therefore is neglected in subsequent calculations.
4. Side losses are neglected.
5. The absorber plate and rear bottom plate have smooth surfaces.
6. The flow is turbulent and fully developed.
7. Negligible shading of collector plate.

2.1 Specifications of SPSAH

The overall dimensions of SPSAH are 2 m × 1 m. The SPSAH is composed of the following components as shown in Fig. 1:

1. Transparent cover or shield of glass.
2. Rear bottom plate.

Fig.1 Schema of single pass solar air heater

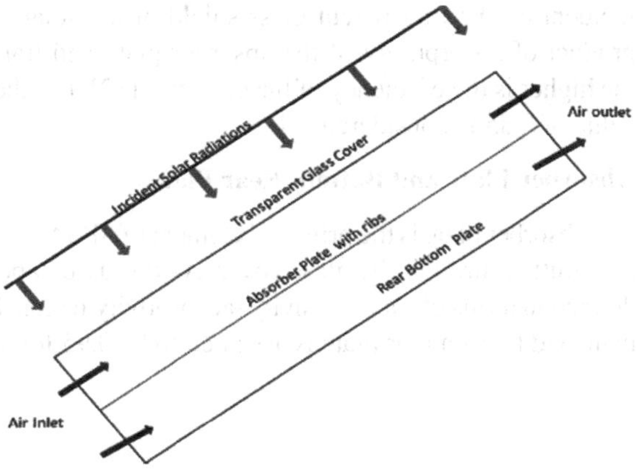

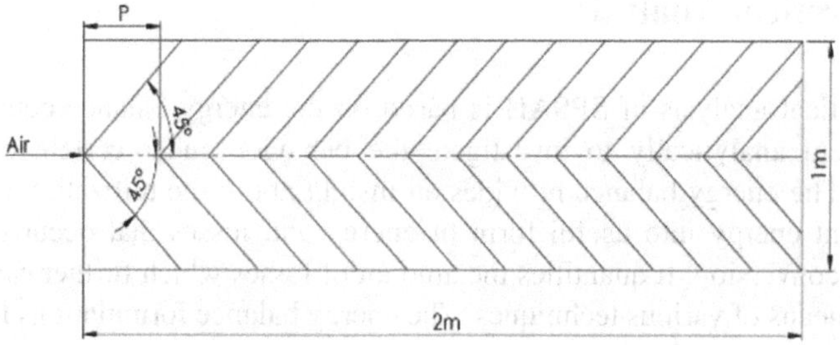

Fig. 2 V-shaped ribbed absorber Plate

3. An absorber plate.

In case of ribbed plate, the absorber plate has been given additional V-shaped ribs at the relative roughness height (e/D_h) of 0.02 with a relative roughness pitch (p/e) of 10. An incident angle of attack of 45° is taken in the present study as shown in Fig. 2.

Glass Cover

The glass shield or cover serves three main purposes. Firstly, it reduces the heat loss from the absorber, both convective and radiant. Secondly, it allows the transmission of the incoming solar radiation to the absorber plate with least loss. And thirdly, it serves as a protection or shield to the absorber plate against ambient environmental conditions [10, 11].

Transmissivity–Absorptivity Factor

The solar radiations incident on the glass cover are partly reflected, absorbed, and much of them are transmitted to the absorber plate. The absorber plate absorbs much part of radiant energy transmitted through the glass cover. The transmissivity–absorptivity product $(\tau\alpha)_{av}$ is used for the purpose of analysis. It is computed by dividing the absorbed energy flux in the absorber plate by the energy flux which is incident on the transparent glass shield of the solar air heater [6]. The greater the product of absorptivity of the absorber plate and transmissivity of glass covering, the higher is the efficiency of the collector [12]. For the purpose of analysis, average value of 0.85 is considered [8].

Absorber Plate and Bottom Rear Plate

The absorber plate is the principal element of SPSAH that absorbs the solar radiations transmitted through the glass cover. As discussed above, the absorbing capacity is defined in terms of transmissivity–absorptivity factor. The emissivity of the absorber plate and bottom rear plate is assumed to be 0.95 for analysis [8].

2.2 Properties of Air

The mean fluid temperature is assumed to be 55 °C [8], and properties are specified at mean temperature of fluid. The properties of air are listed below [4]:

$\rho = \frac{1.077\,\text{kg}}{\text{m}^3}$; $C_p = 1.005\,\text{kJ/kg} - \text{k}$; $\mu = 19.85 \times 10^{-6}\,\text{N} - \text{s/m}^2$; $K = 0.0287\,\text{W/m K}$

For the purpose of analysis, the following ambient conditions are assumed:

Air inlet temperature: 50 °C; Ambient temperature: 20 °C.

3 Calculation of Nusselt Number

Using the assumption that the surfaces are smooth for the first case and the correlation proposed by Kays [13], Nusselt number can be calculated by the following correlations:

$$\text{Nu} = 0.0158\text{Re}^{0.8} \tag{1}$$

In the second case, the absorber plate with V-shaped ribs is considered. The Nusselt number is obtained by below correlations [14]:

$$N_{ur} = 0.067 \times (R_e)^{0.888} \times \left(\frac{e}{D_h}\right)^{0.424} \times (\alpha/60°)^{-0.077} \times \exp\left[-0.782 \times (\ln\alpha/60°)^2\right] \tag{2}$$

Figures 3 and 4 show the Nusselt number variation as a function of Reynolds number for plane and rubber absorber plates, respectively. It can be seen that the Nusselt number varies from 14.38 to the maximum value of 43.59 for first case and it varies from 28.74 to 98.42 for ribbed absorber plate.

3.1 Calculation of Coefficient of Heat Transfer

For the absorber plate, the coefficient of heat transfer is assumed to be equal to coefficient of heat transfer of the rear bottom plate which is given by the following correlation [6]:

$$h_{fp} = h_{pb} = \text{Nu} \times \frac{K}{d_e} \tag{3}$$

where K represents the thermal conductivity while d_e denotes the equivalent diameter which is given by

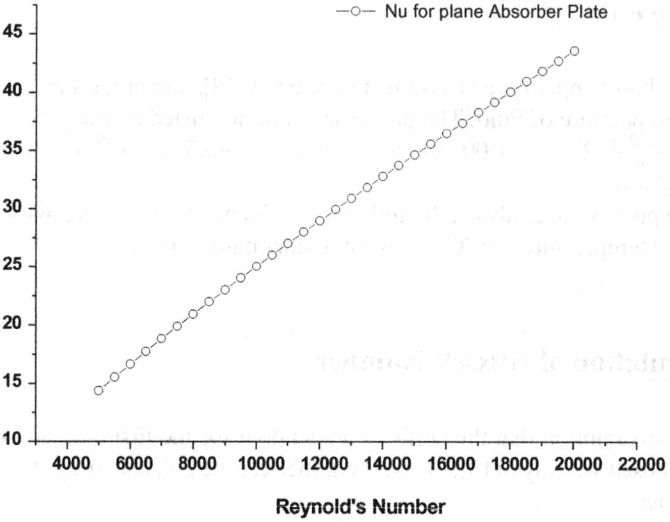

Fig. 3 Nusselt number for plane absorber plate

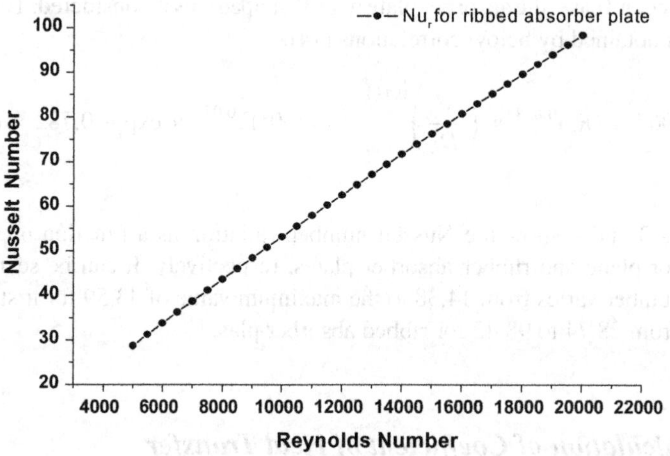

Fig. 4 Nusselt number for ribbed absorber plate

$$d_e = \frac{4 \times \text{cross sectional area of duct}}{\text{wetted perimeter}} \tag{4}$$

3.2 Calculation of Radiative Heat Transfer Coefficients

The coefficient of radiative heat transfer is determined as below [8]:

$$h_r\left(T_{pm} - T_{bm}\right) = \frac{\sigma L dx\left({T_{pm}}^4 - {T_{bm}}^4\right)}{\frac{1}{\varepsilon_p} + \frac{1}{\varepsilon_b} - 1} \tag{5}$$

3.3 Determination of Coefficient of Effective Heat Transfer

The coefficient of effective heat transfer between air stream over the absorber plate is given by the below equation [5]:

$$h_e = \left[h_{fp} + \frac{h_r h_{fb}}{h_r + h_{fb}}\right] \tag{6}$$

4 Loss Coefficient of SPSAH

It is useful to express the overall loss coefficients of the air heaters defined by

$$q_l = U_l A_p\left(T_{pm} - T_a\right) \tag{7}$$

where $U_l = U_b + U_t + U_s$.

Since side losses are neglected, only top and bottom losses are considered. The top loss coefficient is calculated by taking into consideration, the reradiation losses as well as the convection losses in upward direction in the absorber plate and comes out to be 6.2 W/m^2 – K. Similarly, the bottom loss coefficients are calculated by taking into consideration, the convection and conduction losses in downward direction from the absorber plate, and come out to be 0.8 W/m^2 – K. Typical values range from 2 to 10 W/m^2 – K [8]. Hence, our assumptions are correct.

5 Energy Balance for SPSAH

The total energy incident on the glass cover is transmitted to the absorber plate and subsequently to the air stream. A desirable property of the glass material is that it transmits up to 90% of the inbound short-wave rays, while almost zero percent of the returning long wave radiation from the absorber plate can transmit outwards.

The energy balance for each component is given by the equations [8].
Energy balance for absorber plate:

$$SLdx = U_t Ldx(T_{pm} - T_a) + h_{fp} Ldx(T_{pm} - T_f) + \frac{\sigma Ldx({T_{pm}}^4 - {T_{bm}}^4)}{\frac{1}{\varepsilon_p} + \frac{1}{\varepsilon_b} - 1} \quad (8)$$

Energy balance for rear bottom plate:

$$\frac{\sigma Ldx({T_{pm}}^4 - {T_{bm}}^4)}{\frac{1}{\varepsilon_p} + \frac{1}{\varepsilon_b} - 1} = U_b Ldx(T_{pm} - T_a) + h_{fb} Ldx(T_{pm} - T_f) \quad (9)$$

Energy balance for air stream:

$$\dot{m} C_p dT_f = h_{fb} Ldx(T_{pm} - T_f) + h_{fp} Ldx(T_{pm} - T_f) \quad (10)$$

6 Performance Parameters of SPSAH

6.1 Collector Efficiency Factor

The collector efficiency takes into account the various losses including the top and bottom loss coefficients. The top loss coefficient is calculated by taking into consideration, the reradiation losses as well as the convection losses in upward direction in the absorber plate. Similarly, the bottom loss coefficients are calculated by taking into consideration, the convection and conduction losses in downward direction from the absorber plate. The side losses have been neglected for the purpose of analysis. The efficiency factor of the collector is defined as the ratio of real collector output energy to the output energy when absorber plate is at the same mass flow rate and average fluid temperature. The collector efficiency factor is determined for different solar fluxes with varying Re.

6.2 Collector Heat Removal Factor

It is an important parameter for design purposes as it measures and indexes thermal resistance of solar radiation on the absorber in reaching the collector fluid and has been evaluated for the solar radiations ranging from 950 to 1650 W/m^2 for varying Re.

6.3 Useful Heat Gain Rate

The useful heat gain rate for SPSAH is the measure of net amount of energy acquired from incident solar radiations. The useful heat gain is obtained by Hottel–Whiller–Bliss equation and is given below:

$$q_u = F_R A_p\left[S - U_l\left(T_{fi} - T_a\right)\right] \tag{11}$$

7 Results and Discussions

The SPSAH is investigated theoretically for the instantaneous efficiency and temperature of air at outlet. The air heaters are influenced by various parameters including the solar insolation, mass flow rate, quality of air, ambient temperature, material of absorber plate and bottom plate transmissivity of glass cover. In the current study, SPSAH is investigated for variation of Re, and the following results have been obtained.

7.1 Effect of Variation in Re and Solar Incident Radiations on Efficiency

The most important parameter for the measure of performance of SPSAH is instantaneous efficiency. The instantaneous efficiency provides an overview about the performance of solar air heater. It takes into the consideration the overall losses and net useful gain in SPSAH. It is the ratio of useful energy gain to the incident solar radiation over an area for a unit time [9].

Figure 5 represents the effect on efficiency for different incident solar radiations with varying the Reynolds number. For lower Reynolds number, the efficiency changes rapidly as compared to higher Reynolds number. The slope of efficiency curve decreases as Reynolds number increases. Further, the efficiency curves have a common trend as Reynolds number changes from lower values to higher values. As the Reynolds number increases, the efficiency increases less rapidly as compared to the lesser Reynolds number. An increase of 7.75–18.33% in the thermal efficiency is obtained for the V-shaped ribbed solar air heater. The efficiency curves particularly for higher solar insolation 1600 and 1650 W/m^2 show slight variation in the nature of curves. The efficiency increases for increasing values of solar radiations incident on the surface for a particular Reynolds number.

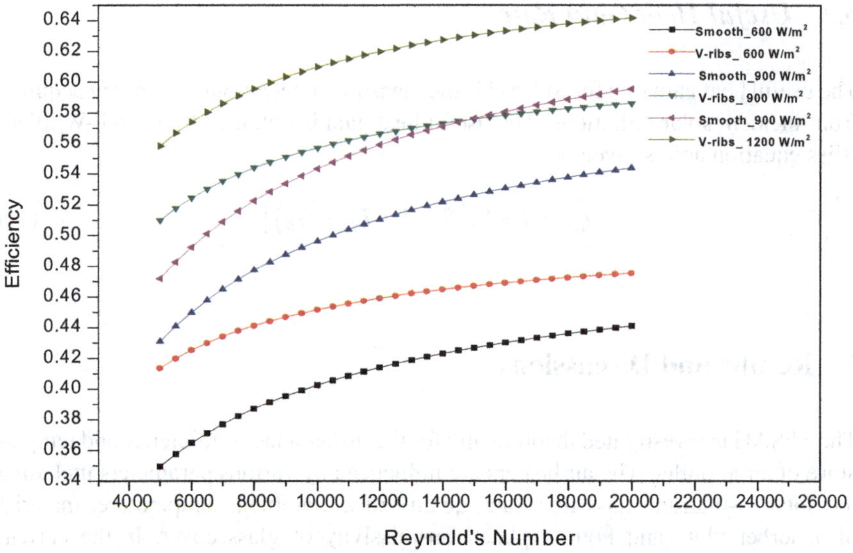

Fig. 5 Efficiency variation as a function of Reynolds number at different solar insolation

7.2 *Effect of Variation in Reynolds Number and Solar Incident Radiations on Outlet Temperature*

Figure 6 shows the trend of outlet temperature (T_{fo}) as a function of Reynolds number. At low solar fluxes, the variation in outlet temperature with increasing Reynolds number is less significant as compared to higher solar fluxes. It was established that the outlet temperature increases rapidly for lower Reynolds number. With the increase in Reynolds number, the slope of the curve shows a decreasing trend indicating fewer rise in outlet temperature. The decrease is more in case of absorber plate without ribs.

At R_e = 5000, the difference in outlet temperature for solar flux of 950 and 1650 W/m^2 is 1.5 °C which is less than the difference in outlet temperature 1.8 °C at R_e = 20,000, thereby indicating higher temperature differences are attained with increasing Reynolds number in both cases.

7.3 *Effect of Variation in Reynolds Number and Incident Solar Radiations on Outlet Useful Heat Gain*

Figure 7 shows the variation. The useful heat for SPSAH increases with increasing Reynolds number with more significant increase for lower values of Reynolds number over higher values of Reynolds number. For a particular solar insolation, the useful

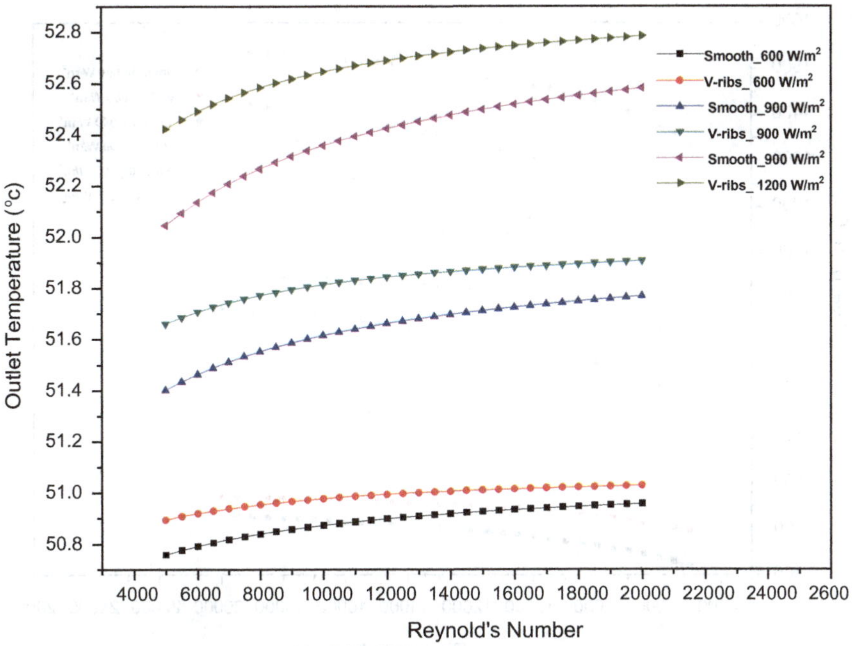

Fig. 6 Outlet temperature at different solar insolation

heat gain changes less rapidly after Reynolds number of around 12,000. The useful heat gain increases with increasing solar insolation. At higher solar insolation, the useful heat gain changes more rapidly as compared to the lower values of solar insolation which is more predominant in case of ribbed absorber plate. For lower values of solar insolation, the difference in the useful heat gain for ribbed absorber plate and plane absorber plate at lower Reynolds number is higher than the difference at higher values of solar insolation.

8 Conclusion

The SPSAH investigated theoretically in the present study has an increasing efficiency with increasing Reynolds number and solar flux over the collector. The SPSAH is the conventional design, and the use of fins and baffles can further increase the efficiency. The V-corrugated plate and double pass finned solar air heaters give higher efficiency as compared to SPSAH. More importantly, the effect of variation in Reynolds number and solar insolation on efficiency, outlet temperature and useful heat gain based on various correlations needs to be given due consideration for optimum design of SPSAH, and this study is a useful step in achieving the optimum design of SPSAH based on specific performance parameters. Based on the results

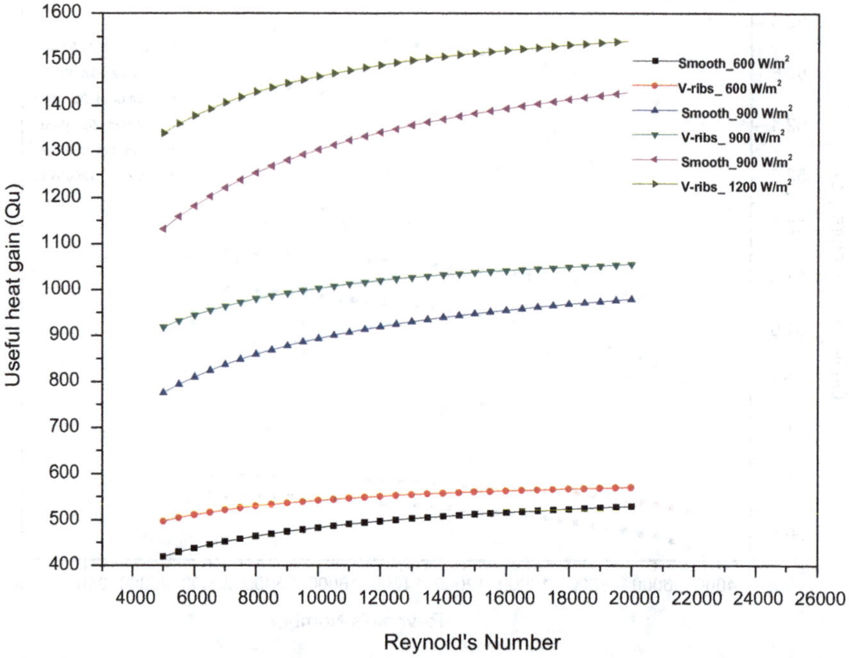

Fig. 7 Useful heat gain at different solar insolation

obtained from the theoretical analysis that the SPSAH with fins and baffles increases the efficiency, the work can be extended to the experimental analysis with different configurations and arrangements of fins. The top surface of V-shaped ribs can be modified to squared or triangular configurations leading to increase in the turbulence. To further enhance the efficiency, baffles may be introduced in the path to make efficient heat transfer. This arrangement not only increases turbulence but also increases the path such that the air remains in contact with the hot surface for longer time and is discharged at higher temperatures as compared to the other configurations.

References

1. Vajjarapu H, Verma A, Gulzar S (2019) Adaptation policy framework for climate change impacts on transportation sector in developing countries. Transp Dev Econ 5(1):3
2. Baba ZU, Shafi WK, Haq MIU, Raina A (2019) Towards sustainable automobiles-advancements and challenges. Prog Ind Ecol Int J 13(4):315–331
3. Kumar R, Ul Haq MI, Raina A, Anand A (2019) Industrial applications of natural fibre-reinforced polymer composites–challenges and opportunities. Int J Sustain Eng 12(3):212–220
4. Lemmon EW (2000) Thermodynamic properties of air and mixtures of nitrogen, argon, and oxygen from 60 to 2000 K at pressures to 2000 MPa. J Phys Chem Ref Data 29(3):331
5. Yadav AS, Bhagoria JL (2013) A CFD (computational fluid dynamics) based heat transfer and fluid flow analysis of a solar air heater provided with circular transverse wire rib roughness on

the absorber plate. Energy 55:1127–1142
6. Gulzar O, Qayoum A, Gupta R (2019) Photo-thermal characteristics of hybrid nanofluids based on therminol-55 oil for concentrating solar collectors. Appl Nanosci 9(5):1133–1143
7. Gulzar O, Qayoum A, Gupta R (2019) Experimental study on stability and rheological behaviour of hybrid Al_2O_3-TiO_2 therminol-55 nanofluids for concentrating solar collectors. Powder Technol 352:436–444
8. Sukhatme (2008) Solar energy: principles of thermal collection and storage. Tata McGraw-Hill Education, New York
9. Duffie JA, Beckman WA (1974) Solar energy thermal processes. Wiley, Hoboken
10. DeWinter F (1990) Solar collectors, energy storage, and materials. MIT Press, Cambridge
11. Perlin J (1999) From space to earth: the story of solar electricity. Earthscan, London
12. Proceedings of ISES World Congress 2007 (vol 1–5): solar energy and human settlement. Springer Science & Business Media, Berlin (2009)
13. Kays W, Crawford M, Weigand B (2005) Convective heat & mass transfer w/engineering subscription card. McGraw-Hill Companies, Incorporated, New York
14. Lanjewar A, Bhagoria JL, Sarviya RM (2011) Heat transfer and friction in solar air heater duct with W-shaped rib roughness on absorber plate. Energy 36(7):4531–4541

[illegible] plate. Sol Energy 85:1127–1142
6. Gupta HK, Das Agrawal G, Mathur J (2015) Investigation for effect of Al_2O_3–H_2O nanofluid flow rate on the efficiency of direct absorption solar collector. Case Stud Therm Eng 5:70–78 [illegible]
7. Gautam G, Dasgupta A, Gupta R (2019) Experimental study on stability and rheological behaviour of hybrid Al_2O_3-TiO_2 thermal oil nanofluids for concentrating solar collectors. Powder Technol 352:[illegible]
8. Sukhatme (2008) Solar energy: principles of thermal collection and storage, 3rd edn. McGraw-Hill Education, New York
9. Duffie JA, Beckman WA (1974) Solar energy thermal processes. Wiley, Hoboken
10. De Winter F (1990) Solar collectors, energy storage, and materials. MIT Press, Cambridge
11. Parfitt [illegible] (1930) From space to earth: the story of solar electricity. Earthscan, London
12. Proceedings of ISES World Congress 2007 (vol. I–vol. V): solar energy and human settlement. Tsinghua University Press, [illegible], Berlin (2009)
13. Kays W, Crawford M, Weigand B (2005) Convective heat and mass transfer, 4th edn. McGraw-Hill, [illegible], New York
14. Hosseini [illegible], Bahrehmand [illegible] (2012) Heat transfer and friction in solar air heater duct with W-shaped rib roughness on absorber plate. Energy [illegible]:[illegible]

Modelling of Ambient Noise Levels in Urban Environment

S. K. Tiwari, L. A. Kumaraswamidhas, and N. Garg

Abstract The study conducts time-series approach for analysing one year noise monitoring data. Support vector machine (SVM) technique is used as a modelling technique for time-series approach. The noise data is trained using tenfold cross-validation to get optimum hyperparameters (γ, ε, C). The performance and accuracy of model are determined by statistical parameters like MSE, RMSE, MAPE in %, R^2. The paper predicts an error of ± 2 dB(A) with the implementation of support vector machine (SVM).

Keywords Noise monitoring · Support vector machine · Ambient noise levels · Tenfold cross-validation · Day noise level · Night noise level

1 Introduction

Noise pollution has significantly increased especially in the urban areas in Indian scenario. With advancement of vehicles and urbanization in cities, there has been a quick addition in traffic volume. Regardless of the way that transportation is an essential part of urban society, its superiority is obscured by its negativity. Inappropriate placement of vehicles at different locations nearby roads is one of the major cause of traffic jam. Some studies affirm that noise pollution has adverse influence on human health [1, 2]. It comprises slant stress impact, sleeping disturbances which clearly cause 'prompt effect' on mental and physical perspective. The Central Pollution Control Board has directed noise levels for different zones, i.e. silence, industrial, commercial, residential zones, and carried many studies for noise monitoring in Indian scenario [3]. Garg et al. [4] discussed the pilot project on the establishment of National Ambient Noise Monitoring Network (*NANMN*) at 35 locations across

S. K. Tiwari (✉) · L. A. Kumaraswamidhas
Indian Institute of Technology (ISM), Dhanbad 826001, India
e-mail: shashikanttiwari61@gmail.com

N. Garg
CSIR-National Physical Laboratory, New Delhi 110012, India

© The Author(s), under exclusive license to Springer Nature Singapore Pte Ltd. 2021
R. M. Singari et al. (eds.), *Advances in Manufacturing and Industrial Engineering*,
Lecture Notes in Mechanical Engineering,
https://doi.org/10.1007/978-981-15-8542-5_70

the seven major cities of the country. The European Environmental Noise Directive 2002/49/EC [5] gives direction for noise mapping that include future plans with financial information and cost-effective assessment. European Directive permits to survey and to look at, inside EU Member States, noise exposure data, particularly for the future execution steps, when noise maps should be drawn up with the basic evaluation strategies.

There are various techniques used for noise assessment and monitoring. Some uses long-term noise monitoring strategy, while Garg et al. [6] emphasized on short-term noise monitoring strategy as a reliable strategy within an accuracy of ±2 dB(A). The high costs of installing and maintaining permanent networks are primarily the main reason for analysing the suitability of short-term strategies to ascertain whether they can provide a suitable and reliable alternative or not as compared to the long-term noise monitoring. There are some illustrations whereby extensive networks have been installed [7]. Morillas and Gajardo [8] evaluated 90% probability interval for random 9 days data to measure L_{den}. Hence, there is a need of alternative approach to predict and forecast ambient noise level by using time-series approach. DeVor et al. [9] used autoregressive moving average (*ARMA*) and Garg et al. [10] used autoregressive integrated moving average (*ARIMA*) model to predict and forecast noise level in time-series analysis. This study explores SVM technique for forecasting and determining the accuracy and performance of the predicted noise level.

2 Methodology

2.1 Ambient Noise Level Calculation

The present study is helpful to opt an optimized strategy for forecasting with the help of a time-series method (*SVM*) in Indian scenario using the 3-year database with minimum error.

The value of A-weighted day noise level (L_{day}) and night noise level (L_{night}) is calculated:

$$L_{\mathrm{day},n} = 10\log\left[\frac{1}{n}\sum_{i=1}^{n} 10^{0.1(l_{\mathrm{day},i})}\right] \tag{1}$$

$$L_{\mathrm{night},n} = 10\log\left[\frac{1}{n}\sum_{i=1}^{n} 10^{0.1(l_{\mathrm{night},i})}\right] \tag{2}$$

where n denotes the numbers of days and nights in long-term noise monitoring strategy. The error is calculated as difference of observed noise level with the predicted noise level for the commercial site for a particular period of data.

2.2 Support Vector Machine

The idea of support vector machine (*SVM*) is mapping a nonlinear dataset. The approach focuses to solve a regression using linear function. The hyperplane is also known as classifier separates classes to get an optimal solution. The data is spitted into training and testing data. Suppose X_i represents input data ($i = 1, \ldots, n$) where n is the number of training data points. The function of hyperplane is as [11]:

$$Y(x) = w^{\mathrm{T}} X_i + b \tag{3}$$

where w is the orientation and b is the position of hyperplane classifying the training data into two classes. C_1 is the positive class, and C_2 is the negative class. The main focus of SVM is to find a new classifier.

$$Y(x_1) = w^{\mathrm{T}} x_i + b > 0 \tag{4}$$

x_1 □ C_1, if x_1 lies on the positive side of the hyperplane.

$$Y(x_2) = w^{\mathrm{T}} x_i + b < 0 \tag{5}$$

x_2 □ C_2, if x_1 lies on the positive side of the hyperplane.

2.3 Nonlinearity in Data

For a nonlinear separable data, SVM can be a optimized time-series technique to have a good solution. The *w* and *b* are determined by two mathematical formulations of a regularized function (*R*(*c*)) in *SVM* [11–13].

$$R(c) = \frac{1}{2}\|w\|^2 + \frac{c}{n}\sum_{i=1}^{n} L_s(d_i, y_i) \tag{6}$$

$$L_s(d_i, y_i) = \begin{cases} |d_i - y_i| - \varepsilon & 0 \\ |d_i - y_i| & \text{otherwise} \end{cases} \tag{7}$$

$\frac{c}{n}\sum_{i=1}^{n} L_s(d_i, y_i)$ is empirical error.

The dot product of two input $\psi(x)$ and $\psi(y)$ vectors is kernel function that should affirm Mercer's condition. Mainly, four kernels are utilized in support vector machine (SVM) modelling which are as follows [12]:

$$\text{Linear Kernel: } K(x, y) = x^{\mathrm{T}} y \tag{8}$$

$$\text{Radial Basis Function Kernel: } K(x, y) = \exp(-\gamma/x - y^2), \quad \gamma > 0 \tag{9}$$

$$\text{Polynomial Kernel: } K(x, y) = (\gamma x^T y + r)^d, \quad \gamma > 0 \tag{10}$$

$$\text{Sigmoid Kernel: } K(x, y) = \tanh(\gamma x^T y + r) \tag{11}$$

Here, the kernel parameters are d, r and γ. Kernel parameters have important significance in the performance of support vector machine model. The complexity of best parameter is controlled by the kernel functions. These kernel functions increase the accuracy of the model as compared to the other NN models with the help of other hyperparameters. The main purpose of kernel parameters is to convert a complex data, i.e. in the form of lower-dimensional space to higher-dimensional space. Here for the research work, radial basis function is used because of its higher accuracy in comparison to other kernel functions.

3 Results and Discussion

Figure 1a shows variation plot of L_{day} in dB(A) for 365 days for a residential site in Delhi. Maximum value of L_{day} is 72 dB(A), and minimum value is 60 dB(A). While Fig. 1b shows variation plot of L_{night} in dB(A) for 365 days for a residential site in Delhi. Maximum value of L_{night} is 72 dB(A), and minimum value is 60 dB(A). The noise values are taken in A-weighting because in most of the industrial application, noise levels are taken in the form of A-weighting only in comparison to C-weighting. The C-weighting noise values are used for very tuned and fined noise, while A-weighting noise levels are used for human audible range.

The kernel applied in the study is radial basis function (RBF) kernel. It has better performance in comparison to the others kernel functions. In RBF kernel, three hyperparameters have been used to analyse the performance of SVM model. These three hyperparameters are: Gamma (γ), Epsilon (ε) and Cost (C). The first stage is to find an optimized parametric combination of these hyperparameters. Hit-and-trial approach was attempted to get the optimum value of hyperparameters. The parametric combination (γ, ε, C) of optimized hyperparameters is (2^5, 0.4, 2^5) for both day and night. Figure 2a, b shows the plot of predicted day and night ambient noise level in comparison to observed noise levels.

The input data is divided into testing and training data. 90% of the input data is taken as training data, and 10% of the data is used as testing data. Mean square error (*MSE*), root mean square error (*RMSE*), mean average percentage error (*MAPE* in%), coefficient of determination (R^2) are the parameters that ascertain the efficiency of the model.

Table 1 shows the statistics performance of training data for both day and night noise levels. The maximum error is 4.54 dB(A) for day and 5.37 dB(A) for night, but the MSE and RMSE error lies within ±2 dB(A) which is sought of reliable accuracy.

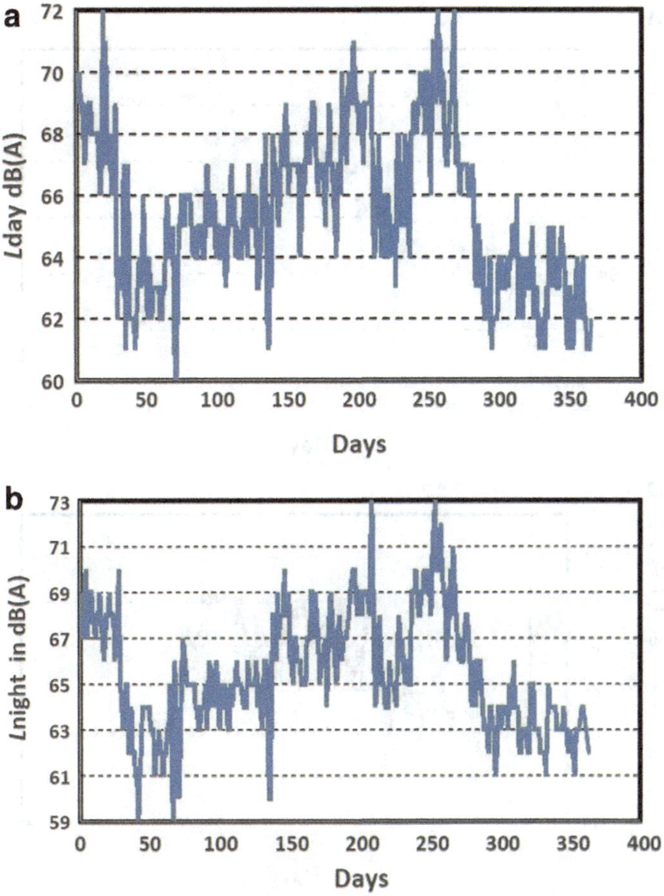

Fig. 1 **a** Time sequence plot of L_{day} in dB(A) for 365 days for a residential site in Delhi. **b** Time sequence plot of L_{night} in dB(A) for 365 days for a residential site in Delhi

The probability of the training data determined from input data can be taken as per the study analysis. There is no particular approach to determine the probability like in the present study, the training data is taken 90%. The determination of coefficient is 0.6 for day and 0.4 for night which implies better performance of the classifier.

To test the model, the testing data is taken 10% of the input data. The SVM model predicts an error of ±2 dB(A) for testing data as well. The determination of coefficient is 0.6 for day and 0.5 for night which implies better performance of the classifier of predicted testing data (Table 2). Hence, it can be observed that the support vector machine is a good approach for forecasting of ambient noise level within an accuracy of ±2 dB(A). Figure 3a, b shows the standardized residual analysis of *SVM* model for L_{day} and L_{night} dB(A) for both training and testing data.

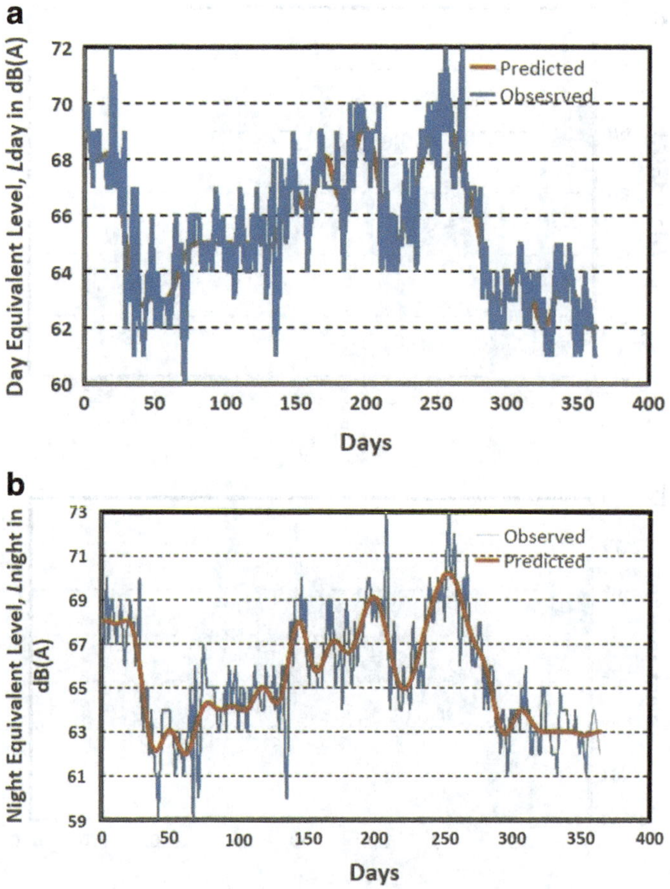

Fig. 2 **a** Comparison of measured (blue line) and predicted (red line) values of L_{day} in dB(A). **b** Comparison of measured (blue line) and predicted (red line) values of L_{night} in dB(A)

Table 1 *SVM* model statistics for training data

Statistical parameter	L_{day} dB(A)	L_{night} dB(A)
MSE in dB(A)^2	1.640	1.664
RMSE in dB(A)	1.281	1.290
MAPE in %	1.47	1.43
Maximum error in dB(A)	4.54	5.37
Minimum error in dB(A)	−4.37	−5.96
R^2	0.6	0.4

Table 2 *SVM* model statistics for testing data

Statistical parameter	L_{day} dB(A)	L_{night} dB(A)
MSE in dB(A)^2	2.316	1.695
RMSE in dB(A)	1.522	1.302
MAPE in %	1.45	1.82
Maximum error in dB(A)	2.47	1.97
Minimum error in dB(A)	−3.20	−3.09
R^2	0.6	0.5

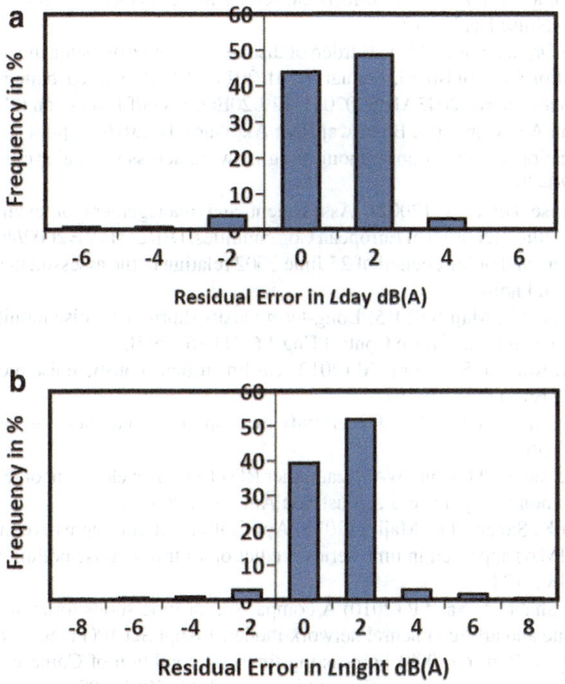

Fig. 3 **a** Standardized residual analysis of *SVM* model for L_{day} dB(A) for both training and testing data. **b** Standardized residual analysis of *SVM* model for L_{night} dB(A) for both training and testing data

4 Conclusion

In the study, *SVM* is used as a time-series modelling technique for the statistical analysis of one-year noise monitoring data set. *SVM* is an outperforming technique that can profoundly predict the ambient noise levels L_{day} and L_{night}. The parametric combination for both day and night is (2^5, 0.4, 2^5). Meanwhile, this is the best set of hyperparameter for classifier which represents a similar trend as the observed pattern.

The application of SVM has rarely been used in the determination of ambient noise level. The result shows that this model can be used as a better fitting model for predicting and forecasting noise levels. The work also emphasizes on the use of RBF kernel to analyse *SVM* model. The performance of model is determined by the statistical parameters like *MSE*, *RMSE*, *MAPE* in % andR^2.

References

1. Ising H, Kruppa B (2004) Health effects caused by noise: evidence in the literature from the past 25 years. Noise Health 6(22):5
2. World Health Organization (2011) Burden of disease from environmental noise. WHO, Geneva
3. Central Pollution Control Board, Annual report, 2011–2012. https://cpcb.nic.in/upload/Annual Reports/AnnualReport%2043AR%202011-12%20English.pdf [access on 02.03.2016]
4. Garg N, Sinha AK, Gandhi V, Bhardwaj RM, Akolkar AB (2016) A pilot study on the establishment of national ambient noise monitoring network across the major cities of India. Appl Acoust 103:20–29
5. European Noise Directive (2002). Assessment and management of environmental noise. 2002/49/EU, Official Journal of European Communities. DIRECTIVE 2002/49/EC of the European Parliament and of the council of 25 June 2002 relating to the assessment and management of environmental noise
6. Garg N, Saxena TK, Maji S (2015) Long-term versus short-term noise monitoring: strategies and implications in India. Noise Control Eng J 63(1):26–35(10)
7. Czyzewski A, Kotus J, Szczodrak M (2012) On-line urban acoustic noise monitoring system. Noise Control Eng J 60(1):69–84
8. Morillas JB, Gajardo CP (2014) Uncertainty evaluation of continuous noise sampling. Appl Acoust 75:27–36
9. DeVor RE, Schomer PD, Kline WA, Neathamer RD (1979) Development of temporal sampling strategies for monitoring noise. J Acoust Soc Am 66(3):763–771
10. Garg N, Soni K, Saxena TK, Maji S (2015) Applications of autoregressive integrated moving average (ARIMA) approach in time-series prediction of traffic noise pollution. Noise Control Eng J 63(2):182–194
11. Samsudin R, Shabri A, Saad P (2010) A comparison of time-series forecasting using support vector machine and artificial neural network model. J Appl Sci 10(11):950–958
12. Ding Y, Song X, Zen Y (2008) Forecasting financial condition of Chinese listed companies based on support vector machine. Expert Syst Appl 34(4):3081–3089
13. Benítez-Peña S, Blanquero R, Carrizosa E, Ramírez-Cobo P (2019) Cost-sensitive feature selection for support vector machines. Comput Oper Res 106:169–178

Development and Characterizations of ZrB_2–SiC Composites Sintered Through Microwave Sintering

Ankur Sharma and D. B. Karunakar

Abstract Zirconium diboride (ZrB_2) has superior properties like high electrical and thermal conductivity, high melting point, high hardness, high elastic modulus, high corrosion, and thermal shock resistance, making ZrB_2 as a potential candidate for hypersonic aerospace vehicles. The inclusion of silicon carbide (SiC) in the ZrB_2 matrix makes it suitable for applications in the ceramic armors, nose caps, and leading edges of atmospheric re-entry vehicles. Microwave sintering is applied to develop ZrB_2–SiC (5–35 vol%) composites and investigate their metallurgical and mechanical properties. The addition of SiC as reinforcement in the ZrB_2 matrix not only improves the densification but also enhances the hardness and fracture toughness of composites. The highest relative density and Vickers hardness are 97.72% and 17.03 GPa for ZrB_2-25 vol% SiC composite. The highest fracture toughness is found as 5.64 MPa $m^{1/2}$ for ZrB_2-15 vol% SiC composite.

Keywords Microhardness · Fracture toughness · ZrB_2–SiC composites · Microwave sintering

1 Introduction

The ultra-high temperature ceramics (UHTC) have applications in the nose caps and leading edges of atmospheric re-entry vehicles due to their intriguing properties such as high hardness, high melting temperatures and high strength at elevated temperatures. Zirconium diboride (ZrB_2) is considered as UHTC due to the properties like high hardness (22 GPa), higher melting point (3040 °C), electrical conductivity (9.2×10^{-6} Ω cm), fracture toughness (4–5 MPa $m^{1/2}$), and thermal expansion coefficient (5.5×10^{-6} K^{-1}) [1–5]. Despite these superior properties of ZrB_2, its densification

A. Sharma (✉) · D. B. Karunakar
Mechanical and Industrial Engineering Department, IIT Roorkee, Roorkee, India
e-mail: nkrsharma9@gmail.com

D. B. Karunakar
e-mail: bennyfme@iitr.ac.in

© The Author(s), under exclusive license to Springer Nature Singapore Pte Ltd. 2021
R. M. Singari et al. (eds.), *Advances in Manufacturing and Industrial Engineering*, Lecture Notes in Mechanical Engineering,
https://doi.org/10.1007/978-981-15-8542-5_71

is quite tricky due to covalent bonding, high vapor pressure, and some amount of impurity concentrations [3, 4]. However, the inclusion of SiC increases the sintering as well as densification of ZrB_2 ceramic. The addition of SiC as reinforcement in the ZrB_2 matrix forms the borosilicate glass layer, thus helps in improving the oxidation resistance of composites at elevated temperatures [4, 5].

Various sintering methods have been adapted like hot isostatic pressing, spark plasma sintering and microwave sintering to develop ZrB_2–SiC composites [1–12]. Microwave sintering has been utilized by numerous researchers from the past few decades for the development of composites based on borides, carbides, and nitrides. Microwave sintering promotes not the only densification but also enhances metallurgical and mechanical properties [5]. ZrB_2 has a little penetration depth of microwaves due to high electrical conductivity [6]. The inclusion of SiC increases the microwave coupling of ZrB_2 material because of better microwave energy absorption of SiC material [3, 4]. Studies have shown improvements in metallurgical properties like microstructure with controlled grain growth and enhancements in mechanical properties like flexural strength and fracture toughness using microwave sintering [3, 4, 6–12].

The microwave sintering process achieves uniform rapid heating without any cracking or thermal stress [6]. The microwave furnace comprises of insulation cavity, magnetron, programmable controller, and heating system. Magnetrons produce the microwaves. The infrared pyrometer detects the temperature generated in the sample. A resistance heating tuner prevents temperature fluctuations. The insulation cavity confines microwaves to the sample [6].

Microwaves propagate outward from the inside, thus heating the sample uniformly which leads to enhanced densification along limited grain growth [6, 8, 10].

The present work aims to develop and characterize the ZrB_2–SiC composites by microwave sintering and identifies the effect of an increase in SiC proportions on densification, hardness, and fracture toughness of developed composites.

2 Experimental Procedure

The experimental procedure includes materials, preparation of the sample for compaction and microwave sintering, metallurgical, and mechanical characterizations.

2.1 Materials and Samples Preparation

ZrB_2 (99.5%, 15 μm size) and SiC (99.9%, 20 μm size) powders purchased from Nanoshel LLC (Wilmington, U.S) are used in the present work. The different proportions of SiC are added as 5 vol%, 15 vol%, 25 vol%, and 35 vol%, respectively. For

Fig. 1 Hydraulic pressing machine

simplification, samples are designated as Z5S for ZrB_2-5 vol% SiC to Z35S for ZrB_2-35 vol% SiC. The starting powders are homogenized in a ball mill for 10 h using ethanol as media to remove large agglomerates. The stainless steel balls of around 6 mm diameters are used. The weight ratio for the ball to powder is kept at 5:1. After ball milling, the powders are mixed in agate mortar by using polyvinyl alcohol (PVA) 5 wt% as the binder. After adding the binder, the powders are dried at 150 °C for 1 h in an infrared oven. The pallets with a diameter of 15 mm and a thickness of 5 mm are formed by compaction using a hydraulic press at a load of 10 tons. The compacts are then sintered in a microwave sintering furnace (2.45 GHz frequency and 3 KW power output). The compacts are placed in a silicon carbide crucible. The infrared pyrometer is used to detect the temperature of compacts. All samples are sintered at a sintering temperature of 1800 °C, a heating rate of 35 °C/min and a holding time of 40 min at a maximum temperature in the flowing argon atmosphere. The hydraulic press and microwave sintering furnace are shown in Figs. 1 and 2. Microwave-sintered ZrB_2–SiC composites with 5 vol% SiC, 15 vol% SiC, 25 vol% SiC, and 35 vol% SiC are shown in Fig. 3.

2.2 Characterizations

Microstructures are evaluated using a field emission scanning electron microscope (FE-SEM, FEI Germany). The elements present in the composites are analyzed by energy-dispersive spectroscopy (EDS). The bulk density of composites is evaluated

Fig. 2 Microwave sintering furnace

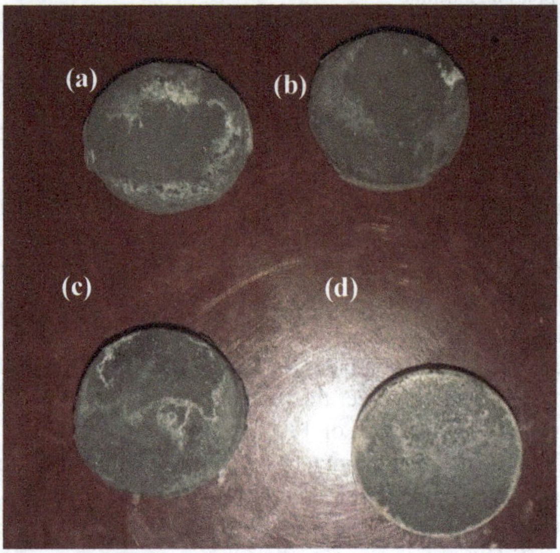

Fig. 3 Microwave-sintered ZrB_2–SiC composites with **a** 5 vol% SiC, **b** 15 vol% SiC, **c** 25 vol% SiC, and **d** 35 vol% SiC

by the Archimedes displacement method. The theoretical density is calculated using the rule of mixture [6], as shown in Eq. (1). The relative density is calculated using Eq. (2).

$$\rho_m = \rho_1 v_1 + \rho_2 v_2 \quad (1)$$

ρ_m density of mixture
ρ_1 density of ZrB_2
ρ_2 density of SiC
v_1 Volume fraction of ZrB_2
v_2 Volume fraction of SiC.

$$\text{Relative density (\%)} = (\text{Sintered density}/\text{Theoretical density}) \times 100 \quad (2)$$

"The microhardness is measured using a Vickers microhardness indentation device at 500-g force load with a dwell time of 10 s as per ASTM E384-17 standards. The five indents are taken per composite sample, and the average value of hardness has been obtained. The fracture toughness is measured by diagonal crack length produced by the Vickers macro-indenter from the macro-indentation tests. The maximum applied indenter load is 150 N for the dwell time of 20 s. Fracture toughness values have been observed with five indents per sample" [6]. "The Antis equation can be used to evaluate fracture toughness if *c*/*a* ratio is higher than 2.5, where '*c*' is radial crack length and '*a*' is the half of indentation diagonal" [6, 13]. The fracture toughness (K_{IC}) is given by Antis equation as:

$$K_{IC} = 0.016\sqrt{\left(\frac{E}{H}\right)}\frac{P}{c^{3/2}} \quad (3)$$

"where E is Young's modulus of the composites by rule of mixtures (Young's modulus of ZrB_2, SiC are 500, 475 GPa, respectively), H is the Vickers hardness (GPa), P is the applied load (N), and c is the diagonal crack length (μm)" [13].

3 Results and Discussions

3.1 Microstructures and EDS Analysis

Microstructures of ZrB_2–SiC composites with different proportions of SiC are shown in Fig. 4. From Fig. 4a, it can be seen that a larger amount of porosity is present for Z5S composite, whereas the very lesser amount of porosity is present for Z25S composites. The increase in SiC proportion reduces the porosity to a more considerable

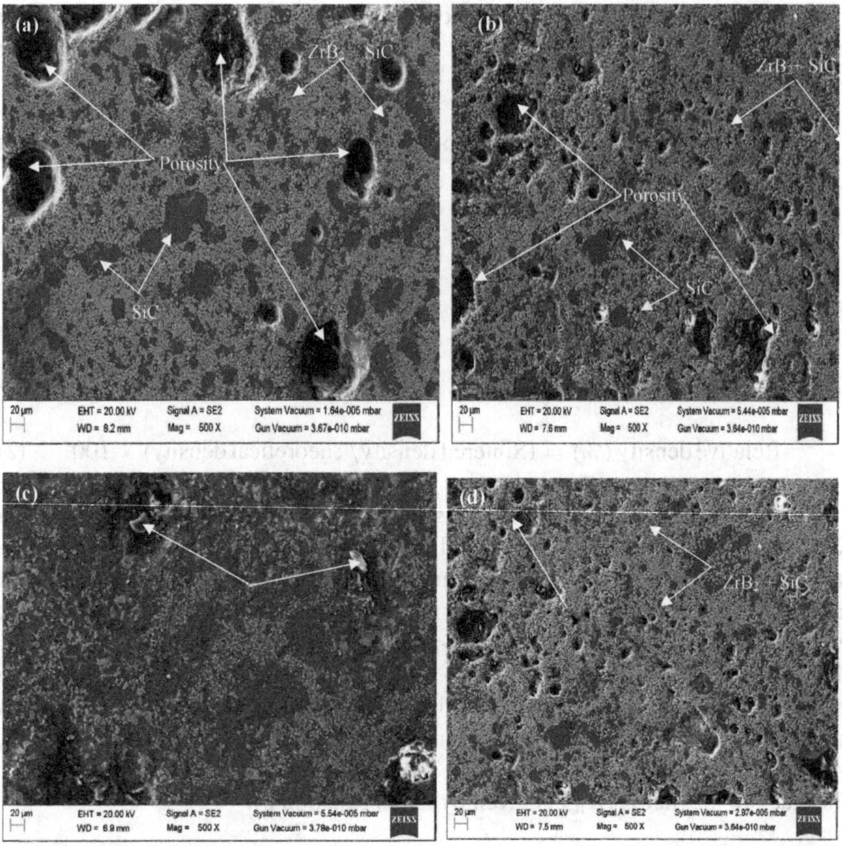

Fig. 4 Microstructures of ZrB_2–SiC composites with **a** 5 vol% SiC, **b** 15 vol% SiC, **c** 25 vol% SiC, and **d** 35 vol% SiC

extent as shown in Fig. 4c, d. The energy-dispersive spectroscopy (EDS) analysis is shown in Fig. 5. The EDS analysis confirms the presence of Zr, B, Si, and C in ZrB_2–SiC composite.

3.2 Relative Density

The relative density of ZrB_2–SiC composites with 5, 15, 25, and 35 vol% of SiC is shown in Fig. 6. The highest relative density of 97.72% has been obtained for Z25S composite, whereas the lowest relative density of 89.1% has been obtained for Z5S composite. The relative density has increased from Z5S to Z25S composite and then decreased. The possible reason for the decrease in relative density for Z35S

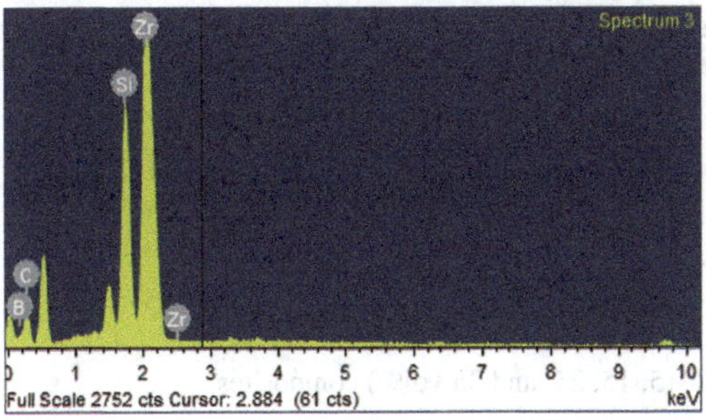

Element	Weight%	Atomic%
B	19.23	26.14
C	50.05	61.24
Si	21.17	11.08
Zr	9.56	1.54

Fig. 5 Energy-dispersive spectroscopy (EDS) analysis of ZrB_2-25 vol% SiC composite

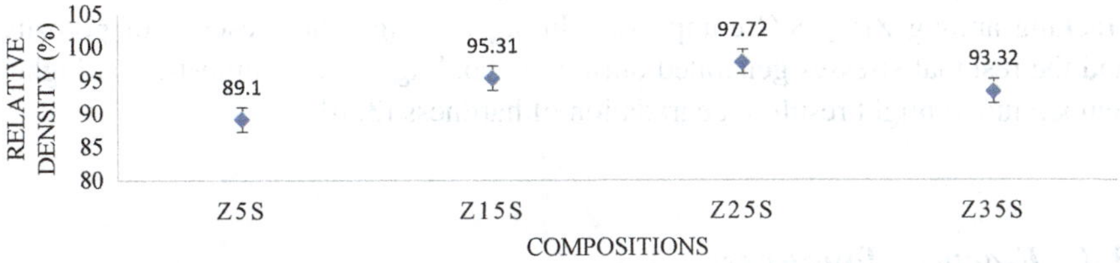

Fig. 6 Relative density of ZrB_2–SiC (5, 15, 25 and 35 vol%) composites

composite is the formation of large agglomerates which results in microcracking during the sintering process [4].

3.3 Vickers Microhardness

The microhardness of ZrB_2–SiC composites with 5, 15, 25, and 35 vol% of SiC is shown in Fig. 7. The highest microhardness of 17.03 GPa has been obtained for Z25S composite, whereas the lowest microhardness of 9.89 GPa has been obtained for Z5S composite. The lower value of microhardness might be due to difference

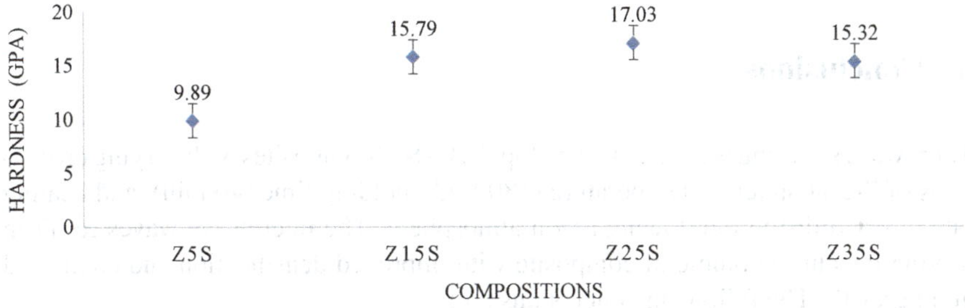

Fig. 7 Microhardness of ZrB_2–SiC (5, 15, 25, and 35 vol%) composites

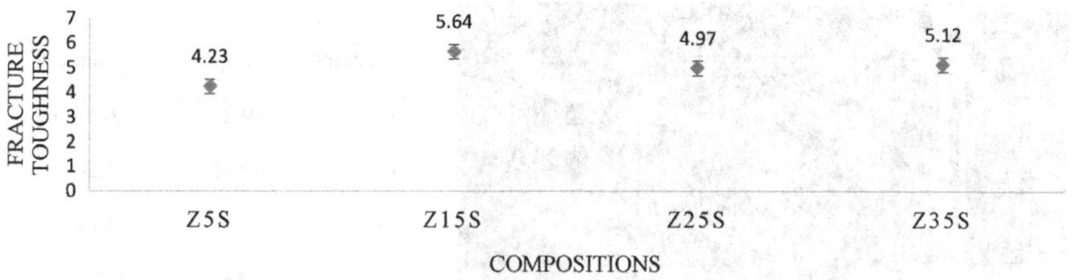

Fig. 8 Fracture toughness of the ZrB_2–SiC (5, 15, 25, and 35 vol%) composites

in coefficients of thermal expansion of ZrB_2 ($\alpha_a = 6.66 \times 10^{-6}$/°C, $\alpha_c = 6.93 \times 10^{-6}$/°C) [13] and SiC ($\alpha = 4.3 \times 10^{-6}$/°C) [14], which might have resulted in microcracking among ZrB_2–SiC composites during cooling. The presence of porosity and the residual stresses generated during the cooling of the composites (at higher temperatures) might result in degradation of hardness [3, 6].

3.4 Fracture Toughness

The fracture toughness of ZrB_2–SiC composites with 5, 15, 25, and 35 vol% of SiC is shown in Fig. 8. The highest fracture toughness of 5.64 MPa $m^{1/2}$ has been obtained for Z15S composite, whereas the lowest fracture toughness of 4.23 MPa $m^{1/2}$ has been obtained for Z5S composite. Figure 9 shows the Vickers macro-indentation and crack length of ZrB_2–SiC composites. Z5S composite exhibits higher radial crack length as shown in Fig. 9a, b, whereas Z15S composite exhibits lower radial crack length, as shown in Fig. 9c, d. The fracture toughness depends upon the homogenous distribution of reinforcement in the matrix. Fracture toughness improves with uniform distribution of SiC in the ZrB_2 matrix [5]. The toughening mechanisms like crack branching, crack deflection, and crack impeding are responsible for improving the fracture toughness. The weak interphase bonding between ZrB_2 and SiC increases the fracture toughness due to crack deflection [6, 14]. Microcracking during sintering also causes stress concentration and reduces fracture toughness [13].

4 Conclusions

Microwave sintering is applied to develop ZrB_2–SiC composites with varying proportions of SiC at sintering temperature (1800 °C), holding time (40 min), and heating rate (35 °C/min) in the flowing argon atmosphere. The microwave waves result in heating the entire volume of composite with improved densification and controlled grain growth. The following conclusions are:

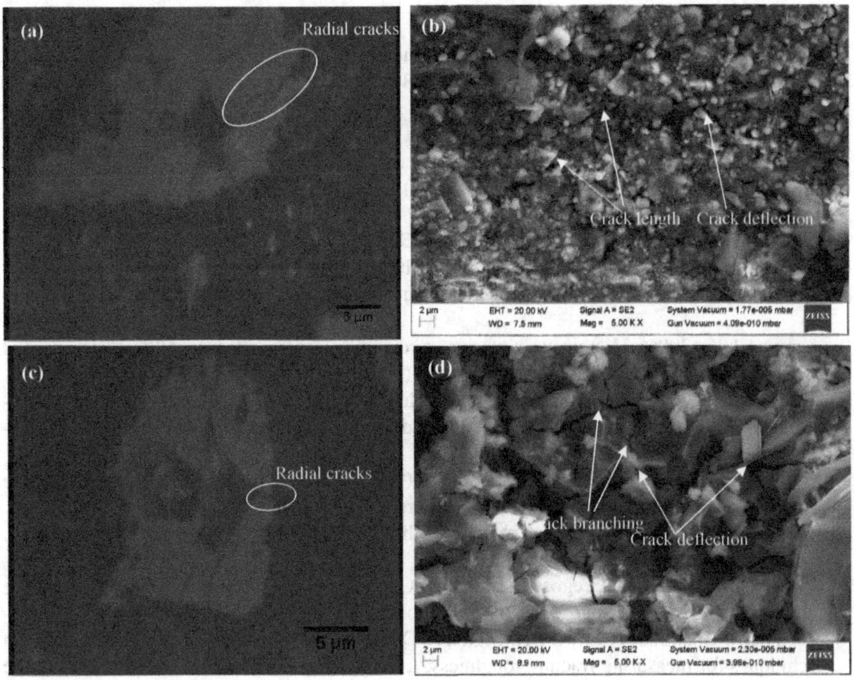

Fig. 9 Fracture toughness of **a** Z5S, **b** crack length of Z5S, **c** Z15S, and **d** crack length of Z15S

- The highest relative density and Vickers hardness are 97.72% and 17.03 GPa for Z25S composite. The highest fracture toughness is 5.64 MPa $m^{1/2}$ for the Z15S composite.
- The microstructural analysis demonstrates the porosity presents in composites, and EDS analysis confirms the elements present in the ZrB_2–SiC composites.
- The inclusion of SiC in ZrB_2–SiC composites improves the densification, metallurgical, and mechanical properties. However, the addition of SiC beyond 25 vol% reduces the mechanical properties due to agglomeration and uneven diffusion of SiC in the ZrB_2 matrix.

References

1. Goldstein A, Geffen Y, Goldenberg A (2001) Boron carbide–zirconium boride in situ composites by the reactive pressureless sintering of boron carbide–zirconia mixtures. J Am Ceram Soc 84(3):642–644
2. Monteverde F, Bellosi A, Guicciardi S (2002) Processing and properties of ZrB_2 based composites. J Eur Ceram Soc 22(3):279–288
3. Wang HL, Wang CA, Chen DL, Xu HL, Lu HX, Zhang R, Feng L (2010) Preparation and characterization of ZrB_2–SiC ultra-high temperature ceramics by microwave sintering. Front

Mater Sci China 4(3):276–280
4. Zhu S, Fahrenholtz WG, Hilmas GE (2007) Influence of silicon carbide particle size on the microstructure and mechanical properties of ZrB_2–SiC ceramics. J Eur Ceram Soc 27(5):2077–2083
5. Sharma A, Karunakar DB (2019) Development and investigation of densification behavior of ZrB_2–SiC composites through microwave sintering. Mater Res Express 6:105072
6. Zhang SC, Hilmas GE, Fahrenholtz WG (2008) Pressureless sintering of ZrB_2–SiC ceramics. J Am Ceram Soc 91(1):26–32
7. Zimmermann JW, Hilmas GE, Fahrenholtz WG, Dinwiddie RB, Porter WD, Wang H (2008) Thermophysical properties of ZrB_2 and ZrB_2–SiC ceramics. J Am Ceram Soc 91(5):1405–1411
8. Wroe FCR, Rowley AT (1995) Microwave enhanced sintered of ceramics. Ceram Trans 59:69–76
9. Agrawal DK (2006) Microwave sintering of ceramics, composites, metallic materials and melting of glasses. Trans Indian Ceram Soc 65(3):129–144
10. Ghasali E, Yazdani RR, Rahbari A, Ebadzadeh T (2016) Microwave sintering of aluminum-ZrB_2 composite: focusing on microstructure and mechanical properties. Mater Res 19(4):765–769
11. Macaigne R, Marinel S, Goeuriot D, Saunier S (2018) Sintering paths and mechanisms of pure $MgAl_2O_4$ conventionally and microwave sintered. Ceram Int 44(17):21107–21113
12. Li Z, Bradt RC (1986) Thermal expansion of the hexagonal (6H) polytype of silicon carbide. J Am Ceram Soc 69(2):863–866
13. Zhu S, Fahrenholtz WG, Gregory EH, Zhang SC, Yadlowsky EJ, Keitz MD (2008) Microwave sintering of a ZrB_2–B_4C particulate ceramic composite. Compos Part A 39(3):449–453
14. Anstis GR, Chantikul P, Lawn BR, Marshall DB (1981) A critical evaluation of indentation techniques for measuring fracture toughness: I, direct crack measurements. J Am Ceram Soc 64(9):533–538

Characterization of Ni-Based Alloy Coating by Thermal Spraying Process

Manmeet Jha, Deepak Kumar, Pushpendra Singh, R. S. Walia, and Qasim Murtaza

Abstract The Ni- and Fe-based coatings have manifested their efficiency to get used in wear-related applications due to their better tribological properties. In the present study, the thermally flame-sprayed wear-resistant Ni–Cr–Fe–Mo coatings were deposited on SS304 substrate. The microstructure analysis of the deposited alloy has been studied with the help of scanning electron microscope (SEM). The coating has solidified successive molten or semi-molten splats, and the morphologies obtained due to deposition are layered. The coating obtained is the amalgamation of different phases such as amorphous, nano-crystalline grains and precipitates which further induces surface roughness. The mechanical tests were then applied for measuring the hardness and surface roughness. The results showed that the hardness and surface roughness were improved due to thermal flame-sprayed alloy coating.

Keywords Thermal spray · Flame spray · Alloy coating · Hardness · SEM

1 Introduction

The production of modern gas-turbine engines has been carried by the desire for greater efficiency and improved performance [1]. Increase in the efficiency and power of gas-turbine engines can be achieved by raising the operating temperature of the turbine. And raising the operating temperature can cause serious repercussions to the turbine blades [2]. Repercussions can be of any degree ranging from minor grain cracks to material fracture, softening of the material, surface roughness and inducement of porosity in the material which further causes oxidation and corrosion.

To overcome the mentioned problems, thermally spray coating process introduced, which proves to be efficient and productive, due to its ability to melt powder

M. Jha · D. Kumar (✉) · P. Singh · Q. Murtaza
Delhi Technological University, New Delhi, Delhi 110042, India
e-mail: deepak.kumar@dtu.ac.in

R. S. Walia
PEC University, Chandigarh 160012, India

© The Author(s), under exclusive license to Springer Nature Singapore Pte Ltd. 2021
R. M. Singari et al. (eds.), *Advances in Manufacturing and Industrial Engineering*, Lecture Notes in Mechanical Engineering,
https://doi.org/10.1007/978-981-15-8542-5_72

particles and deposit it on the substrate, with high velocity, to form a protective film. Of which, high-velocity oxygen-fuel sprayed wear-resistant MCrFeMo (M is a combination of nickel or cobalt or either of them) alloy coatings seems to be of peculiar importance in impeding the blazing hot parts such as static and rotary blades of the turbine from wear, elevated temperature oxidation and hot corrosion [3, 4]. It is well-approved that sprayed nickel-based alloy coatings exhibit better-elevated temperature impediment to corrosion. These nickel-based alloys also increase their refusal to wear after integrating Mo and W elements to the alloy [5].

Many researchers have shown their concomitant views of wear resistance and improving hardness ability of flame-spray used Ni-based coating. Bolelli et al. [6] compared the dry-sliding wear response by ball-on-disc test of HVOF and flame-sprayed Ni–32% Mo–15% Cr–3% Si against different counter bodies at different conditions. Results show that the loss of steel due to wear is very low after heat treatment as compared to preheat test [7]. Fe–Cr–Ni–B–Mo–C is used which is applied by HVOF, which further reaffirms the decrease in wear rate of the coating. A higher value of wear resistance put-forth Fe-based coatings as a potential alternative in future. Fe-based coatings are also economical and environmentally friendly. Ni et al. [8] made amorphous steel coating with constituents of $Fe_{48}Cr_{15}Mo_{14}C_{15}B_6Y_2$ by HVOF thermal spraying. It shows that the wear resistance and microhardness of the coating are worthier to those of electroplated Ni and Cr-based coating.

Apart from the wear resistance, some researchers also measured the hardness of the thermally sprayed coating [9]. Selection of coating and the materials of the valve components determine an important ingredient for the monetary success of gas and oil production activities. An alloy of intermetallic compounds combining Co, Cr, Ni and Mo is used with the help of flame-spray coating. The hardness of the coating was measured with the Vickers Hardness Test. Results show that hardness before corrosion test for Cr_3C_2–NiCr based coating is 928 $HV_{0.300/15}$ and after corrosion test is 905 $HV_{0.300/15}$ [10]. Ni–Mo–Cr–Si–B high-velocity oxy-fuel based coating on BS970 EN8 carbon steel plate have been assessed. The hardness of 612–692 $HV_{0.5}$ was obtained by Vickers Hardness method, which also shows the increase in the hardness of material as compared to the normal substrate. Therefore, in the present study, Ni–Cr–Fe–Mo composite coating on the stainless-steel substrate was applied with the help of thermally sprayed flame-spray process. The microstructure and various mechanical properties of the then material were probed by SEM and EDS processes.

2 Material and Method

2.1 *Substrate Preparation*

Stainless-steel SS304 is an austenitic stainless steel with a minimum of 8% nickel, 18% chromium, combined with a maximum of 0.08% carbon. It is also defined as

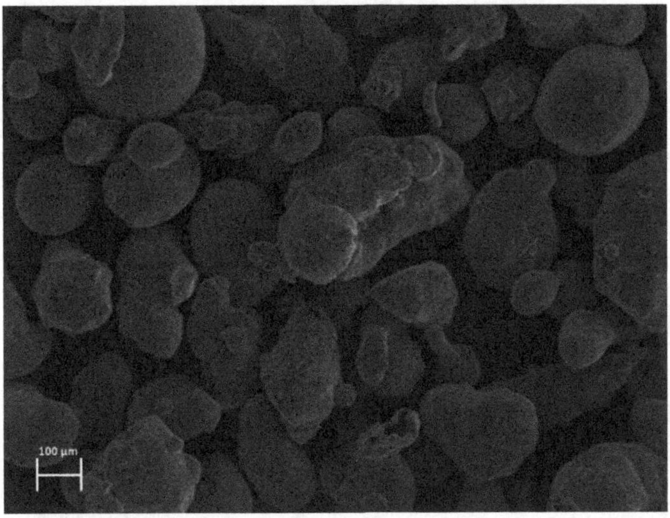

Fig. 1 SEM micro-image of Ni–Cr–Fe–Mo powder

nickel–chromium austenitic alloy taken as a base material for coating deposition. It exhibits good corrosion and oxidation resistance [11]. Against the backdrop of the flame-spray deposition process, the substrate SS304 is first preheated and sandblasted. Sandblasting is done to eliminate oxide to arouse the adherence of the layers of molten powder to be sprayed. The composition of powder includes Ni, Cr, Fe and Mo as shown in Fig. 1. Scanning electron microscope micrograph shows that Mo powder is in a circular shape, and other elements Ni, Cr and Fe are in an irregular shape. Amongst the spraying materials, Ni-based alloys are of significant use due to its greater resistance to wear, elevated temperature corrosion and oxidation [12].

2.2 *Coating Deposition*

Thermal spray is one of the most versatile methods used for coating materials, and its use in manufacturing and industrial application has been exponentially increased [13]. Thermal spraying is essentially a process in which coating powder material is exposed to a heating zone, where it gets melted and then it is propelled onto the surface to be coated [14]. There are numerous methods available for the propelling coating onto the substrate. Flame-spraying is one of the most commonly used amongst them. The schematic representation of the above-discussed flame-spray deposition process is shown in Fig. 2.

Powder-based flame spray is a thermal coating process which produces high quality and calibre surface coating using the thermal energy produced due to the combustion of fuel gas, acetylene in the presence of oxygen.

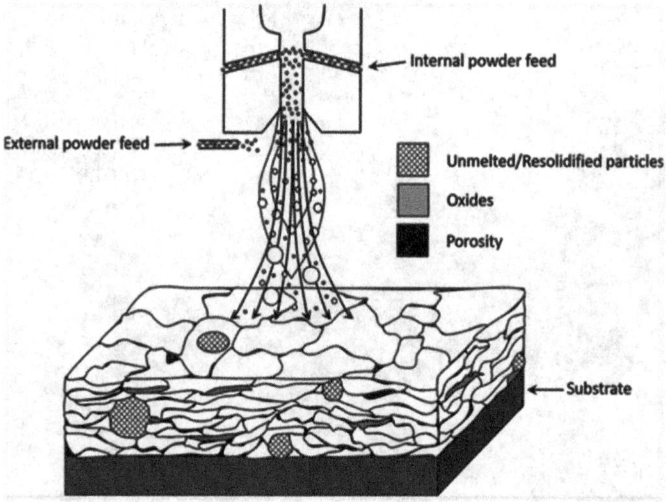

Fig. 2 Schematic representation of flame-spray coating deposition [15]

The combustion of gases takes place in the combustion chamber at a higher pressure of up-to 3 bars of oxygen and 2 bars of acetylene [16]. This heat is used to melt the coating material which then propelled onto the substrate to be coated. The various flame-spray systems differ in the process of powder injection, design and construction of the chamber of combustion and geometry of the nozzle [17]. Heating of particles and the acceleration takes place within the nozzle of the torch, and the free jet outside. Due to its economy and application in a wide range of materials, it is the most preferred one. The deposited desired coating on the stainless-steel substrate is shown in Fig. 3. The parameters used for the flame-spraying process are given in Table 1.

3 Result and Discussion

3.1 Morphological Analysis

Microstructural characterization of flame-spray coatings involves the assessment of geometrical idiosyncrasy, and these geometrical features include porosity (in the form of cracks, voids and certain other defects) and investigation of material aspects of coatings. Material aspects include splat structures, phases, etc. [18].

The microstructure of coated substrate was studied with the help of scanning electron microscope, and micrograph of the coated surface of the specimen is shown in Fig. 4. During the thermal flame-sprayed coating, splats of melted alloy powder are formed on the surface of the SS substrate. These splats then cool at the surface

Fig. 3 Coated substrate

Table 1 Process parameter of coating deposition

Parameter	Range
Nozzle	P7C-M1
Oxygen pressure	2.2 bar
Acetylene pressure	1.0 bar
Oxygen flowmeter reading	45.0
Acetylene flowmeter reading	55.0
Spray distance	205 mm
Spray rate	152 Gm/min

forming adhesion and cohesion contact bonds with the substrate surface. And, the laminar configuration of structure is obtained. The entire process is associated with very low heat transfer to the substrate, and mechanical interlocking is the main adhesion mechanism.

3.2 *Mechanical Properties Analysis*

Microhardness

Hardness is the resistance of a material to penetration, indentation and deformation by means such as drilling, abrasion, impact scratching and/or wear. It is measured by hardness tests such as Rockwell, Brinell, Knoop or Vickers.

Fig. 4 SEM micro-image of Ni–Cr–Fe–Mo coating

In the present study, the microhardness is measured by the operating principle of the instrumented indentation test according to the DIN EN IDO 14577 standard. The instrument used is Fischer indenter as shown in Fig. 5. In the applied instrumentation indentation test, an indenter penetrates the specimen's surface using the load 300 mN/20 s. During the process, the depth of the indentation from the surface is continuously measured. The microhardness is measured at the peak value of force and depth.

$$\mathrm{hm} = \mathrm{hs} + \mathrm{hc} \tag{1}$$

where hs is the displacement of the surface at the contour of the contact and hc is the vertical distance along which the contact is made. hm is the depth at peak load as mentioned in Eq. 1. The hardness is expressed by the ratio between the applied load and the contact area as expressed in Eq. 2. In theory:

$$H = \frac{\mathrm{Fm}}{\mathrm{Ap}} \tag{2}$$

where Fm is the maximum load and Ap is the contact between the specimen and the indenter at the maximum depth by the load as expressed in Eq. 3.

$$Ap = C_0 hc^2 \tag{3}$$

where hc is the contact depth (see Fig. 6) and C_0 depends upon the indenter geometry ($C_0 = 24.5$ for a Berkovich and 6.3 for a cube-corner indenter). The microhardness obtained for the substrate is 281.51 HV and after the propulsion and measuring the indentation at 10 different points it comes out to be 550 HV. This shows that the microhardness of the as-sprayed coating was way greater than as in normal substrate. This is also supported by Bergant and Grum [19]. Bergant

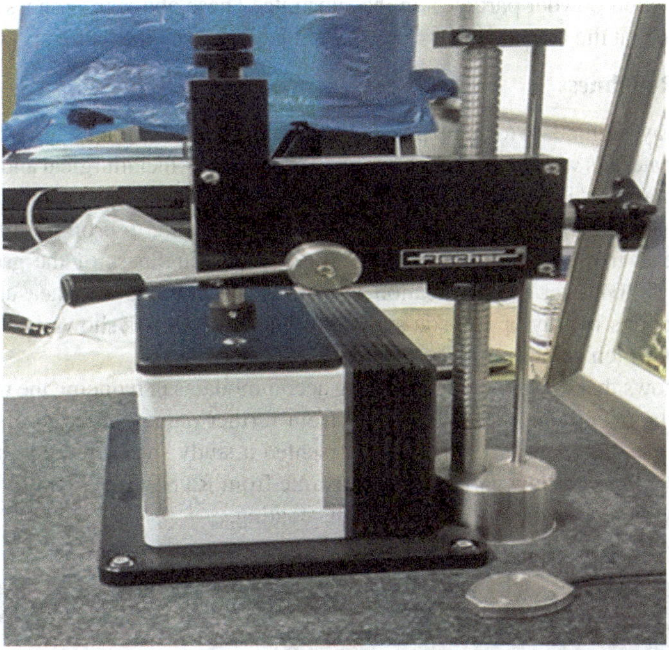

Fig. 5 Fischer indenter

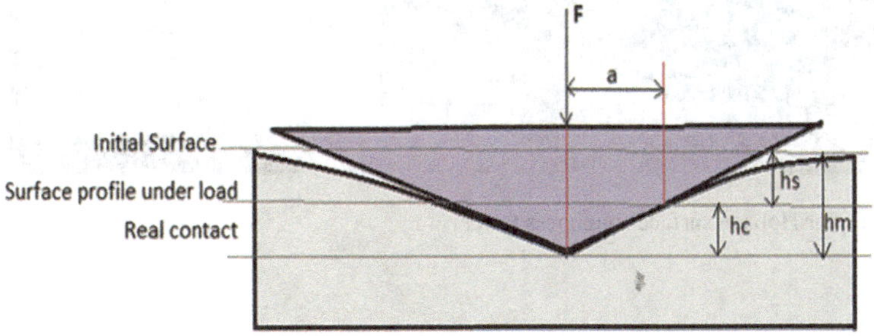

Fig. 6 A schematic representation of the indentation process showing the decreasing of indentation depth during loading (according to Oliver and Pharr)

and Grum [19] studied the hardness of NiCrBSi coatings on cylindrical specimen substrate prepared from structured of carbon steel, whose hardness was between 200 and 240 HV. Coating hardness was measured using Vickers hardness test method, with load of 3 N on a Leitz Wetzlar gauge. The value of coating hardness obtained at 930 °C was 450 $HV_{0.3}$. Further, it was found that the hardness at the substrate-coating interface was higher than that of the hardness in the bulk of coating. Redjdal et al. [20] asserted that the hardness at the substrate-coating interface was 350 HV which is way far than that from the bulk of coating, i.e. 250 HV. The bridging is due to a shot-peening effect resulted due to conundrum blasting and the smashing of the molten or semi-molten powder particles on the substrate. These above inferences show the conclusion that the microhardness is definitely increased by the flame-spray coating.

Surface Roughness

Flame-spray is a method to dump powder on hitherto treated surfaces, non-metallic or metallic materials whose main gluing agency is chemical–metallurgical and mechanical anchorage [21]. The grooming of the substrate SS304 comprises cleansing, the build-out of an irregular surface and sometimes preheating to promise mechanical anchorage at micro-welding sites. In our study, the roughness of the grit-blasted substrate SS304 after the application of NiCrFeMo coating was measured at 10 different points by Taylor Hobson surface roughness tester as shown in Fig. 7, the average of which comes out to be 6.8 μm.

This shows that the flame-spray coating accommodates in reducing the roughness of the surface which in turn can prevent from further damages due to undesirable high roughness. Milanti et al. [7] also presented a study showing the reduction of surface roughness of low carbon steel substrate from Ra 8.3 ± 0.9 μm to Ra 7.1 ± 0.5 μm by the application of FeNiCrMoBC coating.

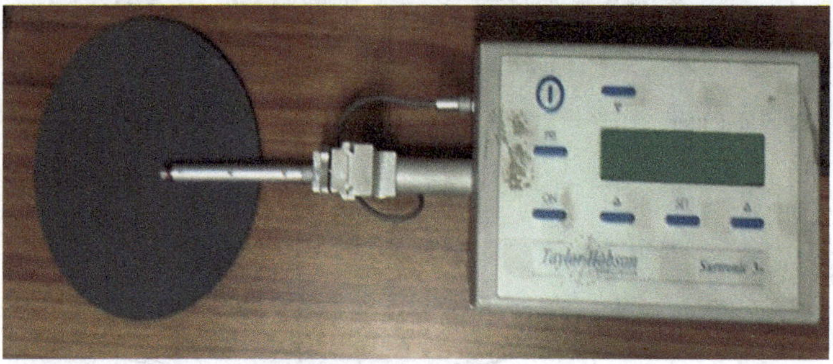

Fig. 7 Taylor Hobson surface roughness tester

4 Conclusion

The flame-spray coating technique was used for the successful deposition of NiCr-FeMo coating on the SS304 substrate material. The morphological analysis of deposited coating was studied with the help of a scanning electron microscope. From which it can be deciphered that Ni-based coatings show better resistance to wear and high-temperature oxidation and corrosion. Coating of high microhardness value 550 HV has been achieved which is far better than the substrate, for which it was 281.51 HV. Further, the surface roughness is also decreased to 6.8 μm due to high temperature and high homogeneity of the thermal-based coating. The SEM analysis of the flame-sprayed coating exhibited a dense, free of snags or deficiencies and well-bounded coating. Moreover, mechanical and microstructural properties of the SS304 substrate were not damaged and remain intact after the flame-spraying thermal process.

References

1. Gurrappa I (2001) Identification of hot corrosion resistant MCrAlY based bond coatings for gas turbine engine applications. Surf Coat Technol 139:272–283
2. Leo CV, Luk-Cyr J, Liu H, Loeffel K, Al-Athel K, Anand L (2014) A new methodology for characterizing traction-separation relations for interfacial delamination of thermal barrier coatings. Acta Mater 71:306–318
3. Vilar R, Santos EC, Ferreira PN, Franco N, Silva RC (2009) Structure of NiCrAlY coatings deposited on single-crystal alloy turbine blade material by laser cladding. Acta Mater 57:5292–5302
4. Tahari M, Shamanian M, Salehi M (2012) Microstructural and morphological evaluation of MCrAlY/YSZ composite produced by mechanical alloying method. J Alloys Compd 525:44–52
5. Tan JC, Looney L, Hashmi MSJ (1999) Component repair using HVOF thermal spraying. J Mater Process Technol 92:203–208
6. Bolelli G, Cannillo V, Lusvarghi L, Montorsi M, Mantini FP, Barletta M (2007) Microstructural and tribological comparison of HVOF-sprayed and post-treated M-Mo–Cr–Si (M=Co, Ni) alloy coatings. Wear 263(7–12):1397–1416
7. Milanti A, Koivuluoto H, Vuoristo P, Bolelli G, Bozza F, Lusvarghi L (2014) Microstructural characteristics and tribological behavior of HVOF-sprayed novel Fe-based alloy coatings. Coatings 4(1):98–120
8. Ni HS, Liu XH, Chang XC, Hou WL, Liu W, Wang JQ (2009) High performance amorphous steel coating prepared by HVOF thermal spraying. J Alloy Compd 467(1–2):163–167
9. Scrivani A, Ianelli S, Rossi A, Groppetti R, Casadei F, Rizzi G (2001) A contribution to the surface analysis and characterisation of HVOF coatings for petrochemical application. Wear 250(1–12):107–113
10. Shrestha S, Hodgkiess T, Neville A (2001) The effect of post-treatment of a high-velocity oxy-fuel Ni-Cr-Mo-Si-B coating. Part I: microstructure/corrosion behavior relationships. J Therm Spray Technol 10(3):470–479
11. Singh SK, Chattopadhyaya S, Pramanik A, Kumar S (2018) Wear behavior of chromium nitride coating in dry condition at lower sliding velocity and load. Int J Adv Manuf Technol 96(5–8):1665–1675

12. Fernández E, Cadenas M, González R, Navas C, Fernández R, Damborenea J (2005) Wear behaviour of laser clad NiCrBSi coating. Wear 259(7–12):870–875
13. Stachowiak G, Batchelor AW (2013) Engineering tribology. Butterworth-Heinemann, Oxford
14. Bhushan B (1999) Principles and applications of tribology. Wiley, Hoboken
15. Davis JR (2004) Introduction to thermal spray processing. In: Handbook of thermal spray technology, pp 3–13
16. Singh SK, Chattopadhyaya S, Pramanik A, Kumar S, Basak AK (2019) Effect of lubrication on the wear behaviour of CrN coating deposited by PVD process. Int J Surf Sci Eng 13(1):60–78
17. Praveen AS, Sarangan J, Suresh S, Channabasappa BH (2016) Optimization and erosion wear the response of NiCrSiB/WC-Co HVOF coating using Taguchi method. Ceram Int 46:1094–1104
18. Deshpande S, Kulkarni A, Sampath S, Herman H (2004) Application of image analysis for characterization of porosity in thermal spray coatings and correlation with small angle neutron scattering. Surf Coat Technol 187:6–16
19. Bergant Z, Grum J (2009) Quality improvement of flame sprayed, heat treated, and remelted NiCrBSi coatings. J Therm Spray Technol 18(3):380–391
20. Redjdal O, Zaid B, Tabti MS, Henda K, Lacaze PC (2013) Characterization of thermal flame sprayed coatings prepared from FeCr mechanically milled powder. J Mater Process Technol 213(5):779–790
21. Culha O, Celik E, Azem NA, Birlik I, Toparli M, Turk A (2008) Microstructural, thermal and mechanical properties of HVOF sprayed Ni–Al-based bond coatings on stainless steel substrate. J Mater Process Technol 204(1–3):221–230

A Review on Solar Panel Cleaning Through Chemical Self-cleaning Method

Ashish Jaswal and Manoj Kumar Sinha

Abstract In last few years, the global coating industries and scientific have introduced superhydrophobic coating with high water repellency. Photovoltaic (PV) panels installation in the dusty regions results in the reduction of its power output because the soil deposition on it resists the conversion of light into power. Thus, in order to prevent the above problem from happening, PV panels must be cleaned in a proper interval of time. However, the cleaning of the solar panel manually is a very lethargic and time-wasting task, and in addition, this cleaning technique can break the PV substrate due to poor brushing which results in unfavourable production of power. Moreover, manual cleaning does not give the best shot in cleaning very small dust particulates. In this manner, analysts around the world are advancing the self-cleaning strategies, viz. electrostatic strategy, mechanical strategy and coating strategy for PV surface cleaning. This paper focuses on the cleaning of a solar PV panel using a superhydrophobic coating.

Keywords Nanoparticle · Superhydrophobic · Contact angle

1 Introduction

In this era, everyone is very much familiar with the environmental pollution which is caused by coal and petroleum industries. Therefore, for diminishing the number of gases like sulphur dioxide (SO_2), carbon dioxide (CO_2), nitrous oxide (NO_2), carbon monoxide (CO), etc., which is discharged by the businesses into the air. Now, companies are heading towards renewable energy sources for power generation because these resources are independent of moving items, and due to these, productions of both air particles and noise intensity do not take place. It is eco-friendly, adaptable in different zones of the environment and geology, and does not require immense but or maybe low-cost support [1]. But one major problem with them is that their surface gets covered with dust after a few days, which should be removed otherwise, it may

A. Jaswal · M. K. Sinha (✉)
National Institute of Technology Hamirpur, Hamirpur 177005, India
e-mail: mksinha@nith.ac.in

© The Author(s), under exclusive license to Springer Nature Singapore Pte Ltd. 2021

R. M. Singari et al. (eds.), *Advances in Manufacturing and Industrial Engineering*,
Lecture Notes in Mechanical Engineering,
https://doi.org/10.1007/978-981-15-8542-5_73

produce small amount of energy from the PV cell. Places with abundant rain (e.g. Japan, India) clean the dust automatically which gathers on the panel. In fashionable, the PV module is operated in an open environment, wherein it stories a considerable variant in environmental parameters, like wind speed, ambient temperature, solar irradiance, humidity and dirt pollution [2]. Those natural parameters have an impact on the overall performance of the PV board, and among those parameters, dust plays a gigantic part in bringing down the overall performance of the PV module. Atmospheric dust is caused by diverse sources, such as soil erosion due to high-speed wind, volcanic activity, automobile activity and pollutants [3]. Size of dirt particles, its shape and elements vary from place to place. In addition, the intensity of the dirt removal rate depends on the surrounding location. But what about those districts where rain barely happens like dry, deserts and semi-arid districts [4].

It has been found tentatively that utilizing nanocoating on PV panels gives an impressive improvement in light transmission and voltage which is not connected to any load in a circuit because it understands the issue of dust electrical losses [5]. Moreover, the cleaning cost of the solar panel depends on a number of repeated cycles. Therefore, cleaning cost by conventional methods is about 2.25 Euros/m^2/year. Whereas, cleaning cost by nanocoating is about 1.89 Euros/m^2/year which is equivalent to 18.900 Euros/MW/year. This shows that the production of nanocoated material is beneficial [6]. In spite of the heavy rains for a few months, decrease in transmission of light was found for uncovered glass samples. So, nanocoated films of various thicknesses were deposited on the glass so that their cleaning performance gets improved. Higher the thickness of the coating, the lower will be the reduction in transmission [7]. Providing nanocoating to PV panels after studying microscopic and macroscopic dust deposition patterns, its density and spectral transmittance give better performance for large tilt angles. As superhydrophobic surfaces have low attachment vitality hence, the deposition of huge clean particles (>20 μm) gets to be difficult [8]. This nanocoating will be resistant to numerous natural components such as scraped area cycle, corrosive rain, saline introduction, antacid arrangement and extraordinary temperature cycling [9].

Another reason that influences the PV cell performance is wind. If the wind speed is greater than 3 m/s, then it causes sand and dust to rise in the dusty regions, which blocks the solar radiation to fall on the solar cell and hence reduces its productivity. If the solar panel is faced in the direction of the wind, then it will remove some of the gathered particles on the uppermost layer, but in addition to that, blowing wind is also carrying very small diameter dust particles which will stick on the uppermost layer of the cell and causes a reduction in the power output [10].

2 Various Techniques for Cleaning Solar Panel

2.1 *Natural Cleaning*

As the name indicates, this technique is based on the combined positions of solar panel and wind speed and the number of rainfalls. But this technique gives better performance for small installations of the solar panel [6].

2.2 *Manual Cleaning*

Manual cleaning can be done, but it will take a lot of time and additionally make fracture on the PV panel surface due to rough brushing. The manual cleaning of the PV panel applies unequal pressure. With too high pressure, the board can be harmed most exceedingly bad cases. On the other hand, in case the weight is too low, the surface will not be cleaned completely. Moreover, the PV panels are at a certain height which makes the cleaner difficult to reach its sides and corners. So, it is not wise to clean the PV panel by hand [4].

2.3 *Mechanical*

This cleaning technique includes robots, manpower, brushes, etc., but these machines or robots are robust. Moreover, they need customization and a lot of construction. Another one is sprinkler-based robots, but the major drawback is that the user needs a large amount of water [1].

Brushing: This mechanical technique is used for small installations but increases energy utilization. This strategy cannot be utilized where dust intensity is exceptionally high because maintenance is required for the module, and also, there is a chance of ruining of PV module [6].

Ultrasonic Vibration: Nowadays, this technique is being used in cleaning the solar panel. In this technique, vibrations in the range of 80–100 kHz are supplied to the PV panel with the help of transducer so that the dust or dirt stick with the surface can be removed. In other words, a piezoelectric effect is used to provide an ultrasonic self-cleaning PV panel [6].

2.4 *Electrical*

The next one is an electrodynamic shield (EDS) which has the ability to get rid of surface dust on the solar panel. In this, an electrodynamic wave is generated due to which dust erosion from the surface takes place [11]. The parallel electrode which is mounted on the PV panel surface generates electrodynamic waves inside the adhesive material and below the dielectric sheet. EDS can use the PV panel power, which will be in a negligible amount [12]. Factors affecting the EDS potential of cleaning an amount of dust from PV panel are EDS electrical operation, properties of electrode, properties of electrically insulating material, properties of dust and circumstances of environment [13].

2.5 *Chemical Self-cleaning*

Super-Hydrophilic Coating: In this, titanium dioxide (TiO_2), as a coating material, is used for making the hydrophilic surface. It destroys organic dust under the photocatalytic effect using ultraviolet (UV) energy and a small quantity of water [6].

Superhydrophobic Coating: In this technique, lotus effect is used in which a water drop takes off a large amount of dust from the surface. Also, the contact angle of water reaches up to 150° [6]. This coating will respond under the ultraviolet (UV) light and split the organic dirt. It has been found that modules coated with the films are having an average daily loss of 2.5%, but on the other hand, modules without coating reach a loss of 3.3% [1].

3 Properties of Superhydrophobic Materials

3.1 *Wettability*

The property of wettability is similar to adsorption phenomena because these both depend upon the interactivity between molecules of different substances. Wettability can be defined as the phenomena in which the adhesion of the liquid is stronger as compared to the cohesion on the uppermost layer of the substrate and this possible only when there is a decrease in free energy of the system [14]. But in the case of water repellent surfaces, the sticking of the liquid molecules on the surface to which they are exposed is weaker than the cohesion of liquid molecules. As a result, liquid drops will not stick to the uppermost layer of the glass or on a solid nonmetal surface [15].

3.2 *Contact Angle*

When the boundary of liquid and vapour meets together, then the angle which is measured through the liquid is called contact angle. It helps in determining the surface tension and surface energy. In terms of contact angle, the superhydrophobic surface can be defined as if the liquid contact angle is between 150° and 180°, and if the water contact angle is between 0° to 90° and 90° to 150°, then the surface is known as hydrophilic and hydrophobic surface, respectively. Tensiometer is a device with the help of which a contact angle of liquid can be measured [15].

3.3 Surface Energy and Surface Tension

Surface tension and surface energy, both are interrelated phenomena. It is well known that the molecules at the centre of the liquid are pulled uniformly towards every direction by other molecules, but this not the case with molecules at the surface of the liquid. Therefore, surface energy is also known as surface tension of a substance. So, from the above concept, it has been found that there is a high energy density at the surface of a liquid but the superhydrophobic surfaces tend to reduce the high energy and the liquid droplet will take the spherical shape, having minimum surface area [15].

3.4 Surface Roughness

There are two different states in which a water droplet can exist when it has a contact with a rough surface. These two models were developed by one Wenzel and other by Cassie and Baxter so that the superhydrophobic effect can be explained with the help of surface roughness. Inside the first model (Wenzel model), there are not any air bubbles beneath the droplet, and the droplet is in absolute touch with the surface. The droplet sticks very well to the surface, it is referred to as a pinned droplet, and as a result, excessive contact angle hysteresis is discovered. In the second version (Cassie–Baxter model), the water droplet sits on the peak of tiny air bubbles. In this model, water droplets will roll off with a purpose to be useful for water repellent and self-cleaning surfaces. A surface can be self-cleaned because any water droplets that touch it will roll off, selecting up and dirt alongside the manner [15].

4 Effect of Different Dust Pollutant on the Performance of Photovoltaic

Efficiency gets reduced when unlike dust pollutants settle on the surface of PV panels, and different geographical sites consist of different types of dust pollutants. Dirt may contain little amount of dust and additionally organisms, green plants, nanofibers, and, maximum generally, natural minerals including sand clay and wore away the

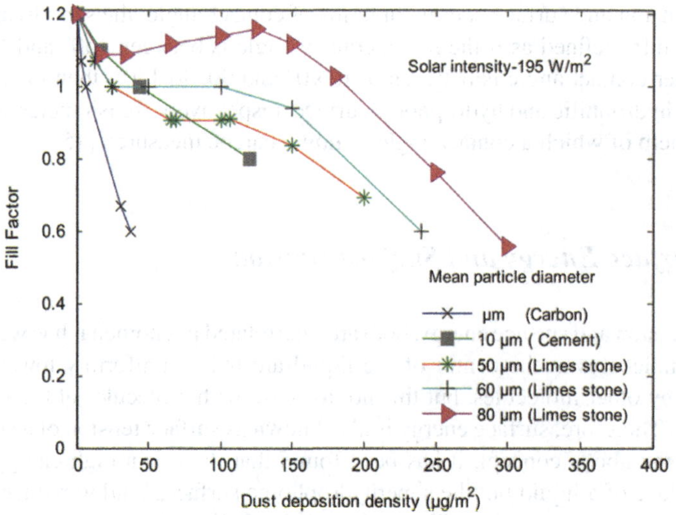

Fig. 1 Showing that fill factor is decreasing with dust collection [18]

limestone [16]. For instance, limestone is worked from (rain, snow, etc.) of (silvery metal) $CaCO_3$, ash is given off from vehicle exhaust, whereas red soil is migrated from African deserts [17].

The impact of polluted air is extreme in cities due to the excessive population density and increases inside the commercial sports, particularly dirt and debris which are created with the help of the combustion of fossil fuel sand creation exercises. Fourteen compounds were found mainly within the dust samples by the researchers with different proportions depending on the area [18] (Fig. 1).

El-Shobokshy and Hussein classified the dust type in terms of their size of those, three had been limestone primarily-based, ground into three unique grades. Cement turned black due to its presence in as a fundamental constructing substances, and it is largely present in the air in maximum populated regions [19, 20] (Fig. 2).

5 The Durability of Superhydrophobic Coating

Upgradation of the durability of superhydrophobic surface is an important point for materialistic and mechanical applications. Thus, the application of water repellent surface in viable utilization is constrained because of their bad mechanical scraped area resistance and unfriendly natural circumstances, in any case, a maximum number of superhydrophobic coating got by the larger part of strategies have limited scraped area resistance. Covalent bonding of the nanoparticles to the adhesive substance was recognized as fundamental for durability improvement. The scratched spot resistance

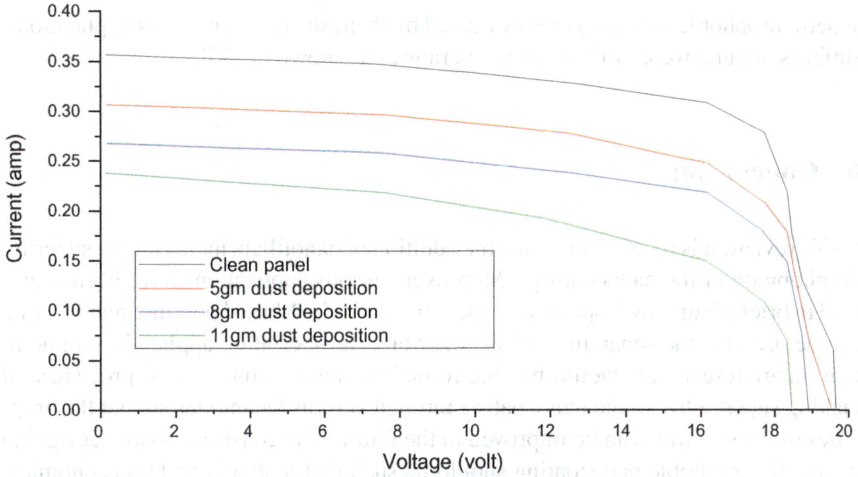

Fig. 2 *I–V* characteristic of the PV module with limestone dust deposition [2]

is chosen by analysing the change in a stable contact point, contact angle hysteresis and the coefficient of outer boundary contact [21–30] (Fig. 3).

Which coated on a glass substrate and found that, that it appears broadband anti-reflection, and the most extreme transmittance comes as large as 97% at the wavelength of 816 nm [31]. The coated glass material transmittance is more important than 95.0% under the wavelength span of 530–1340 nm, in differentiate to that (91.0%) of the uncovered glass substrate [32–43]. Additionally, the thermal stability of the

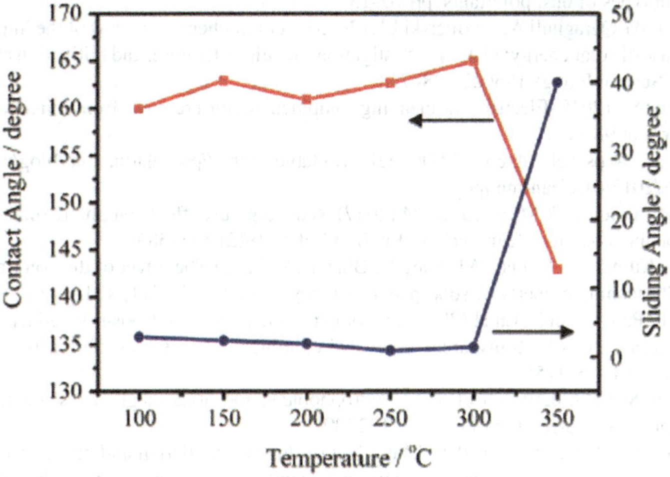

Fig. 3 Contact and sliding angle of static water sample annealed for 2 h at varying temperature [30]

superhydrophobic subtle layer is evaluated by them since, in numerous applications, surfaces are uncovered to hoisted temperatures as shown in [30].

6 Conclusion

In this review, it is recognizable that the addition of nanofillers increases the superhydrophobicity of the nanocoatings. Moreover, various applications have been examined in brief of superhydrophobic nanocoating, and ZnO-based polymer nanocoating can be used for the separation of oil/water and transportation application. In addition, many researchers would be able to understand the concept and procedure of making superhydrophobic nanocoating through various techniques so that the properties of the substrate can be improved in the future. The construction and designing of superhydrophobic nanocoating should be such that it should be less economical and large-scale production can be obtained. Hence, it can be stated that a huge amount of development could be done in this field which could be profit-oriented and lead to mass production.

References

1. Al-Housani M, Bicer Y, Koç M (2019) Assessment of various dry photovoltaic cleaning techniques and frequencies on the power output of CdTe-type modules in dusty environments. Sustainability 11(10)
2. Tripathi AK, Aruna M, Murthy CSN (2018) Performance degradation of PV module due to different types of dust pollutants, pp 10–13
3. Sarver T, Al-Qaraghuli A, Kazmerski LL (2013) A comprehensive review of the impact of dust on the use of solar energy: history, investigations, results, literature, and mitigation approaches. Renew Sustain Energy Rev 22:698–733
4. Kawamoto H (2019) Electrostatic cleaning equipment for dust removal from soiled solar panels. J Electrostat 98:11–16
5. "clean-me @ pscsolaruk.com." [Online]. Available: https://pscsolaruk.com/blog/wp-content/uploads/2017/01/clean-me.jpeg
6. Fathi M, Abderrezek M, Friedrich M (2017) Reducing dust effects on photovoltaic panels by hydrophobic coating. Clean Technol Environ Policy 19(2):577–585
7. Hee JY, Kumar LV, Danner AJ, Yang H, Bhatia CS (2012) The effect of dust on transmission and self-cleaning property of solar panels. Energy Procedia 15(2011):421–427
8. Zhang L, Pan A, Cai R, Lu H (2019) Indoor experiments of dust deposition reduction on solar cell covering glass by transparent super-hydrophobic coating with different tilt angles. Sol Energy 188:1146–1155
9. Maharjan S et al (2020) Self-cleaning hydrophobic nanocoating on glass: a scalable manufacturing process. Mater Chem Phys 239:122000
10. Kazem HA, Chaichan MT (2019) The effect of dust accumulation and cleaning methods on PV panels' outcomes based on an experimental study of six locations in Northern Oman. Sol Energy 187:30–38
11. "how-to-clean-solar-panels @ ecotality.com." [Online]. Available: https://ecotality.com/wp-content/uploads/2019/01/how-to-clean-solar-panels.jpg

12. Panel_Cleaning_Robot_-_CSS @ www.solarpowerportal.co.uk [Online]. Available: https://www.solarpowerportal.co.uk/files/images/Panel_Cleaning_Robot_-_CSS.jpg
13. Chesnutt JKW, Ashkanani H, Guo B, Wu C (2017) Simulation of microscale particle interactions for optimization of an electrodynamic dust shield to clean desert dust from solar panels. Sol Energy 155:1197–1207
14. 0618ctt-self-cleaning-test-lores @ ceramics.org. [Online]. Available: https://ceramics.org/wp-content/uploads/2014/06/0618ctt-self-cleaning-test-lores.jpg
15. Nguyen-Tri P et al (2019) Recent progress in the preparation, properties and applications of superhydrophobic nano-based coatings and surfaces: a review. Prog Org Coatings 132:235–256
16. Mishra A, Rathi V (2017) Super hydrophobic antireflective coating to enhance efficiency of solar PV cells. IJEREEE 3(12):29–34
17. Kaldellis JK, Kapsali M (2011) Simulating the dust effect on the energy performance of photovoltaic generators based on experimental measurements. Energy 36(8):5154–5161
18. Darwish ZA, Kazem HA, Sopian K, Al-Goul MA, Alawadhi H (2015) Effect of dust pollutant type on photovoltaic performance. Renew Sustain Energy Rev 41:735–744
19. El-Shobokshy MS, Hussein FM (1993) Effect of dust with different physical properties on the performance of photovoltaic cells. Sol Energy 51(6):505–511
20. Rajput DS, Sudhakar K (2013) Effect of dust on the performance of solar PV panel. Int J Chem Tech Res 5(2):1083–1086
21. Behniafar H, Haghighat S (2008) Thermally stable and organosoluble one-pot preparation and characterization. Polym Adv Technol 19:1040–1047
22. Bhushan B, Jung YC (2011) Natural and biomimetic artificial surfaces for superhydrophobicity, self-cleaning, low adhesion, and drag reduction. Prog Mater Sci 56(1):1–108
23. She Z, Li Q, Wang Z, Li L, Chen F, Zhou J (2013) Researching the fabrication of anticorrosion superhydrophobic surface on magnesium alloy and its mechanical stability and durability. Chem Eng J 228:415–424
24. Chen Y, Chen S, Yu F, Sun W, Zhu H, Yin Y (2009) Fabrication and anti-corrosion property of superhydrophobic hybrid film on copper surface and its formation mechanism. Surf Interface Anal 41(11):872–877
25. Farhadi S, Farzaneh M, Kulinich SA (2011) Anti-icing performance of superhydrophobic surfaces. Appl Surf Sci 257(14):6264–6269
26. Kulinich SA, Farhadi S, Nose K, Du XW (2011) Superhydrophobic surfaces: are they really ice-repellent? Langmuir 27(1):25–29
27. Dotan A, Dodiuk H, Laforte C, Kenig S (2009) The relationship between water wetting and ice adhesion. J Adhes Sci Technol 23(15):1907–1915
28. Arianpour, F (2010) Water and ice-repellent properties of nanocomposite coatings based on silicone rubber. Université du Québec à Chicoutimi
29. Davis AMJ, Lauga E (2010) Hydrodynamic friction of fakir-like superhydrophobic surfaces. J Fluid Mech 661:402–411
30. Cohen N, Dotan A, Dodiuk H, Kenig S (2016) Superhydrophobic coatings and their durability. Mater Manuf Process 31(9):1143–1155
31. Milionis A, Loth E, Bayer IS (2016) Recent advances in the mechanical durability of superhydrophobic materials. Adv Colloid Interface Sci 229:57–79
32. Xu L, He J (2013) A novel precursor-derived one-step growth approach to fabrication of highly antireflective, mechanically robust and self-healing nanoporous silica thin films. J Mater Chem C 1(31):4655–4662
33. Liu LQ et al (2012) Broadband and omnidirectional, nearly zero reflective photovoltaic glass. Adv Mater 24(47):6318–6322
34. Deng X, Mammen L, Butt HJ, Vollmer D (2012) Candle soot as a template for a transparent robust superamphiphobic coating. Science 335(6064):67–70
35. Janssen RAJ, Hummelen JC, Sariciftci NS (2005) Bulk etherojunction polymer–fullerene solar cells. MRS Bull 30:33–36
36. Deng X et al (2011) Transparent, thermally stable and mechanically robust superhydrophobic surfaces made from porous silica capsules. Adv Mater 23(26):2962–2965

37. Geng Z, He J, Xu L, Yao L (2013) Rational design and elaborate construction of surface nano-structures toward highly antireflective superamphiphobic coatings. J Mater Chem A 1(31):8721–8724
38. Xu L, Geng Z, He J, Zhou G (2014) Mechanically robust, thermally stable, broadband antireflective, and superhydrophobic thin films on glass substrates. ACS Appl Mater Interfaces 6(12):9029–9035
39. Simpson JT, Hunter SR, Aytug T (2015) Superhydrophobic materials and coatings: a review. Rep Prog Phys 78(8)
40. Superhydrophobic-behavior-Water-droplets-1mm-in-diameter-on-treated-surfaces-display @ www.researchgate.net
41. Barati Darband G, Aliofkhazraei M, Khorsand S, Sokhanvar S, Kaboli A (2018) Science and engineering of superhydrophobic surfaces: review of corrosion resistance, chemical and mechanical stability. Arab J Chem
42. Syafiq A, Pandey AK, Adzman NN, Rahim NA (2018) Advances in approaches and methods for self-cleaning of solar photovoltaic panels. Sol Energy 162:597–619
43. Das S, Kumar S, Samal SK, Mohanty S, Nayak SK (2018) A review on superhydrophobic polymer nanocoatings: recent development and applications. Ind Eng Chem Res 57(8):2727–2745

Investigations on Process Parameters of Wire Arc Additive Manufacturing (WAAM): A Review

Mayank Chaurasia and Manoj Kumar Sinha

Abstract Wire arc additive manufacturing (WAAM) is an imperative method for fabricating 3D metallic parts. By the large, additive manufacturing (AM) innovation is utilized to prevail over the restriction of conventional subtractive manufacturing (SM) for manufacturing larger parts with a low buy-to-fly ratio. There are mainly three heat sources utilized in WAAM: metal inert gas welding (MIG), tungsten inert gas welding (TIG) and plasma arc welding (PAW). WAAM is picking up a reputation for the manufacturing of 3D parts using metal as raw material but the method is difficult to control because of its implicit residual stress, low hardness, discontinuous deposition, roughness, low grain growth and distortion. These problems create major issues for WAAM since they affect the part's geometric precision and extremely demean the properties of manufacturing parts. In this review paper, the WAAM process for the manufacturing of 3D metal parts along with its process parameters and their effect is reviewed.

Keywords Additive manufacturing (AM) · Wire arc additive manufacturing (WAAM) · Metal inert gas (MIG)

1 Introduction

Additive manufacturing (AM) is a method used in the creation of 3D parts with layer-by-layer material deposition [1]. AM was earlier known as rapid prototyping (RP), rapid tooling (RT) and layered manufacturing (LM). AM may be able to fabricate useful items directly from computer-aided design (CAD) data [2]. AM is developed in three phases. First phase is manual prototyping where prototypes are not very sophisticated, and level of complexity is simple. Second stage is soft or virtual prototyping begun between 1975 and 1980. In this stage, computerized models can be stressed, modelled and tested by a computer and the complexity of items is twice as complex as in the past stage [3]. The third phase is rapid prototyping, which

M. Chaurasia · M. K. Sinha (✉)
National Institute of Technology Hamirpur, Hamirpur 177005, India
e-mail: mksinha@nith.ac.in

© The Author(s), under exclusive license to Springer Nature Singapore Pte Ltd. 2021
R. M. Singari et al. (eds.), *Advances in Manufacturing and Industrial Engineering*, Lecture Notes in Mechanical Engineering,
https://doi.org/10.1007/978-981-15-8542-5_74

started in the mid-1980s, which results in reducing the manufacturing time by using the hard prototype and increase the complexity almost three times higher than the second phase [3].

Despite the fact that there is a couple of variability inside the mechanical behaviour of AM parts which are routinely demonstrated by the heterogeneity and anisotropy developing from the variety in heat supplied [4]. The primary advantages of AM over the traditional conventional methods are high precision, faster manufacturing rate, and reduction in material waste [5]. Most of the added substances utilized were plastic, polymer and ceramic but presently nowadays metals are also being used as raw material in various AM processes. The usage of manufacturing technologies is subordinate to the raw material status during 3D metal printing.

2 Metal Additive Manufacturing

Metal additive manufacturing is known as a deposition of liquid metal in layers. It may be a direct feeding process including combining electrical sources, a movement framework and feedstock based on the wire. It is the improvised form of the fused deposition modelling (FDM) technique. On the basis of raw materials used, metal AM is classified into wire-based, powder-based and sheet-based processes. Among these technologies, the wire-based method leads in terms of structural efficiency and deposition rates. Wire-based methods also suitable for continuous and uncluttered material flow. Hence, wire-based AM is the most suitable for the production of costly components [6]. This technology can proficiently deliver large-scale metal components. However, with regard to accuracy and surface roughness, it is lower than other AM methods [7, 8]. Uses of metal additive manufacturing in recent times increases because of its ability to deliver metal parts at less cost and low buy-to-fly ratio. On the basis of the power source used, metal AM processes can be also classified as:

2.1 Electron Beam Additive Manufacturing (EBAM)

In this method large size, complex and intricate parts are possible to fabricate by an electron beam which uses as a heat source. Also, the use of vacuum in this technique allows the easy deposition of reactive metals like titanium otherwise for the fabrication of reactive materials we required some separate shield arrangement [6] (Fig. 1).

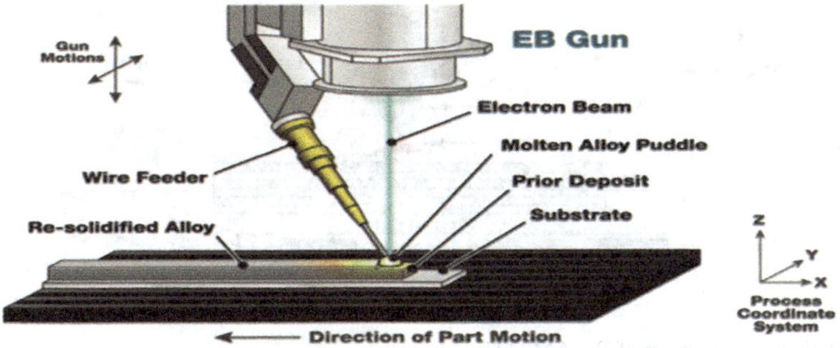

Fig. 1 Schematic diagram of EBAM [9]

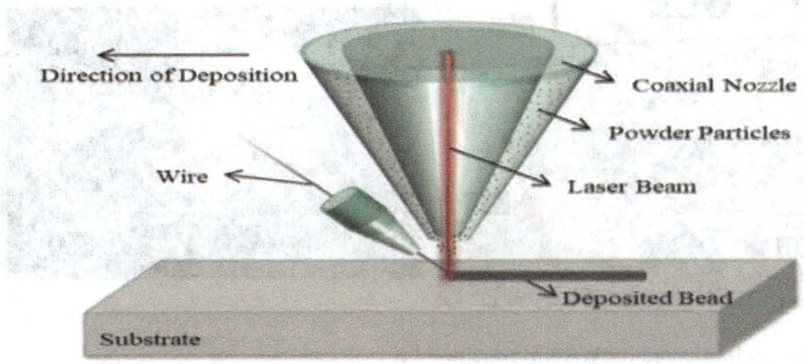

Fig. 2 Schematic diagram of WLAM [10]

2.2 Wire and Laser-Based Additive Manufacturing (WLAM)

WLAM is a process of fabricating 3D complex shape products by continuous feeding of wire into the molten metal pool. Uses of high power density in this technology, small and complex characteristics are easily obtained. High cost and incapability of producing large components are the primary limitation of WLAM [6] (Fig. 2).

2.3 Wire and Arc-Based Additive Manufacturing (WAAM)

In this technique, fabrication of 3D parts was carried out with the help of welding torch to generate weld pool by selecting optimum process parameters and adaptive tool path generating strategy. The main advantage of WAAM over other processes is the high deposition rate which makes this method faster and compatible (Fig. 3).

Metal inert gas (MIG) [12, 13], tungsten inert gas (TIG) [14–16] and plasma arc welding (PAW) [17, 18] are the main source of heat used in the WAAM technology. The afterwards execution of cold metal transfer (CMT) as upgraded MIG it is widely used as the heat source. Due to its capacity to deliver a high rate of deposition with

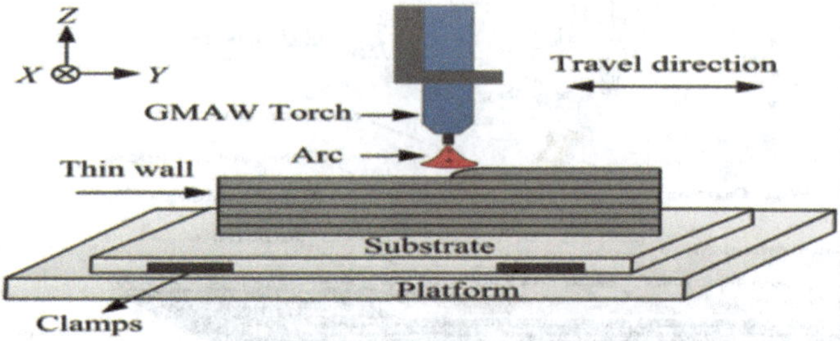

Fig. 3 Schematic diagram of WAAM [11]

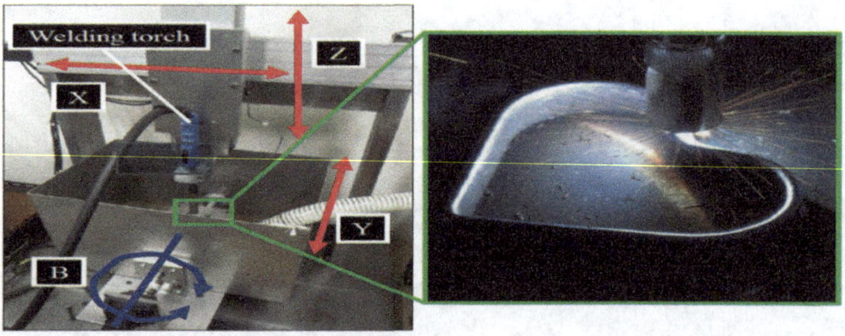

Fig. 4 Specimen fabricated by WAAM machine [26]

lower heat input, CMT has superior execution than MIG [19, 20]. MIG has four welding modes: globular mode, short-circuit mode, splash mode and pulsed-spray mode. In 1920, Shirizly et al. [21] utilized a fusible electrode to form an overlapping metal deposit. Subsequently, Ujiie et al. [22, 23] created the procedure of making a circular cross section through dynamic weld metal deposition. An offline observing framework was created to permitted a computer-aided design (CAD) format to be sliced to encourage weld deposition in layer-by-layer in an indicated to organize [24] (Fig. 4).

3 Parameters Affecting the WAAM Technology

3.1 Wire Feed Speed (WFS)

The study shows that the wire feed rate of Al5Si alloy below 10 m/min and above 45 m/min creates a discontinuous deposit. The best regular bead with a 0.1 mm standard deviation is obtained with 35 m/min wire feed speed [25]. It also shows that bead height of Hastelloy X alloy (nickel-based alloy) is linearly related to wire feed rate [27].

3.2 Travel Speed

Travel speed inversely affects the humping defect, melt through depth and bead width. Study shows that an increase in travel speed of Hastelloy X alloy (nickel-based alloy) results decrease in bead width and melt through depth, due to this the roughness is increased [27]. It also shows that the extreme speed at which humping starts in mild steel wire is 0.6 m/min for 0.8 mm wire diameter [28].

3.3 Heat Input

Results show that the decrease in heat input slows down the grain growth rate of ER5356 (aluminium–magnesium alloy). Also, heat supplied alternatively between layers formed small pores and cracks, large-sized grains and relatively low micro-hardness [29]. Bead structure of Ti6Al4V alloy in the first few layers differs along with the deposition height due to changes in the heat dissipation route but later it is insignificant of heat input [30]. Also, microstructure, grain size and crystalline stage of the Ti6Al4V sheet due to impacts of heat input vary along the construction direction. Optimum temperature to get favourable mechanical properties of Ti6Al4V is 200 °C [31].

3.4 Deposition Direction

The hardness of AA5183 aluminium alloy in the horizontal direction is approximately 75 kg/mm^2, whereas hardness shown in the vertical direction is about 70–75 kg/mm^2. It is also found that the deposition of multilayers from consequent passes gives cracking in weld [32]. Uniform deposition of molten metal depends on the wire feeding angle. A lower value of wire feeding angle (30°–50°) resulting in cracking of deposition of Ti6Al4V alloy, whereas the wire feeding angle of (70°) causes droplets splattering on the side of deposition [33].

3.5 Bulk Deformation

The study shows that the pores larger than 5 μm in diameter of 2319 Al alloy are dispensed with a rolling load of 45 kN [34]. There are two types of roller used in cold working of ER70S-6 steel, slotted roller and profiled roller. Among them, the slotted roller has a high efficiency of distortion elimination and deposition than later one [35].

4 Applications of WAAM

WAAM is suitable for the production of large-sized costly components with high complexity; hence, it is suitable to be used in areas such as aerospace industry, automotive industry, defence industry, naval industry and nuclear industry [36].

The main applications of WAAM technology are given below:

4.1 Aerospace Industries

Manufacturing of titanium and nickel alloys components with high complexity is the main focus in the aerospace industry because subtractive methods seem to be difficult and costly for the production of these material parts [37, 38].

4.2 Nuclear Industry

For manufacturing parts in the nuclear industry, WAAM is an appropriate method because it replaces some less useful nickel parts to stainless steel parts so that cost and weight both are reduced [39].

4.3 Medical Industry

In the medical industry, different alloys of cobalt, titanium and chromium are used for fabricating human vertebra, hip stem implants, dental implants and treatment of bone fracture with the help of WAAM technology [40, 41].

5 Issues and Challenges

5.1 Surface and Material Quality

Excessive heat input is required for achieving a high deposition rate in WAAM but due to high heat input, several challenges like residual stresses and distortions are coming as result, so for the fabrication of large metal components through WAAM these two are the primary concern. To get rid of these problems, post-welding heating is used for releasing residual stress and preheating is used for the problem of surface cracks and distortions.

5.2 *Residual Stress and Distortion*

Residual stress causes component failure because of uneven heat flow, and also, it is responsible for rough tolerance. For reducing the residual stress, post-processing is applied but the problem of reduced tolerance is remaining. Post-weld heat treatment (PWHT) can be used to reduce the residual stress during the process.

6 Conclusion

Numerous theories on WAAM state that WAAM can essentially decrease costs and enhance manufacturing effectiveness in industrial areas, particularly the aerospace, nuclear industry and automotive areas. Subsequently, numerous researches have focused on upgrading the WAAM process through moderation during 3D printing. Several process variations have been evolving in recent times to improving the microstructure and mechanical properties of the manufactured components. Moreover, the maximum number of alloys like aluminium, titanium and steel are utilized previously in fabricating the component using WAAM technology with fabulous outcomes.

References

1. Zhou JG (1999) A new rapid tooling technique and its special binder study. Rapid Prototyp J 5:82–88
2. Shrestha R, Simsiriwong J, Shamsaei N (2019) Fatigue behavior of additive manufactured 316L stainless steel parts: effects of layer orientation and surface roughness. Addit Manuf 4:224–278
3. Chua CK, Leong KF (2003) Rapid prototyping principles and applications. World Scientific Publishing Co., Pte. Ltd., Singapore
4. Nimawat D, Meghvanshi M (2012) Using rapid prototyping technology in mechanical scale models. Int J Eng Res Appl 2:215–219
5. Sodeifian G, Ghaseminejad S, Akbar A (2019) Preparation of polypropylene/short glass fiber composite as fused deposition modeling (FDM) filament. Results Phys 12:205–222
6. Gibson I (2010) Additive manufacturing technologies. In: Rapid prototyping to direct digital manufacturing
7. Szost BA, Terzi S, Martina F, Boisselier D, Prytuliak A, Pirling T, Hofmann M, Jarvis DJ (2015) A comparative study of additive manufacturing techniques: residual stress and microstructural analysis of CLAD and WAAM printed Ti–6Al–4V components. Mater Des 8:684–699
8. Li JZ, Alkahari MR, Rosli NAB, Hasan R, Sudin MN, Ramli FR (2019) Review of wire arc additive manufacturing for 3D metal printing. Int J Autom Technol 13:346–353
9. AM_EBDM_Illustration @ Additivemanufacturing.Com [Online]. Available: https://additivemanufacturing.com/wp-content/uploads/2015/08/AM_EBDM_Illustration.jpg
10. 0104-9224-Si-0104-9224SI220406-Gf01 @ www.Scielo.Br [Online]. Available: https://www.scielo.br/img/revistas/si/v22n4//0104-9224-si-0104-9224SI220406-gf01.jpg

11. 10033_2018_276_Fig1_HTML @ Media.Springernature.Com [Online]. Available: https://media.springernature.com/original/springer-static/image/art%3A10.1186%2Fs10033-018-0276-8/MediaObjects/10033_2018_276_Fig1_HTML.png
12. Zhang Z, Sun C, Xu X, Liu L (2018) Surface quality and forming characteristics of thin-wall aluminium alloy parts manufactured by laser assisted MIG arc additive manufacturing. Int J Light Mater Manuf 6:494–506
13. Xiong J, Lei Y, Chen H, Zhang G (2016) Fabrication of inclined thin-walled parts in multi-layer single-pass GMAW-based additive manufacturing with flat position deposition. J Mater Process Tech 3:278–310
14. Bai JY, Yang CL, Lin SB, Dong BL, Fan CL (2015) Mechanical properties of 2219-Al components produced by additive manufacturing with TIG. Int J Adv Manuf Technol 2:424–484
15. Baufeld B, Van Der Biest O, Gault R (2010) Additive manufacturing of Ti–6Al–4V components by shaped metal deposition: microstructure and mechanical properties. Mater Des 31:106–111
16. Martina F, Ding J, Williams S, Caballero A, Quintino L (2018) Tandem metal inert gas process for high productivity wire arc additive manufacturing in stainless steel. Addit Manuf 4:594–684
17. Aiyiti W, Zhao W, Lu B, Tang Y (2013) Investigation of the overlapping parameters of MPAW-based rapid prototyping. Rapid Prototyp 12:165–172
18. Wu CS, Wang L, Ren WJ, Zhang XY (2014) Plasma arc welding: process, sensing, control and modeling. Manuf Process 16:74–85
19. System MW, Chen X, Su C, Wang Y, Noor A (2018) Cold metal transfer (CMT) based wire and arc additive. Surf Investig 12:1278–1284
20. Cong B, Ding J, Williams S (2015) Effect of arc mode in cold metal transfer process on porosity of additively manufactured Al-6.3% Cu alloy. Adv Manuf Technol 76:1593–1606
21. Shirizly A, Dolev O (2018) From wire to seamless flow-formed tube: leveraging the combination of wire arc additive manufacturing and metal forming. Miner Met Mater Soc 7:585–594
22. Ujiie A (1971) U.S. Patent No. 3,558,846. U.S. Patent and Trademark Office, Washington, DC
23. Ujiie A (1973) U.S. Patent No. 3,746,833. U.S. Patent and Trademark Office, Washington, DC
24. Ribeiro F (1994) Metal based rapid prototyping for more complex shapes. Comput Technol Weld 8:724–735
25. Ortega AG, Galvan LC, Mezrag B (2017) Effect of process parameters on the quality of aluminium alloy Al5Si deposits in wire and arc additive manufacturing using a cold metal transfer process. Sci Technol Weld Join 6:1743–2936
26. Abe T, Mori D, Sonoya K, Nakamura M, Sasahara H (2019) Control of the chemical composition distribution in deposited metal by wire and arc-based additive manufacturing. Precis Eng 55:231–239
27. Dinovitzer AM, Chen X (2018) Effect of wire and arc additive manufacturing (WAAM) process parameters on bead geometry and microstructure. Addit Manuf 34:997–1013
28. Adebayo A, Mehnen J, Tonnellier X (2012) Limiting travel speed in additive layer manufacturing. Trend Weld Res 3:884–892
29. Su C, Chen X, Gao C, Wang Y (2019) Effect of heat input on microstructure and mechanical properties of Al-Mg alloys fabricated by WAAM. Appl Surf Sci 486:431–440
30. Wu B, Ding D, Pan Z, Cuiuri D, Li H, Han J, Fei Z (2017) Effects of heat accumulation on the arc characteristics and metal transfer behavior in wire arc additive manufacturing of Ti6Al4V. J Mater Process Technol 54:112–140
31. Wu B, Pan Z, Ding D, Cuiuri D, Li H (2018) Effects of heat accumulation on microstructure and mechanical properties of Ti6Al4V alloy deposited by wire arc additive manufacturing. Addit Manuf 94:713–740
32. Horgar A, Fostervoll H, Nyhus B, Ren X, Eriksson M, Akselsen OM (2018) Additive manufacturing using WAAM with AA5183 wire. J Mater Process Technol 259:68–74
33. Wu Q, Lu J, Liu C, Shi X, Ma Q, Tang S (2017) Obtaining uniform deposition with variable wire feeding direction during wire-feed additive manufacturing obtaining uniform deposition with variable wire feeding direction during wire-feed additive manufacturing. Mater Manuf Process 6914:1532–2475

34. Gu J, Ding J, Williams SW, Gu H, Ma P (2016) The effect of inter-layer cold working and post-deposition heat treatment on porosity in additively manufactured aluminum alloys. J Mater Process Technol 230:26–34
35. Colegrove PA, Coules HE, Fairman J, Martina F, Kashoob T, Mamash H, Cozzolino LD (2013) Microstructure and residual stress improvement in wire and arc additively manufactured parts through high-pressure rolling. J Mater Process Technol 213:1782–1791
36. Thompson MK, Moroni G, Vaneker T, Fadel G, Campbell RI, Gibson I, Bernard A, Schulz J, Graf P, Ahuja B, Martina F (2016) Design for additive manufacturing: trends, opportunities, considerations, and constraints. CIRP Ann Manuf Technol 65:737–760
37. Merlin PW (2009) Design and development of the blackbird: challenges and lessons learned. In: 47th AIAA aerospace sciences meeting including the new horizons forum and aerospace exposition, pp 1–38
38. DEFACTO_STELIA_PANEL-752x1024 @ 3dprintingindustry.Com [Online]. Available: https://3dprintingindustry.com/wp-content/uploads/2018/02/DEFACTO_STELIA_PANEL-752x1024.jpg
39. Abe T, Sasahara H (2016) Dissimilar metal deposition with a stainless steel and nickel-based alloy using wire and arc-based additive manufacturing. Precis Eng 45:387–395
40. Buchanan C, Gardner L (2019) Metal 3D printing in construction: a review of methods, research, applications, opportunities and challenges. Eng Struct 180:332–348
41. Medical-Implants_feature @ I0.Wp.Com [Online]. Available: https://i0.wp.com/www.pddinnovation.com/wp-content/uploads/2017/11/medical-implants_feature.jpg?fit=584%2C437&ssl=1

A State-of-the-Art Review on Fused Deposition Modelling Process

Kamal Kishore and Manoj Kumar Sinha

Abstract Additive manufacturing (AM) is one of the most emerging and demanding concepts in the field of prototyping. In fact, this technology is now directly used in making the functional products in different sectors such as medical, aerospace, defence and automotive industries. Fused deposition modelling (FDM) is an extrusion-based AM process which is widely used because of process simplicity and associated low cost. It uses mainly thermoplastics in making the prototype and functional parts of complex structure. This paper reviews the different polymers used in FDM and gives detail about the process parameters which must be optimised for better quality of product. At last, research area of FDM is discussed in terms of new materials such as composites, ceramic, metals and other applications.

Keywords Fused deposition modelling · Additive manufacturing · Thermoplastics · Composites

1 Introduction

The manufacturing is the continuous process which is evolving day by day due development of new concepts, methods and tools. Esmaeilian et al. [1] reviewed the tangible and intangible components of the manufacturing system. They described the evaluation of researches in manufacturing system starting from its history, current trends and future developments. They also presented how the manufacturing system had been classified. They covered the eight main areas that include technologies, planning, taxonomies, design, remanufacturing, sustainability, etc. The main focus of the paper was to cover the advance manufacturing techniques which include sustainable manufacturing, nanomanufacturing, social manufacturing and additive

K. Kishore · M. K. Sinha (✉)
National Institute of Technology Hamirpur, Hamirpur 177005, India
e-mail: mksinha@nith.ac.in

K. Kishore
e-mail: kamal@nith.ac.in

© The Author(s), under exclusive license to Springer Nature Singapore Pte Ltd. 2021
R. M. Singari et al. (eds.), *Advances in Manufacturing and Industrial Engineering*, Lecture Notes in Mechanical Engineering,
https://doi.org/10.1007/978-981-15-8542-5_75

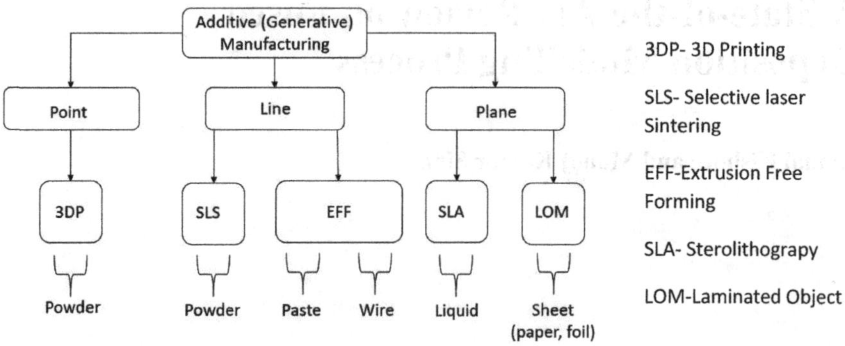

Fig. 1 Classification of AM on their starting materials [3]

manufacturing. The term additive manufacturing (AM) is present around in field of advance manufacturing over the past three decades. But with the development of cheaper computational techniques, designing software and evolution in various new material development methods, AM has shown extraordinary double-digit growth in past 20 years [2]. Additive manufacturing manufactured the real-world object directly from the digitised data using layer-by-layer printing approach. AM is also popular as rapid prototyping, as it is used for quickly development of the prototypes before final production of the items. It finds a bright spot in the field of manufacturing. But today it is popularly known as 3D printing due to its similarity in principle of printing. The tool path and projection pattern or combination of both are used to determine the overall part geometry in AM (Fig. 1).

2 History of Additive Manufacturing

The trace of AM dated back to 1960s and 70s with the development of photopolymerisation, sheet lamination (1979) and powder fusion (1972). With the development of numeric control tools, computer-aided design in 1950s supported the development of AM. However, at that time technology was just for research purpose. The development of 3D printer by MIT in 1989, stereolithography technique in 1988, fused deposition modelling (FDM) in 1989 and laser sintering in 1992 helped in its commercialisation with better control on geometry modelling capabilities [4]. But the high cost of manufacturing, low-dimensional accuracy and less choice of material limit its use to prototyping only. The advancement of technology in late 2000s in fields of solid modelling, networking and reduction in cost of powerful computers helps in cost reduction and accuracy of the product produced and hence makes AM a bright spot in the realm of advance manufacturing [2, 5]. Some of the common file formats of AM are common layer interface (CLI), stereolithography (STL), layer exchange ASCII format (LEAF), additive manufacturing format (AMF) and layer manufacturing interface (LMI).

3 Role of AM in Industries 4.0

The Industries 4.0 is new concept, and we are still on the cusp of fourth industrial revolution. The Industries 4.0 tries to cover all the aspects of manufacturing right from value addition to sustainability. Anderl [6] explained about Industries 4.0, its implementation and technological approaches. The main goal of Industries 4.0 is smart products, smart plants, smart logistics and smart grids. Mehrpouya et al. [7] reviewed the potential of AM in the smart factories. The Industries 4.0 forms an interconnected network of virtual and real world. Ceruti et al. [8] explained the role of augmented reality and AM in the maintenance task and production of spare parts in the aviation industries. Craveiro et al. [9] reviewed the use of AM technology for digital manufacturing under the umbrella of Construction 4.0. They described the methods such as contour crafting and concrete printing technology, which were widely used for construction on small and medium scale. O'Brien [10] wrote the qualification and certification standards that were required for manufacturing of parts for space using AM technologies.

4 Fused Deposition Modelling

The American Society for Testing Material (ASTM) classifies the AM technique according to deposition and solidification method. According to ASTM, material extrusion method such as FDM are the most cost-efficient, widely used extensively researched and commercially used method [11]. The FDM technique was developed by founder of Stratasys Company, Scott Crump in 1992. In the FDM machine, a 3D part is directly printed from CAD model using extrusion-based process. The filament of the thermoplastic in which one is support material along with main material is continuously fed to system in the form of wire from wire spool. The liquefier head converts that solid filament into semisolid state. The nozzle extrudes the semisolid filament, and layer is formed on the printing platform by depositing the material in form of lines adjacent to each other [12]. The layer-by-layer approach is used to make the complete geometry. The important part of the FDM system has liquefier and printing head with nozzle attached to it, printing platform, material feed mechanism, gantry and temperature control chamber [13]. The distribution of temperature during FDM process can be monitored by IR (infrared) camera [14] (Fig. 2).

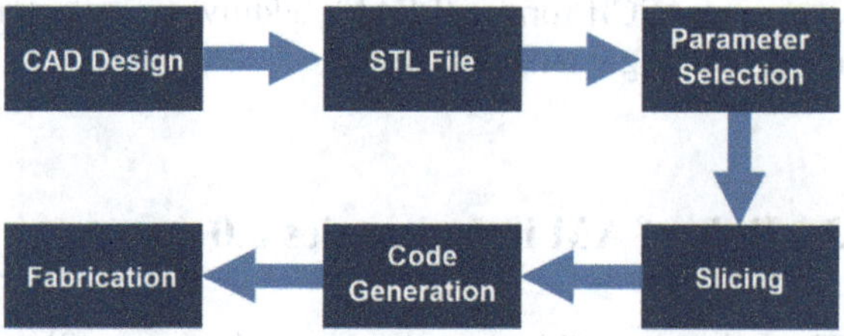

Fig. 2 Steps of FDM

5 Materials in FDM

The FDM process work on simple principle is frequently used extrusion-based technology. In FDM, mostly polymers are used as raw material. Polyether ether ketone (PEEK), acrylonitrile styrene acrylate (ASA), polyether imide (commercially known as ULTEM 9085), polylactic acid (PLA) and acrylonitrile butadiene styrene (ABS) are few widely used polymers.

6 Defects

Although FDM has freedom of design, ability to print complex design and personalisation is in one go. It provides some serious challenges that are the attribute to the nature AM. Agarwala et al. studied the surface quality of the printed part from the FDM. They described the internal and surface defects which must be avoided as they limit the structural properties of part produced [15]. Some of the main challenges are [16]:

6.1 Void Formation

The formation of void is the one of the main drawbacks of the FDM as it not only introduces porosity but also reduced the mechanical strength up to some extent. The amount of void formation is largely depending on the process parameter considered and printing technique. Wang et al. [17] introduced the thermally expandable microsphere in FDM filaments to tackle the problem of the void formation. They suggested an approach of mixing the thermal expendable microsphere with polywax while preparing the filament for the FDM.

6.2 Anisotropic Nature of Structure

Due to layer-by-layer printing approach, the microstructure of each layer is different when compared with the boundaries between the layers. This introduces the anisotropic nature in the component produced. Bellehumer et al. [18] investigated the formation of bonds between the adjacent extruded filaments of ABS. The model printed from FDM process has inferior mechanical properties at the interface (bonding zone) than that of material used. The thermal analysis was performed to model the dynamics of bond formation during the extrusion of semi-melted filament in FDM process. Further sintering experiments were carried out to evaluate the obtained results from thermal analysis, and the results were used to predict the quality of bond formation between the filaments.

6.3 Deviation from Actual Design

Computer-aided design (CAD) software is used to design the 3D component directly. These designs are directly printed, but due to limitation of FDM, the printed part may deviate from the expected solid design element. The concept of tessellation is employed to approximate the models, but inaccuracies and defect generally occurred in case of the curved surfaces. A very fine tessellation may solve this problem to some extent, but this leads to increase in computational time as well as printing time which make the process time consuming and eventually costlier.

6.4 Layer-by-Layer Appearance

The one of the most important challenge in field of FDM is its layer-by-layer appearance. However, appearance may not be an important factor if the final application of the part is hidden such as scaffolds in tissue engineering. But in case of the toys, aerospace, automobile: the appearance matters a lot. This layer-by-layer appearance not only makes the product dull but also increases the surface roughness [19]. The chemical and physical postprocessing methods may be used to reduce this defect, but this will violate the goal of introduction of AM, i.e. to reduce or eliminate the postprocessing of the component produced. However, this defect can be reduced by reducing the number of layers by adopting the adaptive slicing method.

6.5 Warping Defect

The warping is a defect that causes down layer to deform, while top layer is being printed in FDM. This warping defect reduces the overall height of printed object. The main factors that influence the warping defamation are stacking section length, shrinkage coefficient of material and chamber temperature [20]. The most influencing factor among them is stacking section length. An algorithm is developed to reduce the warping by limiting the stacking section length by splitting model into bricks spatially locked form with configurable gaps between them [21].

6.6 Residual Stress and Strain

Due to the rapid heating and cooling of the material in FDM process, residual stress is accumulated during the part building. This residual stress led to the distortion and de-layering of part due to residual strain. Antreas Kantaros and Dimitris Karalekas investigated the magnitude residual strain induced in the FDM processed part during its solidification [22]. They considered layer thickness and deposition orientation as their important parameters. The ABS material is used, and residual strain and coefficient of thermal expansion were recorded. Casavola et al. [23] performed the hole-drilling experiment to measure the residual stress in the part produced. The electronic speckle pattern interferometry is used to measure the displacement of the surface due to stress relaxation.

7 Process Parameters of FDM

Sood et al. [24] performed the parametric appraisal of polymers using FDM-based printing technology considering important factors into consideration. The part orientation, raster angle, raster width, layer thickness, and air gap were taken as the factors influencing the part quality. Few important parameters of FDM are slicing parameter, building orientation and temperature conditions [25]. The some important parameters are described below [12]:

Layer Thickness: It is the slice height at which the cad model is sliced in.stl file for part printing. It generally described as distance travel in the z-direction with respect to previous layer. It has direct impact on surface quality and building time of the part produce. It also depends on nozzle diameter, type of material and machine used for FDM.

Nozzle Tip Diameter: The FDM machines generally consist of a set of tips. A single set only allows specific line width and layer thickness to be printed.

Building Temperature: This temperature generally refers as temperature of the heating element which is inside the liquefier head and which control the viscosity of the molten material that extruded from the nozzle also known as extrusion temperature.

Line Width/Bead Width: Raster width or line width or bead width is the width of single line extruded from the nozzle tip on the platform. The line width differs machine to machine according to the nozzle diameter.

Raster Angle: An angle of tool path deposits with x-axis of the building table 0 to ±90° are typically allowed raster angles in steps of 15°.

Air Gap: There are three types of air gaps found in the FDM-printed object; the first one is raster-to-raster air gap; second one is perimeter to raster air gap and part sparse fill air gap.

Orientation [24]: The term orientation or part building orientation refers to the inclination of part on the building platform with respect to *X*-, *Y*-, *Z*-directions among which it is considered that thickness is provided in *Z*-direction and *X*-axis and *Y*-axis lie in plane of building platform.

Infill Pattern: The infills are used to fill the material between the boundaries. These have different shapes which provide strength at different level to the component.

Kerekes et al. [26] studied the effect of process variables such as infill density and layer thickness on damage and deformation behaviour of 3D-printed specimen by FDM process. Jiang et al. [27] developed an algorithm to minimise the material consumption, production time and energy consumed in FDM process. Although additive manufacturing was eco-friendly process than the other conventional subtractive processes, still a lot of work has to be done to achieve sustainability in term of material consumption and energy consumption. The support materials that were used in FDM process consumed significant amount of energy and time.

8 Research Area in FDM

The FDM process is widely used in various domains such as automobiles, aerospace, tissue engineering and biomedical. The FDM is commonly used in pharmaceutical applications such as tablet manufacturing, capsule printing and personalised implant. Owida et al. [28] fabricated a coronary artery bypass graft with the help of FDM and electrospinning ground collector method. The mould for electrospinning was developed by FDM using ABS as material. The scaffold of optimum geometry was made with help of above technique. Krause et al. [29] explored the possibility of printing the pressure sensitive drug delivery system through FDM. There are many researches going on to develop materials to have better mechanical and thermal properties and which is suitable for FDM. Dul et al. [30] developed a nanocomposite of graphene-based ABS material by solvent-free method. Zhang et al. [31] developed a water content sensor using fused deposition modelling. Macy [32] showed the feasibility of combining FDM with aerosol technology for printing the sensors, electronic circuits directly onto the complex surface. Ning et al. [33] mixed carbon fibres in ABS for enhancing the mechanical responses of the component produced through FDM. Dong et al. [34] investigated the improvement in the mechanical properties of the component produce through FDM using Kevlar nylon composite. Bellini et al. [35] proposed and developed a new set-up called mini extruder deposition system based on FDM. This system is mounted on the precision positioning system and operates using bulk material in granulation. This specification of system opens gate for using wide variety of engineering materials.

9 Conclusion

The FDM technology is around as from quite few years. There are many authors who optimised the process parameters of FDM-based machines using thermoplastics such as ABS, PLA and PEEK. The research is still continued to develop new material which have enough strength for the functional components. The composites developed by mixing of material like carbon fibres, jutes, etc., provide enough strength to material; however, there is compromise in the surface finish and accuracy of the model. This paper reviews the various defects and material presents for fused deposition modelling. There are many composites which are developed in recent times for FDM, but little work has been done regarding change in structure of machine which needs to be done in future.

References

1. Esmaeilian B, Behdad S, Wang B (2016) The evolution and future of manufacturing: a review. J Manuf Syst 39:79–100
2. Yan X, Gu P (1996) A review of rapid prototyping technologies and systems. CAD Comput Aided Des 28(4):307–318
3. Travitzky N, Bonet A, Dermeik B, Fey T, Filbert-Demut I, Schlier L, Schlordt T, Greil P (2014) Additive manufacturing of ceramic-based materials. Adv Eng Mater 16(6):729–754
4. Wohlers Associate (2015) Wohlers report 2015. History of additive manufacturing, p 34
5. Thompson MK, Moroni G, Vaneker T, Fadel G, Campbell RI, Gibson I, Bernard A, Schulz J, Graf P, Ahuja B, Martina F (2016) Design for additive manufacturing: trends, opportunities, considerations, and constraints. CIRP Ann Manuf Technol 65(2):737–760
6. Anderl R (2015) Industrie 4.0—technological approaches, use cases, and implementation. At-Automatisierungstechnik 63(10):753–765
7. Mehrpouya M, Dehghanghadikolaei A, Fotovvati B, Vosooghnia A, Emamian SS, Gisario A (2019) The potential of additive manufacturing in the smart factory industrial 4.0: A review. Appl Sci 9(18):3865
8. Ceruti A, Marzocca P, Liverani A, Bil C (2019) Maintenance in aeronautics in an industry 4.0 context: the role of augmented reality and additive manufacturing. J Comput Des Eng 6(4):516–526

9. Craveiro F, Duarte JP, Bartolo H, Bartolo PJ (2019) Additive manufacturing as an enabling technology for digital construction: a perspective on construction 4.0. Autom Constr 103:251–267
10. O'Brien MJ (2019) Development and qualification of additively manufactured parts for space. Opt Eng 58(01):1
11. Rane K, Strano M (2019) A comprehensive review of extrusion-based additive manufacturing processes for rapid production of metallic and ceramic parts. Adv Manuf 7(2):155–173
12. Masood SH (2014) Advances in fused deposition modeling. Elsevier, Amsterdam
13. Turner BN, Strong R, Gold SA (2014) A review of melt extrusion additive manufacturing processes: I. Process design and modeling. Rapid Prototyp J 20(3):192–204
14. Parandoush P, Lin D (2017) A review on additive manufacturing of polymer-fiber composites. Compos Struct 182:36–53
15. Agarwala MK, Jamalabad VR, Langrana NA, Safari A, Whalen PJ, Danforth SC (1996) Structural quality of parts processed by fused deposition. Rapid Prototyp J 2(4):4–19
16. Ngo TD, Kashani A, Imbalzano G, Nguyen KTQ, Hui D (2018) Additive manufacturing (3D printing): a review of materials, methods, applications and challenges. Compos Part B Eng 143:172–196
17. Wang J, Xie H, Weng Z, Senthil T, Wu L (2016) A novel approach to improve mechanical properties of parts fabricated by fused deposition modeling. Mater Des 105:152–159
18. Bellehumeur C, Li L, Sun Q, Gu P (2004) Modeling of bond formation between polymer filaments in the fused deposition modeling process. J Manuf Process 6(2):170–178
19. Ahn D, Kweon JH, Kwon S, Song J, Lee S (2009) Representation of surface roughness in fused deposition modeling. J Mater Process Technol 209(15–16):5593–5600
20. Armillotta A, Bellotti M, Cavallaro M (2018) Warpage of FDM parts: experimental tests and analytic model. Robot Comput Integr Manuf 50:140–152
21. Guerrero-De-Mier A, Espinosa MM, Domínguez M (2015) Bricking: a new slicing method to reduce warping. Procedia Eng 132:126–131
22. Kantaros A, Karalekas D (2013) Fiber Bragg grating based investigation of residual strains in ABS parts fabricated by fused deposition modeling process. Mater Des 50:44–50
23. Casavola C, Cazzato A, Moramarco V, Pappalettera G (2017) Residual stress measurement in fused deposition modelling parts. Polym Test 58:249–255
24. Sood AK, Ohdar RK, Mahapatra SS (2010) Parametric appraisal of mechanical property of fused deposition modelling processed parts. Mater Des 31(1):287–295
25. Popescu D, Zapciu A, Amza C, Baciu F, Marinescu R (2018) FDM process parameters influence over the mechanical properties of polymer specimens: a review. Polym Test 69:157–166
26. Kerekes TW, Lim H, Joe WY, Yun GJ (2019) Characterization of process–deformation/damage property relationship of fused deposition modeling (FDM) 3D-printed specimens. Addit Manuf 532–544
27. Jiang J, Xu X, Stringer J (2019) Optimization of process planning for reducing material waste in extrusion based additive manufacturing. Robot Comput Integr Manuf 59:317–325
28. Owida A, Chen R, Patel S, Morsi Y, Mo X (2011) Artery vessel fabrication using the combined fused deposition modeling and electrospinning techniques. Rapid Prototyp J 17(1):37–44
29. Krause J, Bogdahn M, Schneider F, Koziolek M, Weitschies W (2019) Design and characterization of a novel 3D printed pressure-controlled drug delivery system. Eur J Pharm Sci 140:105060
30. Dul S, Fambri L, Pegoretti A (2016) Fused deposition modelling with ABS-graphene nanocomposites. Compos Part A Appl Sci Manuf 85:181–191
31. Zhang X, Luo W, Zhang J, Lu Z, Hong C (2019) Development of a FBG water content sensor adopting FDM method and its application in field drying-wetting monitoring test. Sens Actuators A Phys 297:111494
32. Zhang B, Seong B, Nguyen V, Byun D (2016) 3D printing of high-resolution PLA-based structures by hybrid electrohydrodynamic and fused deposition modeling techniques. J Micromech Microeng 26(2):025015

33. Ning F, Cong W, Hu Y, Wang H (2017) Additive manufacturing of carbon fiber-reinforced plastic composites using fused deposition modeling: effects of process parameters on tensile properties. J Compos Mater 51(4):451–462
34. Dong G, Tang Y, Li D, Zhao YF (2018) Mechanical properties of continuous kevlar fiber reinforced composites fabricated by fused deposition modeling process. Procedia Manuf 26:774–781
35. Bellini A, Shor L, Guceri SI (2005) New developments in fused deposition modeling of ceramics. Rapid Prototyp J 11(4):214–220

3D Modelling of Human Joints Using Reverse Engineering for Biomedical Applications

Deepak Kumar, Abhishek, Pradeep Kumar Yadav, and Jitendra Bhaskar

Abstract Computer-aided design (CAD) has been supporting engineers to design better products in a very short time frame. Nowadays, reverse engineering and rapid prototyping are available for enhancing the design capabilities of design engineers for reducing product lead time and making possible customized products. Customized products are very much in demand in biomedical applications as implants and products for operation planning. Real 3D models of organs help doctors to observe the physiological changes in the organs of patients for better implant design and operational planning 3D scanners. Computerized axial tomography (CAT) is used for collecting the data of the geometric shape of organs or human joints and developing 3D models. In this paper, efforts have been made to create a 3D geometric model of the knee joint and ankle joint using CT scan data of two patients. There were different challenges both joints in terms of soft tissues, ligaments, and removal of meshes. 3D slicer and Blender methodologies were adopted for creating 3D geometric models. 3D models were fabricated on fused deposition Modelling 3D printer using Polylactic material. It was observed that 3D slicer and Blender were capable of handling soft tissues of knee joints and removal meshes in ankle joints, respectively. 3D-printed models were close to joints of specific patients.

Keywords Reverse engineering · 3D scanner · 3D printing · CT scan data

1 Introduction

Reverse engineering (RE) is a process in which we create a geometrical model of an object by using a point cloud data generated from the object surface. The point cloud data have inaccuracies and noise which is removed in RE software. 3D point cloud data generated is aligned in a single frame to create a 3D CAD model. The data generated from scanning can be used in RE software to design, test, and inspect different

D. Kumar (✉) · Abhishek · P. K. Yadav · J. Bhaskar
Department of Mechanical Engineering, Harcourt Butler Technical University, Kanpur, Uttar Pradesh, India
e-mail: deepakbherwani123@gmail.com

© The Author(s), under exclusive license to Springer Nature Singapore Pte Ltd. 2021

R. M. Singari et al. (eds.), *Advances in Manufacturing and Industrial Engineering*, Lecture Notes in Mechanical Engineering,
https://doi.org/10.1007/978-981-15-8542-5_76

parts of the human body. Geometric models generated from software are also modified for different purposes such as new design and its simulation. RE concept with additive manufacturing creates a physical model of different parts of the body [1, 2]. The generated physical models are customized and also reduce the operation time. Biocompatible implants are also fabricated using biomedical material can be used to support the damaged part of the body. Biocompatible implants materials are environmentally friendly and provide the same strength that is provided by natural implants. In RE software, the noise is removed through various segmentation methods. Image segmentation separates the different regions like bones, ligaments, and cartilages in case of the knee joint. Reverse engineering using CT scanning captures 2D images in DICOM format in the number of slices. This 2D image represents the complex anatomy of a patient joint which is imported in RE software to generate a 3D CAD model. The steps involved in the RE process are as follows:

1. To capture CT scan data of the patient.
2. Segmentation and filling of holes to remove mesh defects.
3. Generation of a 3D CAD model.
4. Fabrication of the geometrical model.
5. Analysis of the geometrical model to study load distribution and strength of each bone.

3D scanners are used for capturing the 3D data. Specialized software converts these data into a geometrical model. This geometrical model is converted into a *.stl file* and *G-codes* for fabricating a 3D model using a 3D printer [3]. In biomedical areas, computerized axial tomography (CT) scan and magnetic resonance imaging (MRI) data are used in place of 3D scanning data for 3D modelling [4]. There are various types of other methods such as laser scanning, ultrasound, and positron emission tomography are also used for obtaining patient data. The *.stl file* of the 3D model is imported into a 3D printer and physical model is fabricated using the 3D printer. Various rapid prototyping techniques like fused deposition modelling (FDM), stereolithography (SLA) and Inkjet 3D printing are popular in medical applications. Nowadays, techniques like Direct Metal Laser Sintering (DMLS), Selective Laser Sintering (SLS), Selective Laser Melting (SLM), and Electron Beam Melting (EBM) are being used to manufacture models and implants with sufficient density in the metallic form [5]. But, Fused Filament Fabrication method is a very simple and cost-effective method. Generally, filaments of Polylactic acid (PLA), Acrylonitrile Butadiene Styrene (ABS), and Polyether Ether Ketone (PEEK) are used in FDM. 3D printing technology is being used as a clinically promising technology for rapid prototyping of surgically implantable products. Additive manufacturing (AM) provides extensive customization as per the individual patient data such as finding musculoskeletal disorders. These 3D-printed models from 3D-reconstructed images are also used as graspable objects. Models can enhance a patient's understanding of their pathology and surgeon pre-operative planning. Customized implants and casts can be made to match an individual's anatomy [6, 7]. 3D-printed models of human body parts are also very useful as teaching material for anatomy subject due to societal controversy in terms of cultural and ethical issues, especially in the context of employing cadaveric materials. Avoiding the use of cadaveric materials is helpful for not only reducing the financial burden on the medical institutions but also it is good for the health and safety of medical students and staff [8]. Models of individual human organs are developed through special software which takes care of a variety of materials available in human bodies such as soft tissue, foreign bodies, and vascular structures.

All these medical applications can be summarized as below:

1. Pre-operative planning
2. Education
3. Custom manufacturing—implants, prosthetics, and surgical guides
4. Biological applications [9].

In this study, reverse engineering of knee joint and ankle joint has been done from CT scan data-keeping view of above various medical applications.

2 Methodology

For this research work, knee joint and ankle joint have been taken as a case study for reverse engineering. Nature of both joints was different in terms of some meshes, the involvement of soft tissues and ligaments. CT scan data of knee joint were taken from a 30-year-old patient having a bone fracture in knee and data of ankle joint were taken from a patient having talus fracture in the foot. Following two open-source software were used for creating a geometrical model of above two joints.

1. 3D slicer
2. Blender

3D slicer was used for knee joint due to its capability handling soft tissues. But it is weak in removing mesh defects in case of ankle joint. Blender was used for ankle joints due to its advantage in removing more number of meshes. But 3D Blender software is not capable of separating data of soft tissues from data of bone. Details of the procedure of using this software are given below:

2.1 3D Slicer

3D slicer software is a platform for the analysis and visualization of medical images [9]. It is open-source software available in the various operating system. It has the capability of modelling complex parts of the human body. It supports for multimodality imaging including, MRI, CT, ultrasound (US), and microscopy [10, 11]. A 3D slicer is also capable of bidirectional interfacing. Procedure for creating a

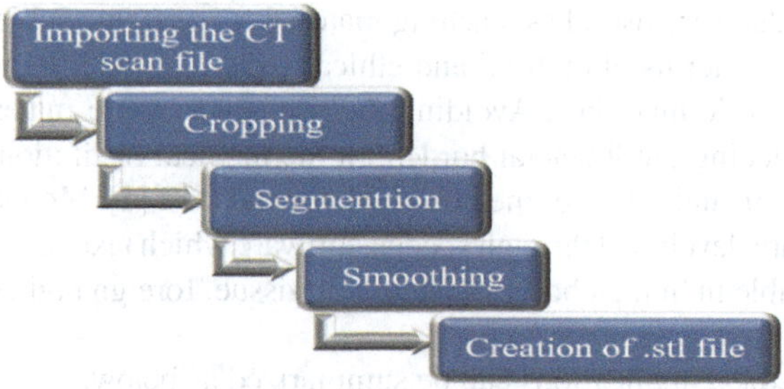

Fig. 1 Steps to create a geometrical model in the 3D slicer software

geometrical model using CT scan data has been explained using the block diagram (Fig. 1).

Details of each step are given below for creating 3D CAD Model of the knee joint.

Importing CT scan files. The CT scan data files were in Digital Imaging and Communications in Medicines (DICOM) format. DICOM is a standard for handling, storing, printing, and transmitting information in medical imaging [12]. The knee data were in 266 two-dimensional images with an image resolution of (512 * 512) pixels. These 2D images were imported into the 3D slicer software. Now, these files have been converted in Nearly Raw Raster Data (NRRD) format in 3D slicer software [13]. NRRD format shows the view of knee in transverse, coronal, and sagittal plane as shown in Fig. 2. Shaded portion shows the bone part which is used for segmentation.

Cropping images. 3D model of the knee in NRRD format was with an undesired portion of the thigh. Cropping of the thigh part was needed to remove an undesired

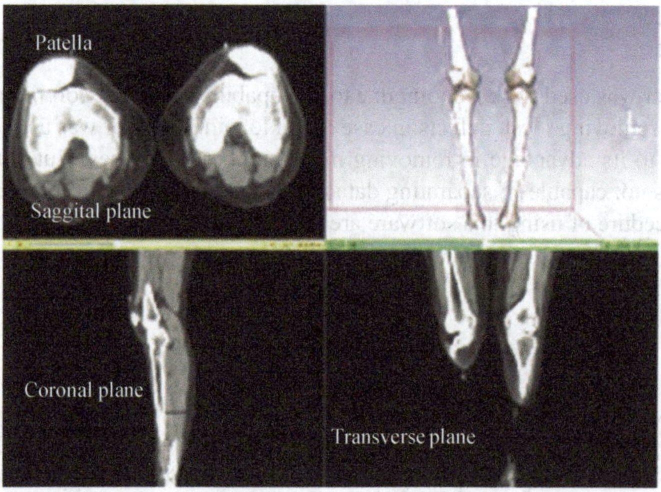

Fig. 2 General interface window of a 3D slicer

part from the model. This thigh part was removed selecting the crop volume option. The portion of the thigh for cropping has been shown in Fig. 3. Cropped model of the knee joint has been shown in Fig. 4.

Segmentation. Cropped model of the knee joint was consisting of soft tissues, ligaments, and cartilages. These are also undesired part for the fabrication of knee joint. Knee joint should consist of an only bone part. So, the 3D-printed model should be

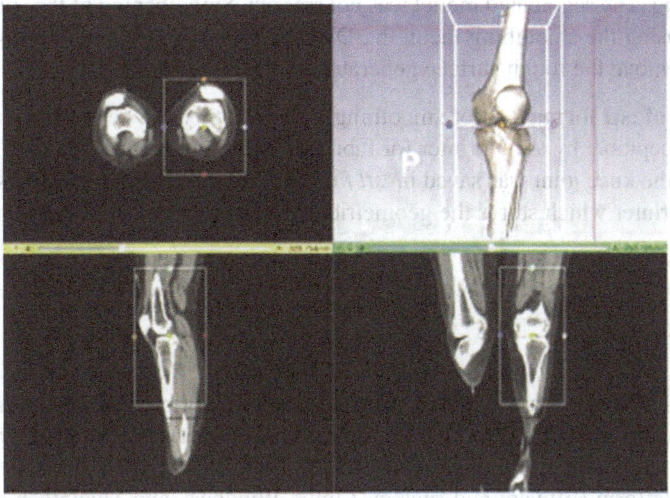

Fig. 3 Thigh portion in boxes for cropping

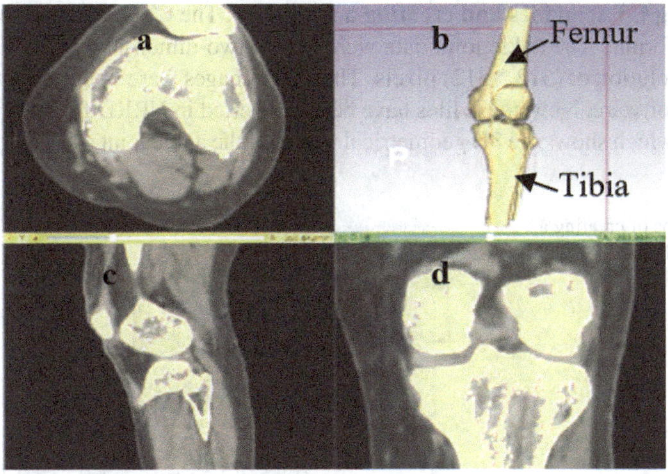

Fig. 4 Views of a cropped model of knee joints with segmentation: **a** Top view of tibia, **b** isometric view of knee, **c** side view of knee, **d** front view of the knee

free from ambiguity. Segmentation was required to remove this undesired part of soft tissues, ligaments, and cartilages [14]. This was done by using *Segmentation tool* in 3D slicer software. Various options such as *Atlas*, *Manual*, and *Threshold* were available for segmentation. *Threshold method* was used for this research work. Hounsfield unit (HU) was taken in the range of 400–1200 HU. Segmentation in the cropped model has been shown in Fig. 4. Segmentation is an important parameter for the accuracy of the 3D CAD model.

Smoothing. The segmented model was not smooth. Smoothening of the surface was done by using the smoothing menu in 3D slicer with a range of 0–1. Smoothing is done to remove the rough surface generated after segmentation.

Creation of .stl format. After smoothing of bones, this had been saved into a file format acceptable by a 3D printer for fabrication and producing G-codes. Final 3D model of the knee joint was saved in *.stl format* [15]. STL format is understandable by a 3D printer which stores the geometrical information of the 3D CAD model.

2.2 *Blender*

Blender is a free and open-source software used for a variety of purposes like the generation of the 3D CAD model, simulation, and animations. It is used when mesh errors are extensive in the geometric model created from CT scan data. Blender is cross-platform software and runs on Linux, Windows, and Macintosh operating system [16, 17]. Procedure for creating a geometrical model using CT scan data has been explained using the block diagram (Fig. 5).

Importing CT scan files and creating a 3D model. The CT scan data files were in DICOM format. The ankle joint data were in 320 two-dimensional images with an image resolution of (512 * 512) pixels. These 2D images were imported into the 3D Blender software. Now, these files have been converted in NRRD format in Blender software which shows the 3D geometrical model of the ankle joint as shown in Fig. 6.

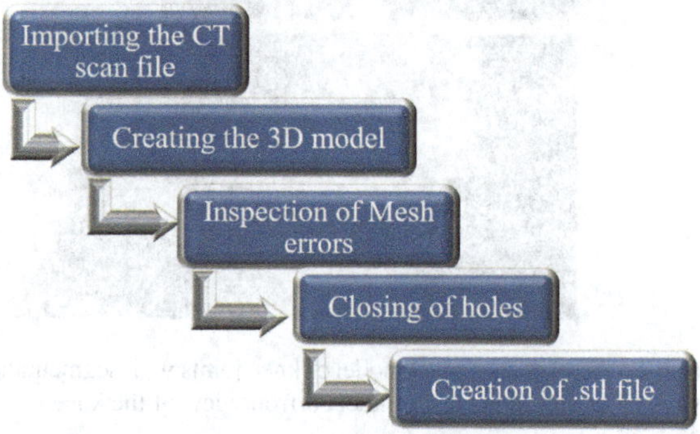

Fig. 5 Steps of creating a geometrical model of ankle joint using blender

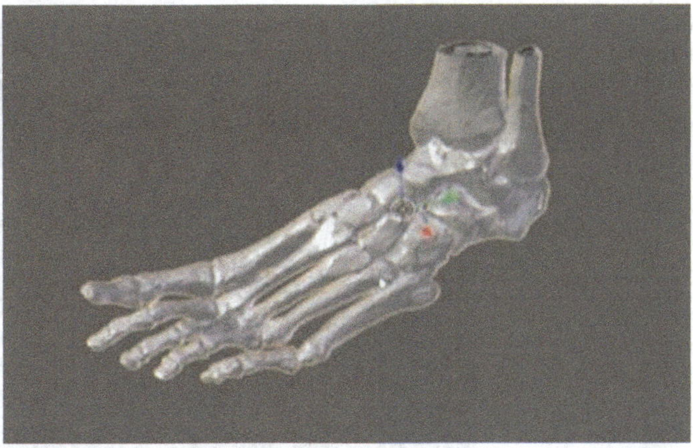

Fig. 6 A 3D geometrical model of ankle joint

Inspection of mesh errors. First meshes were created using a 3D model in NRRD format. It was observed that there were non-manifold defects within the meshes as shown by yellow colour points in Fig. 7. Mesh size was large at some places shown by the circle in ankle joint Fig. 7. Non-manifold means, there are gaps between meshes which can create problem in 3D printing (Fig. 8).

Closing of holes. In this step, a cross-section of bones was visible as holes. So, it was desired to close this. All the vertices were selected from a medullary portion of the tibia bone where non-manifold meshes are available using ctrl-L. An Undesired portion of a medullary mesh of the thin fibula bone was removed. The holes were closed by holding the Alt key and right-click along the edge of the fibula. The face

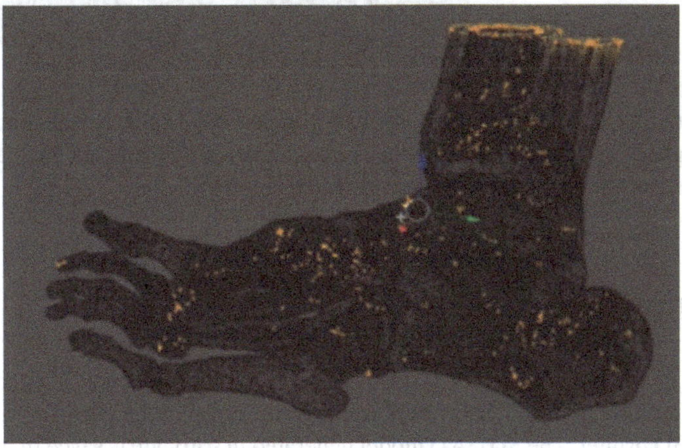

Fig. 7 Non-manifold defects are shown by yellow colour points

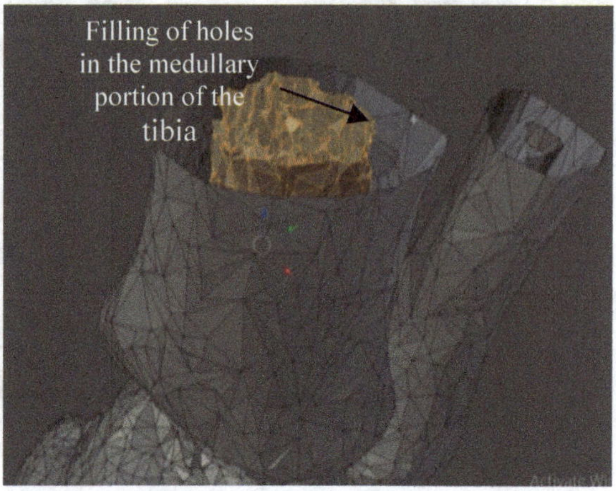

Fig. 8 Holes in the medullary portion of geometric model of tibia

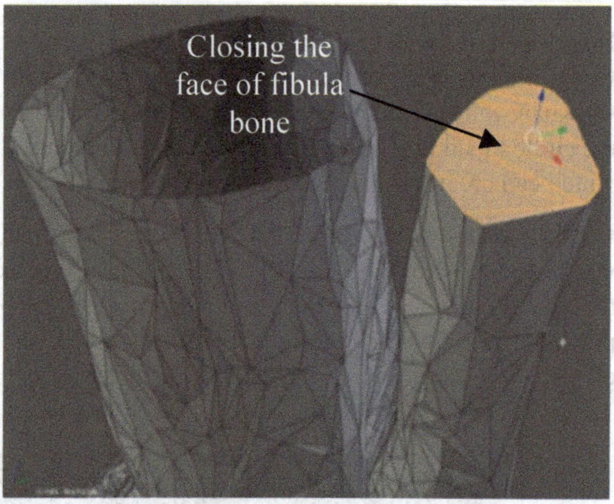

Fig. 9 Bones with closed holes and triangular meshes

was selected by using ctrl-F and converted polygon meshes into triangular meshes by holding ctrl-T. After correcting mesh errors and closing the holes, final error-free file is saved as *.stl* format in Blender software (Fig. 9).

3 Fabrication of Joints

All joints of the human body vary from person to person. So, any 3D geometrical model created by a 3D slicer or Blender Software can only be used for fabrication using the 3D printer. The 3D CAD model generated in software is saved in *.stl* format and imported into 3D printing software. G-codes were generated for 3D printer using open-source software Repetier. This is mention here that all 3D printing machine runs on G-codes. Slic3r in Repetier host 3D printing software calculates the number of layer and printing time to fabricate the 3D CAD model.

Fused deposition modelling (FDM) was used which is very simple and of low cost. It is good for orthopedics application such as fabrication of bone model. FDM-based 3D printer: DESK 200 3D printer was used for this purpose. Joints were 3D printed using filament of polylactic acid (PLA). Parameters of 3D printing for the fabrication of ankle joint and the knee joint were taken as shown below:

Parameters	Details
Nozzle diameter	0.2 mm
Bed temperature	50 °C
Raster angle	45 °C
Rectangular infill%	20
Extruder temperature	220 °C

4 Results and Discussions

Reverse engineering was made possible using CT scan data of both knee joint and ankle joint. 3D geometric models of knee joint and ankle joints were successfully created using 3D slicer and Blender, respectively. The geometrical models were free from soft tissues and cartilages. These models were consisting of only bone data. These models were 3D printed using PLA filament in white colour. This is suitable for clearly visualizing the fractured or deformed part of the bone. These models were very close to the actual models which are very good for teaching and learning resource for medical students (Figs. 10 and 11). The accuracy of 3D-printed model depends on image segmentation. Segmentation using the threshold method generates

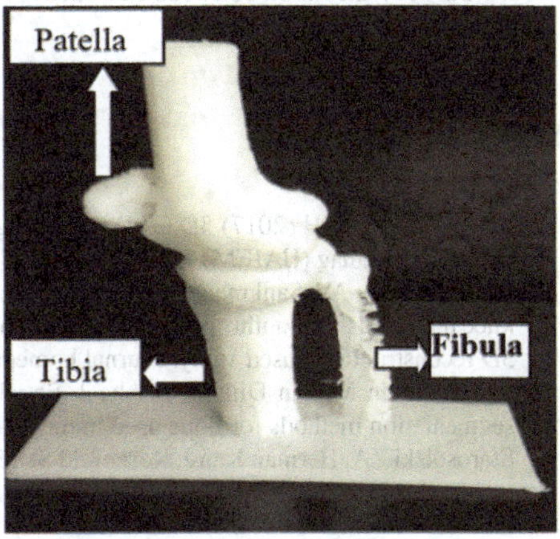

Fig. 10 Fabricated knee joint using FDM-based 3D printer

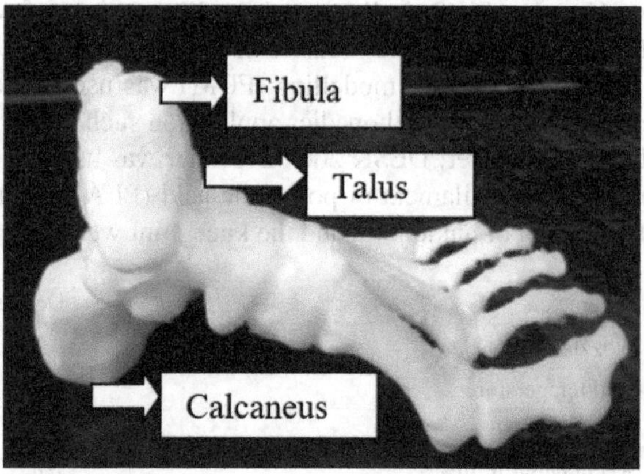

Fig. 11 Fabricated ankle joint using FDM-based 3D printer

3D CAD model which is compatible with 3D printing software and also helps in easy visualization of each bone in knee joint and ankle joint.

5 Conclusions

In this study, efforts were made for the prototype of knee joint and ankle joint. The effort was successful for converting CT scan data into a 3D geometrical prototype of PLA-based material A prototype was free from soft tissues, ligaments, and cartilages. Non-manifold mesh defects and infilling of holes were successfully removed in Blender software. 3D slicer and Blender have been very helpful for converting CT scan data into a prototype. Each bone in the fabricated model of knee and ankle joint was representing the actual bone which allows these 3D-printed models to be used for teaching. Doctors can easily visualize and inspect bone fracture through these 3D-printed models. The fabricated 3D model is beneficial in pre-operative planning before surgery and also reduces the surgery time.

References

1. Patel PG, Patel NH (2017) 3D CAD modelling of human knee joint using mimics. Int J Appl Innov Eng Manag (IJAIEM) 6(6): 256–264, ISSN 2319–4847
2. Gurava Reddy AV, Sankineani SR, Agrawal R, Thayi C (2019) Comparative study of existing knee prosthesis with anthropometry of Indian patients and other races, a computer tomography 3D reconstruction-based study. Journal homepage: www.elsevier.com/locate/jcot
3. van Eijnatten M, van Dijk R, Dobbe J, Streekstra G, Koivisto J, Wolff J (2017) CT image segmentation methods for bone used in medical manufacturing. Med Eng Phys 000: 1–11
4. Starosolski ZA, Herman Kan J, Rosenfeld SD, Krishnamuthy R, Annapragada A (2013) Application of 3-D printing (rapid prototyping) for creating physical models of pediatric orthopedic disorders. Springer, Berlin

5. Soni A, Modi YK, Agrawal S (2018) Computed tomography based 3D modeling and analysis of human knee joint. Mater Today Proc 5:24194–24201
6. Sandeep Kumar Y, Rajeswara Rao KVS, Yalamalle SR, Venugopal SM, Krishna S (2018) Applications of 3D printing in TKR pre surgical planning for design optimization—a case study. Mater Today Proc 5:18833–18838
7. Bedo T, Muntean SI, Popescu I, Chiriac A, Pop MA, Milosan I, Munteanu D (2019) Method for translating 3D bone defects into personalized implants made by additive manufacturing. Mater Today Proc 19:1032–1040
8. Mulford JS, Babazadeh S, Mackay N (2016) Three-dimensional printing in orthopaedic surgery: review of current and future applications. Department of Orthopaedics, Launceston General Hospital, Launceston, Tasmania, Australia
9. Sindhu V, Soundarapandian S (2018) Three-dimensional modelling of femur bone using various scanning systems for modelling of knee implant and virtual aid of surgical planning. Measurement 141:190–208. Journal homepage: www.elsevier.com/locate/measurement
10. Calignano F, Galati M, Iuliano L, Minetola P (2019) Design of additively manufactured structures for biomedical applications: a review of the additive manufacturing processes applied to the biomedical sector. Hindawi J Healthc Eng. Article ID 9748212
11. Hieu LC, Sloten JV, Hung LT, Khan L, Soe S, Zlatov N, Phuoc LT, Trung PD (2010) Medical reverse engineering applications and methods. In: Proceedings of international conference on innovations
12. Martorelli M, Gloria A, Bignardi C, Cali M, Maietta S (2020) Design of additively manufactured lattice structures for biomedical applications. Hindawi J Healthc Eng. Article ID 2707560
13. Javaid M, Haleem A (2017) Additive manufacturing applications in medical cases: a literature based review. Alexandria J Med 54:411–422
14. Aimar A, Palermo A, Innocenti B (2019) The role of 3D printing in medical applications: a state of the art. Hindawi J Healthc Eng. Article ID 5340616
15. Radharamanan R (2017) Additive manufacturing for the production of a low cost knee prototype. J Manage Eng Integr 10(1):1–13
16. Wong TM, Jin J, Lau TW, Fang C, Yan CH, Yeung K, To M, Leung F (2017) The use of three-dimensional printing technology in orthopaedic surgery: a review. J Orthop Surg 25(1)
17. Sivarasu S, Prasanna S, Mathew L (2011) Reverse engineering vs conceptual design principles in making of artificial knee models. Trends Biomater Artif Organs 25(2):60–62
18. Fedorov A, Beichel R, Kalpathy-Cramer J, Finet J, Fillion-Robin J-C, Pujol S, Bauer C, Jennings D, Fennessy F, Sonka M, Buatti J, Aylward SR, Miller JV, Pieper S, Kikinis R (2012) 3D slicer as an image computing platform for the quantitative imaging network. Magn Reson Imaging 30(9):1323–1341. PMID: 22770690

Institutional Distance in Cross-Border M&As: Indian Evidence

Sakshi Kukreja, Girish Chandra Maheshwari, and Archana Singh

Abstract Cross-border mergers and acquisitions (M&A) are inherently afflicted by additional complexities and "liability of foreignness" on account of the home-host distances, often influencing the strategic decisions involving the deals. Extant studies have often alluded to unidimensional reasons like culture and geographic distance, but have not focused enough on disaggregated multi-dimensional institutional distance approach which may expectedly provide a better understanding of the choices being made in the market for corporate control. This research seeks to investigate the impact of home-host distance on the acquirer's ownership structure choices in an emerging market setting using measures of institutional distance including cultural, geographic, financial, administrative, global connectedness, knowledge, economic, demographic and economic distances. For this purpose, 1542 completed M&A deals involving Indian firms, as target or acquirer, constitute the sample for the study. The results confirm to the fact that (multi-dimensional) home-host country distances causing institutional dynamics are factored in choosing a foreign target.

Keywords Mergers and acquisitions · Cross-border deals · Institutional distance · Ownership structure

1 Introduction

Cross-border mergers and acquisitions bring with it a pack of peculiarities owing to home-host country differences and are conceivably more exposed to challenges vis-à-vis their domestic counterparts. Dealing with cross-border targets requires coordination across home-host differences, over a spectrum of dimensions, including but beyond the traditionally utilised geographic and cultural measures. Unfamiliarity with target environment or the lack of requisite skills to manage those may prove to be hazardous. Concerns over establishing legitimacy also increases with the increase in home-host distances, driving the "liability of foreignness" and hence causing

S. Kukreja (✉) · G. C. Maheshwari · A. Singh
Delhi School of Management, Delhi Technological University, New Delhi, Delhi, India
e-mail: kukrejasakshi09@gmail.com

© The Author(s), under exclusive license to Springer Nature Singapore Pte Ltd. 2021
R. M. Singari et al. (eds.), *Advances in Manufacturing and Industrial Engineering*, Lecture Notes in Mechanical Engineering,
https://doi.org/10.1007/978-981-15-8542-5_77

performance difficulties and adding to costs [1–4]. As a strategic response to the external institutional risks and deal peculiarities, a firm may accordingly adjust its level of ownership acquired in the target firm, balancing with the desired level of control, resource commitment and risks [5]. Linked with survival, performance and stability, the choice of ownership level to be acquired represents a crucial decision and does warrant a special attention. The extant research in the area has majorly relied on the unidimensional traditional measures of geographic or cultural distance. The present study aims to provide an emerging market evidence, utilising dual perspective of both inbound and outbound deals, for the impact of institutional distance on the strategic choices involved in cross-border M&A.

Drawing from the institutional and resource-based theory, it is argued that foreign firms either conform to the pressures of local isomorphism or have to import firm-specific capabilities [2]. To meet the host country needs, a firm may choose to imitate the local organisational practices, hence motivating the partial ownership in the target to benefit from the local expertise and knowledge. While if a firm decides to gain competitive advantage through its organisational capabilities, majority or full ownership may seem to be incentivising. The efforts in managing the regulatory, cognitive and cultural differences, attaining local legitimacy and transferring practices also increase as the home-host nations get more distant [2]. Acquirers could comprehend and adjust more easily in case of similar home-host legal institutional environment [4]. In case of large home-host distances, acquirers are able to diversify their risks, while at the same time are exposed to higher uncertainties and lack of knowledge on host environment. The nine institutional distance dimensions, as proposed by Berry et al. [6], are utilised in this research. These include "economic, financial, political, administrative, cultural, demographic, knowledge, and global connectedness as well as geographic distance". This study aims to contribute to the literature in at least three ways. First, utilising a disaggregated institutional distance framework, the study provides a more comprehensive insights on cross-country distance over the much prevalent unidimensional measures [7]. Second, given a significant chunk of extant M&A studies focused on the developed markets, it would be worthwhile to conduct the study in an emerging market context owing to their distinctiveness and growing role in the global landscape [8]. The emerging markets are arguably more prone to challenges posed by distance due to their dynamic and vulnerable home country conditions. While it might be convenient to assume all emerging nations to be facing similar institutional challenges, but practically, significant differences exist across emerging nations demanding a focused study [9]. Being credited with several big-ticket deals and having emerged as a significant player in the global M&A landscape, India warrants a special focus on the institutional determinants of strategic choices. Third, the study presents a broader understanding by providing a comparative view between the inbound and outbound deals involving Indian firm.

The paper proceeds as follows. The following section covers a brief review of the related literature on cross-border M&A, institutional distance and ownership structure. The detailed research methodology is followed by the results, discussions and implications. The final section concludes the paper and provides suggestions for future research.

2 Literature Review

2.1 Institutional Distance

The world is turning into a level playing field with the advancements in information technologies and reducing national barriers [10]. Yet it still would be dangerous to undermine the influence of cross-country distance. While entering a foreign land, companies are faced with challenges caused by distance resulting in increased costs and risks [11]. A range of cross-country distance measures needs to be recognised and used as explanatory variables to gain meaningful insights on distance effect. Despite much criticism, Geert Hofstede's dimensions of cultural distance remain at the heart of cross-country research ever since it was made available through his book "Culture's consequences: International differences in work-related values" in 1980 [12]. A crucial assumption underlying these measures was the highly time unvarying nature of it. Kogut and Singh [13] proposed a method of calculating composite culture distance index using the underlying cultural dimensions by adapting the Euclidean distance metric. The Kogut and Singh index has since gained much popularity, albeit not without criticism [14].

The power of unidimensional approach in studying distance stays limited; hence, a multi-dimensional approach, simultaneously studying multiple distance dimensions, is recommended [15]. Over the time, institutional distance has catapulted as a leading approach for investigating the cross-country differences and often delved in to study its influence on various dependent variables such as strategic decision, costs and performance. It has been propounded as a complementary rather than a replacement to the individual constructs like culture, to capture the broader spectrum of national differences [13]. Multiple attempts have been made at defining and operationalising it. One such momentous attempt was at underlining the three pillars of institutional framework, encompassing normative, regulatory and cognitive pillar [16]. Yet another classification adopted in studies is of informal and formal institutional distance [17]. While providing as a simplistic approach, it misses out on the detailed and comparative analysis for encompassing individual dimensions. The CAGE Framework proposed cultural, administrative, geographic and economic dimensions [11, 18]. Delineating the multi-dimensional nature of distance with a nine-dimensional disaggregated measure, Berry et al. [6] proposed an institutional framework for cross-country distance. It included cultural distance, geographical distance, administrative distance, financial distance, global connectedness distance, knowledge distance, economic distance, demographic distance and political distance. These nine dimensions can be looked at as accommodating and rather extending the previously proposed measures. His contribution has been widely acknowledged and utilised for further studies.

2.2 *Cross-Border M&A, Institutional Distance and Ownership Structure*

The studies have often considered national characteristics as a salient determinant of cross-country deal flows [19]. Supported by theories like "cost of doing business abroad", "Liability of Foreignness", "Liability of origin" and "Liability of multinationality", cross-country institutional distance constitutes a crucial source of risk for companies venturing on to foreign lands. More the host country be institutionally distant from the home country, higher the liability of foreignness faced by a firm and hence higher are the efforts in managing [3]. Also, being a "stranger in a strange land" involves unfamiliarity hazards, discrimination hazards and relational hazards [3, 20]. A possible remedy to these hazards can be continued involvement of the local partner to benefit from its legitimacy, knowledge and expertise [21]. Hence, the possible hazardous institutional pressures may favour a minority stake for the acquirer firm. The strategic choice of the level of ownership stake, minority versus majority ownership, to be acquired in a deal represents one of the most crucial decisions. Minority acquisition represents "distinct organisational strategy" and comes up as a favoured option to gain more insights into the anticipated synergies and keep target managerial incentives intact [22]. Demanding lower levels of resource commitments, it provides an edge over contractual relationships facilitating cooperation between two firms [23]. It is well connected to the level of control desired by the acquirer firm [24] and requires to strike a balance among the levels of resource commitment and risk. Higher stakes may be only preferred when the acquirer is confident about the post-acquisition performance as well as its management abilities. Firms are often argued to prefer a lower ownership strategy as the home-host institutional distance increases, though the results far from being conclusive and varying for each of the distance dimensions [3, 25]. Also, in cases of uncertainty like in case of doubtful target valuation, partial acquisition is often a preferred strategy taking assurance from the targets continued stakes [26].

Underlining the choice of ownership structure as a strategic response to the uncertainties of operating in an institutionally distant country, a recent study by Ferreira et al. [5] investigated the impact of institutional distance on the ownership strategy for the Brazilian acquisitions during the period of 2008–2012. The financial, cultural and geographic distances were found to be significant factors influencing ownership levels. Utilising informal and formal institutional distance dimensions, Ellis et al. [27] examined the institutional determinants of ownership structure in the African context for the period 2008–2014. While the informal distance was found to be negatively related to the ownership position, the formal institutional distance was reported to be positively related. Even while focusing on similar distance dimensions, conflicting results have been reported. Like, of the studies focusing on the cultural distance, a few have reported larger cultural distance to be encouraging full ownership [5, 28], while some found otherwise [25, 29]. Partial ownership in case of culturally distant nations is seemed preferable given the increasing need to preserve the high-powered incentives for integrating target firm managers and incentivising tacit knowledge sharing [30]. However, with the peculiarities of deals involving emerging market firms, the motivations for majority ownership vs. minority ownership may differ [31] and hence may impact the ownership choices differently. M&A deals emerging out of emerging market often aim synergistic gains through taking control of tangible as well as the intangible resources motivating majority or full ownership. Some of the synergistic benefits can only be realised through nothing less than a majority acquisition.

3 Methodology

3.1 Data Collection and Sample

The sample for the study comprises completed Indian cross-border M&A deals announced from 1 January 2010–31 December 2015 as listed on Thompson Reuters Eikon database (now Refinitiv Eikon). For comparison, the sample was bifurcated as inbound and outbound deals. To avoid misleading results, the deals wherein only the remaining interest was being acquired have been eliminated. Further, each deal was individually matched with the set of nine home-host country distance dimensions by the year and country pair. The distance data for eight out of nine measures are sourced from Berry et al. [6], available through their official Website. Due to the limited availability of cultural data in Berry et al. [6], the Traditional Euclidean Hofstede-based cultural distance index, given by Beugelsdijk et al. [32], was utilised for measuring cultural distance. Few observations had to be dropped due to non-availability of data, resulting in the final sample of 1542 deals, of which 422 are of outbound deals and 1120 inbound deals. The top five inbound acquirer countries, for the completed deal count, are the USA, Singapore, Mauritius, Japan and the United Kingdom, whereas the top target countries are the USA followed by the United Kingdom, Germany, Singapore and Australia.

3.2 Measures

The acquirer ownership level post-M&A deal forms the dependent variable of the study. In line with the previous research [22, 33] and the objective of the present research, the dependent variable takes dichotomous form with a cutoff at 50% of the stakes acquired. Majority ownership (more than 50% in target), signifying greater control, is accompanied with higher resource commitment and risk. Taking brackets of equity ownership which defines distinct strategic choices is an appropriate method since the motive of the study is to study the choice of ownership structures (and not predict the exact percentage of stake acquired) [34]. It takes the value of "1" if the more than 50% of the stakes are acquired indicating majority ownership, and "0" otherwise.

The impact of home-host country distance is studied using disaggregated institutional distance measures, thus recognising their varying impacts. The nine-dimensional approach as given by Berry et al. [6] is utilised for the study. These include cultural distance, geographical distance, administrative distance, financial distance, global connectedness distance, knowledge distance, economic distance, demographic distance and political distance. Each of the nine dimensions of institutional distance, presenting the extent of inter-country differences, is studied individually for its impact on the acquirer ownership levels to provide a comparative as well as a broader view. A dummy variable for relatedness is also included in the analysis.

4 Results and Discussion

The dependent variable used for the study is of dichotomous nature, viz. majority ownership vs. minority ownership. Hence, binary logistic regression has been used to meet the requirements of the data [35]. The sample size requirements are satisfied with over fifteen hundred observations included in the analysis. Table 1 presents the collinearity diagnostic test results. Variance inflation factor (VIF) values all below five and none of the variables had a tolerance level below 0.2. Hence, the problem due to multicollinearity could be ruled out [36, 37]. The analysis is carried on in three sections with the sample of cross-border M&A deals involving: inbound deals, outbound deals and all deals combined. Table 2 presents the results for the binary logistic regression.

The results for the sample of inbound deals confirm that the nine institutional distance parameters when considered together are able to significantly predict the choice of ownership structure with Chi-square = 172.369, df = 10, $N = 1120$, $p < 0.001$ exhibiting an excellent level of overall fit. The odds ratio or the exponential of "B" (log-odds) value indicates the change in odds resulting from a unit change in the predictor variable. The odds ratio presented in Table 2 should be interpreted as the change in odds of majority ownership as compared to minority ownership. In the model, financial, cultural, global connectedness, knowledge, economic and administrative distances are found to be significant predictors for the ownership structure choices. Financial, knowledge and cultural distances with lower than one value of odds ratio signify, ceteris paribus, decreasing odds of majority ownership with a unit increase in these predictor variables. The increased distance across these dimensions may motivate shared ownership with the local partners for utilising their

Table 1 Collinearity statistics (VIF)

	Related	Direction	Fin	Culture	Demo	GConnect	Know	Pol	Eco	Admin	Geo
Combined	1.096	1.122	1.713	2.991	2.465	1.684	2.965	1.286	1.185	1.982	3.367
Target	1.110		1.768	4.806	3.426	1.930	3.020	1.353	1.215	3.498	4.429
Acquirer	1.023		1.799	1.637	1.578	1.603	2.994	1.289	1.243	1.215	2.509

Table 2 Logistic regression results for ownership structure

	Combined		95% C.I. for Exp(*B*)			Outbound deals		95% C.I. for Exp(*B*)			Inbound deals		95% C.I. for Exp(*B*)		
	B	Sig	Lower	Exp(*B*)	Upper	*B*	Sig	Lower	Exp(*B*)	Upper	*B*	Sig	Lower	Exp(*B*)	Upper
Related	**−1.182**	**0**	0.238	**0.307**	0.396	−0.45	0.107	0.369	0.638	1.103	**−1.319**	**0**	0.200	**0.267**	0.358
Direction	**1.562**	**0**	3.503	**4.767**	6.488										
Financial	**−0.05**	**0.025**	0.910	**0.951**	0.994	−0.027	0.538	0.891	0.973	1.062	**−0.057**	**0.027**	0.897	**0.944**	0.994
Culture	−0.006	0.512	0.977	0.994	1.011	0.014	0.39	0.983	1.014	1.046	**−0.025**	**0.033**	0.953	**0.975**	0.998
Demographic	0.027	0.125	0.993	1.027	1.063	−0.005	0.888	0.927	0.995	1.068	0.02	0.354	0.978	1.02	1.064
Global connectedness	**0.087**	**0.034**	1.007	**1.091**	1.182	−0.042	0.567	0.831	0.959	1.107	**0.143**	**0.007**	1.040	**1.154**	1.280
Knowledge	**−0.026**	**0.003**	0.958	**0.974**	0.991	−0.006	0.798	0.953	0.994	1.038	**−0.032**	**0.001**	0.950	**0.968**	0.987
Political	−0.009	0.355	0.972	0.991	1.010	−0.038	0.096	0.921	0.963	1.007	−0.005	0.651	0.974	0.995	1.016
Economic	**0.02**	**0.008**	1.005	**1.02**	1.035	−0.002	0.892	0.965	0.998	1.032	**0.021**	**0.009**	1.005	**1.022**	1.038
Administrative	0.003	0.26	0.998	1.003	1.009	**−0.011**	**0.012**	0.981	**0.989**	0.998	**0.011**	**0.007**	1.003	**1.011**	1.019
Geographic	**0**	**0**	1.000	**1**	1.000	0	0.592	1.000	1	1.000	**0**	**0**	1.000	**1**	1.000
Constant	−0.558	0.203		0.573		2.588	0.014		13.297		−0.418	0.424		0.658	
N	1542								422					1120	
Pseudo *R*-Sq	0.264								0.067					0.19	

Figures in bold are statistically significant ($p < 0.05$)

expertise and overcoming the liability of foreignness. The political distance measure constituting of home-host democracy difference, stability and trade bloc memberships is reported to be insignificant and very close to zero indicating its indifference in influencing ownership structure choices for Indian inbound deals. India represents the largest democracy in the world with a high level of political stability as well as cordial trade relations with most of the nations around the globe. Robust Indian business environment and favourable policy regimes have ensured foreign capital inflows in the country (IBEF 2019), whereas with a greater-than-one and significant odds ratio, the likelihood of majority acquisition increases in case of increase in global connectedness, economic and administrative distance. This implies that when economic, administrative and global connectedness distance is higher between India and the acquirer nation, acquirer firms have a higher likelihood of majority acquisition. Correspondingly, a majority acquisition is much more likely in case of related acquisitions. Also, for all the three sample groups, none of the significant predictor variables had the threshold value of one included in the odds ratio (exponential of log odds) confidence interval (C.I.). Hence, the observed direction of relationships can be said to be true for the population [37].

The complete model for the sample, including both Indian acquirer and target firms, is also found to be significant with Chi-square $= 336.717$, df $= 11$, $N = 1542$, $p < 0.001$. A dummy variable for the direction of deal as either inbound or outbound was included for the analysis. Global connectedness and economic distance have a significant positive association with the majority acquisition, increasing the odds of majority acquisition with every unit increase in distance, whereas an increase in the financial and knowledge distance between the home-host nation decreases the likelihood of majority acquisition. Unrelated acquisitions are associated with significantly lower odds of majority ownership. The majority and minority ownership structure choices are equally likely in case of geographic distance with an odds ratio of one. Hence, home-host geographic distance is not an influencing factor for deciding upon the ownership structure choices in cross-border deals involving Indian firms.

However, in case of Indian acquirer firms only administrative distance, constituted of differences across colonial ties, language, religion and legal system, is found to be a significant predictor for the choice of ownership structure. The results support decreasing odds of majority ownership with an increase in administrative distance between India and the host nation. Given the differences across administrative parameters, Indian firms may be apprehensive about going in for a majority acquisition for the involved post-acquisition difficulties involved in managing the target operations. The overall model for Indian acquiring firms, including all predictor variables, was also not found to be significant.

5 Conclusion

The study commenced with a view to understand the impact of institutional distance on the strategic choices involved in cross-border M&A. Building upon the institutional theory, this research examines the impact on the acquirer's choice of ownership structure in the emerging market setting of India. The country pair and year-wise disaggregated measures for nine-dimensions of institutional distance, involving cultural, geographical, administrative, financial, global connectedness, knowledge, economic, demographic and political distance, were utilised for empirically testing the impact of cross-country distance.

For the firms entering a foreign land, liability of foreignness poses a critical challenge. Adopting an optimal level of ownership level, balancing the risks and returns, the company might well be able to successfully enter, sustain and reap the benefits in distant nations. The dimensions of cross-country distance may have differing influence, the knowledge of which shall be crucial for the managers involved in potential cross-border deals involving Indian firms for managing the institutional uncertainties. By structuring ownership level choices for managing the cross-country differences, a firm may establish competitive advantage in foreign land or choose to leverage on the local partner's knowledge, getting an edge over contractual relationships.

The study on the impact of cross-country distance on the ownership decisions has both strong theoretical contribution and managerial implication. Interestingly, the results suggest that the home-host geographic distance does not influence the choice of ownership structure in any of the three sub-samples. The present era backed by advanced information technology developments has virtually shrunken the geographic barriers across nations. While geographical proximity may mean more face-to-face contact, but being geographically distant does not impede knowledge acquisition nor may be related to relational ties or social proximity [38, 39]. Hence, the strategic decisions of a firm are more influenced by factors other than geographic proximity. Further, for the inbound deals, most of the institutional distance dimensions (financial, cultural, global connectedness, knowledge, economic and administrative distances), except the demographic and political distance, are found to be significant predictors for ownership structure choices. The acquirer-target relatedness is also found to be significant predictor favouring majority acquisition. Further, institutional proximity does not largely impact the level of ownership to be acquired in the case of Indian companies targeting foreign firms. For Indian acquirer companies, only administrative distance was found to be significant predictor, indicating distinct motives and other potential factors apart from cross-country distance. Thereafter, in the sample including both inbound and outbound cross-border M&A deals, global connectedness, financial, knowledge and economic distances along with firm relatedness were found to be statistically significant predictors. The future research efforts may be directed toward linking the extant knowledge with deal motivations to espouse deeper insights.

References

1. Wan F, Williamson P, Pandit NR (2020) MNE liability of foreignness versus local firm-specific advantages: the case of the Chinese management software industry. Int Bus Rev 29:101623
2. Zaheer S (1995) Overcoming the liability of foreignness. Acad Manag J 38:341–363
3. Eden L, Miller SR (2004) Distance matters: Liability of Foreignness, Institutional Distance and Ownership Strategy. In: Hitt MA, Cheng JLC (eds) Theories of the Multinational Enterprise: Diversity, Complexity and Relevance. Adv Int Manage, vol. 16, Emerald Group Publishing Limited, Bingley, pp 187–221
4. Kostova T, Zaheer S (1999) Organizational legitimacy under conditions of complexity. Acad Manag Rev 24:64–81
5. Ferreira MASPV, Vicente SCDS, Borini FM, Almeida MIRD (2017) Degree of equity ownership in cross-border acquisitions of Brazilian firms by multinationals: a strategic response to institutional distance. Rev Adm 52:59–69
6. Berry H, Guillén MF, Zhou N (2010) An institutional approach to cross-national distance. J Int Bus Stud 41:1460–1480
7. Dong L, Li X, McDonald F (2019) Distance and the completion of Chinese cross-border mergers and acquisitions. Balt J Manag 14:500–519
8. Lebedev S, Peng MW, Xie E, Stevens CE (2015) Mergers and acquisitions in and out of emerging economies. J World Bus 50:651–662
9. Adeola O, Boso N, Adeniji J (2018) Bridging institutional distance: an emerging market entry strategy for multinational enterprises. In: Aggarwal J, Wu T (eds) Emerging issues in global marketing: a shifting paradigm. Springer, Cham, pp 205–230
10. Friedman TL (2005) The world is flat: a brief history of the twenty-first century. Farrar, Straus and Giroux, New York
11. Ghemawat P (2001) Distance still matters. Harv Bus Rev 79:137–147
12. Hofstede G (1980) Culture's Consequences: International Differences in Work-Related Values. Sage, Beverly Hills, CA
13. Kogut B, Singh H (1988) The effect of national culture on the choice of entry mode. J Int Bus Stud 19:411–432
14. Konara P, Mohr A (2019) Why we should stop using the Kogut and Singh index. Manag Int Rev 59:335–354
15. Sousa CMP, Bradley F (2008) Cultural distance and psychic distance: refinements in conceptualisation and measurement. J Mark Manag 24:467–488
16. Scott WR (1995) Institutions and organizations. Sage Publications, Thousand Oaks
17. North DC (1991) Institutions. J Econ Perspect 5:97–112
18. Ghemawat P (2007) Redefining Global Strategy: Crossing Borders in a World Where Differences Still Matter. Harvard Business School Press, Boston, Massachusetts, pp 33–64
19. Luong TA (2018) Picking cherries or lemons: a unified theory of cross-border mergers and acquisitions. World Econ 41:653–666
20. Zaheer S (2002) The liability of foreignness, redux: a commentary. J Int Manage 8:351–358
21. Makino S, Delios A (1996) Local knowledge transfer and performance: implications for alliance formation in Asia. J Int Bus Stud Spec Issue 905–927
22. Ouimet PP (2013) What motivates minority acquisitions? The trade-offs between a partial equity stake and complete integration. Rev Financ Stud 26:1021–1047
23. Fee CE, Hadlock CJ, Thomas S (2006) Corporate equity ownership and the governance of product market relationships. J Financ 61:1217–1251
24. Erramilli MK (1996) Nationality and subsidiary ownership patterns in multinational corporations. J Int Bus Stud 225–248
25. Malhotra S, Lin X, Farrell C (2016) Cross-national uncertainty and level of control in cross-border acquisitions: a comparison of Latin American and U.S. multinationals. J Bus Res 69:1993–2004
26. Chen SFS, Hennart JF (2004) A hostage theory of joint ventures: why do Japanese investors choose partial over full acquisitions to enter the United States? J Bus Res 57:1126–1134

27. Ellis KM, Lamont BT, Holmes RM (2018) Institutional determinants of ownership positions of foreign acquirers in Africa. Glob Strateg J 8:242–274
28. Padmanabhan P, Cho KR (1996) Ownership strategy for a foreign affiliate: an emperical investigation of Japanese firms. Manag Int Rev 36:45–65
29. Brouthers KD, Brouthers LE (2001) Explaining the national cultural paradox. J Int Bus Stud 32:177–189
30. Chari MD, Chang K (2009) Determinants of the share of equity sought in cross-border acquisitions. J Int Bus Stud 40:1277–1297
31. Liou R, Lee K, Miller S (2017) Institutional impacts on ownership decisions by emerging and advanced market MNCs. Cross Cult Strateg Manag 24:454–481
32. Beugelsdijk S, Kostova T, Roth K (2017) An overview of Hofstede-inspired country-level culture research in international business since 2006. J Int Bus Stud 48:30–47
33. Makino S, Beamish PW (1998) Performance ventures and survival of joint with ownership structures. J Int Bus Stud 29:797–818
34. Prashant K, Puranam P (2004) Choosing equity stakes in technology-sourcing relationships: an integrative framework. Calif Manage Rev 46
35. Leech NL, Barrett KC, Morgan GA (2005) SPSS for intermediate statistics: uses and interpretation. LAWRENCE ERLBAUM ASSOCIATES, London
36. Landau S, Everitt BS (2004) A handbook of statistical analyses using SPSS. Chapman & Hall/CRC, UK
37. Field A (2018) Discovering statistics using IBM SPSS statistics. SAGE Publications Inc, New York
38. Ganesan S, Malter AJ, Rindfleisch A (2005) Does distance still matter? Geographic proximity and new product development. J Mark 69:44–60
39. Ben Letaifa S, Rabeau Y (2013) Too close to collaborate? How geographic proximity could impede entrepreneurship and innovation. J Bus Res 66:2071–2078

Synthesis and Characterization of PVDF/PMMA-Based Piezoelectric Blend Membrane

Ashima Juyal and Varij Panwar

Abstract The present paper represents the dielectric and piezoelectric properties of the poly (vinyl fluoride (PVDF) and poly-methylmethacrylate (PMMA) blend membrane. The PVDF/PMMA blends were fabricated using the solvent casting technique. The piezoelectricity of the PVDF/PMMA (85/15 ratios) showed highest sensing voltage of 31.4 mV with bending strain of 0.0075 as compared to the pure PVDF (sensing voltage 7.6 mV) cause of the high dielectric constant (ε) of the PVDF/PMMA (85/15) blend membranes. Properties like dielectric analysis, AC conductivity, dissipation factor and piezoelectric effects were experimentally investigated using impedance analyzer in the frequency range 20 Hz–10 MHz. It is observed that the dielectric constant of the blend membrane falls with the rise in frequency. As a result, the AC conductivity increases with increasing frequency. Also, the dissipation factor increases. The piezoelectric effects increase on the addition of PMMA to PVDF as PVDF has a negative piezoelectric coefficient.

Keywords PVDF · PMMA · DMF · TFT · LCD

1 Introduction

In today's era, there is an immense need of bio-degradable polymers. Anything that is bio-degradable has the capability of being decomposed by bacteria or other living organisms, thereby avoiding pollution. In turn, these polymers will minimize the requirement of man-made polymers and eventually lead to less pollution caused by them.

In regard to the types of materials used in this particular work, PMMA is a transparent thermoplastic and is extremely bio-compatible [1, 2]. It is used as a lighter,

A. Juyal · V. Panwar (✉)
Polymer Sensors and Actuator Lab, Department of Electronics and Communication Engineering, Graphic Era (Deemed to be University), Dehradun, Uttarakhand, India
e-mail: varijpanwarcertain@gmail.com

A. Juyal
e-mail: juyalaashima@gmail.com

© The Author(s), under exclusive license to Springer Nature Singapore Pte Ltd. 2021
R. M. Singari et al. (eds.), *Advances in Manufacturing and Industrial Engineering*, Lecture Notes in Mechanical Engineering,
https://doi.org/10.1007/978-981-15-8542-5_78

resistant material as a substitute to glass in aquariums, windows, etc. Therefore, it is easily processed, low-cost and an adaptable material. Also, the temperature at which the transition of PMMA exists is 105 °C. It has low resistance due to the presence of the ester groups in it. It is environmentally stable and that is superior to most of the other plastics. PMMA is used as panels for building windows, signs and displays, as a light guide in thin-film transistor (TFT), liquid crystal display (LCD), etc.

PVDF is a thermoplastic fluoropolymer that is generally non-reactive. It has a low density (1.78 g/cm^3). PVDF is used as a binder component for the carbon electrode in supercapacitors and for other electrochemical applications. It has a glass transition temperature of about −35 °C. PVDF offers strong piezoelectric effect [3]. It has a negative piezoelectric coefficient value which generally states that it compresses instead of expanding or vice versa when subjected to some electric field. When poling is done, PVDF behaves as a ferroelectric polymer, possessing some pyroelectric properties [4]. Hence, this makes it useful in sensor and battery applications. It also offers some more properties like high strain rate, good workability, chemically resistant, etc. [5, 6]. PVDF, at a temperature under 150 °C, behaves as a ferroelectric material. Hence, it is a polymer with both pyroelectric and piezoelectric properties [7]. It also possesses high mechanical strength, is thermally stable and is chemically resistant [8].

In this particular paper, membranes using PVDF and PMMA are developed and are studied by varying their concentrations. Also, a pure PVDF membrane is developed for a means to differentiate. PVDF/PMMA blends have been studied extensively, mainly in relation to piezoelectric properties [9, 10].

2 Experimental Setup

2.1 Materials

The materials used here are polyvinylidene fluoride (PVDF) and poly-methyl-methacrylate (PMMA). N, N-Dimethylformamide (DMF) was made use of as a solvent to serve the purpose of dissolving PMMA into PVDF. The membranes were developed by mixing all the ingredients together.

3 Method of Preparation

PVDF and poly-methyl-methacrylate were taken in a beaker and mixed together in the ratio 90/10 to develop a solution of 20 wt.%.

The concluding amalgam was made by combining together the concentration of 10% solute and 90% solvent which resulted into a 100% solution. This was followed by the addition of 18 ml of the solvent used, i.e., N, N-Dimethylformamide (DMF)

Table 1 Different concentrations of the membrane

S. No	Abbreviations Abbreviations used	Ratio	Polyvinylidene fluoride (PVDF) (gm)	Poly-methyl-methacrylate (PMMA) (gm)	N,N-Dimethylformamide (DMF) (ml)
[5]	AJ-00	100/0	1		9
[6]	AJ-05	90/10	1.8		18
[7]	AJ-06	85/15	1.7		18

to the beaker. The solution was then mixed on a magnetic stirrer using a magnetic bead to ensure uniformity for around 3 h at 80 °C at 400 rpm.

Later, the concentrations of both PVDF and PMMA were varied and membranes were developed. After the solution got uniformly mixed, it was poured onto a glass petri dish followed by its heating for 3 h. When the solvent completely evaporates, the petri dish is made to cool down and the membrane is successfully striped off from the dish (Fig. 1).

Different concentrations varied while developing the membranes are listed below (see Table 1).

4 Characterization

Properties like dissipation factor and capacitance were measured by making use of Impedance analyzer (Key sight E4990A) in the frequency range (20 Hz–10 MHz). Also, the conductivity as well as dissipation factor was measured using the impedance analyzer. The conductivity, energy losses and dielectric constant are important parameters since they are used to decide on the appropriateness for application.

5 Results and Discussions

i. **Dielectric analysis**

This basically reveals that dielectric constant (ε) reduces with increasing frequency (see Fig. 2). This is because with increasing frequency, the total polarization of the material drops down, and hence, its dielectric constant drops too. At peak frequencies, no abundance of the molecules exists.

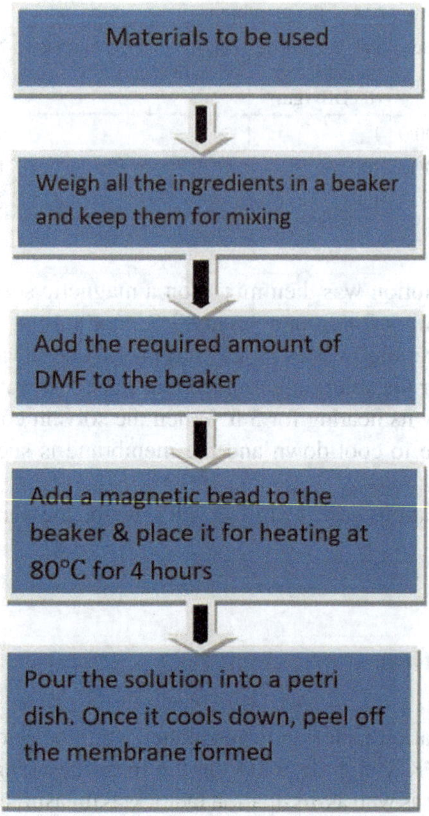

Fig. 1 Synthesis chart

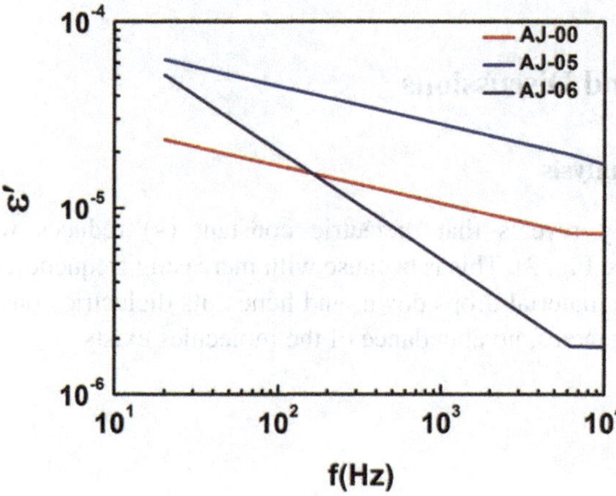

Fig. 2 Dielectric constant of various concentrations of PVDF and PMMA existing in the membrane at frequency ranging from 20 Hz to 10 MHz

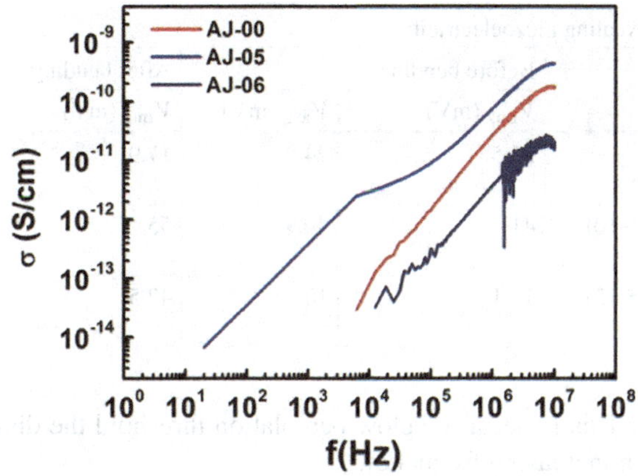

Fig. 3 Conductivity of different concentrations of PVDF and PMMA existing in the membrane at frequency ranging from 20 Hz to 10 MHz

Conductivity analysis

This reveals that the conductivity grows as the frequency is increased (see Fig. 3). This is due to the fact that the electrons mobility enhances in the film or due to the rise in the density of the carrier. Also, at a frequency range, conductivity increases due to the possible release of space charge.

Dissipation factor analysis

This reveals the dissipation factor of different concentrations of PVDF and PMMA (see Fig. 4). From the figure, it is clear that the dissipation factor rises with the rise

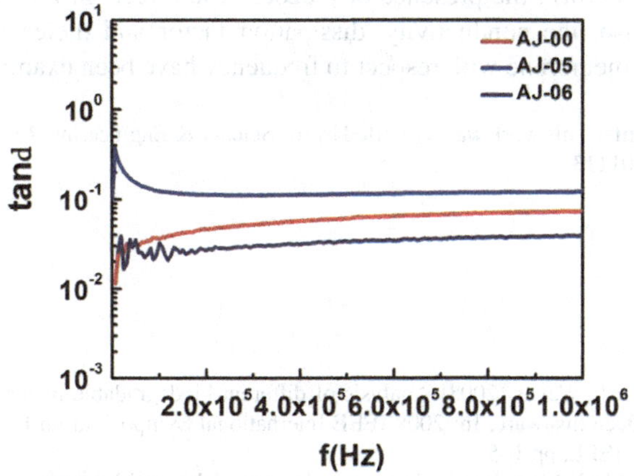

Fig. 4 Dissipation factor of various concentrations of PVDF and PMMA in the frequency range 20 Hz–10 MHz

Table 2 Film's revealing piezoelectricity

Membrane	Before bending		After bending	
	V_{rms} (mV)	V_{avg} (mV)	V_{rms} (mV)	V_{avg} (mV)
Pure PVDF i.e. AJ-00	25.5	24.9	17.9	17.7
PVDF/PMMA (90/10) i.e. AJ-05	41	40.9	55	53.2
PVDF/PMMA (85/15) i.e. AJ-06	11.1	10.9	42.5	38.6

in frequency. This is because below percolation threshold the dissipation factor increases with increasing frequency.

Piezoelectric analysis

Table 2 shows the voltage developed on the effect of force applied on the membranes (see Table 2).

Radius of the film (r) = 0.01 m.

Width of the film (L) = 0.000075 m.

Therefore, Bending force = ($L/2r$) = 0.0075.

This indicates that the membranes show piezoelectric effects [11]. So, these membranes can also be used for application in pressure sensors like blood pressure sensors [11, 12].

6 Conclusion

The final results affirm the presence of piezoelectric effects in PVDF and PMMA membranes. Also, the conductivity, dissipation factor and dielectric analysis of PVDF/PMMA membrane with respect to frequency have been examined.

Acknowledgements This work was supported by the Science & Engineering Research Board, File No.- ECR/2016/001113.

References

1. Jia X, Li J, Li F, Wei B (2008) Analysis of different biodegradable materials and its technique to produce dishware. In: 2008 IEEE International Symposium on Electronics and the Environment. IEEE, pp 1–5
2. Irimia-Vladu M (2014) "Green" electronics: biodegradable and biocompatible materials and devices for sustainable future. Chem Soc Rev 43(2):588–610
3. Takahashi Y, Zakoh T, Hanatani N (1991) Molecular mechanism for elongation: poly (vinylidene fluoride) form II. Colloid Polym Sci 269(8):781–784

4. Takahashi Y, Tadokoro H (1980) Crystal structure of form III of poly (vinylidene fluoride). Macromolecules 13(5):1317–1318
5. Qin Y, Peng Q, Ding Y, Lin Z, Wang C, Li Y, Li Y et al (2015) Lightweight, superelastic, and mechanically flexible graphene/polyimide nanocomposite foam for strain sensor application. ACS Nano 9(9):8933–8941
6. Hu N, Fukunaga H, Atobe S, Liu Y, Li J (2011) Piezoresistive strain sensors made from carbon nanotubes based polymer nanocomposites. Sensors 11(11):10691–10723
7. Zhen Y, Arredondo J, Zhao GL (2013) Unusual dielectric loss properties of carbon nanotube—polyvinylidene fluoride composites in low frequency region (100 Hz< f< 1 MHz). Open J Organic Polym Mater 3(04):99
8. Kang GD, Cao YM (2014) Application and modification of poly (vinylidene fluoride) (PVDF) membranes—a review. J Membr Sci 463:145–165
9. Elashmawi IS, Hakeem NA (2008) Effect of PMMA addition on characterization and morphology of PVDF. Polym Eng Sci 48(5):895–901
10. Saito K, Miyata S, Wang TT, Jo YS, Chujo R (1986) Ferroelectric properties of a copolymer of vinylidene fluoride and trifluoroethylene blended with poly (methyl methacrylate). Macromolecules 19(9):2450–2452
11. Panwar LS, Kala S, Panwar V, Panwar SS, Sharma S (2017) Design of MEMS piezoelectric blood pressure sensor. In: 2017 3rd International Conference on Advances in Computing, Communication & Automation (ICACCA) (Fall), pp 1–7. IEEE, September
12. Panwar LS, Kala S, Panwar V, Sharma S (2017) Modelling of different MEMS pressure sensors using COMSOL multiphysics. Int J Curr Eng Technol 7(1):243–247

Comparative Study of Retrofitted Columns Using Abaqus Software

Geeta Singh, Tarun Shokeen, and Vidrum Gaur

Abstract In the present study, coating of different materials has been done on a concrete column. The coating is being carried out using the ABAQUS software. The displacement values are obtained after applying 100 kN of force. In this study, we observed that how ABAQUS performs analyses of retrofitting column under transverse load and seismic load. The size of column used for analysis is 300 mm × 300 mm × 1200 mm. The load used is 100 kN. In this study, we design the columns of different materials and compare them with standard RCC column using ABAQUS software. The work involves calculating the maximum deflection produced in different columns and determining the most suitable material for retrofitting. Materials used for coating are carbon fibre, steel, aluminium, e-fibre glass, Douglas fir wood, Teflon. ABAQUS modelling and analysis involve pre-processing, simulation, post-processing. According to the results obtained from ABAQUS, the load resisting property observed in decreasing order is Douglas fir wood > E-fibre glass > Aluminium > Steel > Teflon > Carbon fibre > Composite of Aluminium and Steel. Douglas Fir wood is found to have best load resisting property with minimum deflection of 5.422 mm. E-fibre glass is found to have the second best load resisting property with deflection of 6.536 mm.

Keywords Retrofitted · Columns · Douglas fir wood · ABAQUS

1 Introduction

The buckling of column can be reduced by retrofitting column by the carbon fibre. Carbon fibre winding method is superior to steel plate jacketing in cost and simplicity of structures, and carbon coating method is superior to carbon fibre winding method.

G. Singh · T. Shokeen (✉) · V. Gaur
Delhi Technological University, Delhi, India
e-mail: tarunshokeen@yahoo.in

G. Singh
e-mail: geeta.singh@dce.ac.in

© The Author(s), under exclusive license to Springer Nature Singapore Pte Ltd. 2021
R. M. Singari et al. (eds.), *Advances in Manufacturing and Industrial Engineering*, Lecture Notes in Mechanical Engineering,
https://doi.org/10.1007/978-981-15-8542-5_79

Carbon fibre has used in aircraft and sports goods because of high strength, high elastic modulus, lightweight and high durability. However, transverse reinforcement of ordinary concrete members does not need to be ductile. So that brittle carbon fibre can be applied for transverse reinforcement.

The merits of coating are.

1. The coating protects the column from rusting.
2. Increase the life and load-carrying capacity of the column.
3. Also enhance the aesthetic value of the structure.

Considering the above-mentioned facts in view a study was being planned to study the buckling of column by retrofitting the column with various materials by fixing the column from both ends and using 100 KN transverse uniform distributed loading on 300 × 300 mm size column.

2 Objective of the Study

To design the columns by retrofitting different materials and compare it with standard RCC column using ABAQUS software and to determine the most suitable material for retrofitting.

3 Literature Review

Abrams [2] studied the effect of axial load on the reversed lateral cyclic loading of columns and found that the additional axial loading creasers the stiffness, flexural strength and shear capacity. He found that the flexural capacity of columns increases with axial load but ductility reduces. Mo and Wang [3] conducted experiments on reinforced cement column by doing reverse cyclic loading. They introduced alternate ties to enhance seismic performance. He concluded that the lateral displacement of reinforced cement column is directly proportional to spacing of transverse reinforcement. Montes et al. [4] conducted experiment to calculate the ductility capacity of the column. He introduced reinforcement sizing diagram to determine acceptable reinforcement. He also introduced conjugate gradient search method for finding optimal reinforcement of RCC column.

4 Materials Used

In this study, six types of materials were used for retrofitting the column and the property of each material is given below.

4.1 *Carbon Fibre*

- Poisson's ratio: 0.24
- Young's Modulus: 228,000 MP

4.2 *Steel*

- Poisson's Ratio: 0.28
- Young's Modulus: 210,000 MPA

4.3 *Aluminium*

- Density is 2.7 g/cm^3.
- It increases aesthetic value.

4.4 *E-Fibre Glass*

- Tensile strength—3445 MPA
- Poisson's ratio—0.22
- Density: 258 g/cc
- Compressive strength—1080MPA
- Young's Modulus—78,500

4.5 *Douglas Fir Wood*

- It is used in veneer, plywood and construction lumber.
- It is found in Western North America
- Young's modulus: 13,000 MPA
- Specific Gravity (Basic, 12% MC): 0.45, 0.51
- Poisons Ratio—0.376
- Modulus of Rupture: 86.2 MPa
- Elastic Modulus: 12.17 GPa
- Crushing Strength: 47.9 MPa

4.6 *Polytetrafluoroethylene (Teflon)*

- Young's Modulus: 450 MPA
- Poisson's Ratio: 0.46
- Melting point—327 °C
- Density—2200 kg/m^3

5 Methodology

ABAQUS completes all its process in three phases. Pre-processing, simulation and post-processing are the three phases of ABAQUS.

5.1 *Design Steps in ABAQUS [1]*

ABAQUS consists of the following ten steps in complete analysis.

5.1.1 Creating Parts

- This is the first step in designing of column.
- It involves creating the column and its shell.

5.1.2 Assigning Property

- This step involves assigning properties to the created column.
- The properties assigned are elasticity of the material, Young's modulus and Poisson's ratio of the material.

5.1.3 Assembly of Parts

- In this step, both column and shell are assembled together.
- Type of interaction between the parts is provided.

5.1.4 Creating Steps

- Creation of steps occurs here.
- Steps are created to increase the efficiency of the output.

5.1.5 Creating Interaction (i.e. Assigning Friction Coefficient)

- Interaction between the column and shell is given to provide the amount of friction between the two surfaces.
- Here interaction property-1 is given to the column.

5.1.6 Assigning Boundary Condition

- Encastre boundary condition is applied to the column, i.e. (U1 = U2 = U3 = UR1 = UR2 = UR3 = 0).

5.1.7 Assigning Loads

- In this step, loads are assigned to the column.
- 100 kN load is applied to the column which is uniformly distributed to the column.

5.1.8 Creating Mesh

- Meshing is done to reduce the size of the smallest unit of the column.
- It is done to increase the efficiency and accuracy of the output.
- The approximate global size provided here is 100.

5.1.9 Creating and Submitting Job

- This is the final step in designing of column.
- The created work is submitted to check the errors.

5.1.10 Stress–Strain Curve

- Stress–strain diagram of the chosen node is displayed in the result.
- The red colour shows maximum deflection, and blue colour shows minimum deflection.

6 Results and Analysis

6.1 Standard Column

- Maximum deflection in standard column of Length = 300 mm, Breadth = 300 mm, Height = 1200 mm is 1.563e + 4 mm (Fig. 1).

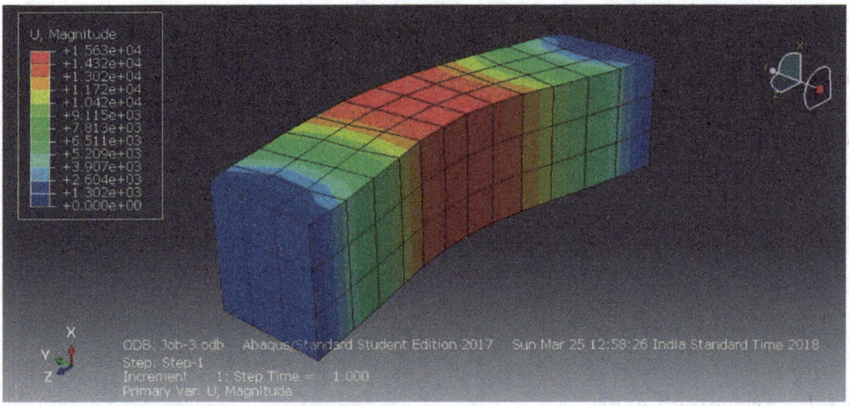

Fig. 1 Image of deflection of standard column in ABAQUS

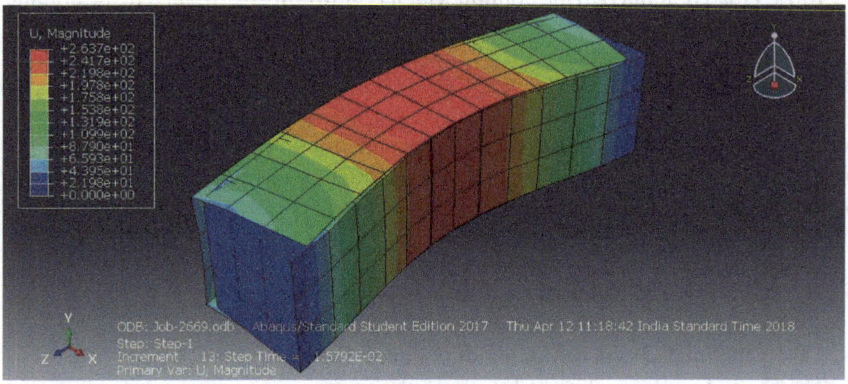

Fig. 2 Image of deflection of carbon fibre column in ABAQUS

6.2 Carbon Fibre

- The maximum deflection observed in the column after coating with carbon fibre is 2.637e + 2 mm.
- The reduction in deflection observed is 1.53e + 4 mm (Fig. 2).

6.3 Steel

- The maximum deflection observed in column after coating with steel is 2.063e + 1 mm.
- The reduction in deflection observed is 1.56e + 4 mm (Fig. 3).

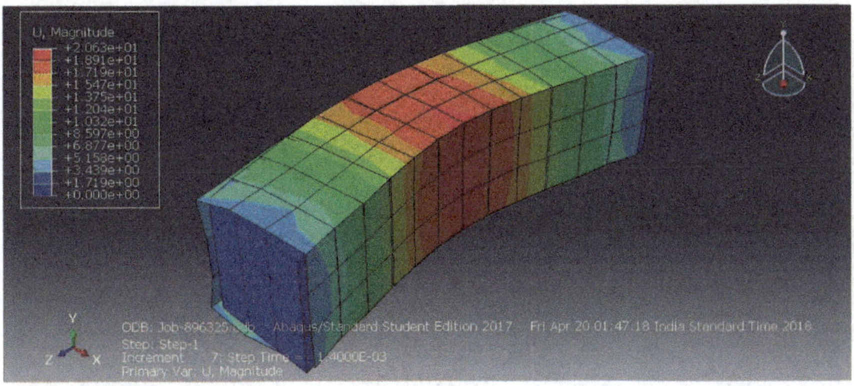

Fig. 3 Image of deflection of steel column in ABAQUS

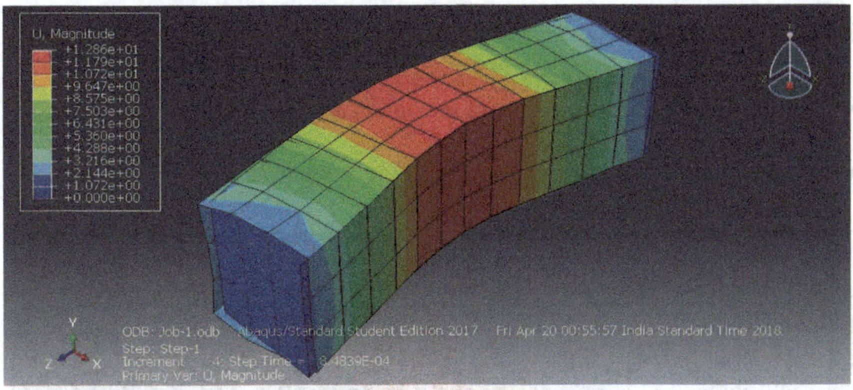

Fig. 4 Image of deflection of aluminium column in ABAQUS

6.4 Aluminium

- The maximum deflection observed in column after coating with aluminium is 1.286e + 1 mm.
- The reduction in deflection observed is 1.56e + 4 mm (Fig. 4).

6.5 Composite of Aluminium and Steel

- The maximum deflection observed in column after coating with composite of aluminium and steel is 1.344e + 4 mm.
- The reduction in deflection observed is 2.1e + 3 mm (Fig. 5).

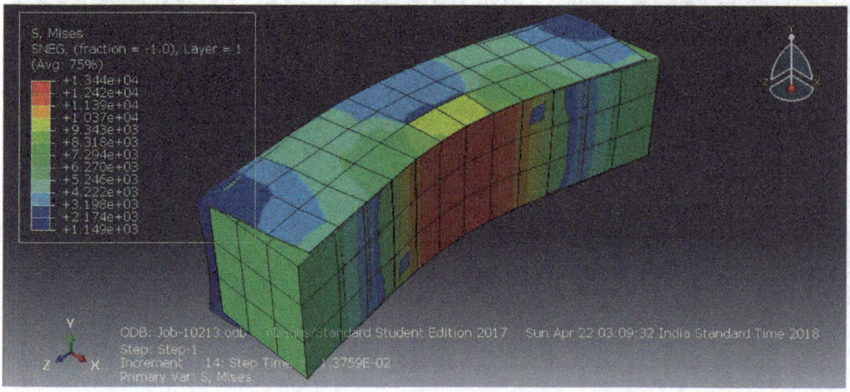

Fig. 5 Image of deflection of composite of aluminium and steel column in ABAQUS

6.6 *E-Fibre Glass*

- The maximum deflection observed in column after coating with e-fibre glass is 6.536 mm.
- The reduction in deflection observed is 1.56e + 4 mm (Fig. 6).

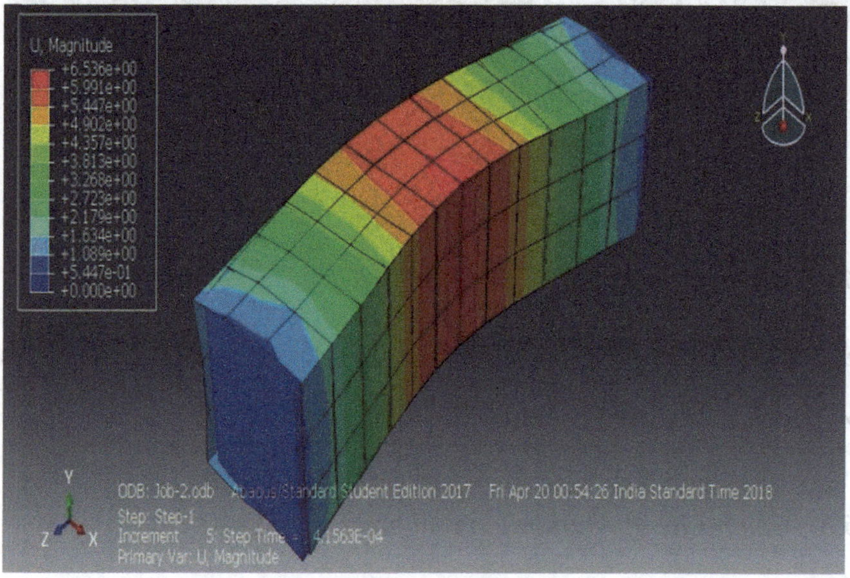

Fig. 6 Image of Deflection of E-Fibre Glass Column in ABAQUS

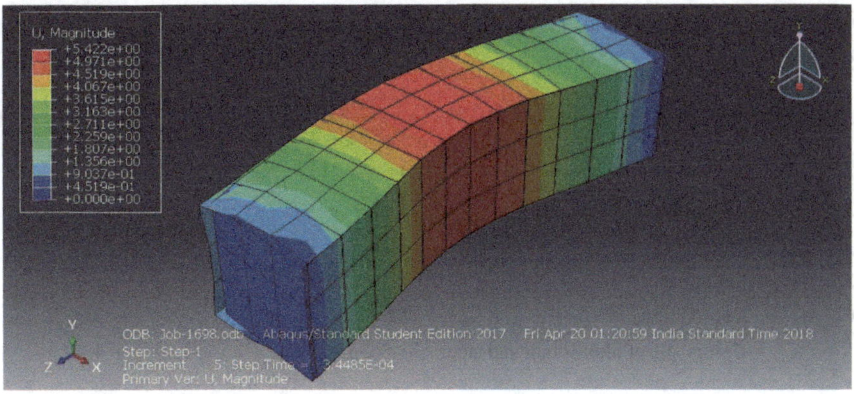

Fig. 7 Image of Douglas Fir wood column in ABAQUS

6.7 Douglas Fir Wood

- The maximum deflection observed in column after coating with Douglas fir wood is 5.422 mm.
- The reduction in deflection observed is 1.56e + 4 mm (Fig. 7).

6.8 Polytetrafluoroethylene (Teflon)

- The maximum deflection observed in column after coating with polytetrafluoroethylene is 6.278e + 1 mm.
- The reduction in deflection observed is 1.55e + 4 mm (Fig. 8).

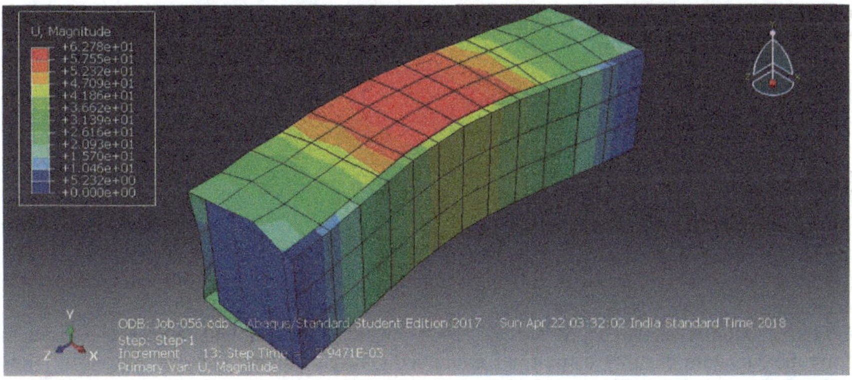

Fig. 8 Image of deflection of polytetrafluoroethylene column in ABAQUS

Table 1 Decreasing order of load resisting property of materials

S. No.	Material	Displacement before Coating (mm)	Displacement after Coating (mm)
1.	Douglas FIR wood	1.563e + 4	5.422
2.	E-Fibre glass	1.563e + 4	6.536
3.	Aluminium	1.563e + 4	1.286e + 1
4.	Steel	1.563e + 4	2.063e + 1
5.	Teflon	1.563e + 4	6.278e + 1
6.	Carbon fibre	1.563e + 4	2.637e + 2
7.	Composite of aluminium and steel	1.563e + 4	1.344e + 4

The columns of same dimensions were analysed with shells of different materials by ABAQUS software. The variations were observed in terms of deflection of columns. The comparison of deflections observed is given in Table 1.

7 Conclusion

The conclusions drawn from the present study are.

- According to the results obtained from ABAQUS, the load resisting property observed in decreasing order is

Douglas Fir Wood > E-Fibre Glass > Aluminium > Steel > Teflon > Carbon Fibre > Composite of Aluminium and Steel.

- **Douglas Fir Wood** is found to have best load resisting property with minimum deflection of 5.422 mm.
- **E-fibre glass** is found to have the second best load resisting property with deflection of 6.536 mm.

Reference

1. ABAQUS/CAE manual
2. Abrams DP (1987) Influence of axil force variation on flexural behaviour of reinforced concrete columns. ACI Struct J 84(3):246–254
3. Martin LMG, Aschheim M, Montes ES (2004) Impact of optimal longitudinal reinforcement on the curvature ductility capacity of reinforced concrete column sections. Mag Concr Res 56(9):499–512. 10.1680/macr.2004.56.9.499
4. Mo YL, Wang SJ (2000) Seismic behaviour of RC columns with various tie configurations. J Struct Eng 126(10). 10.1061/(ASCE)0733-9445(2000)126:10(1122)

Optimal Pricing and Procurement Decisions for Items with Imperfect Quality and Fixed Shelf Life Under Selling Price Dependent and Power Time Pattern Demand

Sonal Aneja and K. K. Aggarwal

Abstract Products with a fixed shelf life pose a challenge to the retail industry. The consumers restrain to buy products that are nearing their expiry date. It is, therefore, of utmost importance to effectively manage the inventory of such products whose demand is influenced by their age. This research paper develops an inventory model for products with fixed shelf life. As the production process is not always perfect so the items received by the retailer may contain a proportion of defectives. Thus, the products go through an inspection process to separate out the defective items. Initially, the demand for the product depends on its selling price and shelf life but as the product approaches its expiry date the retailer offers a price discount before the cycle time to boost the demand of the product and during this period the product follows a power time pattern demand. A mathematical model is developed to estimate the optimal values of selling price and replenishment cycle time so as to maximize the total profit per unit time. Numerical examples are presented to show the validity of the model. Also, a sensitivity analysis is performed on different inventory parameters and their impact on the optimal values of decision variables is observed.

Keywords Inventory · Imperfect quality · Shelf life · Power pattern demand · EOQ

S. Aneja (✉) · K. K. Aggarwal
Department of Operational Research, Faculty of Mathematical Sciences, University of Delhi, Delhi, India
e-mail: sonal.aneja@gmail.com

K. K. Aggarwal
e-mail: kkaggarwal@gmail.com

© The Author(s), under exclusive license to Springer Nature Singapore Pte Ltd. 2021
R. M. Singari et al. (eds.), *Advances in Manufacturing and Industrial Engineering*, Lecture Notes in Mechanical Engineering,
https://doi.org/10.1007/978-981-15-8542-5_80

1 Introduction

Items with a fixed shelf life constitute a significant part of the retail industry. As the life of a product nears its expiry date, it becomes less attractive and less useful to the customers which adversely affect the sale of the product. There are a variety of factors that influence the demand of a product of which price is one. It has been largely observed in grocery retail division that customers are reluctant to buy products that are about to expire. Thus, the retailers need to mark down the price of such products substantially way before their expiry date (or shelf life) so as to attract the customers to buy the product.

The impact of price on demand has been captured by numerous research articles but very few researchers have taken the influence of shelf life on demand. Avinadav et al. [2] addressed the problem of finding an optimal order quantity for items with expiry date. They assumed demand as a declining function of shelf life and developed a profit maximizing problem. Demirag et al. [5] examined an inventory problem for items with demand dependent on the remaining life of the item. They formulated a demand function which was only influenced by the residual life of the item as they argued that reducing the price of the product that are approaching their expiry date to inflate declining demand would also affect the demand of fresh products. Lu et al. [11] introduced a problem for perishable items with a specific age. They argued that items on shelf can be differentiated by their age and age of an item reflects its quality level which has a direct influence on the customer's willingness to pay which resultantly affects the demand of the item. They included all these factors into their model and developed a model to obtain optimal price of the item. Sharma et al. [19] formulated a problem for items that deteriorate with time and have an expiry date. They considered demand as a function of shelf life and price. They developed their model for two cases: (a) shortages at a constant rate, (b) shortages at a variable rate. Shah et al. [18] discussed an inventory control optimization problem for items with imperfect quality and fixed lifetime. They considered demand to be a quadratic function of time and inspection rate as a linear expression of time. The items under consideration also deteriorated with time. Yavari et al. [23] developed a joint optimization inventory model for perishable items with limited shelf life to determine price and order quantity. They considered dynamic pricing and partial backlogging under the condition of time value of money.

Naddor [13] introduced a specific type of demand function and termed it as power pattern of demand in his book titled 'Inventory Systems.' He argued that the demand of a product is dependent upon time as well as cycle length. Later, Datta and Pal [4] formulated an inventory problem with demand following a power time pattern. They developed an ordering quantity model for a product that followed a varying rate of deterioration. Lee and Wu [10] studied an EOQ problem for deteriorating goods with power pattern demand with deterioration rate following Weibull distribution. They developed their model under the assumption that shortages were completely backlogged. Dye [6] amended the model of Lee and Wu [10] by considering partially backlogged shortages with time proportional rate of backlogging. Singh et al. [22]

examined an inventory problem for perishable products. They assumed a power time pattern of demand and developed their model under partial backlogged shortages. Sicilia et al. [20] presented an inventory problem to optimize cycle length and order quantity. They considered demand as a power function of time and analyzed the model under two situations: without shortages and with shortages. Later, Sicilia et al. [21] studied an EPQ model for items whose demand followed a power pattern of time. They considered the rate of production was proportional to rate of demand. They also permitted occurrence of shortages and developed a model with an objective to achieve minimum cost. San-Jose et al. [17] examined an inventory model with demand as an additive function of selling price and power time function. They also assumed holding cost as a power function of time and solved the problem by using the technique of sequential optimization.

Also as the manufacturing process is not always error-free so the presence of imperfect items in the received stock of goods cannot be ignored. There is a vast literature on inventory models that have captured the imperfect nature of items either at the manufacturer or the retailer end. The introductory work in this very area was by Porteus [14] and Rosenblatt and Lee [15]. Later, Salameh and Jaber [16], Cardenass Barron [3], Khan et al. [9], Hsu and Hsu [7], Jain and Aggarwal [8], Mussawai-Haider et al. [12], Aggarwal and Aneja [1], etc., have done considerable work in the field of inventory management of imperfect items.

In the present study, a deterministic economic order quantity model has been developed for products that contain a known proportion of defectives. The price of the product is assumed to be a declining function of shelf life. As the product nears its shelf life, it is offered at a discounted price and after that there is no further reduction in price. The demand of the product varies with price during the pre-discount period, whereas it follows a power-time pattern during the post-discount period. A mathematical problem is formulated by taking all these factors into consideration, and the model is solved to find optimal values of initial selling price, cycle duration, and order quantity.

The remaining paper is organized as follows. The notations and assumptions are mentioned in Sect. 2 which follows with problem definition and model formulation (Sect. 3). Section 4 defines the solution procedure while the numerical illustration and sensitivity analysis is included in Sects. 5 and 6, respectively. The paper ends with Sect. 7 which summarizes the model along with mentioning the possible extension of the present paper.

2 Notations and Assumptions

Following are the notations used in the model:

S cost of ordering ($/order)
c cost of purchase ($/unit)
i cost of inspection ($/unit)

S_L shelf life of the product
t_d time epoch at which discount is offered and $t_d = kS_L$ where $(0 < k < 1)$
I_0 order quantity
t_i inspection time
h cost of holding inventory ($/unit/unit time)
d percentage discount offered on price prevailing at time epoch t_d
α percentage defective
λ inspection rate
p_0 selling price($/unit) (decision variable)
T cycle time (decision variable).

Following are the assumptions used in the model:

1. The rate of replenishment is infinite.
2. The product has a fixed shelf life.
3. A fixed percentage of defectives are present in the received lot and are sold at a price vp_0 where $0 < v < 1$.
4. The inspection process is error-free.
5. A dynamic pricing policy is adopted, and the price at any time is a function of product's shelf life. The price of the product is thus defined as $p(t) = p_0\left(1 - \frac{t}{S_L}\right)$ where p_0 is the initial selling price.
6. The demand of the product is deterministic and is defined as

$$D(t, p(t)) = \begin{cases} D_1(t, p(t)) & , 0 \le t \le t_d \\ D_2(t, (1-d)p(t_d)) & , t_d \le t \le T \end{cases}$$

where $t_d = kS_L$ is the time at which discount is given and it is assumed that $T \le S_L$. The functions $D_1(t, p(t))$ and $D_2(t, (1-d)p(t_d))$ are defined as follows:

(a) For the interval $[0, t_d]$, the demand function is $D_1(t, p(t)) = a_0 - \beta p(t) = a_0 - \beta p_0\left(1 - \frac{t}{S_L}\right)$ which is a linearly decreasing function of selling price $p(t)$ where a_0 is the scale parameter of demand and β is the sensitivity of demand to price such that $a_0 > 0$, $\beta > 0$ and $p_0 \le \frac{a_0}{\beta}$. Also, it is assumed that the selling price is at least equal to the cost of purchase, i.e., $p_0 \ge c$.

(b) For the interval $[t_d, T]$, the demand function is

$$D_2(t, (1-d)p(t_d)) = (a_0 - \beta(1-d)p(t_d))\left(\frac{t^{\frac{1}{n}-1}}{nT^{\frac{1}{n}}}\right)$$

$$= \left(a_0 - \beta(1-d)p_0\left(1 - \frac{t_d}{S_L}\right)\right)\left(\frac{t^{\frac{1}{n}-1}}{nT^{\frac{1}{n}}}\right)$$

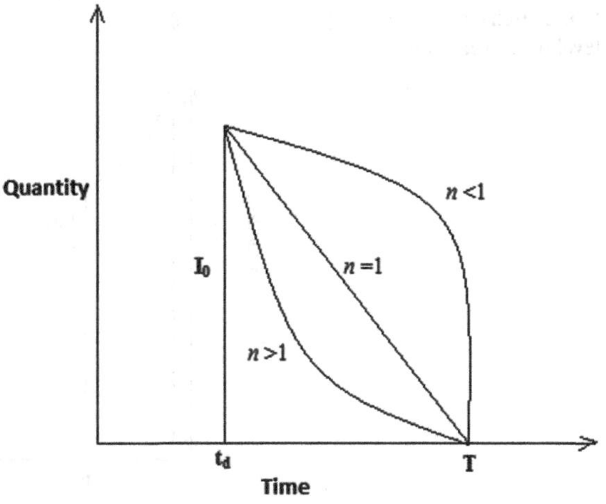

Fig. 1 Behavior of inventory level corresponding to different values of n

which is a function of linearly decreasing discounted selling price and power pattern of time with n as the index of power time pattern of demand such that $n > 0$. The power demand pattern for different values of n is described as:

(i) If $n > 1$, a substantial proportion of demand occurs toward the beginning of the time period.
(ii) If $n = 1$, the demand remains uniform throughout the time period.
(iii) If $0 < n < 1$, a large proportion of demand occurs toward the end of the time period.

This type of demand pattern is suitable to model the demand for different types of product in practical situations.

The following graph (Fig. 1) shows the behavior of inventory in the interval $[t_d, T]$ when the demand follows power pattern of time, and it is drawn corresponding to the functional form of Eq. (6).

3 Problem Definition and Model Formulation

Consider a retailer who receives an order of size I_0 units at the beginning of the cycle, i.e., at $t = 0$. As the retailer knows from past experience that the lot size contains a fixed proportion α of faulty items, thus, the lot passes through an inspection process which is assumed to be 100% perfect. The demand is fulfilled alongside from the stock of perfect items, and the faulty items segregated by inspection are stocked separately. At time t_i, the inspection process finishes and the defectives accumulated till t_i are sold as a separate batch at a reduced price. From Fig. 2, it can be observed that in the interval $[0, t_i]$ the demand is met at the rate $D_1(t, p(t))$ and after withdrawing αI_0 items from the inventory at time t_i, the demand continues to be fulfilled at the same

Fig. 2 Behavior of inventory level over cycle time

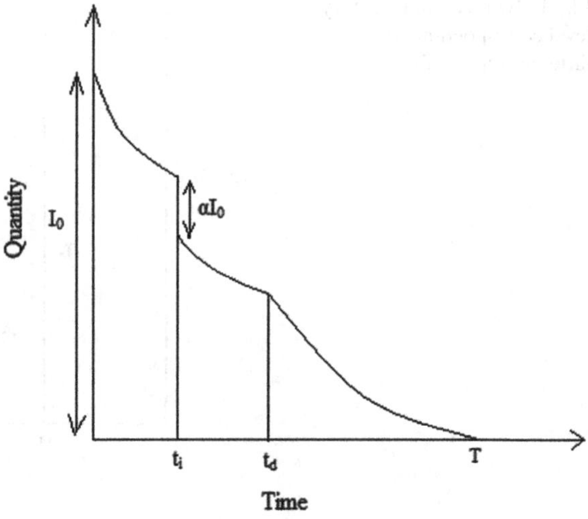

rate, i.e., $D_1(t, p(t))$ till time t_d. At time t_d, the product is offered at a discounted price and the price of the product is no more dynamic. Since the product has a fixed shelf life S_L so it is assumed that $t_d = kS_L$, i.e., a fixed fraction of the shelf life of the product where $0 < k < 1$ and $kS_L \leq T$. During the time period $[t_d, T]$, the product is demanded at the rate $D_2(t, (1-d)p(t_d))$ and at time T the inventory level becomes zero and the cycle ends.

The differential equations representing the rate of change of inventory over different time intervals during the cycle time are as follows:

For the interval $[0, t_i]$,

$$\frac{\mathrm{d}}{\mathrm{d}t}I_1(t) = -\left[a_0 - \beta p_0\left(1 - \frac{t}{S_L}\right)\right] \tag{1}$$

For the interval $[t_i, t_d]$,

$$\frac{\mathrm{d}}{\mathrm{d}t}I_2(t) = -\left[a_0 - \beta p_0\left(1 - \frac{t}{S_L}\right)\right] \tag{2}$$

For the interval $[t_d, T]$,

$$\frac{\mathrm{d}}{\mathrm{d}t}I_3(t) = -\left[a_0 - \beta(1-d)p_0\left(1 - \frac{t_d}{S_L}\right)\right]\left(\frac{t^{\frac{1}{n}-1}}{nT^{\frac{1}{n}}}\right) \tag{3}$$

The solution of Eqs. (1), (2), and (3) with the boundary conditions, $I_1(0) = I_0$, $I_2(t_i) = I_1(t_i) - \alpha I_0$ and $I_2(t_d) = I_3(t_d)$, respectively, is

$$I_1(t) = I_o - a_o t + \beta p_o\left(t - \frac{t^2}{2S_L}\right) \tag{4}$$

$$I_2(t) = (1-\alpha)I_o - a_o t + \beta p_o\left(t - \frac{t^2}{2S_L}\right) \tag{5}$$

$$I_3(t) = (1-\alpha)I_o - a_o t_d + \beta p_o\left(t_d - \frac{t_d^2}{2S_L}\right) + \left(a_0 - \beta(1-d)p_0\left(1 - \frac{t_d}{S_L}\right)\right)\left[\left(\frac{t_d}{T}\right)^{\frac{1}{n}} - \left(\frac{t}{T}\right)^{\frac{1}{n}}\right] \tag{6}$$

As $t_d = kS_L$, so Eq. (6) can be written as

$$I_3(t) = (1-\alpha)I_o - a_o kS_L + \beta p_o kS_L\left(1 - \frac{k}{2}\right) + (a_0 - \beta(1-d)(1-k)p_0)\left[\left(\frac{kS_L}{T}\right)^{\frac{1}{n}} - \left(\frac{t}{T}\right)^{\frac{1}{n}}\right] \tag{7}$$

Also at $t = T, I_3(T) = 0$. Thus, from Eq. (7), we have

$$I_0 = \left(\frac{1}{1-\alpha}\right)\left[a_o kS_L - \beta p_o kS_L\left(1 - \frac{k}{2}\right) + (a_0 - \beta(1-d)(1-k)p_0)\left(1 - \left(\frac{kS_L}{T}\right)^{\frac{1}{n}}\right)\right] \tag{8}$$

As this paper develops a profit maximization model so first we need to formulate the total revenue and total cost components of the underlying problem.

The total revenue (TR) is the revenue obtained from selling perfect (non-defective) and imperfect (defective) items. Thus, we have

Total revenue (TR) = Revenue earned from selling perfect items (R_{Per}) + Revenue earned from selling imperfect items $\left(R_{\text{Imper}}\right)$

$$R_{\text{Per}} = \int_0^{t_d} p(t)D_1(t, p(t))\,\mathrm{d}t + \int_{t_d}^{T} (1-d)(p(t_d))D_2(t, (1-d)p(t_d))\,\mathrm{d}t$$

$$= a_0 p_0 k\left(1 - \frac{k}{2}\right)S_L + \beta p_0^2\{(1-k)^3 - 1\}S_L$$

$$+ (1-d)(1-k)p_0\{a_0 - \beta(1-d)(1-k)p_0\}(nT)\left(1 - \left(\frac{kS_L}{T}\right)^{\frac{1}{n}}\right)$$

$$R_{\text{Imper}} = v\alpha p_0 I_0$$

The total cost (TC) is the sum total of ordering cost, purchase cost, inspection cost, and holding cost. Each of these components is discussed below.

(i) Ordering cost $= S$
(ii) Purchase cost $= cI_0$
(iii) Inspection cost $= iI_0$
(iv) Holding Cost $= h\left[\int_0^{t_i} I_1(t)\,dt + \int_{t_i}^{t_d} I_2(t)\,dt + \int_{t_d}^{T} I_3(t)\,dt\right]$

$$= h\left[\begin{array}{l}(1-\alpha)I_0T + \frac{\alpha I_0^2}{\lambda} + \frac{a_0}{2}k^2S_L^2 - a_0kS_LT \\ + \beta p_0\left(\frac{k}{3} - \frac{1}{2}\right)k^2S_L^2 + \beta p_0 k\left(1 - \frac{k}{2}\right)S_LT \\ + \{a_0 - \beta(1-d)(1-k)p_0\}\left\{\begin{array}{l}\left(\frac{kS_L}{T}\right)^{\frac{1}{n}}T - \left(\frac{n}{n+1}\right)T \\ -\left(\frac{1}{n+1}\right)kS_L\left(\frac{kS_L}{T}\right)^{\frac{1}{n}}\end{array}\right\}\end{array}\right]$$

Thus, total profit (*TP*) is given as:

$$\begin{aligned} TP &= TR - TC \\ &= a_0p_0k\left(1 - \frac{k}{2}\right)S_L + \beta p_0^2\{(1-k)^3 - 1\}S_L + v\alpha p_o I_o - S - (c+i)I_0 \\ &\quad + (1-d)(1-k)p_0\{a_0 - \beta(1-d)(1-k)p_0\}\left(1 - \left(\frac{kS_L}{T}\right)^{\frac{1}{n}}\right) \\ &\quad - h\left[\begin{array}{l}(1-\alpha)I_0T + \frac{\alpha I_0^2}{\lambda} + \frac{a_0}{2}k^2S_L^2 - a_0kS_LT \\ + \beta p_0\left(\frac{k}{3} - \frac{1}{2}\right)k^2S_L^2 + \beta p_0 k\left(1 - \frac{k}{2}\right)S_LT \\ + \{a_0 - \beta(1-d)(1-k)p_0\}\left\{\begin{array}{l}\left(\frac{kS_L}{T}\right)^{\frac{1}{n}}T - \left(\frac{n}{n+1}\right)T \\ -\left(\frac{1}{n+1}\right)kS_L\left(\frac{kS_L}{T}\right)^{\frac{1}{n}}\end{array}\right\}\end{array}\right]\end{aligned}$$

The total profit per unit time $(\overline{TP})$ is

$$\overline{TP}(p_0, T) = \frac{TP}{T}$$

$$= \frac{1}{T}\left[\begin{array}{l} = a_0 p_0 k\left(1-\frac{k}{2}\right)S_L + \beta p_0^2\left\{(1-k)^3-1\right\}S_L \\ +v\alpha p_o I_o - S - (c+i)I_0 \\ +(1-d)(1-k)p_0\{a_0-\beta(1-d)(1-k)p_0\}\left(1-\left(\frac{kS_L}{T}\right)^{\frac{1}{n}}\right) \\ -h\left[\begin{array}{l} (1-\alpha)I_0T + \frac{\alpha I_0^2}{\lambda} + \frac{a_0}{2}k^2S_L^2 - a_0kS_LT \\ +\beta p_0\left(\frac{k}{3}-\frac{1}{2}\right)k^2S_L^2 + \beta p_0 k\left(1-\frac{k}{2}\right)S_LT \\ +\{a_0-\beta(1-d)(1-k)p_0\}\left\{\begin{array}{l}\left(\frac{kS_L}{T}\right)^{\frac{1}{n}}T - \left(\frac{n}{n+1}\right)T \\ -\left(\frac{1}{n+1}\right)kS_L\left(\frac{kS_L}{T}\right)^{\frac{1}{n}}\end{array}\right\}\end{array}\right]\end{array}\right] \tag{9}$$

Using Eq. (8), the above function can be written as:

$$\overline{TP}(p_0, T) = \frac{1}{T}\left[\begin{array}{l} = a_0 p_0 k\left(1-\frac{k}{2}\right)S_L + \beta p_0^2\left\{(1-k)^3-1\right\}S_L \\ + v\alpha p_o I_o - S - (c+i)I_0 \\ +(1-d)(1-k)p_0\{a_0-\beta(1-d)(1-k)p_0\}\left(1-\left(\frac{kS_L}{T}\right)^{\frac{1}{n}}\right) \\ -h\left[\begin{array}{l} \left\{\begin{array}{l} a_okS_L - \beta p_o kS_L\left(1-\frac{k}{2}\right) \\ +(a_0-\beta(1-d)(1-k)p_0)\left(1-\left(\frac{kS_L}{T}\right)^{\frac{1}{n}}\right)\end{array}\right\}T \\ +\frac{\alpha}{(1-\alpha)^2\lambda}\left\{\begin{array}{l} a_okS_L - \beta p_o kS_L\left(1-\frac{k}{2}\right) \\ +(a_0-\beta(1-d)(1-k)p_0)\left(1-\left(\frac{kS_L}{T}\right)^{\frac{1}{n}}\right)\end{array}\right\}^2 \\ +\frac{a_0}{2}k^2S_L^2 - a_0kS_LT + \beta p_0\left(\frac{k}{3}-\frac{1}{2}\right)k^2S_L^2 \\ +\beta p_0 k\left(1-\frac{k}{2}\right)S_LT \\ +\{a_0-\beta(1-d)(1-k)p_0\}\left\{\begin{array}{l}\left(\frac{kS_L}{T}\right)^{\frac{1}{n}}T - \left(\frac{n}{n+1}\right)T \\ -\left(\frac{1}{n+1}\right)kS_L\left(\frac{kS_L}{T}\right)^{\frac{1}{n}}\end{array}\right\}\end{array}\right]\end{array}\right] \tag{10}$$

The objective of our study is to maximize the total profit per unit time, and thus, the optimization problem is formulated as:

$$\left.\begin{array}{l} \underset{p_0, T}{Max}\ \overline{TP}(p_0, T) \\ \text{s.t.} \\ T \le S_L,\ kS_L \le T,\ c \le p_0 \le \frac{a_0}{\beta} \\ p_0, T \ge 0. \end{array}\right\} \quad (11)$$

4 Solution Methodology

Due to the complex nature of total profit function (10), a closed-form solution cannot be obtained for the problem (11), and also, the concavity of the $\overline{TP}$ function cannot be derived analytically. Thus, the nonlinear programming problem (11) has been solved on LINGO to find optimal values of p_0 and T, and the concavity of the $\overline{TP}$ function is established graphically by using MATLAB.

5 Numerical Examples

This section presents numerical illustrations to show the application of model developed in Sect. 4. The optimal values of initial selling price $\left(p_0^*\right)$, cycle length (T^*), order quantity $\left(I_0^*\right)$, and maximum profit per unit time $\left(\overline{TP}^*\right)$ are obtaining by assigning different values to the parameters of the model.

The following numerical examples are solved using different parameters:

Example 1 $S = \$250/\text{order}$, $c = \$\,100/\text{unit}$, $h = \$\,15/\text{unit/yr}$, $i = \$\,10/\text{unit}$, $\alpha = 0.06$, $\beta = 0.7$, $d = 0.3$, $v = 0.6$, $k = 0.45$, $S_L = 0.6$ yr, $n = 0.5$, $a_0 = 2000$, $\lambda = 20,000$ units/yr. By using the software LINGO, optimal solution is obtained as: $p_0^* = \$\,1076$, $T^* = 0.4838$ yr, $I_0^* = 1605.025$ units and $\overline{TP}\left(p_0^*, T^*\right) = \$\,704408.4$. Also shown below (Fig. 3) is the three-dimensional concave graph for the $\overline{TP}$ function in p and T corresponding to the given data.

Example 2 $S = \$250/\text{order}$, $c = \$\,120/\text{unit}$, $h = \$\,10/\text{unit/yr}$, $i = \$\,10/\text{unit}$, $\alpha = 0.06$, $\beta = 0.7$, $d = 0.3$, $v = 0.6$, $k = 0.35$, $S_L = 0.6$ yr, $n = 1$, $a_0 = 1800$, $\lambda = 20000$ units/yr. By using LINGO, the optimal solution is obtained as: $p_0^* = \$\,1069.8$, $T^* = 0.4503$ yr, $I_0^* = 1092.54$ units and $\overline{TP}\left(p_0^*, T^*\right) = \$\,579432.3$. Also shown below (Fig. 4) is the three-dimensional concave graph for the $\overline{TP}$ function in p and T corresponding to the given data.

Example 3 $S = \$200/\text{order}$, $c = \$\,140/\text{unit}$, $h = \$\,15/\text{unit/yr}$, $i = \$\,10/\text{unit}$, $\alpha = 0.06$, $\beta = 0.65$, $d = 0.4$, $v = 0.9$, $k = 0.40$, $S_L = 0.7$ yr, $n = 1.5$, $a_0 = 2200$, $\lambda = 20000$ units/yr. By using LINGO, the optimal solution is obtained as: $p_0^* = \$\,1086.089$, $T^* = 0.5037$ yr, $I_0^* = 1157.66$ units and $\overline{TP}\left(p_0^*, T^*\right) =$

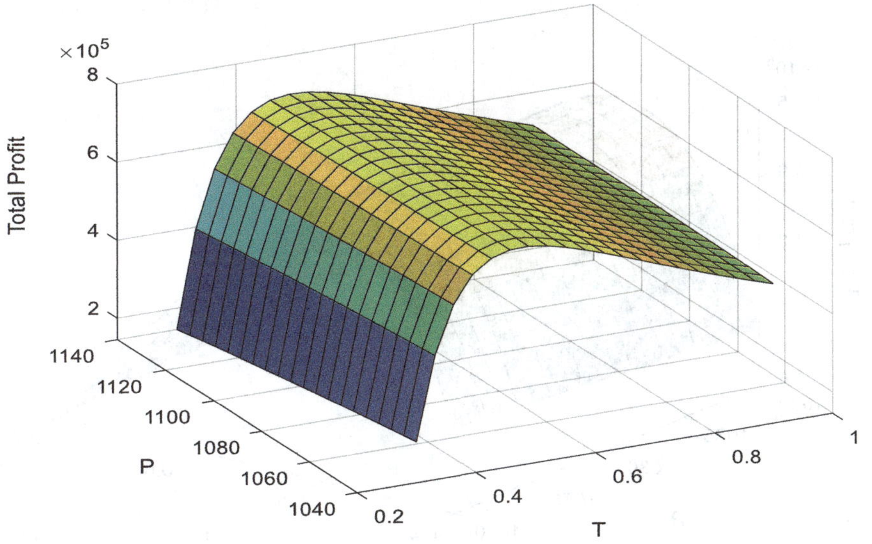

Fig. 3 Concavity of total profit function in p and T

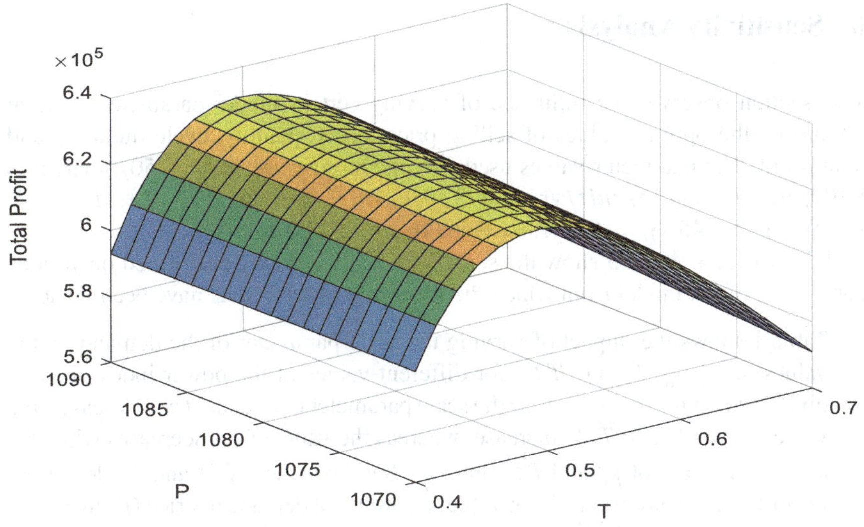

Fig. 4 Concavity of total profit function in p and T

\$ 496764.8. Also shown below (Fig. 5) is the three-dimensional concave graph for the $\overline{TP}$ function in p and T corresponding to the given data.

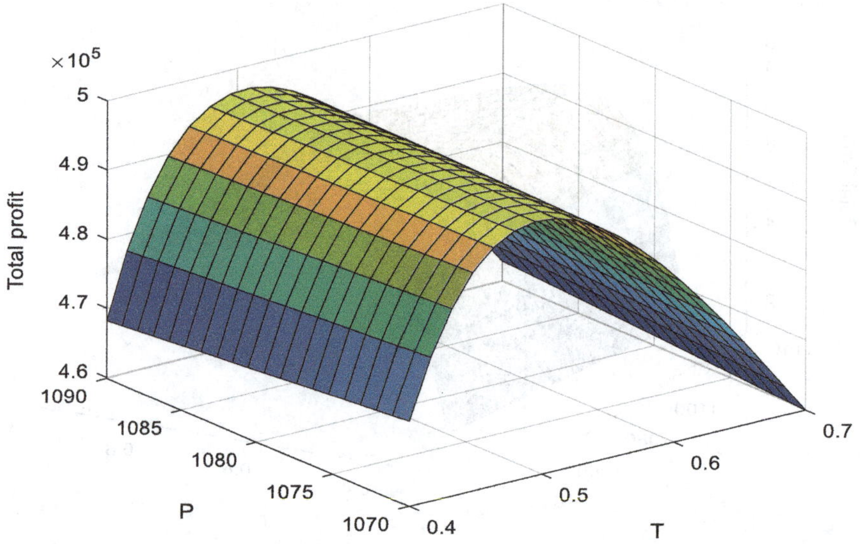

Fig. 5 Concavity of total profit function in p and T

6 Sensitivity Analysis

This section observes the influence of varying certain model parameters (one at a time) on the optimal values of selling price, order quantity, cycle duration, and total profit. The parameter values used for this purpose are: $S = \$250/\text{order}$, $c = \$100/\text{unit}$, $h = \$15/unit/yr$, $i = \$10/unit$, $\alpha = 0.06$, $\beta = 0.8$, $d = 0.3$, $v = 0.6$, $k = 0.45$, $S_L = 0.6$ yr, $a_0 = 2000$, $\lambda = 20000$ units/yr.

Tables 1, 2, 3, 4 and 5 show the sensitivity analysis being performed on various parameters of the model from which the following observations have been made:

1. Table 1 shows the impact of varying the scale parameter of the demand on the values of p_0^*, I_0^*, T^* and $\overline{TP}^*$ for different values of the power index $'n'$. It is observed that for $0 < n \leq 1$, as the scale parameter of demand (a_0) increases, the values of p_0^*, I_0^* and $\overline{TP}^*$ increase, whereas the value of T^* decreases while for $n > 1$ the values of p_0^* and $\overline{TP}^*$ increase but the values of I_0^* and T^* decrease.
2. From Table 2, it is observed that as the sensitivity of demand to price (β) increases the values of p_0^* and $\overline{TP}^*$ both decrease, whereas the values of I_0^* and T^* increase for all values of $n > 0$.
3. Table 3 shows the impact of varying discount rate (d) on the optimal values of decision variables. As the value of $'d'$ increases for $0 < n \leq 1$, there is a decrease in the values of p_0^*, T^* and $\overline{TP}^*$ but the optimal order quantity $\left(I_0^*\right)$ increases, whereas for $n > 1$ the optimal values of all decision variable along with corresponding total profit decrease.

Table 1 Impact of scale parameter of demand (a_0) on optimal values of decision variables

	a_0	p_0^*	T^*	I_0^*	$\overline{TP}^*$
$n = 0.5$	1200	679.32	0.5189	1013.94	163501.5
	1600	876.48	0.4942	1324.35	390116.5
	2000	1076	0.483775	1605.025	704408.4
	2400	1276.449	0.478017	1957.516	1106146
	2800	1477.354	0.474382	2276.112	1595259
$n = 1$	1200	610.27	0.6	828.1488	96184.43
	1600	778.5173	0.552141	1060.361	243259.5
	2000	949.5571	0.530874	1298.017	449778.3
	2400	1122.48	0.519765	1539.626	715389.5
	2800	1296.301	0.512975	1783.118	1039989
$n = 1.5$	1200	552.59	0.6	1279.557	67631.84
	1600	700.8886	0.543923	1157.743	180966.4
	2000	850.1303	0.51553	1070.74	341704.7
	2400	1002.285	0.501552	1004.56	549434.3
	2800	1155.768	0.493313	947.365	804025.7

Table 2 Impact of shape parameter of demand (β) on optimal values of decision variables

	β	p_0^*	T^*	I_0^*	$\overline{TP}^*$
$n = 0.5$	0.48	1745.613	0.471245	1586.633	1429917
	0.64	1326.651	0.476984	1595.051	976142.5
	0.8	1076	0.483775	1605.025	704408.4
	0.96	909.6159	0.491909	1616.835	523765.8
	1.12	791.499	0.501788	1630.822	395256.3
$n = 1$	0.48	1528.857	0.507276	1265.721	938586.4
	0.64	1165.909	0.517832	1280.385	632748.2
	0.8	949.5571	0.530874	1298.017	449778.3
	0.96	806.7677	0.547255	1319.284	328347.5
	1.12	706.2867	0.56821	1344.986	242206
$n = 1.5$	0.48	1361.578	0.486616	1036.369	729639.5
	0.64	1040.632	0.499208	1140.991	486877.8
	0.8	850.1303	0.51553	1070.74	341704.7
	0.96	725.3449	0.537177	1275.062	245453.7
	1.12	638.6308	0.566581	1332.367	177319.4

Table 3 Impact of discount factor (d) on optimal values of decision variables

	d	p_0^*	T^*	I_0^*	$\overline{TP}^*$
$n = 0.5$	0.1	1112.662	0.485325	1521.803	926177.7
	0.2	1099.375	0.485316	1563.547	819098
	0.3	1076	0.483775	1605.025	704408.4
	0.4	1040.688	0.479605	1642.834	583347.7
	0.5	991.0218	0.470423	1669.212	457872.1
$n = 1$	0.1	996.5375	0.546199	1260.611	586970.7
	0.2	977.764	0.540898	1282.564	520109.9
	0.3	949.5571	0.530874	1298.017	449778.3
	0.4	909.2953	0.51253	1298.502	377027.8
	0.5	850.5935	0.476639	1299.09	303813.2
$n = 1.5$	0.1	906.7385	0.552814	1096.02	437886.6
	0.2	883.4397	0.538713	1082.241	390533
	0.3	850.1303	0.51553	1070.74	341704.7
	0.4	801.5085	0.475366	1029.434	292799.7
	0.5	711.1361	0.387834	868.51	247999.6

Table 4 Impact of shelf life (S_L) on optimal values of decision variables

	S_L	p_0^*	T^*	I_0^*	$\overline{TP}^*$
$n = 0.5$	0.4	1285.981	0.335939	1478.181	1150145
	0.5	1165.426	0.411878	1563.858	872734.2
	0.6	1076	0.483775	1605.025	704408.4
	0.7	1006.867	0.551824	1709.339	593548.7
	0.8	951.715	0.616261	1773.885	516145
$n = 1$	0.4	1145.292	0.389483	1189.528	703640.2
	0.5	1032.857	0.464871	1248.678	544900
	0.6	949.5571	0.530874	1298.017	449778.3
	0.7	884.993	0.588343	1340.125	388063.2
	0.8	833.199	0.638137	1376.482	345780.6
$n = 1.5$	0.4	1032.596	0.4	998.9262	508256.4
	0.5	929.6158	0.466959	1130.074	403285.4
	0.6	850.1303	0.51553	1070.74	341704.7
	0.7	787.873	0.551644	1272.873	302985.5
	0.8	737.1292	0.576986	1314.014	277657.5

Table 5 Impact of time to discount (t_d) on optimal values of decision variables

	k	$t_d = kS_L$	p_0^*	T^*	I_0^*	$\overline{TP}^*$
$n = 0.5$	0.35	0.21	1150.845	0.385763	1492.43	1073565
	0.40	0.24	1111.531	0.43612	1568.717	864351.8
	0.45	0.27	1076	0.483775	1605.025	704408.4
	0.50	0.30	1042.621	0.52794	1705.275	577704.4
	0.55	0.33	1009.825	0.567388	1762.804	474329.1
$n = 1$	0.35	0.21	1027.492	0.441866	1202.997	666626
	0.40	0.24	986.7193	0.489995	1254.988	543313.7
	0.45	0.27	949.5571	0.530874	1298.017	449778.3
	0.50	0.30	913.8186	0.562409	1329.032	376383
	0.55	0.33	876.8656	0.581038	1341.276	317354.4
$n = 1.5$	0.35	0.21	931.9109	0.453934	1019.368	488027
	0.40	0.24	889.6665	0.49093	1051.89	404327.2
	0.45	0.27	850.1303	0.51553	1070.74	341704.7
	0.50	0.30	809.9682	0.523911	1068.797	293676
	0.55	0.33	763.2703	0.508249	1027.815	256837.5

4. Table 4 shows the impact of varying shelf life (S_L) on optimal values of decision variables. As the value of S_L increases, the values of p_0^* and $\overline{TP}^*$ decrease, whereas there is an increase in the values of I_0^* and T^* for all values of $n > 0$.
5. Table 5 shows the influence of varying time to discount (t_d) on optimal values of decision variables. As t_d increases and moves closer to the shelf life the values of p_0^* and $\overline{TP}^*$ decrease, whereas the values of I_0^* and T^* increase for all values of $n > 0$

7 Conclusion

The present work addresses the problem of optimizing the total profit by finding the optimal values of selling price and cycle time. The model developed considers a different pre-discount and post-discount demand function. The price of the product is assumed to be dynamic and is a declining function of shelf life. The demand is defined as a linear function of price till a specific period of time which is the pre-discount period. However, as the product has a fixed shelf life so the product draws less customer attention as it nears its shelf life. In order to keep up the demand of the product, the retailer offers a discount on the price at a time which is a fixed proportion of the product's shelf life and is less than the cycle time. This time period is called the post-discount period, and during this period, demand is multiplicative function of discounted selling price and power-time pattern. A mathematical model is developed for this problem, and an optimal solution is obtained. Also by varying certain parameters and keeping other fixed at a time, the behavior of optimal values of selling price, cycle time, order quantity, and total profit has been discussed. This model will help managers to take decision in setting price and finding optimal inventory level for fixed shelf life items. The model can be further extended to include stochastic demand rate, credit policy, shortages, time value of money, etc.

References

1. Aggarwal KK, Aneja S (2016) An EOQ model with inspection error, rework and sales return. Int J Adv Oper Manage 8(3):185–199
2. Avinadav T, Arponen T (2009) An EOQ model for items with a fixed shelf-life and a declining demand rate based on time-to-expiry: technical Note. Asia-Pacific J Oper Res 26(6):759–767
3. Cárdenas-Barrón LE (2000) Observation on: economic production quantity model for items with imperfect quality. Int J Prod Econ 67(2):201
4. Datta TK, Pal AK (1988) Order level inventory system with power demand pattern for items with variable rate of deterioration. Int J Pure Appl Math 19(11):1043–1053
5. Demirag OC, Kumar S, Rao KSM (2017) A note on inventory policies for products with residual-life-dependent demand. Appl Math Modell 43:647–658
6. Dye CY (2004) A note on: an EOQ model for items with Weibull distributed deterioration, shortages and power demand pattern. Inf Manage Sci 15(2):81–84
7. Hsu JT, Hsu LF (2013) An EOQ model with imperfect quality items, inspection errors, shortage backordering, and sales returns. Int J Prod Econ 143(1):162–170
8. Jain D, Aggarwal KK (2012) The effect of inflation-induced demand and trade credit on ordering policy of exponentially deteriorating and imperfect quality items. Int Trans Oper Res 19(6):863–889
9. Khan M, Jaber MY, Bonney M (2011) An economic order quantity (EOQ) for items with imperfect quality and inspection errors. Int J Prod Econ 133(1):113–118
10. Lee WC, Wu JW (2002) An EOQ model for items with Weibull distributed deterioration, shortage and power demand pattern. Int J Information Manage Sci 13(2):19–34
11. Lu J, Zhang J, Lu F, Tang W (2017) Optimal pricing on an age-specific inventory system for perishable items. Oper Res Int J 1–21
12. Moussawi-Haidar L, Salameh M, Nasr W (2014) Effect of deterioration on the instantaneous replenishment model with imperfect quality items. Appl Math Modell 38(24):5956–5966
13. Naddor E (1966) Inventory systems. Wiley, New York
14. Porteus EL (1986) Optimal lot sizing, process quality improvement and setup cost reduction. Oper Res 34(1):137–144
15. Rosenblatt MJ, Lee HL (1986) Economic production cycles with imperfect production process. IIE Trans 18(1):48–55
16. Salameh MK, Jaber MY (2000) Economic production quantity model for items with imperfect quality 64(1):59–64
17. San-Jose LA, Sicilia J, Cardenas-Barron LE, Gutierrez JM (2019) Optimal price and quantity under power demand pattern and non-linear holding cost. Comput Industrial Eng 129:426–434
18. Shah NH, Chaudhari U, Jani MR (2018) Optimum inventory control for imperfect quality item with maximum life-time under quadratic demand and preservation technology investment. Int J Appl Eng Res 13(16):12475–12485
19. Sharma S, Singh S, Singh SR (2018) An inventory model for deteriorating items with expiry date and time varying holding cost. Int J Procurement Manage 11(5):650–666
20. Sicilia J, Febles-Acosta J, González-De-la-Rosa M (2012) Deterministic inventory systems with power demand pattern. Asia-Pacific J Oper Res 29(5), article no. 1250025 (28 pages)

21. Sicilia J, González-De-la-Rosa M, Febles-Acosta J, Alcaide-López-de-Pablo D (2014) Optimal policy for an inventory system with power demand, backlogged shortages and production rate proportional to demand rate. Int J Prod Econ 155:163–171
22. Singh JJ, Singh SR, Dutt R (2009) An EOQ model for perishable items with power demand and partial backlogging. Int J Prod Econ 15(1):65–72
23. Yavari M, Zaker H, Emamzadeh ESM (2019) Joint dynamic pricing and inventory control for perishable products taking into account partial backlogging and inflation. Int J Appl Comput Math 5(1) (28 pages)

Experimental Analysis of Portable Optical Solar Water Heater

Hasnain Ali, Ovais Gulzar, K. Vasudeva Karanth, Mohammad Anaitullah Hassan, and Mohammad Zeeshan

Abstract The unchecked consumption of natural resources, to meet our daily requirements, has led to a crisis situation. Natural resources are depleting at alarming levels. At the other extreme, even when the total solar energy absorbed by Earth's atmosphere, oceans and land masses is approximately 3,850,000 exajoules (EJ) per year, barely a fraction has been exploited to our benefit. To a generation seeking solutions to its energy needs, solar energy offers tremendous potential. Solar hot water heater can save billion thermos of natural gas a year. Solar hot water systems capture energy from the sun to heat water for homes and businesses, thereby displacing the use of natural gas, or in some cases electricity, with free and limitless solar energy. In this study, we have designed and fabricated a solar collector employing plano-convex lens to extract maximum solar energy in limited space. We have also benchmarked the performance of the fabricated solar collector with the existing conventional collectors. Since in the current setup, the average temperature rise of 14 °C is observed for a concentrator surface area of 0.031 m^2 (446 °C/m^2) against the temperature rise of 20.5 °C for a given concentrator area of 0.216 m^2 (95 °C/m^2) in case of a flat plate collector type heater, the present system is found to be more efficient when less concentrator surface area is available for water heating applications. The present system offers the desired prospects of lesser space acquisition and portability benefits at lesser material and labor cost (about INR 2000) which renders the system effective for basic household usage. However, the present system has its own inherent shortcomings of higher convective heat losses due to which lesser temperature rise is achieved when compared to existing conventional collectors.

H. Ali (✉)
Department of Civil Engineering, Indian Institute of Technology Delhi, Delhi, India
e-mail: hasnain.mit14@gmail.com

O. Gulzar
Department of Mechanical Engineering, Islamic University of Science and Technology, Awantipora, J&K, India
e-mail: bhatovais@gmail.com

K. V. Karanth · M. A. Hassan · M. Zeeshan
Department of Mechanical & Mfg. Engineering, Manipal Institute of Technology, Manipal, Karnataka, India

© The Author(s), under exclusive license to Springer Nature Singapore Pte Ltd. 2021
R. M. Singari et al. (eds.), *Advances in Manufacturing and Industrial Engineering*, Lecture Notes in Mechanical Engineering,
https://doi.org/10.1007/978-981-15-8542-5_81

Keywords Solar water heater · Portable · Plano-convex lens

1 Introduction

Solar water heater system captures energy from the sun to heat water for domestic use and industrial applications, thereby conserving natural gas, and electricity and is available for free in its unlimited form. Solar water heater could save some billion thermos of natural gas a year [1, 2]. With climate change happening, it is more imperative to shift toward solar-based technologies, especially in developing countries [3], the Earth receives 174 peta watts (PW) of incoming solar radiation (insolation) at the upper atmosphere. Approximately 30% is reflected back to space while the rest is absorbed by clouds, oceans and land masses. Though the total solar energy absorbed by Earth's atmosphere, oceans and land masses is approximately 3,850,000 exajoules (EJ) per year. We have energy crisis in many parts of world and India is no exception. In crisis conditions, energy usage should be appropriate and meticulous. Our technology for energy harvest, conversion and efficient application is still lacking behind with that of developed world. We should give emphasis for the development of indigenous technology, where solar water heating for domestic as well industrial usage could be an effective way of saving conventional energy [4–6]. It could be a contributing factor for the environment. Utilization of solar thermal energy is relatively in a nascent state in our country. Therefore, before taking any wider initiative in this field smaller units should be tried first in terms of performance evaluation. Further, new technologies, optimization algorithms and nanotechnologies can help improve our standard of living and reduce our dependence on non-renewable sources of energy [7–9, 14–17].

In this paper, we present an experimental study which is carried out with three objectives in mind. First, to design a solar collector for extracting maximum solar energy within a limited space and convert the same into useful form of energy such as heating of water, using suitable heat absorbing and heat retaining materials. Second, to fabricate the designed solar collector and optimize it for the experimental study. Third, to benchmark the performance of the fabricated solar collector with the existing conventional collectors.

2 Literature Review

Solar flat plate collectors have been in common use for both domestic and industrial purposes. This may be due to a simple design and low cost of maintenance of these collectors. Solar collectors are a special kind of heat exchangers that transform radiant energy from the sun to the internal energy of the transport media. However, flat plate solar collectors are associated with higher heat losses and hence lower thermal performance. A large number of research investigations have been undertaken both numerically and experimentally to enhance the thermal performance of flat plate solar collectors. Kumar and Rosen [1] investigated the thermal performance of an integrated solar water heater with a corrugated absorber plate. They found that the collector produced higher temperatures for longer time compared to the plain surface. Prasad [4] experimentally determined that the use of solar tracking in flat plate collectors can achieve 21% higher thermal efficiency. Gao et al. [9] experimentally found that the cross-corrugated collectors have higher thermal performance as compared to simple flat plate collectors. V-groove collectors also have higher efficiency as reported by Karim et al. [10]. Saini and Verma [11] used the concept of dimple-shaped roughness geometry on the absorber plate to enhance surface area for heat absorption as well as to provide turbulence for air heating applications. Introducing turbulence in the water flow path is another way of enhancing the thermal performance of solar collector. Kumar and Prasad [12] found that the use of twisted tape inserts significantly increases the thermal performance of solar water heater.

It is noted from the literature that the thermal performance of flat plate collector over an optical solar water heater has not been the focus of study. Hence, an attempt has been made in this project to explore the effectiveness of using an optical solar water heater using a large size lens.

3 Methodology and Experimental Setup

The following methodology was adopted in this study.

- Building the framework/structure to support a large size lens and optimize the heat extraction device by using copper pipes and heat absorbing and heat retaining materials.
- Recording the temperature rise of the fluid that is achieved by the experimental test rig for different operating conditions.

3.1 Components of Experimental Test Rig

The experimental setup consists of a large size optical lens, an absorber copper vessel and an adjustable framework which supports the lens along with the copper vessel, a storage tank, connecting pipes which are insulated, and temperature and heat flux measuring devices.

Fig. 1 Optical lens used

Optical Lens: Optical lens as shown in Fig. 1 has a diameter of 230 mm and a thickness of 8 mm at its periphery. It is flat on one side and a convex on the other side to provide a focal length of 270 mm, and the lens is made up of prismatic glass material. The lens is the most critical component of the experimental test rig as the title of the project indicates.

Absorber copper vessel: Absorber copper vessel is made up of a closed cylindrical container of 30 mm diameter and height 30 mm with an inlet pipe of 6 mm diameter at the bottom for water inlet and a 6 mm diameter pipe for the water to exit at its upper end as shown in Fig. 2. The copper vessel has a sand bath enclosing it and helps in retaining the heat that is absorbed by the copper vessel due to radiation and convection. **Wooden structure**: The wooden structure consists of an upper block and a lower block. The upper block supports the lens and the copper vessel. It has adjustable screws so that the copper vessel can be moved vertically to adjust the top surface of the absorber plate to the focus point of the lens. The lens on both sides is sandwiched between two wooden plates for the purpose of rigidity while tilting the same. The upper block is supported by the lower block which rests on the ground and has a tilting arrangement incorporated for the purpose of positioning the lens

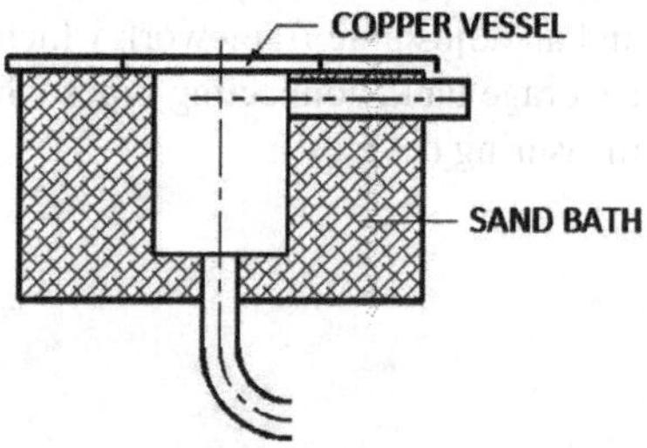

Fig. 2 Schematic diagram of the copper vessel

Fig. 3 Frame

in its normal position to the sun's rays. To provide inclination for the structure, two frames of 30 mm thickness and length of 120 mm and width of 300 mm is created. The structure is suspended between two frames through a 6 inch bolt on the upper side of the frame. A wing nut made of brass is used to lock the bolts at specific angles (Fig. 3).

Storage tank: The storage tank is made up of a 1 L insulated plastic container which is enclosed on the upper side with a provision to mount a thermometer. It has holes on the bottom and upper side to accommodate the insulated pipes. A schematic of the same is shown in Fig. 4. The storage tank is supported on a wooden stand so as to maintain the tank well above the lens structure. The connecting pipes from the storage tank to the copper vessel are of 6 mm diameter and are insulated so as to minimize the heat loss. **Measuring devices**: Pyranometer is used to measure the heat flux due to solar radiation at the place of experimentation. Figure 5 shows a view of the pyranometer used in the experimentation.

Fig. 4 Storage tank

Fig. 5 Pyranometer sensor

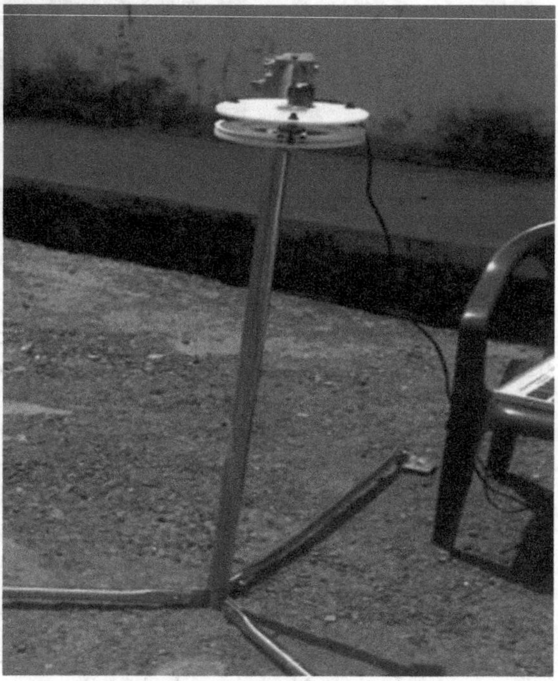

Fig. 6 Copper vessel

4 Fabrication of Experimental Test Rig

4.1 Design and Fabrication of the Frame

The design of frame which holds the lens is done so as to restrict the radial motion of lens and firmly fitted with the structure. The optical lens has the diameter of 230 mm. A square-shaped ply of wood of side 250 mm and thickness of 5 mm is taken and center is marked by drawing the diagonals. A circle of 200 mm is drawn using the center found. A vertical saw machine is used to cut the hole at the center using the circle drawn. At the adjacent sides of optical lens, a square-shaped packing is done of thickness 10 mm and above that a 250 mm square-shaped ply of wood with a hole of 200 mm is nailed. The absorber copper vessel is placed in the wooden container with sand packing around it. Figure 6 shows the copper vessel enclosed in a wooden container with sunrays getting focused on it.

4.2 Design and Fabrication of the Storage Tank and Piping

The storage tank is designed in so as to have an optimum head for flow of water and the high degree of insulation to avoid heat losses. A storage tank of 1 L capacity is taken. It has inlet at the top and outlet at the bottom with diameter of 4 mm. A square plywood of 100 mm side with a hole of 60 mm diameter is taken. This plywood is supported by the 4 wooden logs of 700 mm length at the 4 corners of plywood. These

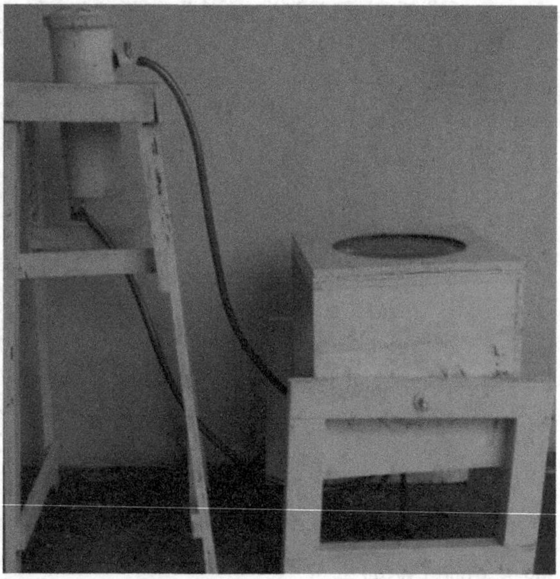

Fig. 7 Assembled experimental setup

logs are nailed to the ply and thus building a firm structure. Insulated pipes of 5 mm internal diameter with the thickness of 2 mm fitted to storage tanks. The length of pipes is kept as small as possible so that to reduce heat losses. Figure 7 shows the complete assembled view of the experimental setup.

5 Experiments

The experimental setup consisting of the wooden structure (which houses the optical lens), storage tank, absorber copper vessel and pipes is assembled. The priming of the setup is done to allow the air bubbles present in the setup to escape. The value of the room temperature and the initial water temperature are noted. Calculated amount of water is filled in the storage tank with the help of a 2 L measuring flask. The storage tank is provided with a thermometer for measuring the rise in the water temperature. A thermocouple is installed on the surface of absorber copper vessel to note the temperature rise occurring in the copper vessel. A pyranometer is employed to measure the heat flux of the sun (in W/m^2). The setup is kept under sunshine and is oriented in such a way that the optical lens focuses the sun rays on the absorber copper vessel. The adjustable screws provided at the bottom of the vessel holder can be used to adjust the distance between the optical lens and the absorber copper vessel. The absorber copper vessel top temperature as indicated by thermocouple is noted after every 5 min. Simultaneously, the thermometer reading of storage tank and pyronometer reading indicating the heat flux is also noted down after every 5 min. The process is continued for time duration of 90 min such that 18 readings are obtained. The readings are tabulated on MS-Excel sheet and graphs are generated using the software Origin Pro.

6 Analytical Computation

6.1 Calculation of Ideal Temperature Rise Due to Optical Solar Collector

Assuming 10% heat loss to the surrounding

$$Q(\text{out}) = 0.9Q(\text{in}) \tag{1}$$

$$Q(\text{in}) = FA_L \tag{2}$$

$$Q_{\text{out}} = KA_c \frac{\mathrm{d}T}{\mathrm{d}x} \tag{3}$$

where $Q(\text{out})$ is heat output. $Q(\text{in})$ is heat input, F is average flux and A_L is surface area of lens. K is conduction heat transfer coefficient. A_c is copper vessel plate area. $\mathrm{d}T$ is temperature difference between vessel top($T(t)$) and ideal water temperature ($T(i)$) and $\mathrm{d}x$ is plate thickness. Taking K(copper) as 386 W/m k and $A = \pi d^2/4$ and dimensions of copper vessel surface plate: diameter = 0.06 m; thickness of copper plate = 0.001 m

$$A_c = \frac{\pi \times 0.06 \times 0.06}{4} = 2.83 \times 10^{-3}\,\text{m}^2 \tag{4}$$

Dimensions of lens: lens diameter = 0.2 m

$$AL = \frac{\pi \times 0.2 \times 0.2}{4} = 0.031\,\text{m}^2 \tag{5}$$

By (1) (2) (3) (4) and (5)

$$\boxed{T_{(i)} = T_{(t)} - 2.59 \times 10^{-5} \times F}$$

7 Results and Discussion

The temperature of water in the storage tank gradually increases with time due to concentration of solar radiation on the absorber copper vessel. Convective heat transfer takes place inside the copper vessel and the water starts circulating due to thermosyphon effect and eventually increasing the temperature of the water in the storage tank. Figure 8 shows the representative plot of water temperature with respect to time for experiments carried out for 6 days. The heat flux kept varying with time due to the variation of solar heat flux with the solar heat flux intensity sometimes falling sharply due to the cloudy conditions. The average heat flux varied from 500 to 600 W/m^2 on a sunny day. This can be inferred from Fig. 9 that the temperature of

Fig. 8 Variation of water temperature

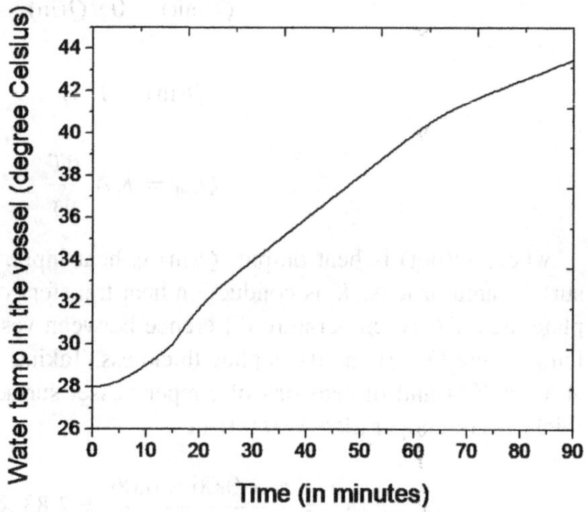

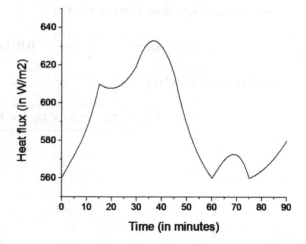

Fig. 9 Variation of heat flux

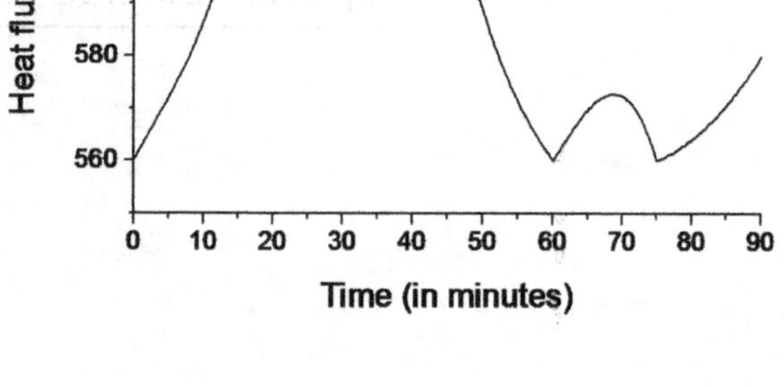

the copper vessel top increased with time due to the concentration of solar radiation on the copper vessel by the plano-convex lens which caused the temperature of the vessel top to rise. The average rise in temperature of the vessel top was observed to be 63 °C. This can be inferred from Fig. 10.

In the **present setup,** an average temperature rise of **14^{0}C** is observed for a concentrator surface area of **0.031 m^2,** whereas the temperature rise in the solar water heater that employs **flat plate collector** was observed to be **20.5^{0}C** for a given concentrator area of **0.216m^2** (Figs. 11 and 12).

Fig. 10 Plot of absorber plate temperature

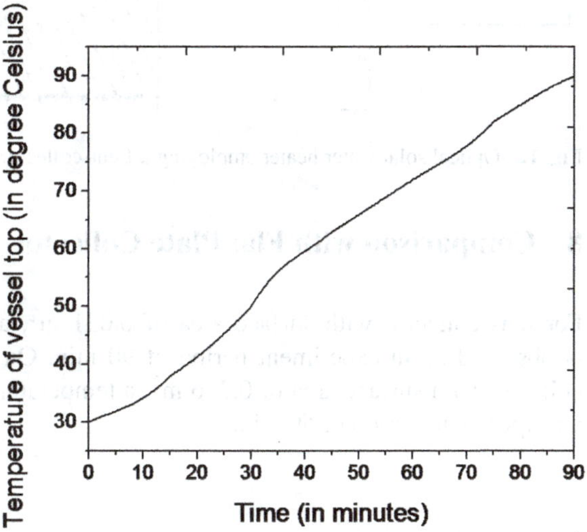

Fig. 11 Variation of ideal and actual temperature

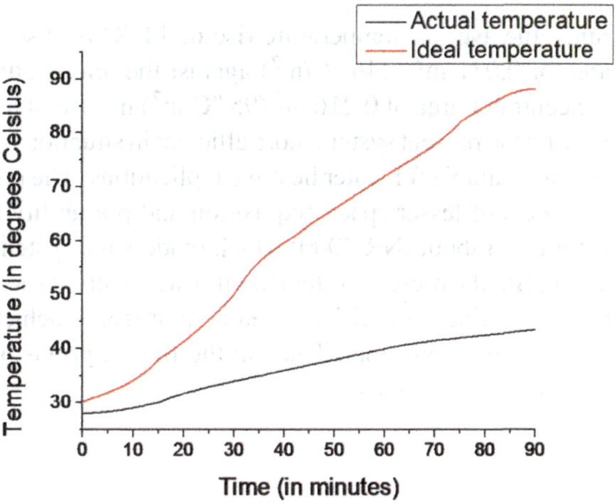

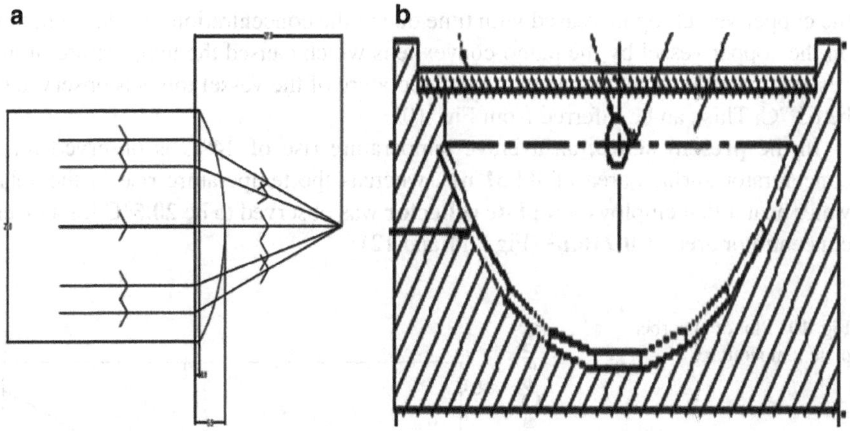

Fig. 12 Optical solar water heater employing a Lens collector and b flat plate collector

8 Comparison with Flat Plate Collector

For lens collector with surface area of 0.031 m^2, a temperature rise of 13.5 °C is observed at an experiment period of 90 min. On the other hand, for flat plate collector with surface area of 0.216 m^2, a temperature rise of 22 °C is observed at an experiment period of 50 min.

9 Conclusion

Since the average temperature rise of 14 °C is observed for a concentrator surface area of 0.031 m^2 (446 °C/m^2) against the temperature rise of 20.5 °C for a given concentrator area of 0.216 m^2 (95 °C/m^2) in case of a flat plate collector type heater, making the present system more efficient in situations where less concentrator surface area is available for water heating applications. The present system offers the desired prospects of lesser space acquisition and portability benefits at lesser material and labor cost (about INR 2000) which renders the system effective for basic household usage. But the present system has its own inherent shortcomings of higher convective heat losses due to which lesser temperature rise is achieved when compared to existing conventional collectors. Thus, in the future, provisions can be made to reduce the convective heat losses.

References

1. Kumar R, Rosen MA (2010) Thermal performance of integrated collector storage solar water heater with corrugated absorber surface. Appl Thermal Eng 1–5
2. Baba ZU, Shafi WK, Ul Haq MI, Raina A (2019) Towards sustainable automobiles-advancements and challenges. Int J Prog Industrial Ecol 13(4):315–331
3. Vajjarapu H, Verma A, Gulzar S (2019) Adaptation policy framework for climate change impacts on transportation sector in developing countries. Transp Dev Econ 5(1):3
4. Prased PR, Bryegowda HV, Gangavati PB (2010) Experimental analysis of flat plate collector and comparison of performance with tracking collector. Eur J Sci Res 40(1):144–155
5. Gulzar O, Qayoum A, Gupta R (2019a) Photo-thermal characteristics of hybrid nanofluids based on therminol-55 oil for concentrating solar collectors. Appl Nanosci 9(5):1133–1143
6. Gulzar O, Qayoum A, Gupta R (2019b) Experimental study on stability and rheological behaviour of hybrid Al_2O_3-TiO_2 Therminol-55 nanofluids for concentrating solar collectors. Powder Technol 352:436–444
7. Ali H, Kar AK (2018) Discriminant analysis using ant colony optimization–an intra-algorithm exploration. Proc Comput Sci 132:880–889
8. Ali H, Guleria Y, Alam S, Duong VN, Schultz M (2019) Impact of stochastic delays, turnaround time and connection time on missed connections at low cost airports. ATM R&D Seminar—Vienna, Austria, June 17–21, 2019
9. Gao W, Lin W Liu T, Xia C (2007) Analytical and experimental studies on the thermal performance of cross-corrugated and flat plate solar air heaters. Appl Energy 84:425–441
10. Karim A, Hawlader MNA (2004) Performance investigation of flat plate, v-corrugated and finned air collectors. Energy 31:452–470
11. Saini M, Verma J (2008) Heat transfer and friction correlations for a duct having dimple shaped artificial roughness for solar air heaters. Energy 33:1277–1287
12. Kumar A, Prasad BN (2010) Investigation of twisted tape inserted solar water heaters-heat transfer, friction factor and thermal performance results. Renew Energy 19:379–398
14. Gulzar S, Ali H, Doddamna C (2018) A conceptual framework for introducing 'Mobility as a Service' in India: opportunities & challenges. ASCE India Conference, Indian Institute of Technology Delhi, India
15. Gulzar S, Ali H, Doddamna C (2018) Ant colony optimization in pavement asset management. ASCE India Conference, Indian Institute of Technology Delhi, India
16. Gulzar S, Paktin H (2017) Scope of using nanomaterials in pavement engineering. In: Proceedings of Nano Indian 2017, pp 1–22
17. Gulzar S, Underwood S (2019) Use of polymer nanocomposites in asphalt binder modification. In: Advanced functional textiles and polymers: fabrication, processing and applications, p 405

CO_2 Laser Micromachining of Polymethyl Methacrylate (PMMA): A Review

Shrikant Vidya, Reeta Wattal, Lavepreet Singh, and P. Mathiyalagan

Abstract Laser micromachining has become a prominent tool for fabricating microstructures over polymers. Among different polymers such as polyethylene, polypropylene and polycarbonate, polymethyl methacrylate (PMMA) also known as acrylic has gained importance, specially in medical applications due to its low cost and ability to get machined at low power with good accuracy. The purpose of this paper is to present the different aspects and research trends and developments in laser micromachining of PMMA and provide a basis for follow-up research leading to the evolution of superior microfluidic devices. The review begins with the introduction to laser ablation and its principle, followed by the description of micromachining of PMMA and then review of the studies done with respect to the fabrication of microstructures over PMMA.

Keywords Laser ablation · CO_2 laser · PMMA · Review

1 Introduction

There has been an increasing interest in the production of PMMA microfluidic devices because they have lower consumption rate of reagents, higher heat transfer rate, better reaction kinetics, simpler fabrication methods and better suitability for single use, especially in medical applications. Among various microfabrication methods for PMMA microfluidic systems, such as hot embossing [1], injection moulding [2] and micromilling [3], infrared laser ablation techniques have gained much importance due to their property of evaporating substrate material with the application of heat with laser beam only. The high fabrication speed and the peculiar

S. Vidya (✉) · R. Wattal
Department of Mechanical Engineering, Delhi Technological University, New Delhi, India
e-mail: skvrsm@gmail.com

S. Vidya · L. Singh · P. Mathiyalagan
School of Mechanical Engineering, Galgotias University, Uttar Pradesh, India

© The Author(s), under exclusive license to Springer Nature Singapore Pte Ltd. 2021
R. M. Singari et al. (eds.), *Advances in Manufacturing and Industrial Engineering*,
Lecture Notes in Mechanical Engineering,
https://doi.org/10.1007/978-981-15-8542-5_82

ability of infrared laser systems to change design makes it a standout tool for micro-manufacturing. The high radiance of laser light [4] together with the characteristics of PMMA create the possibilities for the production of microfluidic devices by laser ablation.

2 Principle of Laser Machining

LASER stands for 'light amplification by stimulated emission of radiation.' It is typically a high-intensity, coherent and amplified beam of electromagnetic radiation. On the basis of sources of lasers, lasers can be classified as solid, gas and semiconductor as well as on the basis of wavelength: ND: YAG Laser, Excimer Laser and CO_2 Laser.

Laser is generated on the principle of stimulated emission. When a surface is striked by focused beam of laser radiation, laser photons excite the electrons present in the substrate [5] due to which heat is generated by absorption of photon energy, and the amount of energy depends on the material thickness and the light intensity. Due to the heat generated, the macroscopic materials get melted or evaporated from the substrate.

With respect to machining over PMMA, when the PMMA surface gets striked by the focussed laser beam, there is rapid heating of the PMMA material in the focussed spot. Due to the heat generated by direct laser heating, photon absorption and photochemical as well as photothermal processes, all chemical bonds are broken (Fig. 1).

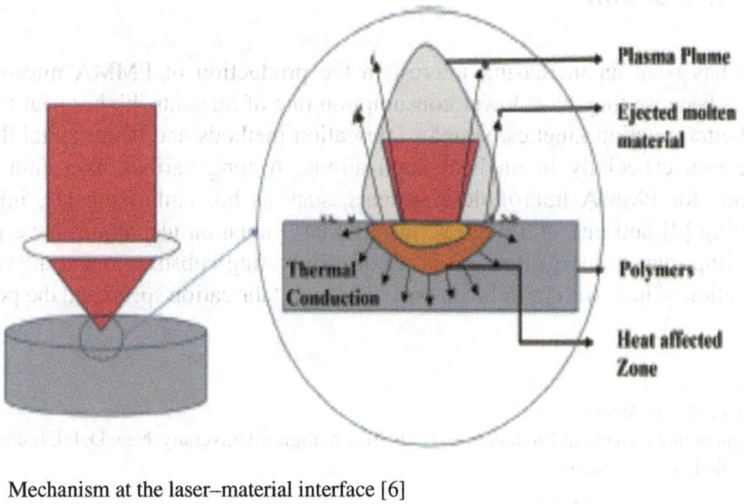

Fig. 1 Mechanism at the laser–material interface [6]

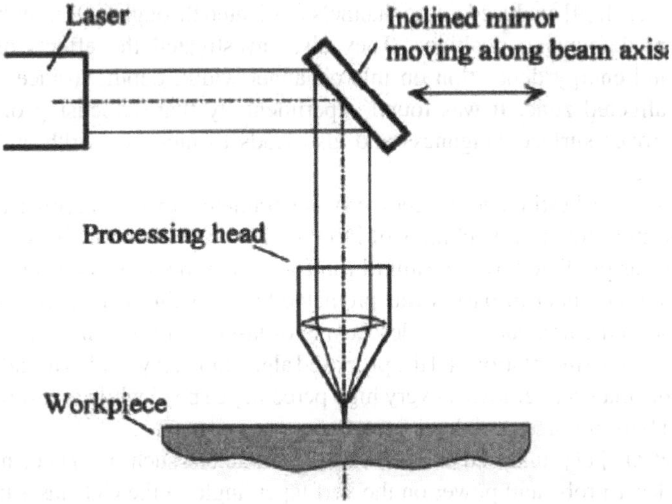

Fig. 2 Schematic CO_2 laser beam cutting system [7]

2.1 Carbon Dioxide (CO_2) Laser

CO_2 lasers typically emit light with a wavelength of 10.6 μm having overall efficiencies of approx. 10–13%. CO_2 lasers utilize continuous laser beam. In CO_2 lasers, a gas mixture of helium, nitrogen, CO_2 and other additives stimulated by electrical gas discharge is utilized to generate the laser beam. During this process, energy gets transferred from nitrogen molecules to the CO_2 molecules leading to transition of CO_2 molecules from upper energy level to a lower energy accompanied by photon release, ultimately resulting in laser beam emission. When the CO_2 molecules collide with the helium atoms, they return to the ground state and become ready to be used for another cycle (Fig. 2).

3 CO_2 Laser Micromachining of PMMA

Nasser et al. [8] fabricated a low-cost microfluidics chip using a universal laser direct writing machine and PMMA material as substrate for use in medical analysis. They also studied the influence of the channel geometry of the inlet channels intersection (DTJ and MDTJ) on the droplet frequency, droplet size and distance between droplets at different flow rate of water and oil. It was found that the droplet generation frequency is more by in case of MDTJ than the DTJ channel by 8–25%, and the droplet was more uniformly distributed inside the MDTJ channel as compared to the DTJ by 3–20%.

Prakash et al. [9] analysed microchannels fabricated through CO_2 laser by etching at different defocusing positions. They also investigated the effects of defocus distance and energy deposition on microchannel width, depth, surface roughness and heat-affected zone. It was found experimentally that defocusing of the laser beam improves surface roughness and also leads to increase width and HAZ of microchannel.

Varsi et al. [10] estimated the geometrical parameters by developing a computer-based algorithm for laser machining of PMMA. The algorithm I predicts the depth of the cavity was predicted by algorithm I for a given number of passes, and algorithm II predicts the number of passes was predicted by algorithm II for a required depth of cut. Due to the preheating and defocusing of laser beam, the divergence appears between the experimentation and the proposed algorithms. It was also found that there is negligible material removal at very high percentage speed while at low percentage speed, the burning of material occurs.

Varsi et al. [11] analysed the effects of parameters such as speed, number of passes of laser probe and power on the kerf taper angle in the CO_2 laser machining of PMMA using regression analysis and found that there is a significant effect of the above-mentioned parameters on the kerf taper angle. It was reported that low kerf taper is formed at lower speed, higher laser power and higher number of passes of laser probe.

Stan et al. [12] fabricated microchannels in PMMA using tip-based technique and analysed their dimensional characteristics. Researchers also investigated the effects of speed and normal on the geometry of microchannels, channel depth as well as friction. It was reported that the normal load value selection serves as a critical factor in the fabrication of desired microchannel and its value must be greater than 4.5 N to achieve a finite depth of microchannel in PMMA.

Imran et al. [13] employed continuous wave CO_2 laser of low power to fabricate microchannels by inscribing periodic wavelet structures on a PMMA substrate. The studies showed that the interaction time of the laser beam and input parameters can be controlled to create fine and smooth wavelet structures. It was concluded that the depth of the depressions increase with increase in laser power. Apart from this, scanning speed and input power of laser affect the formation of bulge on the outer sides of the depressions.

Prakash et al. [14] fabricated microchannels of several aspect ratios over PMMA using CO_2 laser and investigated different characteristics like microchannel width, depth and softened zone. With the evidence of spectroscopic analysis, they reported that there is absence of residual stress or cracks in CO_2 laser machining of PMMA. The microhardness test confirmed the occurrence of softened zone just adjacent to microchannels, which plays a vital role in the bulge formation and requires to be minimized.

Mohammed et al. [15] fabricated microfluidic devices for bioreactor and capillary action using multiple engraving passes and a laminate bonded layer cutting for shallow and larger depths, respectively. They reported that microfluidic channels of depths ranging 50–470 μm having low surface roughness can be produced by multiple engraving process rapidly.

Chung et al. [16] utilized metal film protection technique to minimize defects like clogging, bulges, heat-affected zones and re-solidification around the rim of trenches in CO_2 laser ablation of PMMA. As a result, it was found that the height of bulge got reduced and there was no clogging at the junction of the channels. They also simulated the laser beam interaction with PMMA in ANSYS analyse the temperature distribution.

Yue et al. [17] machined microchannels over PMMA of different molecular weights using a CO_2 laser. Laser cut was performed at low power of 0.25–2.5 W and the cutting speeds ranging from 7.0 to 64 mm/s. After studying the effects of the molecular weight on the width and depth of channels, it was reported that the depth and width of the machined microchannels varied from 18 μm to 660 μm and 110 μm to 450 μm, respectively. Laser ablation at lower laser power ratings provides precise control of the width and depth of channels. It was demonstrated that CO_2 laser is capable of fabricating micochannels with smooth surfaces and the smoothness of microchannels depends on the polymer molecular weight.

Chung et al. [18] fabricated cross-microchannels in PMMA using foil-assisted CO_2 laser micromachining with an aim to investigate the channel's feature sizes and bulges. They reported that the feature size of the cross-channel decreases to 63.6 μm from 229.1 μm with the absence of any clogging effect and the height of the bulge decreases to 0.2 μm from 8.2 μm.

Chen et al. [19] used CO_2 laser to produce microchannels over PMMA and conducted orthogonal experiments to analyze the effects of parameters on processing quality. Four-layer sheets were bonded thermally to build 3D microchannel networks. It was found that the complete bonding takes place at 103 °C and 1.2 MPa. The properties of the four-layer PMMA microchip were evaluated by conducting the concentration gradient generating experiments.

Hashemzadeh et al. [20] cut PMMA using continuous wave CO_2 laser of low power and investigated the effects of workpiece vibration on the machining performance. Researchers reported that with increase in the frequency of workpiece vibration, the cutting speed increases, but the extent of heat-affected zone also increases.

Monfard et al. [21] presented a laser drilling technique to fabricate porous PMMA scaffold of high interlinked porous structure, pore size, regenerative porosity and mechanical properties. In addition, this study illustrated that composite coating enhanced the cell compatibility of PMMA scaffolds and the chitosan/β-TCP layer coating over PMMA scaffold is highly recommended for bone tissue engineering applications.

Prakash et al. [22] fabricated microchannels over PMMA substrate using multi-pass CO_2 laser processing technique. Researchers performed a number of experiments to investigate the effects of scanning speed and laser power on width, depth, surface roughness, surface profiles and heat-affected zone. It was found that multi-pass processing channels have better aspect ratio, finer microchannel walls, lower heat-affected zone and decreased tapering as compared to single-pass processing channels.

Klank et al. [23] produced channels over PMMA using CO_2 laser of varying power and number of passes and power of the beam and analysed performance by evaluating width and depth of channels formed. They studied and analysed different methods of bonding for microstructured polymethyl methacrylate like laminating, melting, surface activation and solvent-assisted glueing using a plasma asher. Among which, a thermal bonding process assisted with solvent was reported as the most efficient with respect to time. A polymer microstructure of three layers including optical fibres was also fabricated by combining laser micromachining as well as bonding.

Huang et al. [24] used various parameters to fabricate chambers of the polymethyl methacrylate microfluidic chip by CO_2 laser cutting, and the surface roughness of the sections cut was also evaluated with the help of a non-contact 3D surface profiler. Authors preheated the polymethyl methacrylate sheet to an adequate ambient temperature during laser processing with an aim to decrease the surface roughness and then it was reported that the best value achieves at 85 °C ($R_a = 100.86$ nm).

4 Conclusions

Microfluidic structures and devices in PMMA can be easily fabricated by CO_2 laser micromachining. CO_2 laser is capable of producing residual stress and crack-free structures, specially in PMMA which has vast applications in medical and engineering fields as it can transfer swift energy to the substrate flexibly and works on the thermal theory of heat transfer.

This paper presented the current research aspects and trends in the CO_2 laser micromachining of polymethyl methacrylate. From the research review, it can be concluded that the performance of CO_2 laser micromachining such as width and depth of cut, smoothness of structures, kerf width, heat-affected zone and bulge formation can be controlled by controlling input parameters like scanning speed, laser power, polymer molecular weight and number of machining passes.

This current review is an attempt to present a quick overview of the latest researches in the CO_2 laser micromachining of PMMA and will act as a platform and open new areas of concern for laser micromachining of PMMA and enhancing the performance of laser machining of PMMA.

References

1. Gerlach A, Knebel G, Guber AE, Heckele M, Herrmann D, Muslija A, Schaller Th (2002) Microfabrication of single-use plastic microfluidic devices for high-throughput screening and DNA analysis. Microsyst Technol 7:265–268
2. R¨otting O, Ropke W, Becker H, Gartner C (2002) Polymer microfabrication technologies. Microsyst Technol 8:32–36
3. McKeown P (1996) From micro- to nano-machining-towards the nanometre era. Sensor Rev 16:4–10

4. Ready JF (1997) Industrial application of lasers, 2nd edn. Academic, San Diego
5. Brown MS, Arnold CB (2010) Fundamentals of laser-material interaction and application to multiscale surface modification. Laser Precis Microfabrication Springer Ser Mater Sci 135:91–120
6. Ravi-Kumar S, Lies B, Zhang X, Lyub H, Quin H (2019) Laser ablation of polymers: a review. Polym Int 68:1391–1401
7. Sadegh AM (2016) An intelligent knowledge based system for CO_2 laser beam machining for optimization of design and manufacturing. Int J Adv Des Manuf Technol 9:39–50
8. Nasser GA, Fath El-Bab AMR, Abdel-Mawgood AL, Mohamed H, Saleh AM (2019) CO_2 laser fabrication of PMMA microfluidic double T-Junction device with modified inlet-angle for cost-effective PCR application. Micromachines 10:678
9. Prakash S, Kumar N, Kumar S (2017) CO_2 laser microchanneling on PolymethylMethacrylate (PMMA) at different defocusing. In: Proceedings of 6th International & 27th All India Manufacturing Technology, Design and Research Conference (AIMTDR-2016), vol 27, pp 1433–1437
10. Varsi AM, Shaikh AH (2018) Developing an algorithm for predicting depth as well as number of passes during CO_2 laser machining on thermoplastic material. J Laser Appl 30:042007
11. Varsi AM, Shaikh AH (2019) Experimental and statistical study on kerf taper angle during CO_2 laser cutting of thermoplastic material. J Laser Appl 31:032010
12. Stan F, Fetecau C, Stanciu NV (2017) Fabrication of micro-channels in PMMA by tip-based microfabrication technique—depth and friction analysis. In: Proceedings of the ASME, 12th International Manufacturing Science and Engineering Conference (MSEC 2017)
13. Imran M, Rahman RA, Ahmad M, Akhtar MN, Usman A, Sattar A (2016) Fabrication of microchannels on PMMA using a low power CO_2 laser. Laser Phys 26:096101
14. Prakash S, Kumar S (2015) Fabrication of microchannels on transparent PMMA using CO_2 laser (10.6 μm) for microfluidic applications: an experimental investigation. Int J Precis Eng Manuf 16(2):361–366
15. Mohammed MI, Alam MNHZ, Kouzani A, Gibson I (2017) Fabrication of microfluidic devices: improvement of surface quality of CO_2 laser machined poly(methylmethacrylate) polymer. J Micromech Microeng 27:015021
16. Chung CK, Tan TK, Lin SL, Tu KZ, Lai CC (2012) Fabrication of sub-spot-size microchannel of microfluidic chip using CO_2 laser processing with metal-film protection. Micro Nano Lett 7(8):736–739
17. Yue CY, Lam YC (2010) Morphology and geometry of CO_2-laser machined PMMA microchannels: influence of molecular weight and number of cut passes. Int J Nanomanufacturing 6:85–98
18. Chung CK, Lin SL (2011) On the fabrication of minimizing bulges and reducing the feature dimensions of microchannels using novel CO_2 laser micromachining. J Micromech Microeng 21:065023
19. Chen X, Shen J, Zhou M (2016) Rapid fabrication of a four-layer PMMA based microfluidic chip using CO_2-laser micromachining and thermal bonding. J Micromech Microeng 26:107001
20. Hashemzadeh M, Voisey KT, Kazerooni M (2012) The effects of low-frequency workpiece vibration on low-power CO_2 laser cutting of PMMA: an experimental investigation. Int J Adv Manuf Technol 63:33–40
21. Monfard KR, Fathi A, Rabiee SM (2016) Three-dimensional laser drilling of polymethyl methacrylate (PMMA) scaffold used for bone regeneration. Int J Adv Manuf Technol 84:2649–2657
22. Prakash S, Kumar S (2017) Experimental investigations and analytical modeling of multi-pass CO_2 laser processing on PMMA. Precis Eng 49:220–234
23. Klank H, Kutter JP, Geschke O (2002) CO_2 Laser micromachining and back-end processing for rapid production of PMMA based microfluidic systems. Lab Chip 2:242–246
24. Huang Y, Liu S, Yang W, Yu C (2010) Surface roughness analysis and improvement of PMMA-based microfluidic chip chambers by CO_2 laser cutting. Appl Surf Sci 256:1675–1678

Design of Delay Compensator for a Selected Process Model

Oumayma Benjeddi, M. Chaturvedi, P. K. Juneja, G. Yadav, V. Joshi, and R. Mishra

Abstract The dead time in a complex industrial system is a critical problem to address in process control. The recompense of the negative effects of delay time has been the subject of several studies throughout the history of the control systems. The comparison between the PID controllers designed will enable to determine the influence of the different parameters of the PID correctors on the delay time and also on the other characteristics of the system. Dead time compensation technique is applied to pay off the undesirable effects of delay on the controllers designed for process model.

Keywords Dead time · IMC · Compensator

1 Introduction

Plants with large dead time always present a challenge in front of the control engineers as they change the dynamics of the plant. Also, this a widely acceptable fact that every practical process inherently has delays. A first-order process model having an intrinsic dead time can be given as follows:

$$G(s) = \frac{Ke^{-\theta s}}{(1 + \tau s)} \tag{1}$$

The original version of this chapter was revised: The author name "Vivek Joshi" has been changed to "V. Joshi". The correction to this chapter is available at https://doi.org/10.1007/978-981-15-8542-5_108

O. Benjeddi
E.N.S.I.S.A, Universite de Haute-Alsace, Mulhouse, France

M. Chaturvedi (✉) · P. K. Juneja · G. Yadav · V. Joshi · R. Mishra
Graphic Era Deemed to be University, Dehradun, India
e-mail: mayankchaturvedi.geit@gmail.com

©The Author(s), under exclusive license to Springer Nature Singapore Pte Ltd. 2021, corrected publication 2021
R. M. Singari et al. (eds.), *Advances in Manufacturing and Industrial Engineering*, Lecture Notes in Mechanical Engineering,
https://doi.org/10.1007/978-981-15-8542-5_83

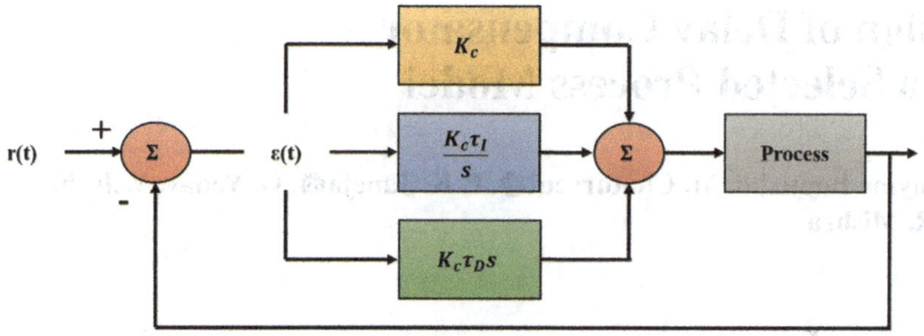

Fig. 1 Closed-loop system with PID Controller

where K is gain, θ represents the transportation lag, and τ is the time constant. In the same way, a second-order process having time constants τ_1 and τ_2 and with a natural lag can be formulated as follows:

$$G(s) = \frac{Ke^{-\theta s}}{(1+\tau_1 s)(1+\tau_2 s)} \quad (2)$$

When the process has a noteworthy transportation lag, the performance of the closed-loop system can be enhanced by using a predictor assembly. Pemberton proposed a tuning scheme with three formulas which allow the passage of the parameters of the transfer function of the system to the parameter of the corrector [1].

There are many fresh advances in the process control engineering; however, PID controller remains the primary solution of the researchers to control the complex processes. This is due to its simple structure and wide application on most of the processes which exists in the industry [2]. Figure 1 represents the closed-loop system, in which the process is being controlled by a PID controller. The transfer function of a PID controller can be characterized by

$$G_c(s) = K_c\left[1 + \frac{\tau_I}{s} + \tau_D s\right] \quad (3)$$

Skogestad PID tuning method gives formulas for the integral time, which are supposed to avoid slow disturbance compensation [3]. The IMC design procedure is a two-step design process that aims to offer an appropriate trade-off among performance and robustness [4]. Smith predictor, proposed in 1957 by Smith, is the highest prevalent algorithm used to compensate dead time in a process industry [7, 8].

2 Methodology

The dead time is an imperfection of the vast majority of industrial systems, so several studies have tried to provide a solution to this problem in order to reduce its negative influence on the system and not to disrupt the functioning of the industrial system. In process of treating this negative effect, various correctors with several methods have been used, such as: Skogestad, IMC, Pemberton. Each time after application of different correctors, system's response to a step is analyzed and visualized, to gauge the performance improvement. Analyzing and visualizing the system's response to a step, by applying all the different correctors.

The selected process can be represented by the following equation:

$$G(s) = \frac{5e^{-4s}}{(1+20s)(1+8s)} \tag{4}$$

It can be noted from the above equation that, the dead time present in the process is 4 s which is significant. It is highly desirable to compensate it, otherwise it can have adverse effects on the performance of the closed loop.

The most promising corrector is decided for a selected process model with a delay. Considering the possibility compensating the effects of dead time, Smith predictor compensation technique is implemented. The systematic implementation strategy in the form of flowchart is shown in Fig. 2.

3 Results and Discussion

As shown in Fig. 3, the tuning of all correctors with three different Skogestad, Pemberton and fine-scale methods of internal model control with three different values for the time constant (T_c = 1, 3, 5) is carried out. Table 1 represents the characteristics of the different correctors represented.

From Fig. 3, it is evident that the optimized response is achieved in case of controller designed using Skogestad PID controller design method.

After the application of the various correctors, the presence of imposing oscillations is still there, which influences its functioning negatively. In order to optimize the response of the system the maximum possible, the Smith predictor compensation technique is used as a delay compensator, as shown in Fig. 4. By comparing the responses in Fig. 5, it is evident that the Smith predictor has suppressed all the oscillations of the system and also the overshoot.

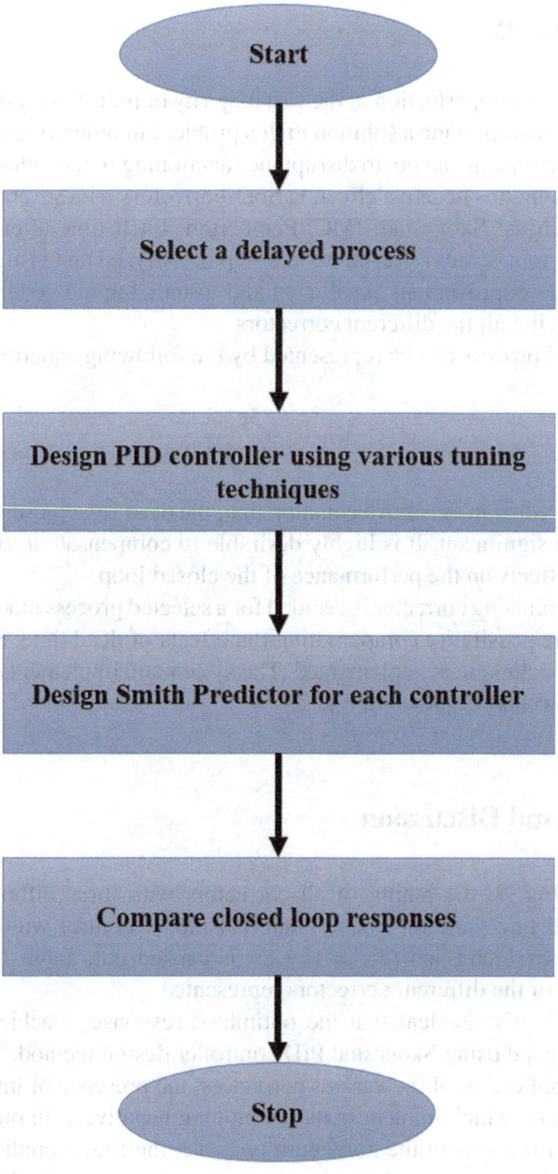

Fig. 2 Methodology

4 Conclusion

A comparative study of controllers designed for the selected process model with delay is implemented, and optimized performance is decided based on the performance

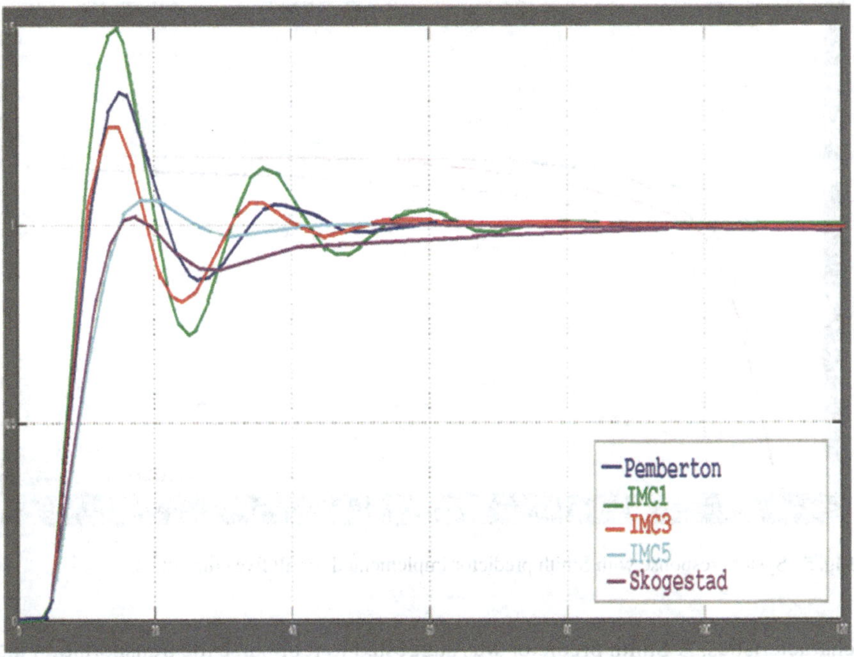

Fig. 3 Comparative responses of five controllers for same process model

Table 1 Important characteristics

Tuning Technique	Rise time (s)	Settling time (s)	Overshoot (%)	Peak	Gain margin (dB)	Phase margin (°)	Closed-loop stability
PEMBERTON	4.63	51.4	33.9	1.34	5.09	42.3	Stable
IMC ($T_c = 1$)	3.9	70.8	50.9	1.51	3.51	31.2	Stable
IMC ($T_c = 3$)	4.42	47.9	26.5	1.26	4.72	48.7	Stable
IMC ($T_c = 5$)	7.54	35.8	6.39	1.06	8.61	61	Stable
Skogestad	7.25	81.3	1.91	1.02	8.02	65.2	Stable

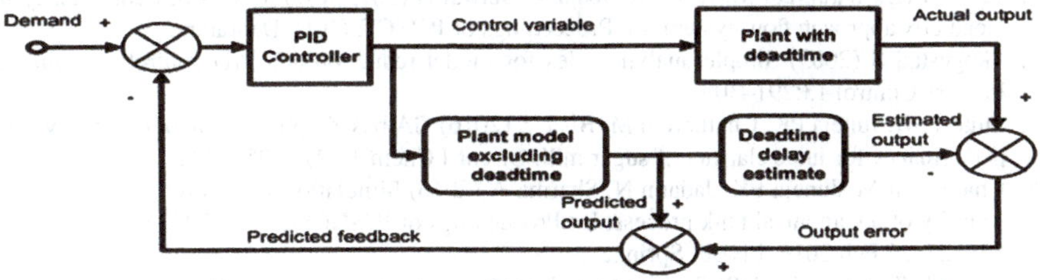

Fig. 4 Block diagram of Smith predictor

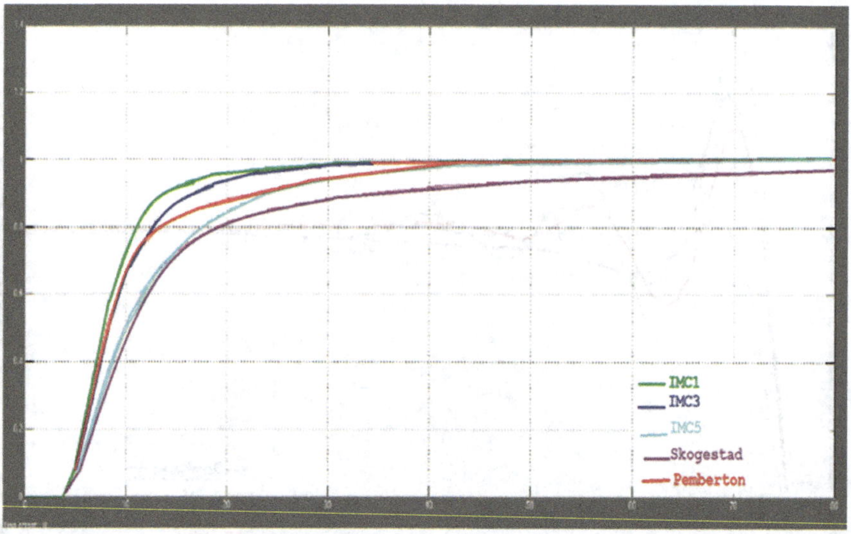

Fig. 5 System response with Smith predictor implemented for all five controllers

characteristics. A Smith predictor was suggested to reimburse the transportation lag for a carefully chosen process model. The PID controllers have been implemented to generate control signal for the selected process by means of different tuning techniques. Closed-loop time constants for IMC technique have been varied to cover a wider range. The suggested IMC controllers and smith predictor have proved to deliver an outstanding functionary of the delayed process model under consideration.

References

1. O'Dwyer A (2009) Handbook of PI and PID controller tuning rules. World Scientific
2. Sunori SK, Juneja PK, Chaturvedi M, Mittal J (2017) Dead time compensation in sugar crystallization process. In: Proceedings of ICICCD-2017, India, AISC, vol 479. Springer, Singapore, pp 375–381
3. Juneja PK, Chaturvedi M, Ray AK, Joshi V, Belwal N (2019) Control of stock consistency in Head box approach flow system. In: Proceedings of ICISCT, GEU, Dehradun
4. Skogestad S (2003) Simple analytic rules for model reduction and PID controller tuning. J Process Control 13:291–309
5. Sunori SK, Juneja PK, Chaturvedi M, Bisht S (2016) GA based optimization of control system performance for juice clarifier of sugar mill. Orient J Chem 32(4):2205–2208
6. Chaturvedi M, Juneja PK, Jadaun N, Sharma A (2016) Simulation studies for delay effect on stability of a Canonical tank process. In: Proceedings of ICSNCS-2016, JNU, New Delhi, vol 396, 25–27 Feb 2016, LNEE. Springer
7. Zhang B, Tan W, Li J (2019) Tuning of Smith predictor based generalized ADRC for time-delayed processes via IMC. ISA Trans
8. Singh S, Chaturvedi M, Juneja PK (2016) Design of modified Smith predictor for dead time compensation for SOPDT process. In: Proceedings of ICCCS 2016, 9–11 Sept 2016, Gurgaon, India, pp 935–940

Fly Ash, Rice Husk Ash as Reinforcement with Aluminium Metal Matrix Composite: A Review of Technique, Parameter and Outcome

Jagannath Verma and Harish Kumar

Abstract Aluminium metal matrix composite refers to the category of that class that has properties like better strength to weight ratio, it is of very less density and lighter in weight, and it is suitable in many fields and commonly used in the automotive, aerospace and many other industries such as marine, agro and military fields. As per the market trend, we have design reinforced composite material that suits our recent requirement of demand. Researchers have noticed that the inclusion of ash of rice husk which is agrowaste and fly ash to create aluminium metal matrix composite results in enhanced properties like physical and better wear resistance ability. Here, in this paper, primary objective is to evaluate suitable technique such as stir casting to fabricate various forms of aluminium alloys with the given reinforcements. This article review the dry sliding behaviour at high temperature, friction and wear rate variation with different composition (wt%) of agrowaste such as rice husk as first constituent and fly ash as second in aluminium matrix composite by making use of pin and disc setup. The rate of volume loss, frictional force and also the effects of the size of particles of agrowaste ash on tribological behaviour. Further, the machinability has been investigated by conducting experiments on electrical discharge machining (EDM). To investigate the results of scanning electron micrograph of agrowaste ash, fly ash distribution inside aluminium composite. An attempt has also been made to investigate its mechanical and corrosion behaviour of composites.

Keywords Rice husk ash · Pin-on-disc device · Fly ash

J. Verma (✉) · H. Kumar
Department of Mechanical Engineering, National Institute of Technology Delhi, New Delhi 110040, India
e-mail: jagan.verma90@gmail.com

© The Author(s), under exclusive license to Springer Nature Singapore Pte Ltd. 2021
R. M. Singari et al. (eds.), *Advances in Manufacturing and Industrial Engineering*, Lecture Notes in Mechanical Engineering,
https://doi.org/10.1007/978-981-15-8542-5_84

1 Introduction

AMCs have been designed to enhance aluminium alloy properties that can meet the requirements of modern engineering products. In various application purposes, better strength to weight ratio is suitable in many fields. It has incredible properties like good enough specific strength, thermal expansion coefficient is very low and heat resistant, and good in vibration neutraliser and high rigidity.

On milling 1 ton kg of rice in rice field for rice manufacturing, nearly 22% of husk of rice was obtained and approximately 25% of rice husk ash made after treating it in the boiler [1, 2]. This matrix utilised to manufacture a number of components in the aero field, auto vehicles field, sea vehicles and in nuclear field. This agrowaste is a cheaper relative to other reinforcement material such as silicon carbide, alumina and titanium carbide. Rice husk is the milled paddy waste. These rice husks are utilised as source of energy by various rice firms to form steam to handle boiling process [3]. Properties of material with reinforcement improved with the exception of flexural rigidity [4]. Millions of tons of RHA are produced worldwide each year. RHA dumping pollutes the surrounding area and the ground. Recent studies have confirmed that comprise of around ninety per cent silica [5]. Fly ash is produced by burning crushed coal at thermal power plants as a by-product. Once inhaled through the skin, it is harmful so, it was planned to use fly ash in a beneficial way to upgrade the desired mechanical properties of aluminium-based matrix [6].

2 Review of Experimental Analysis

The wear characteristics coated with rice field waste and power plant waste fly ash was observed on the pin's dry sliding test. Load range, sliding velocities and rotational speed of disc are variable parameter. There are various mechanisms of wear, with the help of the electronic microscope: abrasive wear, oxidative wear, delamination wear, adhesive wear and melting wear [7]. Wear volume with respect to time decreases with rise in the content of reinforcements. Main reasons of strain of this composite are due to melting of composite and it get thermally very soft [8]. We noted that the rise in weight content of rice field waste (RHA) particles modified the wear causing factor, initially, it was micro-cutting but later changed to abrasion and caused to reduce structure and shape of wear particles of remains [9]. At the compo-casting temperature, RHA homogenously distribute in matrix, which upgrades the characteristics like more resistant to indentation and also its ultimate tensile strength nature [10]. With rise in spread of interaction, surface between RHA content matter and matrix particulates resulting in a significant enhance its strength. With the rise in the weight proportion of ash material, it further enhanced all type of strengths. Due to higher density of dislocation, it is utilised in improvement of various physical properties [11]. At temperature above room temperature, volume rate loss in sliding behaviour with fly ash, as a consequence composite produced using technique known as stir casting [12]. The strengthening of fly ash content matter is obtained in various temperatures, which is examined to improve wear resistance of compo-cast matrix composite are most prominent wear method like adhesion, metal flow, delamination, abrasion and oxidation [13]. Volume rate loss in a one loading depends on the amount of reinforcement, temperature and sliding velocity. At fixed temperature, with the mixing of industrial waste like fly ash rate of volume loss decreases [14]. The transparent boundary and good binding which give obstruction to fly ash particle loosening and crack propagation. Because of the variation in coefficients of thermal expansion, stretching occurs on all the direction of the particles of FA [15]. Such stretching fields accumulate dislocations that offer resistance to crack propagation by [16].

3 Technique of Fabrication and Wear Mechanism

Powder metallurgy, which is one of the best and the most known techniques used in Al-MMNC manufacturing. Investigations of the dry wear behaviour of cast AA7075 + titanium carbide MMC by making use of Taguchi method [17]. Composite's wear rate increases linearly with higher sliding speed, sliding length and standard composite load. Delamination is considered as the main wear process at higher sliding speed along with ploughing and other wear mechanisms [18]. The wear resistance improved by RHA particles [19].

4 About Aluminium Alloy

It has very good inner strength and easier to handle as it is not much heavy and it is in limelight in modern-day technology since threshold areas are the concern of environment and energy. It has strength and wear resistance which is equivalent to cast iron (Table 1).

5 About Agro-Generated Waste

Ash of rice husk is a costless stuff for reinforcement compared with other reinforcements such as alumina, silica and so on. Rice husk is actually known for cheaply available agrowaste [21].

It consists of around 15–20% of stuff derived from non-living and the remaining is made up of natural substance [22]. This waste is utilised to create steam at steam-making factories. Put husk into a flame and extract substance which is highly flammable and turn it into ash [23] (Table 2).

Table 1 Elements proportion in alloy 61S [20]

Elements	wt%
Magnesium	1.079
Iron	0.169
Silicon	0.629
copper	0.319
Manganese	0.519
Vanadium	0.009
Titanium	0.019
Aluminium	Remainder

Table 2 Percentage proportion in rice husk ash [2]

Additives	wt%
Silica	94.039
Alumina	0.2499
Iron	0.1359
Calcium oxide	0.6219
Magnesium oxide	0.4419
Nitrous oxide	0.0229
Kallium oxide	2.489
Lithium oxide	3.519

6 About Fly Ash (FA)

A cenosphere is not much heavy, empty round solids, largely consists of silica and alumina which is empty inside, usually generated as a by-product in coal combustion. The spherical shape produces a low volume-to-surface area ratio. This round solid matter has chemical compositions derive from the product of the coal supply and the combustion process. Advantage of it includes light in weight, thermally reflective, spherical in shape, strength and refractory.

7 Rice Husk to Ash Preparation Processing

Wash this agrowaste to remove foreign particles and dehumidify it for one day in a normal temperature. After that, it is warmed to two hundred degree for an hour to dehumidify it completely [24]. Because of the burning of organic matter. The extraction of the carbonaceous material is further heated to 600 °C for 12 h [25]. Remaining ash comprises of silica and ready for reinforcement [26].

8 Technique to Fabricate Compo-Composite

Stir casting is primarily utilised for manufacturing a composite metal matrix with reinforcement particles. Then, the reinforcement is stirred in the molten metal [27]. Then, to solidify the melt, the mould is placed below the furnace. Such type of setup improves particulate absorption and reduces the time molten metal pouring [28, 29]. It further stabilise, time reduction of pouring and the mould's composite characteristics. The composite matrix subjected to high temperature crucible. Stirrer is connected to motor. Using gear connection, the stirrer can be pushed up and down [30]. Mixing techniques are used to introduce and disperse the discontinuous phase homogeneously [31, 32].

9 Effect on Mechanical Properties of Compo-Composite on Addition of Rice Husk Ash and Fly Ash

This agrowaste particle reinforcement greatly improves AMC's hardness and its UTS. Aluminium alloy (6061) with 8% agrowaste ash (wt fraction) displays approx 167% better anti-indentation property and approx 57% better UTS with respect to parent aluminium alloy without any reinforcement [3]. During phase transformation in solidification process, due to difference of the thermal expansive coefficient establishes stretching region near to ash of the agrowaste particles [29, 33]. Fields of strain avoid the dislocation during the loading of the tensile. To transfer the dislocation around the strain fields, higher load is required [30].The granular size is inverse proportion by taking tensile strength on other side. Now granular entities refine the area which is resisting applied loading and periphery and hence this area increases [34]. Due to the dislocation, it increases micro-hardness and UTS of the composite [14, 35] (Figs. 1 and 2).

10 AA6061/RHA Aluminium Matrix Composite Sliding Wear Behaviour

After researching literature reviews, it can be observed that volume rate loss is reduced with rise in proportion of rice husk in one fixed loading condition [37]. The improved hardness on addition of RHA provides resistance to counter-asperities slicing action is strengthening [38]. It is a solid round in appearance that prevent accumulation of stress and delays premature particle detachment [39].

It leads to a wear rate reduction (Table 3).

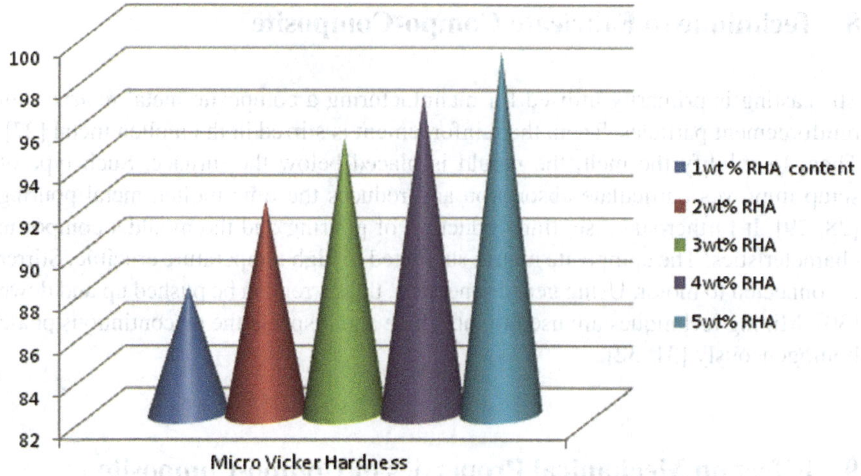

Fig. 1 Experimental result of variation of anti-indentation property with respect to RHA content [36]

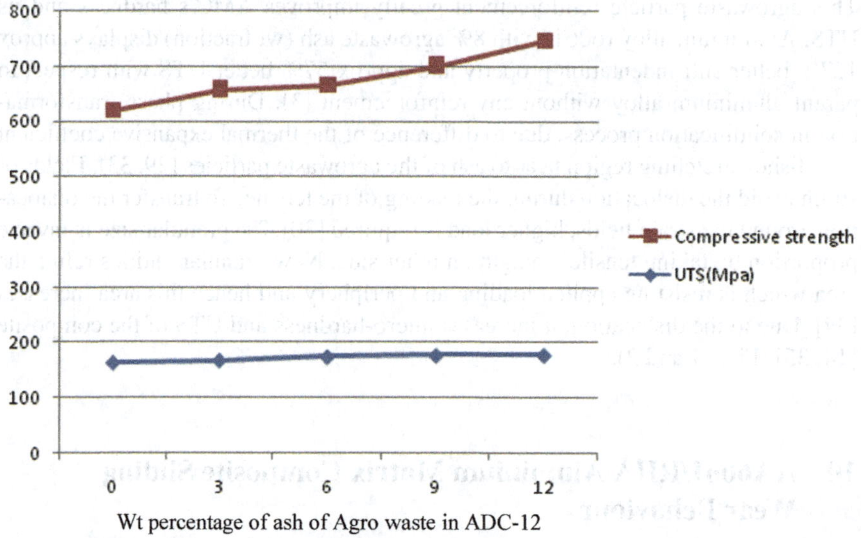

Fig. 2 Compressive and ultimate tensile strength versus weight % of rice husk ash [36]

Table 3 Agricultural waste (organic powders) reinforcements in friction stir processing [20]

Aluminium alloy type and its dimensions	Geometry of the tool	Type of reinforced matter	Dimensions of the groove	Various variable parameters	Outcomes	References
AA6061 ((l = 100 mm) × (b = 50 mm) × (h = 10 mm))	Pin tool threaded profile which is made of steel with diameter of shoulder, length of pin and diameter of pin having values 17.9, 5.9 and 5.79 mm	Fly ash, (0, 6, 12, and 18.00 vol. %).	Deep = 5.50 mm and wide = (0.40, 0.80, and 1.19 mm)	10 KN of axis load, and1600 rpm and 60 mm/min = traverse speed	Micro-hardness and wear increases	[40]
Aluminium alloy 6061	Chromium-molybdenum hot work steel tool H13	Rice husk ash of 5 weight%	NA	Rotational speed = 800 rpm, 1000 rpm and 1200 rpm	At 1200 rpm higher micro-hardness visualised	[41]
Aluminium alloy 1100 6 mm thickness of plate	HSS tool, diameter of the pin 5 mm, length of the pin is 5 mm shoulder diameter(20 mm)	Rice husk ash and SiO_2	Depth is 0.5 mm and 1 mm of width	Rotational speed = 600, 865,1140 and 1500 rpm where as traverse speed = 45 mm/min	Higher hardness, fine grains and wear resistance observed	[42]
Aluminium alloy 606, diameter of pin = (6 mm)	Chromium-molybdenum hot work steel tool H13 and & steel HCHCr tool of shoulder diameter (18 and 24 mm), length of pin (5.8, 5 and 4.5 mm)	5 μm of fly ash	NA	RPM = 1000, 1200 rpm and traverse speed = 41 to 65 mm/min	Micro-hardness stabilised due to fly ash particles.	[43]
Aluminium alloy 5083 ((l = 100 mm) × (b = 100 mm) × (h = 5 mm^3))	NA	Powder of fly ash	Wide = 1.00 mm deep = 2.00 mm (grooves dimensions)	Rotational speed = 1100, 1400 rpm and traverse speed = 21.99, 24.990 mm/min at rotational speed = 1450 rpm and traverse speed = 25.00 mm/min	Fly Ash worked like anti-corrosion inside aluminium alloy 5083	[44]

11 Conclusion

Regardless of the volume fraction, RHA content matter spread in equal proportion in Al alloy. RHA particles were not aggregated or separated. Most of the molecules were within the borders of the grain. During the process, rice husk ash particles experienced fracture due to plastic strain [20].

If the proportion of RHA rises above a certain limit, its impact has also been reduced due to poor aluminium alloy wettability. RHA content matter behaved as a location of formation of new nucleus for grain, it further purified aluminium matrix grain. Development of Al grains was resisted by RHA content matter. RHA particles improve surface wear resistance and mitigated the plastically strain caused by the remains of particles after wear. Enhanced wear resistance has been due to increased stiffness, stretched field development, uniformly throughout distribution and round in shape of RHA content matter. With rise in rice husk content matter wear mechanism changed, initially, it was adhesive later it changed to abrasive. It is observed that, the hardness of alloy rises with the ageing temp. maximum range to 175 °C after that it declines [45].

References

1. Alaneme KK, Bodunrin MO (2011) Corrosion behavior of alumina reinforced aluminium (6063) metal matrix composites. J Minerals Mater Characterization Eng 10(12):1153–1165
2. Tiwari S, Pradhan MK (2017) Effect of rice husk ash on properties of aluminium alloys: a review. Mater Today Proc 4(2):486–495
3. Ayswarya EP et al (2012) Rice husk ash—a valuable reinforcement for high density polyethylene. Mater Des 41:1–7
4. Mukesh K, Kumar A (2019) Sliding wear performance of graphite reinforced AA6061 alloy composites for rotor drum/disk application. Mater Today Proc
5. Gupta VK, Gupta M, Sharma S (2001) Process development for the removal of lead and chromium from aqueous solutions using red mud—an aluminium industry waste. Water Res 35(5):1125–1134
6. Sharma VK, Singh RC, Chaudhary R (2017) Effect of fly ash particles with aluminium melt on the wear of aluminium metal matrix composites. Eng Sci Technol Int J 20(4):1318–1323
7. Das S, Dan TK, Prasad SV, Rohatgi PK (1986) Aluminium alloy—rice husk ash particle composites. J Mater Sci Lett 5(5):562–564
8. Alaneme KK, Olubambi PA (2013) J Mater Res Technol 2(2):188–194
9. Bodunrin MO, Alaneme KK, Chown LH (2015) J Mater Res Technol 4(4):434–445
10. Chawla N, Shen Y-L (2001) Adv Eng Mater 3(6):357–370
11. Tony TA, Parameshwaran R, Muthukrishnan A, Arvind Kumaran M (2014) Development of feeding and stirring mechanisms for stir casting of Aluminium metal matrix composites. In: Proceedia materials science, vol 5, pp 1182–1191
12. Kalaiselvan K, Murugan N, Parameswaran S (2011) Production and characterization of AA6061–B4C stir cast composite. Mater Des 32:4004–4009
13. Rosenberger MR, Schvezov CE, Forlerer E (2005) Wear of different Aluminum matrix composites under conditions that generate a mechanically mixed layer. Wear 259(1–6):590–601
14. Jerome S, Ravisankar B, Kumar Mahat P et al (2010) Synthesis and evaluation of mechanical and high temperature tribological properties of insitu Al–TiC composites. Tribol Int 43:2029–2036

15. Bastwros MM, Esawi AM, Wifi A (2013) Friction and wear behavior of Al–CNT composites. Wear 307:164–173
16. Abdollahi A, Alizadeh A, Baharvandi HR (2014) Dry sliding tribological behavior and mechanical properties ofAl2024–5wt.% B4C nanocomposite produced by mechanical milling and hot extrusion. Mater Des 55:471–481
17. Dehnavi MR, Niroumand B, Ashrafizadeh F, Rohatgi PK (2014) Effects of continuous and discontinuous ultrasonic treatments on mechanical properties and micro-structural characteristics of cast Al413/SiCnp nano composite. Mater Sci Eng, A 617:73–83
18. Carreño-Gallardo C et al (2014) Dispersion of silicon carbide nanoparticles in a AA2024 Aluminum alloy by a high-energy ball mill. J Alloys Compounds 586:S68–S72
19. Srikanth N, Thein MA, Gupta M (2004) Effect of milling on the damping behaviour of nano-structured copper. Mater Sci Eng A 366:38e44
20. Dinaharan I, Kalaiselvan K, Murugan N (2017) Influence of rice husk ash particles on microstructure and tensile behavior of AA6061 Aluminum matrix composites produced using friction stir processing. Compos Commun 3:42–46
21. Deiana C et al (2008) Industrial Eng Chem Res 47(14):4754–4757
22. Gowda H, Rajendra Prasad P (2017) Influence of heat treatment on corrosion resistance of A356/RHA/Al_2O_3 based hybrid composites. Mater Today Proc 4(10):10870–10878
23. Rozainee M et al (2008) Bioresour Technol 99(4):703–713
24. Sun L, Gong K (2001) Industrial Eng Chem Res 40(25):5861–5877
25. Huang Y, Bird RN, Heidrich O (2007) A review. Resour Conserv Recycling 52(1):58–73
26. Siriwardena S, Ismail H, Ishiaku US (2001) Polym Int 50(6):707–713
27. Soltani BN, Pech-canul MI, Gutierrez CA (2016) Crit Rev Environ Sci Technol 46:143–207
28. Raviraj MS, Sharanprabhu CM, Mohankumar GC (2014) Experimental analysis on processing and properties of Al-TiC metal matrix composites. Proc Mater Sci 5:2032–2038
29. Tony TA, Parameshwaran R, Muthukrishnan A, Arvind Kumaran M (2014) Development of feeding and stirring mechanisms for stir casting of Aluminium metal matrix composites. Proc Mater Sci 5:1182–1191
30. Balasivanandha Prabu S, Karunamoorthy L, Kathiresan S, Mohan B (2006) Influence of stirring speed and stirring time on distribution of particles in cast metal matrix composite. J Mater Process Technol 171:268–273
31. Venkat Prasad S, Subbramanian R, Rahika N, Anandavel B, Arun L, Praveen N (2011) Influence of parameter on dry sliding behaviour of aluminium/fly ash/graphite hybrid metal matrix composite. Euro J Sci Res 53(2):280–290
32. Siva Prasad A, Rama Krishna (2011) Production and mechanical properties of A356.2/RHA composites. Int J Adv Sci Technol 33 (August)
33. Alaneme KK, Ajayi OJ (2015) Microstructure and mechanical behaviour of stir cast Zn-27Al based composites reinforced with rice husk ash, Silicon carbide and graphite. J King Saud Univ Eng Sci
34. Looney HL, Hashmi MSJ (1999) Metal matrix composites: production by the stir casting method. J Mater Process Technol 92–93:1–7
35. Rajaram G, Kumaran S, Rao TS et al (2010) Studies on high temperature wear and its mechanism of Al–Si/graphite composite under dry sliding conditions. Tribol Int 43:2152–2158
36. Muralimohan R, Kempaiah UN (2018) Seenappa influence of rice husk ash and B4C on mechanical properties of ADC 12 alloy hybrid composites. Mater Today Proc 5:25562–25569
37. Gladston J, Kingsly A et al (2015) Production and characterization of rich husk ash particulate reinforced AA6061 aluminium alloy composites by compocasting. Trans Nonferrous Metals Soc China 25(3):683–691
38. Krishnarao RV, Godkhindi MM (1992) Effect of Si3N4 additions on the formation of SiC whiskers from rice husks. Ceram Int 18(3):185–191
39. Wu GH, Dou ZY, Jiang LT, Cao JH (2006) Damping properties of aluminium matrix fly ash composites. Mater Lett 60:2945e2948
40. Fatchurrohman N, Farhana N, Marini CD (2018) Investigation on the effect of friction stir processing parameters on micro-structure and micro-hardness of rice husk ash reinforced Al6061 metal matrix composites. IOP Conf Ser Mater Sci Eng 319(1):0–6

41. Dinaharan, Nelson R, Vijay SJ, Akinlabi ET (2016) Microstructure and wear characterization of aluminium matrix composites reinforced with industrial waste fly ash particulates synthesized by friction stir processing. Mater Charact 118:149–158
42. Zuhailawati H, Halmy MN, Almanar IP, Seman AA, Dhindaw BK (2016) International journal of mechanical properties of friction stir processed 1100 Aluminum reinforced with rice husk ash silica at different rotational speeds. Int J Metall Mater Eng 2(120):1–6
43. Dinaharan, Akinlabi ET (2018) Low cost metal matrix composites based on aluminium, magnesium and copper reinforced with fly ash prepared using friction stir processing. Compos Commun 9:22–26
44. Prabhakar GVNB, Kumar NR, Sunil BR (2018) Surface metal matrix composites of Al5083—fly ash produced by friction stir processing. Mater Today Proc 5(2):8391–8397
45. Mishra P, Mishra P, Rana RS (2018) Effect of rice husk ash reinforcements on mechanical properties of Aluminium alloy (LM6) matrix composites. Mater Today Proc 5(2):6018–6022

Optimization of CNC Lathe Turning: A Review of Technique, Parameter and Outcome

Vivek Joshi and Harish Kumar

Abstract Computer numeric control (CNC) machine are electro-mechanical device that uses computer program as an input for performing the desired machining. Various machine tools that can be numerically controlled are mills grinders and lathe. Conventional machining to impart complex geometries on blank requires complex jig to control cutting tool motion. But since CNC tool path is digitally programmed so there is no need of jig in CNC machine. In addition to this, CNC offers various advantages over conventional machine-like product can easily replicated thousands of time, less labor needed to operate on CNC; CNC software increases production options, more accuracy, etc. In turning, the cutting mainly affects the MRR and surface finish. Increasing struggle for higher production with high quality surface finish has forced the production industry to use quality machining tools. The sundry parameter turning process which effects surface qualities are spindle velocity, cutting depth, feed and cutting velocity. The present work is to review the work done by the researcher in area of the optimization of CNC lathe turning. Due to its wide spread availability and its capability of performing various tasks without altering its setup, the lathe machine was chosen for parameter optimization. As turning operation provides various benefits such as it can be used for machining large variety of the material, and it is one of the cheapest machining processes that is why turning operation was explicitly chosen.

Keywords CNC · Turning · Cutting parameter · Optimization

V. Joshi (✉) · H. Kumar
National Institute of Technology Delhi, New Delhi 110040, India
e-mail: vivekjoshinitcadcam@gmail.com

© The Author(s), under exclusive license to Springer Nature Singapore Pte Ltd. 2021

R. M. Singari et al. (eds.), *Advances in Manufacturing and Industrial Engineering*,
Lecture Notes in Mechanical Engineering,
https://doi.org/10.1007/978-981-15-8542-5_85

1 Introduction

In metal working industry, the turning on lathe is prevailing machining process. To bring more flexibility in the production, the advancements in the lathe were done by controlling the lathe through computer that is CNC. CNC lathe provides better control over cutting parameter like feed, cutting depth and cutting velocity. Costly turning on CNC lathe and better control provided by CNC over the process parameter, forces the industries to go for further advancement that is nothing but the optimization of CNC turning. The various input parameter in turning is feed, cutting velocity, spindle speed, cutting depth and tool-nose radius while the output parameter is surface quality and rate of metal removal. Taguchi, genetic algorithm, particle swarm optimization, response surface methodology and artificial neural network are various optimization techniques. Miroslav [1] describe gave the various steps for carrying out the optimization. First step is identification of objectives, constraints and parameters. Second step is describing the problem that needs to be optimized. Selecting and constructing the mathematical model of optimization as the third step. Fourth step is choosing the optimizing method. Finally, solving the constructed optimization problem as fifth step.

2 Literature Review

Ajaja [2] carried out the optimization of turning of high strength steel. Different optimization methods used were grey relation analysis (GRA), Taguchi, proportion quality loss reduction (PQLR). At low cutting velocity (50 m/min), the surface qualities achieved with carbide inserts was higher than that was obtained with blended ceramics and cubic boron nitride (PCBN). At high velocity (100 m/min), the surface roughness obtained with blended ceramics and polycrystalline cubic boron nitride was better than that obtained with ceramics (Table 1).

Table 1 Contribution of control parameter [2]

Control parameter	Percentage contribution		
	R_z	R_a	R_v
Tool	19	32	20
Speed	10	7	7
Feed	57	41	48
Depth of cut	14	20	25

Optimum condition found by GRA and PQLR results in enhanced surface condition but the GRA method is better than PQLR as optimum condition obtained from GRA resulted in reduction of R_Z by 44% while optimum obtained from PQLR resulted in reduction of R_Z by 32%. Good surface quality was attained with cutting velocity = 50 m/min, feed = 0.051 mm/revolution, cutting depth = 0.762 mm. Ashish George [3] studied tool approach and rake angle effect on tool vibration and surface roughness. Increment in feed resulted in increment in R_a value while decrement in R_a was reported with spindle speed increment. Also, the decrement in roughness was noted with reduction in approach angle. Roughness was highly influenced by spindle velocity and feed, whereas approach angle effect on roughness was more pronounced as compared to rake angle. Optimum condition reported at approach angle = 75°. Spindle speed = 1200 rpm, cutting depth = 0.50 mm, rake angle = 9° and feed = 0.05 mm/rev. Bagaber [4] carried out dry turning. High feed and high cutting depth resulted in generation of huge temperature at tool-chip interface which ultimately resulted in sticking of chip on tool followed by abrasion of tool. The abraded tool resulted in high surface roughness. Surface roughness was not much affected by speed. Optimal solution was obtained at cutting velocity = 110 m/min, cutting depth = 1.33 mm and feed = 0.159 mm/rev with power consumption of 25.91 Wh. Garcia[5] carried out turning of aluminum alloy with dry condition and reduced quantity lubrication. Higher depth of cut resulted in larger cutting forces that produce vibration and this vibration led to high value of peak to valley surface roughness. Dry machining brings waviness over the surface but this waviness was not generated when RQL was used. Optimality was achieved by RSM in alliance with Behnken design (Table 2).

Ender Hazir [6] reported that spindle speed was the highest influencing factor that controls the surface roughness. Minimum surface roughness that can be achieved by integrated approach of RSM and DFM was 3.519 μm while for integrated approach of RSM, DF and GA, the minimum surface roughness achieved was 3.512 μm. Laghari [7] reported that for tool life maximization optimal value of cutting parameter were 0.01 mm/rev as feed, 6.283 m/min as velocity, 0.2 mm as cutting depth. Similarly, to achieve better surface finish the cutting speed should be 18.85 m/min, feed should be 0.015 mm/revolution, cutting depth should be 1.5 mm. The minimum surface finish obtained was 0.044 μm. For tool life cutting, velocity and feed are most dominating factor. Aissalaouissi [8] used neural network and response surface methodology to established mathematical model and used genetic algorithm (GA) technique for

Table 2 Optimum values of cutting parameter and obtained roughness under dry and RQL [5]

Cutting parameter (optimum)	Dry	RQL
Velocity (m/min)	851	403
Feed (mm/rev)	0.07	0.05
Cutting depth (mm)	2	0.5
Obtained R_a (μm)	0.44 ± 0.05	0.18 ± 0.01
Obtained R_z (μm)	2.73 ± 0.20	0.96 ± 0.06

Table 3 Percentage contribution of parameter on roughness [8]

Parameter	Uncoated tool		Coated tool	
	Minimum contribution (%)	Maximum contribution (%)	Minimum contribution (%)	Maximum contribution (%)
Feed	65	74	73	86
Cutting speed	12	15	3	10
Depth of cut	2.5	4.5	3.3	7.7

optimization. Uncoated ceramic tool imparted poor surface finish as compared to coated ceramic tool (Table 3).

By using genetic algorithm for achieving the best reconciliation among surface quality, cutting force, energy consumed in cutting and MRR the cutting condition should be 299.525–512.571 m/min for cutting velocity, 0.8 to 0.121 mm/revolution for feed, 0.251 to 0.586 mm for cutting depth. Nataraj [9]used RSM to create mathematical model and done optimization by graphical analysis. Feed was found to be most influencing parameter for vibration while cutting depth and cutting velocity was the most influencing parameter for surface roughness. Chew Ying Nee [10] used differential evolution as method of optimization for minimizing surface roughness. As compared to RSM, regression modeling and experiments, the use of differential evolution optimization method resulted in decrease in surface roughness by 30%, 72% and 81%. Salman [11]reported that to obtain good surface quality without affecting rate of metal removal and the cutting velocity should be high and both feed and cutting depth should be low. Serra [12] used genetics algorithm as optimization technique and reported that feed rate and interaction (depth and feed) had very profound impact on surface roughness while surface roughness was unaffected by cutting depth and cutting velocity in the studied range therefore to improve productivity these parameter that is cutting speed and cutting depth can be increased as they do not have any significant impact on surface quality. Power consumed in cutting is highly dependent on interaction of feed and cutting velocity followed by cutting depth. Nan Xie [13] carried out energy-based modeling of turning and reported that there existence of nonlinear dependency between the surface quality and energy consumption per unit volume of material removal (SCEC). For a given value of cutting depth, surface roughness reduced with SCEC. As depth of cut increases, there will be decrease in SCEC which results in bad surface quality.

Year	Author	Work material	Cutting tool	Remark
2018	Saidi [14]	Stellite	CNGG-coated tool	Depth of cut and feed effects tangential force
2017	Chabbi [15]	Polyoxymethylene	Cemented carbide	ANN outperform RSM, feed is the most influencing parameter for roughness
2017	Xiaol [16]	AISI 1045	YT5	Feed greatly influences surface roughness
2016	Park [17]	AISI 1045	Carbide	Multi-energetic criteria adversely affect cutting zone temperature
2017	Rocha [18]	AISI H13	PCBN	For tool life, cutting speed is most influencing parameter while feed is most influencing for roughness
2018	Camposeco [19]	AISI 1018 steel	DCMT carbide inserts	Same MRR do not have same power consumption as power consumption depends on cutting parameter values
2015	Kim [20]	AISI 1045	WC inserts	Depth of cut independently did not effects roughness, instead of cutting speed spindle speed influences the roughness
2014	Zebala [21]	Sintered carbide(wc–co)	PCD	Sintered carbide having lower Co content results in high surface roughness

(continued)

(continued)

Year	Author	Work material	Cutting tool	Remark
2014	Batish [22]	Steel	Cubic boron nitride	Machining force highly depend upon feed rate and hardness, surface roughness is highly depend upon feed
2014	Bouzid [23]	Stainless steel	Coated carbide	Feed rate significantly effects surface roughness, MRR significantly affected by depth of cut
2014	Rao [24]	Hardened steel	Carbide	Increment in cutting depth and feed of cut increases MRR, with spindle velocity roughness decreases
2014	Santos [25]	Aluminum alloy	Carbide	Required machining force in turning Al alloy is high at low velocity due to reduced heat production
2014	Shihab [26]	Hardened alloy steel	Coated carbide	Cutting speed affects micro hardness,feed affects surface roughness
2014	Jafarian [27]	Inconel 718	Cemented carbide	Medium cutting depth and feed with higher cutting velocity results in improved surface finish
2014	Homami [28]	Inconel 718	TiAlN-coated tungsten carbide	Cutting velocity, feed, tool-nose radius affects flank wear

(continued)

(continued)

Year	Author	Work material	Cutting tool	Remark
2014	Wang [29]	Carbon steel	Carbide	Surface finish optimization effect is limited as cutting factor is in reasonable range before optimization
2013	Xiong [30]	Heavy duty CNC machine tool	Coated tool	Multi-objective optimization is superior to single objective optimization
2011	Saeedy [31]	Stainless steel	Tic-coated inserts	Dominant factor for tool wear and roughness are cutting speed and feed, respectively
2011	Munawar [32]	AISI 1018	TiAln-coated carbide	High feed with low cutting velocity results in good surface finish
2011	Natarajan [33]	Brass	Carbide	With increment in feed roughness increases and then decreases
2011	Pawade [34]	Inconel 718	Cubic boron nitride	Cutting force and surface quality is highly depend on feed
2011	Ranganathan [35]	Stainless steel (316)	WC inserts	Feed has dominant affect while workpiece temperature has least effect on surface roughness

(continued)

(continued)

Year	Author	Work material	Cutting tool	Remark
2012	Senthikumaar [36]	Inconel 718	Uncoated carbide	Interaction of cutting velocity and cutting depth affects flank wear, while feed alone influences roughness
2011	Raja [37]	Brass, aluminum, copper, mild steel	Tungsten carbide	Low feed, high cutting velocity with low cutting depth gives better surface quality
2010	Bhushan [38]	Al alloy (7075) sic composite	Tungsten carbide, polycrystalline diamond	Aluminum has better surface roughness than turned Al alloy composite
2010	Yang [39]	EN 24 steel	Tungsten carbide	DE-based ANN model is easy, swift and robust
2010	Pytlak [40]	18 HGT steel	Cubic boron nitride	Feed accompanied by depth of cut influences roughness

3 Conclusions

1. Several surface integrity characteristics simultaneous optimization by single response Taguchi method is not good because of there exist interaction between the control parameter.
2. Multi-objective genetic algorithm (MOGA) technique is not authentic optimization as it is almost impossible to substantiate optimality of solution from MOGA.
3. ANN predictive capacity is better than that of RSM, but RSM method performs better when interaction between cutting parameter is needed to be evaluated. So these methods are complement to each other.
4. Slight variation in surface finish with time is not detected by neural network model.
5. Adverse effect on cutting zone temperature, productivity and tool life may occur due to multi-energetic optimizing criteria so an integrated optimization method with more objectives should be used.

References

1. Radovanović M (2019) Multi-objective optimization of multi-pass turning AISI 1064 steel. Int J Adv Manuf Technol 100:87–100. https://doi.org/10.1007/s00170-018-2689-z
2. Ajaja J, Jomaa W, Bocher P et al (2019) Hard turning multi-performance optimization for improving the surface integrity of 300M ultra-high strength steel. Int J Adv Manuf Technol 104:141–157. https://doi.org/10.1007/s00170-019-03863-3
3. Ashish George J, Lokesha K (2019) Optimisation and effect of tool rake and approach angle on surface roughness and cutting tool vibration. SN Appl Sci 1. 10.1007/s42452-019-1175-z
4. Bagaber SA, Yusoff AR (2018) Multi-responses optimization in dry turning of a stainless steel as a key factor in minimum energy. Int J Adv Manuf Technol 96:1109–1122. https://doi.org/10.1007/s00170-018-1668-8
5. Garcia RF, Feix EC, Mendel HT et al (2019) Optimization of cutting parameters for finish turning of 6082-T6 aluminum alloy under dry and RQL conditions. J Brazilian Soc Mech Sci Eng 41. 10.1007/s40430-019-1826-4
6. Hazir E, Ozcan T (2019) Response surface methodology integrated with desirability function and genetic algorithm approach for the optimization of CNC machining parameters. Arab J Sci Eng 44:2795–2809. https://doi.org/10.1007/s13369-018-3559-6
7. Laghari RA, Li J, Xie Z, Wang S, Qi (2018) Modeling and optimization of tool wear and surface roughness in turning of Al/SiCp using response surface methodology. 3D Res 9. https://doi.org/10.1007/s13319-018-0199-2
8. Laouissi A, Yallese MA, Belbah A et al (2019) Investigation, modeling, and optimization of cutting parameters in turning of gray cast iron using coated and uncoated silicon nitride ceramic tools. Based on ANN, RSM, and GA optimization. Int J Adv Manuf Technol 101:523–548. https://doi.org/10.1007/s00170-018-2931-8
9. Nataraj M, Balasubramanian K (2017) Parametric optimization of CNC turning process for hybrid metal matrix composite. Int J Adv Manuf Technol 93:215–224. https://doi.org/10.1007/s00170-016-8780-4

10. Nee CY, Saad MS, Mohd Nor A et al (2018) Optimal process parameters for minimizing the surface roughness in CNC lathe machining of Co28Cr6Mo medical alloy using differential evolution. Int J Adv Manuf Technol 97:1541–1555. https://doi.org/10.1007/s00170-018-1817-0
11. Salman KH, Elsheikh AH, Ashham M et al (2019) Effect of cutting parameters on surface residual stresses in dry turning of AISI 1035 alloy. J Brazilian Soc Mech Sci Eng 41. https://doi.org/10.1007/s40430-019-1846-0
12. Serra R, Chibane H, Duchosal A (2018) Multi-objective optimization of cutting parameters for turning AISI 52100 hardened steel. Int J Adv Manuf Technol 99:2025–2034. https://doi.org/10.1007/s00170-018-2373-3
13. Xie N, Zhou J, Zheng B (2018) An energy-based modeling and prediction approach for surface roughness in turning. Int J Adv Manuf Technol 96:2293–2306. https://doi.org/10.1007/s00170-018-1738-y
14. Saidi R, Fathallah B Ben, Mabrouki T et al (2019) Modeling and optimization of the turning parameters of cobalt alloy (Stellite 6) based on RSM and desirability function. Int J Adv Manuf Technol 100:2945–2968. https://doi.org/10.1007/s00170-018-2816-x
15. Chabbi A, Yallese MA, Nouioua M et al (2017) Modeling and optimization of turning process parameters during the cutting of polymer (POM C) based on RSM, ANN, and DF methods. Int J Adv Manuf Technol 91:2267–2290. https://doi.org/10.1007/s00170-016-9858-8
16. Xiao Z, Liao X, Long Z, Li M (2017) Effect of cutting parameters on surface roughness using orthogonal array in hard turning of AISI 1045 steel with YT5 tool. Int J Adv Manuf Technol 93:273–282. https://doi.org/10.1007/s00170-016-8933-5
17. Park HS, Nguyen TT, Dang XP (2016) Multi-objective optimization of turning process of hardened material for energy efficiency. Int J Precis Eng Manuf 17:1623–1631. https://doi.org/10.1007/s12541-016-0188-4
18. Rocha LCS, de Paiva AP, Rotela Junior P et al (2017) Robust multiple criteria decision making applied to optimization of AISI H13 hardened steel turning with PCBN wiper tool. Int J Adv Manuf Technol 89:2251–2268. https://doi.org/10.1007/s00170-016-9250-8
19. Camposeco-Negrete C, de Dios Calderón Nájera J, Miranda-Valenzuela JC (2016) Optimization of cutting parameters to minimize energy consumption during turning of AISI 1018 steel at constant material removal rate using robust design. Int J Adv Manuf Technol 83:1341–1347. https://doi.org/10.1007/s00170-015-7679-9
20. Kim YS, Kwon WT (2015) The effect of cutting parameters for the finest surface roughness during the turning of AISI 1045 with a WC insert. J Mech Sci Technol 29:3437–3445. https://doi.org/10.1007/s12206-015-0742-5
21. Wojciech Z, Kowalczyk R (2014). Estimating the effect of cutting data on surface roughness and cutting force during WC-Co turning with PCD tool using Taguchi design and ANOVA analysis. https://doi.org/10.1007/s00170-014-6382-6
22. Batish A, Bhattacharya A, Kaur M, Cheema MS (2014) Hard turning : parametric optimization using genetic algorithm for rough/finish machining and study of surface morphology †. 28:1629–1640. https://doi.org/10.1007/s12206-014-0308-y
23. Bouzid L, Boutabba S, Girardin F (2014) Simultaneous optimization of surface roughness and material removal rate for turning of X20Cr13 stainless steel. https://doi.org/10.1007/s00170-014-6043-9
24. Babu T, Gopala RA (2014). Modeling and multi-response optimization of machining performance while turning hardened steel with self-propelled rotary tool. https://doi.org/10.1007/s40436-014-0092-z
25. Article O (2014) Multi-objective optimization of cutting conditions when turning aluminum alloys (1350-O and 7075-T6 grades) using genetic algorithm. https://doi.org/10.1007/s00170-014-6314-5
26. Shihab SK, Khan ZA, Mohammad AAS, Siddiquee AN (2014) Optimization of surface integrity in dry hard turning using RSM
27. Jafarian F, Amirabadi H, Fattahi M (2014) Improving surface integrity in finish machining of Inconel 718 alloy using intelligent systems, pp 817–827. https://doi.org/10.1007/s00170-013-5528-2

28. Homami RM, Tehrani AF (2014) Optimization of turning process using artificial intelligence technology, pp 1205–1217. https://doi.org/10.1007/s00170-013-5361-7
29. Wang Q, Liu F, Wang X (2014) Multi-objective optimization of machining parameters considering energy consumption, pp 1133–1142. https://doi.org/10.1007/s00170-013-5547-z
30. Xiong Y, Wu J, Deng C, Wang Y (2013). Machining process parameters optimization for heavy-duty CNC machine tools in sustainable manufacturing. https://doi.org/10.1007/s00170-013-4881-5
31. Mahdavinejad RA, Saeedy S (2011) Investigation of the influential parameters of machining of AISI 304 stainless steel. Sadhana Acad Proc Eng Sci 36:963–970. https://doi.org/10.1007/s12046-011-0055-z
32. Munawar M, Chen JCS, Mufti NA (2011) Investigation of cutting parameters effect for minimization of sur face roughness in internal turning. Int J Precis Eng Manuf 12:121–127. https://doi.org/10.1007/s12541-011-0015-x
33. Natarajan C, Muthu S, Karuppuswamy P (2011) Prediction and analysis of surface roughness characteristics of a non-ferrous material using ANN in CNC turning. Int J Adv Manuf Technol 57:1043–1051. https://doi.org/10.1007/s00170-011-3343-1
34. Pawade RS, Joshi SS (2011) Multi-objective optimization of surface roughness and cutting forces in high-speed turning of Inconel 718 using Taguchi grey relational analysis (TGRA). Int J Adv Manuf Technol 56:47–62. https://doi.org/10.1007/s00170-011-3183-z
35. Ranganathan S, Senthilvelan T (2011) Multi-response optimization of machining parameters in hot turning using grey analysis. Int J Adv Manuf Technol 56:455–462. https://doi.org/10.1007/s00170-011-3198-5
36. Senthilkumaar JS, Selvarani P, Arunachalam RM (2012) Intelligent optimization and selection of machining parameters in finish turning and facing of Inconel 718. Int J Adv Manuf Technol 58:885–894. https://doi.org/10.1007/s00170-011-3455-7
37. Bharathi Raja S, Baskar N (2011) Particle swarm optimization technique for determining optimal machining parameters of different work piece materials in turning operation. Int J Adv Manuf Technol 54:445–463. https://doi.org/10.1007/s00170-010-2958-y
38. Bhushan RK, Kumar S, Das S (2010) Effect of machining parameters on surface roughness and tool wear for 7075 Al alloy SiC composite. Int J Adv Manuf Technol 50:459–469. https://doi.org/10.1007/s00170-010-2529-2
39. Yang SH, Natarajan U, Sekar M, Palani S (2010) Prediction of surface roughness in turning operations by computer vision using neural network trained by differential evolution algorithm. Int J Adv Manuf Technol 51:965–971. https://doi.org/10.1007/s00170-010-2668-5
40. Pytlak B (2010) Multicriteria optimization of hard turning operation of the hardened 18HGT steel. Int J Adv Manuf Technol 49:305–312. https://doi.org/10.1007/s00170-009-2375-2

Innovations and Future of Robotics

Ayush Kumar Agrawal, Pritam Pidge, Manisha Bharti, M. Prabhat Dev, and Prashant Kaduba Kedare

Abstract Next-generation applications in robotics and mechatronics will necessitate basic, flexible, compact, and affordable technologies. Technologies could be easy due to the virtual elimination of small-level feedback control, detectors, cabling, and computers. Traditional electric motors, including linear motors and drivetrain, are not really suitable for Boolean technology, though, because they become complicated, bulky as well as costly. Such automation was not well recognized and thus was encountered with main issues of usability. So thus in the next generation, the latest trend which is the Internet of things will be used with that of the devices and technologies which we use in our day-to-day life to do our work more efficiently and with very ease. Two-way communication which may be stated as full-duplex communication, where the device communicates with another device of its own and work accordingly, is been termed as the Internet of things and can be understood as stating it as an interconnection of things or devices.

Keywords Internet of things · Robotics · Automated devices · Automation · Mechatronics

1 Introduction

The turn of the period witnessed the development of manufacturing processes and their large acceptance. Industrialization was in middle at the turn of the millennium, and by the time we also created the automobile and were kind enough to display supersonic ideas. The effect on people's lives was enormous; socioeconomic laws are updated that regulated traveling, medical care, production, employment conditions, and domestic life. Each cycle was replicated with that of the change in innovation in the past century though at a quicker pace. Development relocated into the household from the workshop and research center [1–4].

A. K. Agrawal (✉) · P. Pidge · M. Bharti · M. P. Dev · P. K. Kedare
National Institute of Technology Delhi, New Delhi 110040, India
e-mail: ayush6295@gmail.com

© The Author(s), under exclusive license to Springer Nature Singapore Pte Ltd. 2021

R. M. Singari et al. (eds.), *Advances in Manufacturing and Industrial Engineering*, Lecture Notes in Mechanical Engineering,
https://doi.org/10.1007/978-981-15-8542-5_86

The guiding powers became the modern fields of computing, networking, robotics, and processing, instead of the electrical components of the Middle Ages. There have been virtually no cellphones in the late 1800s, but mobiles were a regular sight at the turn of the decade; machines were mostly dreamed of a fifty years earlier, but they have become ubiquitous. We are here at the precipice of an almost important modern industrial change: the Technology Era. The transition would put the twenty-first millennium at a crucial point in time. More specifically, it would have an irrevocable effect on the futures of both our families and generations to come [5–7].

In comparison, automation uses a combination of electronic systems, power systems, and modern computer and technology approach. It is by integrating the potential of several current and emerging technologies that a very impressive variety of machines and autonomous frameworks has been built and will therefore be.

2 Next-Generation Future Robotics

"Robotics" is sometimes described over its capabilities—it is really a system that can efficiently execute a complicated set of things, particularly, that can be programmed by software. It is a literal depiction that includes a huge percentage of the kind of traditional robots you see everyday in movies. This description, and the influence of existing social conceptions of how a machine is, is affecting our perceptions of what being a machine maybe. By analyzing societal views toward bots all over the globe, we could see the strongest evidence of this. The effect of such social and cultural disparities throughout the creation in technology is deep: Global automation is deeply involved with military science; whereas, common robots are centered on assistance, medical services, and business. It also reinforces our skewed expectations about what a bot will look or even how it will act. We also have a chance to move freely in certain protocols. A machine is not realistic, get arms, move, or speak. We could have a somewhat narrower understanding of how a device is, instead. There is a redrawing of distinctions among smart structures, nanotechnology, abstraction, genetics, and automation. That is how robots over the next twentieth to fifty decades can truly impact humankind [8–11].

Including bots capable of tracking and restoring a physical world to micro-machines to detect and destroy tumors, as well as from machines capable of guiding the path to global exploration to device compañeros to save us all from isolation in older years. There really is no aspect of our culture or life which modern robots will not influence. They will be commonplace, in fact.

To attain mechanical prevalence, we also need to learn and reproduce the features of the design, but also to move above them with accelerated growth (surely accelerated than developmental timespans!) and also more common and versatile technology. As living beings, one way of talking about potential robotics is just too. Rather of a traditional bot that can be broken down into physical, electromagnetic, and technological realms, we should conceive of a bot in terms of its human equivalent as they have three central parts: a frame, ahead as well as an abdomen. The potential energy in the abdomen of species then spreads throughout the organ to fuel the body and then mind that in effect regulates the humans. Therefore, there is a conceptual distinction between certain model entity and the human world: The mind is equal to the machine or linear actuator; the heart is equal to the machine's physical framework; as well as the abdomen is identical to the machine's electricity supply, whether a generator, solar panel, or another energy source. The strength of both the synthetic system models was that we were able to use all the attributes in species, but move through them. These support virtues still unresolved by existing nanotechnology science, like activity under diverse and difficult environments, beneficial incorporation with the climate, regeneration, decay, and decay. Those are necessary for the creation of omnipresent automated species [12–15].

Solving these problems will only be feasible by concentrated work in the fields of intelligent structures, nanotechnology, computing, and adjustment. There we could concentrate on the creation of innovative intelligent technologies components, and so we can see if the production of components does not take place in parallel from the other critically required engineering fields.

3 Intelligent Devices

An intelligent substance is something that once induced by some other area experiences any measurable impact in this field. This encompasses certain areas from electronic, electromagnetic, biological, electronic, atmospheric, etc. In example, until warm, a thermochromic substance experiences some colored shift; whereas, when thermally activated, an electroactive plastic produces a physical production. Quick technologies may bring new electronics skills and particularly synthetic entities. An intelligent substance that adjusts mechanical properties once subjected to a solvent can also be used. The author already has enough indications of what could be achieved with natural fibers in existence, and we would like to leverage those strengths in developing devices. Now let us glance then at some of the innovations that can provide this functionality. That state-of-the-art soft robotics technology may be classified across several communities: (1) flexible kinetic and motorized structures; (2) intelligent actuators and control structures; or (3) devices that alter rigidity. On another side, covalent molecular hydraulics are intelligent technologies that work by a specific electrical concept. The IPA is constructed out of a single osmotic connector sheet, still wedged by two charging connectors; however, the electrical potential is somewhat smaller (kV/m) in comparison with Rud, and thus the membranes will be more heat-resistant. If adding the electrical force, released atoms inside the covalent resistor travel into the edges which they accumulate. Thanks to localized charged particle effects, the substantial population with atoms at just the interfaces allows themselves to extend while as though-charge repels [16–19].

4 Use of the Robotics

These have focused on either the developments that could send a golden age of automation earlier, and let us now look at how so many machines may function in human worlds and whether we can communicate with them rather than function. Adherence to weak technologies enables their best design for immediate human cell interactions. Necessarily, the medium-soft experiences between a flexible bot and person are far better than that of a hard force design enforced by traditional stiff bots. Most research has been conducted on clever products, namely electromagnetic contacts and computer parts, for immediate surface-to-skin touch and application on latex. By that, the need for medicines, the fundamental part of these hides resides in smarter rubber gloves to facilitate curing but reduces the incidence of infectious tolerance microbes. Cosmetics may, of reality, cover garments we are much distance toward societal recognition of third-skins as just a supplement for traditional attire. If, on the other side, we take advantage of spongy soft rotor developments including the silicone coils control module and type titanium polymer components for information. These intelligent fabrics often deliver a remarkable new campus: Since the intelligent fabric has been in close touch with both the body which has rotor capability, the surface may be actively activated by the process. We should incorporate physical touch into the garments in this manner. Another sense had mostly left that visual medium of contact behind. Take the current smartphones for example; it has a large throughput of both auditory inputs with almost semi-existent relaxing contact functionality. They may produce normal "effective" sensations of contact through contact-enabled clothes, thereby bringing us just a possibly groundbreaking new medium of interaction. We may brush, lick, or inflict fun sensory sensations rather than a rugged magnetic mechanism (as it is used in cellphones). When the above intelligent garments can produce greater forces, it may be used not only for correspondence but for the support structure. A potential alternative is in the form of wealth-assistant apparel which will improve autonomy for individuals who seem to be weak, injured, or aged. Maintaining flexibility might have a significant effect on the user's well-being and could also encourage them to move to economic life, hence benefiting the overall economy. That problem for such an assertion was the strength capacity within the support unit of the actuation technological. Unless the person is low, for once since they have reduced body fat, we would require substantial strength, so it may be extremely costly to calculate this spare strength. The help system would also be as small and convenient as feasible with either the helical gears providing a strength capacity that is considerably greater than that of the physiological tissues. This is above the latest technology at a moment. Essentially, portable support systems can make obsolete traditional assistance tools. They may expand that chem-integration onto the skin, as demonstrated by portable devices mentioned above. Since hard robots are too ideal for physiological organ interaction, this is normal to conceive of a system that could be inserted within the anatomy or that could literally communicate with inner structures. They could then create endoscopic pharmaceuticals that might regain the functions of an organ or tissues that are disfigured and harmed, taking cartilage cancers, of instance, which may involve areas varying from both the bladder or kidney and nasal cavity or windpipe. Traditional diagnosis of these conditions includes the therapeutic tumor amputation as well as the control of the accompanying infection. The person of the laryngeal tumor can undergo colon resection, but subsequently will not be able to communicate and might have to have a lifelong tracheostomy. We will regain usable functions by creating and transplanting soft mechanical substitute organs to encourage the individual to talk, drink, swallow to enjoy their lives once again. These thermo-integrating flexible robots are also under progress and are projected to surface throughout the hospital within the next 10 to 15 decades [20–22].

5 Conclusion and Future Scope

Among this paper, we just gotten started about what a bot is, how it can be conceived of as a flexible robotic body, as well as how intelligent technologies can further understand and fundamentally change potential technologies. That individual influence was addressed but the real magnitude of this effect is one we can just hint at. Much as it has become difficult to foresee the effect of the Internet and the World Wide Web, we could not anticipate how potential technology would lead us. As we stroll the technology revolution path, we are going to look back on just this century as the day that robotics really took out and laid the groundwork for our futuristic society.

References

1. Rossiter J (2016) Robotics, smart materials, and their future impact for humans. In: The next step. Exponential life. BBVA, Madrid
2. Albrecht T, Bührer C, Fähnle M, Maier K, Platzek D, Reske J (1997) First observation of ferromagnetism and ferromagnetic domains in a liquid metal. Appl Phys Mater Sci Process 65(2):215
3. Amend JR, Brown E, Rodenberg N, Jaeg HM, Lipson H (2012) A positive pressure universal gripper based on the jamming of granular material. IEEE Trans Rob 28(2):341–350
4. Bar-Cohen Y (ed) (2004) Electroactive Polymer (EAP) actuators as artificial muscles—reality, potential, and challenge, 2nd edn. SPIE Press, Bellingham, WA.
5. BBC News (2003) Octopus intelligence: jar opening. https://news.bbc.co.uk/1/hi/world/europe/2796607.stm. 2003-02-25. Retrieved 2016-10-10
6. Cao W, Cudney HH, Waser R (1999) Smart materials and structures. PNAS 96(15):8330–8331
7. Curie J, Curie P (1881) Contractions et dilatations produites par des tensions dans les cristaux-hémièdres à faces inclines (Contractions and expansions produced by voltages in hemihedral crystals with inclined faces). Comptesrendus (in French) 93:1137–1140
8. Haines CS et al (2014) Artificial muscles from fishing line and sewing thread. Science 343(6173):868–872
9. Ilievski F, Mazzeo AD, Shepherd RF, Chen X, Whitesides GM (2011) Soft robotics for chemists. AngewandteChemie 123:1930–1935

10. Jahromi SS, Atwood HL (1969) Structural features of muscle fibres in the cockroach leg. J Insect Physiol 15(12):2255–2258
11. Jin S, Koh A, Keplinger C, Li T, Bauer S, Suo Z (2011) Dielectric elastomer generators: How much energy can be converted? IEEE/ASME Trans Mechatron 16(1)
12. Keplinger C et al (2013) Stretchable, transparent, ionic conductors. Science 341(6149):984–987
13. Kier WM, Smith KK (1985) Tongues, tentacles and trunks: The biomechanics of movement in muscular-hydrostats. Zoological J Linnean Soc 83:307–324
14. Kim DH et al (2011) Epidermal electronics. Science 333(6044):838–843
15. Knoop E, Rossiter J (2015) The tickler: a compliant wearable tactile display for stroking and tickling. In: Proceedings of the CHI 2015, 33rd Annual ACM Conference on Human Factors in Computing Systems, pp 1133–1138
16. Lendlein A, Kelch S (2002) Shape-memory polymers. Angew Chem Int Ed 41:2034–2057
17. Mather JA (2006) Behaviour development: a cephalopod perspective. Int J Comparative Psychol 19(1)
18. Meller MA, Bryant M, Garcia E (2014) Reconsidering the McKibben muscle: energetics, operating fluid, and bladder material. J Intell Mater Syst Struct 25:2276–2293
19. Morin SA, Shepherd RF, Kwok SW, Stokes AA, Nemiroski A, Whitesides GM (2012) Camouflage and display for soft machines. Science 337(6096):828–832
20. Pelrine R, Kornbluh R, Pei Q, Joseph J (2000) High-speed electrically actuated elastomers with strain greater than 100%. Science 287(5454):836–839
21. Pfeifer R, Gómez G (2009) Morphological computation—connecting brain, body, and environment. Creating Brain-Like Intell, Lect Notes Comput Sci 5436:66–83
22. Polygerinos P, Wang Z, Galloway KC, Wood RJ, Walsh CJ (2015) Soft robotic glove for combined assistance and at-home rehabilitation. Robot Autonomous Syst 73:135–143

Optimization of EDM Process Parameters: A Review of Technique, Process, and Outcome

Akash Gupta and Harish Kumar

Abstract Electric discharge machining (EDM) is a vital non-traditional process, which is extensively used for machining a range of difficult-to-machine material such as heat-treated tool steels, composites, ceramics, carbides, heat-resisting steels with convoluted shape. The various process of EDM: die sinking EDM, wire EDM (WEDM), EDM milling, and wire electrical discharge grinding (WEDG). Principle behind EDM process is thermo-electric source between the job and an electrode. Majorly electrical discharge machining parameters are usually evaluated on the basis of surface roughness (SR), MRR (MRR), cutting speed (V), and tool wear rate (TWR). The vital electrical discharge machining parameter affecting to the characteristic measures of the course is Ip, arc gap, Ton, and Toff. Numerous approaches are proposed in the past to resolve the problems such as to optimize the condition of higher MRR, and improvement in surface finish and lower SR with low tool wear rates are achieved with various optimization techniques. A substantial work has been reported by the researchers on the EDM performance for various materials on the basis of parameters. It is observed that a review of the diverse approaches developed previously would aid to assess their foremost prospects and their advantages or boundaries in which the majority of materials have to be machined so as to choose the most appropriate approach for a meticulous application based on the material and also tells about the prospects that need further consideration. In this paper, we will review research work done in past regarding optimization of EDM process parameters by different techniques.

Keywords Electric discharge machining · Process parameters · Taguchi methodology

---

A. Gupta (✉) · H. Kumar
Department of Mechanical Engineering, National Institute of Technology Delhi, New Delhi, India
e-mail: akashgupta0304@gmail.com

© The Author(s), under exclusive license to Springer Nature Singapore Pte Ltd. 2021

R. M. Singari et al. (eds.), *Advances in Manufacturing and Industrial Engineering*,
Lecture Notes in Mechanical Engineering,
https://doi.org/10.1007/978-981-15-8542-5_87

Nomenclature

MRR	Material removal rate
SR	Surface roughness
TWR	Tool wear rate
KW	Kerf width
CS	Cutting speed
Ip	Peak current
SV	Servo voltage
Toff	Pulse off time
Ton	Pulse on time
GV	Gap voltage
DE	Dimensional error
RLT	Thickness of recast layer
W_T	Wire tension
MH	Microhardness
F	Frequency of vibration
SG	Spark gap
WS	Wire speed
WF	Wire feed

1 Introduction

EDM was primarily invented in 1770 by Joseph Priestly. Though, EDM is not entirely used for machining in anticipation of 1943 after Russian scientists B R and N I Lazarenko used it for machining purposes. It has now recognized worldwide a best machining process for production of forming tools to generate plastics molding, die casting, forging die, machining for heat-treated tool steel, super alloy, ceramics, and MMC requiring high exactness, complicated shapes with high surface finish, etc., because of its unique machining characteristics and high precision [1]. The removal of material in EDM is based on erosion of electric sparks taking place amid two electrodes. There are numerous theories in attempts to clarify the complicated phenomenon of "erosive spark" [2]

The following are various theories:

- **Electro-mechanical theory**: It is based on fact that abrasive effect of material particles takes place as action of the intense electric field. This theory proposes that electric field separate the object particles of the work piece as it exceeds the cohesive force in the structure of material. Investigation lacks support this theory.
- **Thermo-mechanical theory**: It is based on fact that MRR in EDM operation occurs due to melting of material due to "flame jets." The flame jets are created

as a consequence of diverse electrical effects of discharge.This does not consent through experiments and fail to furnish a realistic justification of spark erosion.

- **Thermo-electric theory**: Theory, best-support investigational evidence, suggests that MRR in EDM takes place by generation of extremely high temperature generated by elevated intensity of discharge current. Though well supported, but it cannot be considered as definite and absolute because of difficulty in explanation.

Material Removal in the form of small sphere [3]:

At place wherever the field of electricity is the strong, a discharge will be started; this being in fact the results of a whole process which is described in Fig. 1, and the stages of removal process are shown below.

A. Under the effect of strong electric field, the positive ions and electrons are accelerated getting a high velocity, and swiftly they form an electrically conductive ionized channel.
B. At these stages, current flows through channel a spark are initiated between two electrodes causing an infinite number of collisions between particles. At the same time, a gas bubble develops due to vaporization of dielectric, and its pressure goes up gradually to a very high value.

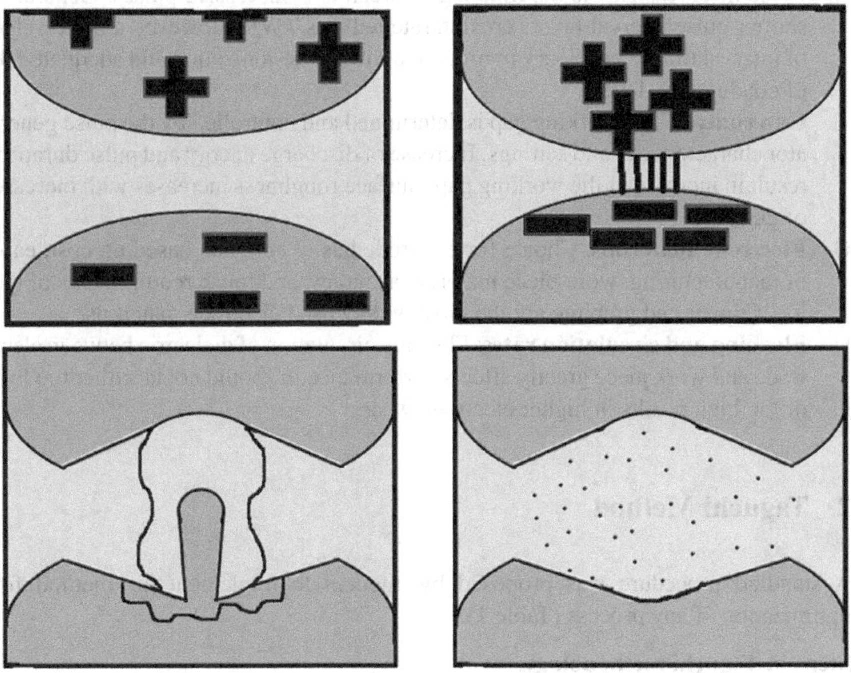

Fig. 1 Shows process of material removal

C. After current is switched off, abrupt decline in temperature origin implosion of bubble, giving rise toward dynamic forces that have effect of projecting melted material out of crater.
D. The eroded material then re-solidifies in form of small spheres, and it is removed by circulating dielectric flow.

Process parameters and controlling factors in EDM In general, EDM performance depends upon the following parameters [4]:

A. **Polarity**: It is selected based upon the arrangement of electrode as well as work materials in order to achieve required MRR with minimum TWR.
B. **No-load voltage**: This is the open circuit voltage during delay time prior to current starts flowing. It is elected by manufacturers. Usually, it varies from 60 to 120 V.
C. **Discharge current**: This influences the MRR explicitly, provides high erosion rates with reduced amperage, and vice versa. Moreover, wear rates often rise with that amperage in the case of copper electrodes, and wear levels do not escalate beyond certain amperage in the case of graphite electrodes.
D. **Pulse duration**: It is the period of discharge or the time of passing of current in a single pulse. It is in the microsecond order. It influences corrosion rate, wear of the electrode, and finish of surface.
E. **Pulse interval**: This is the time lag between two successive pulses. Generally, shorter pulse interval favors erosion rate reduces TWR. However, a finite value of interval time is necessary in order to allow for de-ionization and adequate fall of conductivity level.
F. **Gap control**: The working gap is determined and controlled by the pulse generator characteristics and settings. Increase of discharge energy and pulse duration result in increase of the working gap. Surface roughness increases with increase of gap.
G. **Electrode materials**: Choice for electrode has to be made based on cost, ease of manufacturing, work piece material, accuracy, and finish requirements of the job. Copper and graphite are the most widely used electrode materials.
H. **Flushing and circulation rates**: Flushing circulation of dielectric between electrode and work piece greatly affects performance. It should not be either too low or too high results in higher electrode wear.

2 Taguchi Method

A standard procedure was proposed by Taguchi to implement his method for optimization of any process (Table 1).

Steps in Taguchi methodology:

A. Identifying the most important function and failure method.
B. Identifying noise-related factors.

Table 1 Taguchi method for improvement and optimization of performance parameters for EDM process

Author	Work piece material	Electrode material	Machining parameters (level)	Performance parameters	OA	Remarks
Manjaiah and Laubscher [5]	AISI D2 Steel	Brass	Ton, Toff, SV, WF	MRR, SR	L27	Ton and SV are most important variables affecting MRR and SR
Subrahmanyam and Nancharaiah [6]	Inconel 625	Brass wire	Ton, Toff, Ip	MRR, SR	L9	For MRR, Ton and Toff were major factor. Toff has highest contribution. For SR, Ton and Toff were major factor. Ton has the highest contribution
George et al. [7]	Carbon–carbon composites	Copper	Ip, GV, Ton	MRR, TWR	L8	Parameters affecting TWR and MRR, according to significance, are GV, Ip, and Ton. Among various process parameters, Ton was insignificant
Dharmendra and Shyam [8]	Inconel 800	Copper	Ton, Toff, Ip	MRR, TWR	L9	The developed empirical relations can used to estimate MRR and TWR
Muthu Ramalingam [2]	Alpha-beta titanium alloy	Brass, tungsten carbide, copper	Ton, Toff, Ip	MRR, SR	L9	For SR, Ton and Toff are major factors. Ton has highest contribution. For MRR, Ton and Toff were major factors. Toff has highest contribution

(continued)

Table 1 (continued)

Author	Work piece material	Electrode material	Machining parameters (level)	Performance parameters	OA	Remarks
Singh et al. [9]	Al 10%SiCP composite (Al MMC)	Brass	Ip, Ton, flushing pressure	MRR,TWR, SR, taper, radial overcut	L27	Ip have major effect on TWR trail by voltage and Ton, and Ton have maximum effect on SR trail by voltage and current.
Huu-Phan et al. [10]	SKD61 die steel	Copper	Ton, Toff, Ip, F	MRR	L9	Low frequency of vibration is incorporated in work piece to improve surface quality. Ip and F have major influence on MRR. Pulse parameter is less prominent than frequency of vibration.
Palanisamy et al. [11]	LM6-alumina (MMC)	Copper	Ton, Toff, Ip	MRR, TWR, SR	L27	It is revealed that Ip is most influencing parameter trailed by Ton and Toff
Amit Kumar et al. [12]	HSS M2	Molybednum	Ton, Toff, Ip WF	MRR, SR, KW	L27	The characteristics such as MRR, SR, and KW were improved

(continued)

Table 1 (continued)

Author	Work piece material	Electrode material	Machining parameters (level)	Performance parameters	OA	Remarks
Dhruv Bhatta and Ashish Goyal [13]	AISI-304	Copper	Ton, Toff Ip, GV, W_T, WS, SG	MRR, SR	L27	The increase in Ton and SG results in increased MRR, whereas increasing Toff resulted in decrease of MRR. The high value Toff, W_T, and GV results in increased SR
Nayak and Mahapatra [14]	Inconel 718	Deep cryo-treated bronco wire	Part thickness, taper angle, Ton, Ip, WS, W_T	SR, angular error, CS	L27	The consequence of parameters on various performance characteristics such as angular error, SR, and CS is analyzed after taper cutting operation in deep cryo-treated Inconel 718
Ravinder et al. [15, 16]	Carbon fiber-reinforced plastics (CFRPs)	Tungsten carbide	Voltage, capacitance, tool speed	MRR, TWR	L9	It is found that inappropriate placement of tool electrode above conducting fiber layer produces non-circular holes.

(continued)

Table 1 (continued)

Author	Work piece material	Electrode material	Machining parameters (level)	Performance parameters	OA	Remarks
Satish Kumar et al. [17]	Inconel 800	Copper, copper–chromium, graphite	Ton, Toff, Ip, tool material, powder material	MRR, TWR	L27	Ip, Ton, and tool are found most important for MRR, Ip, Ton, tool, and powder for TWR
Maneswar and Patowari [18]	Aluminum	Copper	Compact load, Ip, Ton	TWR, SR, material transfer rate	L16	Greater Ton that results larger will be contact to high temperature of masking material, which outcome damage of edge of masking and deposition
Tripathy and Tripathy [19]	H-11 die steel	Copper	Powder concentration, Ip, Ton, GV duty cycle	MRR,TWR SR, electrode wear ratio	L27	Ip is parameter which has significant involvement toward enhancement in value of preference solution while for GV is trivial.
Prakash et al. [20]	Hybrid 356/B4C/fly ash	Brass	Ton, Toff GV,WF % reinforcement	MRR	L27	GV has largest statistical control on MRR of composites then followed by Ton and Toff
Lin et al. [21]	Al_2O_3–TiC ceramics	Electrolytic copper	Polarity Ip, GV Ton, SV	MRR, TWR, SR	L18	MRR enlarged with peak current also with pulse duration

(continued)

Table 1 (continued)

Author	Work piece material	Electrode material	Machining parameters (level)	Performance parameters	OA	Remarks
Singh et al. [22]	Super Co 605	Graphite	Flushing pressure, GV, Ton, Toff, Ip	MRR, TWR, SR	L18	Ip and Ton found most important factors, MRR and Ip, Toff for TWR
Nikalj et al. [23]	Maraging Steel 300	Copper	Ip, Ton, Toff	TWR, MRR, SR, relative wear ratio	L9	It was established that optimum levels of factors for SR and TWR are similar but differ from optimal level of parameters for MRR and relative WR
Kamal et al. [24]	Ti-6Al-7Nb	Graphite, CuW	Ip, Ton, Toff, work height	TWR, drilling rate	L16	Increasing trend in DR and TWR with rise in values of Ip and Ton. With increase in Toff, value of DR increases, but TWR decreases. With increasing work-height, characteristics see a decline
Carmita Camposeco-Negrete [25]	AISI-O1 tool steel	Brass	Ton, Toff, SV, WS	Machining time, SR	L9	SV was most significant for minimizing machining time, followed by Ton and WS. For SR, Toff was most

(continued)

Table 1 (continued)

Author	Work piece material	Electrode material	Machining parameters (level)	Performance parameters	OA	Remarks
Aliakbari and Baseri [26]	X210Cr12 (SPK)	Electrode without hole, one eccentric hole, two eccentric symmetric holes	Ip, Ton, electrode rotation	MRR, SR, TWR, overcut	L9	Input parameters of Ip, Ton, rotation, and geometry are most efficient on MRR, TWR, and SR. By geometry of electrode, rotation speed of electrode has diverse effect on outputs. This is due to geometry of electrode which is affected on flushing quality when electrode is rotary. Rotary EDM with electrical base; electrical characters are more effective than nonelectric
Ramakrishnan and Karunamoorthy [27]	HT-tool steel	Zinc-coated brass	Ton, W_T delay time, WF, WS, Ip	MRR, SR, TWR	L16	It was identified that Ton and Ip have influenced more than other parameters considered in this

(continued)

Table 1 (continued)

Author	Work piece material	Electrode material	Machining parameters (level)	Performance parameters	OA	Remarks
Kao et al. [28]	Ti-6Al-4 V alloy	Electrolytic copper	Ip, GV, Ton, duty factor	MRR, TWR, SR	L9	Ton was chief influencing feature contributing to characteristic measures, followed by Toff, Ip, and GV
Vikas et al. [29]	En 41	Copper	Ton, Toff, Ip, GV	SR	L27	It was found out that Ip had a great impact over SR. The effect of other characteristic was insignificant
Lin et al. [30]	SKD 61	Electrolytic copper	Ton, Toff, Ip, polarity, GV, SV	MRR, SR	L18	Magnetic force able EDM had a better machining stability. Moreover, number of effective discharge waveforms obtained by magnetic force-assisted EDM. MRR of magnetic force-assisted EDM is thrice of standard EDM

(continued)

Table 1 (continued)

Author	Work piece material	Electrode material	Machining parameters (level)	Performance parameters	OA	Remarks
Amitesh Goswami and Jatinder Kumar [31]	Nimonic-80	Brass	Ton, Toff, Ip	MRR, SR	L27	Contributions of Ton and Toff have been found to dominate other factors. All two factors (Ton-Toff, Ton-Ip, Toff-Ip) in study also significant for MRR
Chander Prakash et al. [32]	Biodegradable Mg-Zn-Mn alloy	Mg–Ca alloy	Ton, Toff, Ip hydroxyapatite powder	SR, RLT, MH	L27	SR and RLT increased through increase in Ip, Ton and decreased with powder, Ton
Murahari Kolli and Adepu Kumar [33]	Ti-6Al-4 V	Copper	Ton, Ip electrode diameter, powder concentration	MRR, SR, TWR, RLT	L9	It is observed from experiments that graphite powder with surfactant added dielectric significantly improves MRR and reduces SR, TWR, and RLT
Zahid and Arshad [34]	SS 304	Brass	Ton Toff, Ip	SR, KW	L9	The Ton is found to most influential factor for both SR and KW
Dabadea and Karidkar [35]	Inconel 718	Zinc-coated brass	Ton, Toff Ip, WF W_T, SV	MRR, SR, KW, DE	L8	Ip is significant parameter for KW and DE, whereas for MRR and SR, SV was observed to significant parameter

(continued)

Table 1 (continued)

Author	Work piece material	Electrode material	Machining parameters (level)	Performance parameters	OA	Remarks
Rahul et al. [36]	Inconel 601, 625, 718, 825	Copper	GV, Ip Ton, duty factor, flushing pressure	MRR, TWR, SR, surface crack density	L16	Increase in Ip results in improved MRR for different grades. An increasing value of SR observed increment of Ip. In surface crack density, with increase in Ip, crack decreases, again increases, and finally constant

C. Testing environment and quality characters.
D. Identifying the objective purpose of optimization.
E. Identifying of control factor and their respective level.
F. Selecting the appropriate orthogonal array (OA) for matrix experiment.
G. Conducting matrix for experiment.
H. Analysis of data; predicting optimum levels and their respective performance.
I. Performing the authentication trial and planning future course of action.

3 Conclusion

- The chief advantage of Taguchi method is that it helps to improve quality of product by providing values that are similar to the desired value mean rather than a value within the statistically defined maximum.
- MRR for materials increases with increase in Ip, Ton, duty cycle, tool shift material from copper to brass, whereas decreases with increase in voltage.
- TWR increases by Ip, Ton, and tool shift material from copper to brass while it decreases with increase with gap voltage and duty cycle.
- SR increases by Ton and duty cycle as it decreases through increase in gap voltage. Though, SR decreases with Ip initially, then increases afterwards.
- The orthogonal array often leads to decreased number of experiments conducted for a specific no. of device parameter with limited number of levels.
- The Taguchi approach is a fast, simple, and easy technique for optimization. The gain is improved productivity with decreased cycle time.
- The key limitation of the Taguchi process is that Taguchi tests. The Taguchi method does not specify exactly which parameter has maximum effect on value of performance features. The orthogonal array does not analyze the arrangement of individual variables as this approach is not capable of establishing a relationship between those variables.
- Further, Taguchi approach does not refer to systems that change over time, dynamic processes.

References

1. Ho KH, Newman ST (2003) State of the art electrical discharge machining (EDM). Int J Mach Tools Manuf 43(13):1287–1300
2. Muthuramalingam T (2019) Effect of diluted dielectric medium on spark energy in green EDM process using TGRA approach. J Clean Prod 238:117894
3. Ojha K, Garg RK, Singh KK. Parametric optimization of PMEDM process using Chromium powder mixed dielectric and triangular shape electrodes. J Miner Mater Charact Eng 10(11)
4. Deshmukh SS, Zubair AS, Jadhav VS, Shrivastava R (2019) ScienceDirect optimization of process parameters of wire electric discharge machining on AISI 4140 using Taguchi method and grey relational analysis. Mater Today Proc 18:4261–4270

5. Manjaiah M, Laubscher RF, Kumar A, Basavarajappa S (2016) Parametric optimization of MRR and surface roughness in wire electro discharge machining (WEDM) of D2 steel using Taguchi-based utility approach. Int J Mech Mater Eng
6. Subrahmanyam M, Nancharaiah T (2019) Materials today: proceedings optimization of process parameters in wire-cut EDM of Inconel 625 using Taguchi's approach. Mater Today Proc
7. George PM, Raghunath BK, Manocha LM, Warrier AM (2004) EDM machining of carbon–carbon composite—a Taguchi approach, vol 145, pp 66–71
8. Dharmendra BV, Kodali SP, Rao BN (2019) Heliyon A simple and reliable Taguchi approach for multi-objective optimization to identify optimal process parameters in nano-powder-mixed electrical discharge machining of INCONEL800 with copper electrode. Heliyon 5:e02326
9. Singh PN, Raghukandan K, Pai BC (2004) Optimization by grey relational analysis of EDM parameters on machining Al—10% SiC P composites, vol 156, pp 1658–1661
10. Nguyen H-P, Pham V-D (2019) Engineering sciences single objective optimization of die-sinking electrical discharge machining with low frequency vibration assigned on workpiece by Taguchi method. J King Saud Univ Eng Sci 1–6
11. Palanisamy D, Devaraju A, Manikandan N, Balasubramanian K, Arulkirubakaran D (2019) Materials today : proceedings experimental investigation and optimization of process parameters in EDM of aluminium metal matrix composites. Mater Today Proc
12. Kumar A, Soota T, Kumar J (2018) Optimisation of wire-cut EDM process parameter by grey-based response surface methodology. J Ind Eng Int 14(4):821–829
13. Bhatt D, Goyal A (2019) ScienceDirect Multi-objective optimization of machining parameters in wire EDM for AISI-304. Mater Today Proc 18:4227–4242
14. Nayak BB, Mahapatra SS (2015) Optimization of WEDM process parameters using deep cryo-treated Inconel 718 as work material. Eng Sci Technol Int J
15. Kumar R, Agrawal PK, Singh I (2018) Fabrication of micro holes in CFRP laminates using EDM. J Manuf Process 31:859–866
16. Khanna R, Kumar A, Pal M, Ajit G (2015) Multiple performance characteristics optimization for Al 7075 on electric discharge drilling by Taguchi grey relational theory. J Ind Eng Int 11(4):459–472
17. Kumar S, Dhingra AK, Kumar S (2017) Parametric optimization of powder mixed electrical discharge machining for nickel-based superalloy Inconel-800 using response surface methodology. Mech Adv Mater Mod Process 3(1)
18. Rahang M, Patowari PK (2015) Parametric optimization for selective surface modification in EDM using Taguchi analysis
19. Tripathy S, Tripathy DK (2015) Multi-attribute optimization of machining process parameters in powder mixed electro-discharge machining using TOPSIS and grey relational analysis. Eng Sci Technol Int J
20. Prakash JU, Juliyana SJ, Pallavi P, Moorthy TV (2018) ScienceDirect optimization of wire EDM process parameters for machining hybrid composites (356/B4C/Fly Ash) using Taguchi technique. Mater Today Proc 5(2):7275–7283
21. Taylor P et al (2013) Materials and manufacturing processes machining performance and optimizing machining parameters of Al_2O_3–TiC ceramics using EDM based on the Taguchi method machining performance and optimizing machining parameters of Al_2O_3–TiC ceramics using EDM, pp 37–41
22. Taylor P et al (2014) Materials and manufacturing processes optimization of parameters using conductive powder in dielectric for EDM of super Co 605 with multiple quality characteristics optimization of parameters using conductive powder in dielectric for EDM of super Co 605 w, August, pp 37–41
23. Nikalje AM, Kumar A, Srinadh KVS (2013) Influence of parameters and optimization of EDM performance measures on MDN 300 steel using Taguchi method
24. Kumar K, Singh V, Katyal P, Sharma N (2019) EDM μ-drilling in Ti-6Al-7Nb : experimental investigation and optimization using NSGA-II
25. Camposeco-negrete C (2019) Prediction and optimization of machining time and surface roughness of AISI O1 tool steel in wire-cut EDM using robust design and desirability approach

26. Aliakbari E, Baseri H (2012) Optimization of machining parameters in rotary EDM process by using the Taguchi method, pp 1041–1053
27. Karunamoorthy RRL (2006) Multi response optimization of wire EDM operations, pp 105–112
28. Kao JY, Tsao CC, Wang SS, Hsu CY (2010) Optimization of the EDM parameters on machining Ti–6Al–4 V with multiple quality characteristics, pp 395–402
29. Kumar A, Kumar K (2014) Effect and optimization of various machine process parameters on the surface roughness in EDM for an EN41 material using Grey-Taguchi. Proc Mater Sci 6:383–390
30. Lin Y, Chen Y, Wang D, Lee H (2008) Optimization of machining parameters in magnetic force assisted EDM based on Taguchi method, vol 9, pp 3374–3383
31. Goswami A, Kumar J (2014) Engineering science and technology, an international journal optimization in wire-cut EDM of Nimonic-80A using Taguchi's approach and utility concept. Eng Sci Technol Int J 17(4):236–246
32. Prakash C, Singh S, Singh M, Verma K, Chaudhary B, Singh S (2018) Multi-objective particle swarm optimization of EDM parameters to deposit HA-coating on biodegradable Mg-alloy. Vacuum
33. Kolli M, Kumar A (2015) Engineering science and technology, an international journal effect of dielectric fluid with surfactant and graphite powder on electrical discharge machining of titanium alloy using Taguchi method. Eng Sci Technol Int J 18(4):524–535
34. Khan ZA, Siddiquee AN, Zaman N, Khan U, Quadir GA (2014) Multi response optimization of Wire electrical discharge machining process parameters using Taguchi based Grey relational analysis. Proc Mater Sci 6:1683–1695
35. Dabade UA, Karidkar SS. Analysis of response variables in WEDM of Inconel 718 using Taguchi technique. In: Procedia CIRP, vol 41
36. Datta S, Bhusan B, Sankar S (2019) Machinability analysis of Inconel 601, 625, 718 and 825 during electro-discharge machining: on evaluation of optimal parameters setting. Measurement 137:382–400

Impact Behavior of Deformable Pin-Reinforced PU Foam Sandwich Structure

Shivanku Chauhan, Mohd. Zahid Ansari, Sonika Sahu, and Afzal Husain

Abstract Present work studies the effect of deformable silicone rubber pin reinforcement on impact and energy absorption characteristics of polymer sandwich structure. Low-density polyurethane foam is used as core material and acrylic sheets as the two skin. The size of the sandwich structure is 50 × 50 × 35 mm. Material properties of core and skin were determined from experiments and were used later in the numerical analysis software ABAQUS. Results for deformation and stresses produced inside the sandwich for 4 J of impact energy. Results show that by reinforcing the sandwich structure with deformable silicone pins, the maximum deflections are reduced to about half, and its post-impact recovery is improved.

Keywords Polyurethane foam · Energy absorption · Silicone rubber · Sandwich composite · Finite element analysis

1 Introduction

Sandwich structures are basically a class of composites in which a soft core material is confined by two hard outer phases. The advantageous feature of sandwich structure includes better flexural rigidity and energy absorption capacity with increased strength-to-weight ratio. Other features of sandwich structure include excellent thermal insulation, acoustic damping, fire retardation, etc. The sandwich composites are basically designed to support the out-of-plane compressive load and bending load [1]. A major problem with the use of sandwich is their susceptibility to damage

---

S. Chauhan · Mohd. Z. Ansari (✉)
Mechanical Engineering Discipline, PDPM-Indian Institute of Information Technology, Design and Manufacturing, Airport Road, Dumna, Jabalpur, MP 482005, India
e-mail: zahid@iiitdmj.ac.in

S. Sahu
School of Automation, Banasthali Vidyapith, Niwai, Rajasthan 304022, India

A. Husain
Department of Mechanical and Industrial Engineering, Sultan Qaboos University, PC-123, Muscat, Oman

© The Author(s), under exclusive license to Springer Nature Singapore Pte Ltd. 2021

R. M. Singari et al. (eds.), *Advances in Manufacturing and Industrial Engineering*, Lecture Notes in Mechanical Engineering,
https://doi.org/10.1007/978-981-15-8542-5_88

due to impact loading. There are many practical accidents like tool drops, runway debris, bird strikes, hailstorms, ballistic loading, and hammer impact which causes damage to the sandwich structures. Low-velocity impact is supposed to be dangerous because the surface may appear to be undamaged after strike and the damage might be left undetected. The low-velocity impact can cause the damage of facings, core material, and the core-facing interface of the sandwich structure. Damage initiation energy and size mainly depend on the properties of the material of core and face sheets. They also depend on the relationship between the properties of the cores and the facings, size, and shape of the structure.

Kulkarni et al. [2] analyzed the crack growth phenomenon in the sandwich structure subjected to the bending. It has been found that fracture or damage starts in the form of the core and skin material debonding parallel to the beam axis. This debonding crack grew parallel to the top interface and then eventually kinked in to the core foam in form of the shear crack, which resulted to the failure. Burman et al. [3] have also done the research in the same field and revealed that the initiation of damage starts at the middle of the core with high shear stress and as soon as the micro crack developed, its crack front kinked away and propagate toward the interface of the sandwich constituents. Long et al. [4] analyzed the failure of foam core composite under the different drop weight energy. The sandwich with a stiffer core has been found to be more susceptible to delamination. Foams are suitable for impact protection of structural components due to their specific energy absorption capability under compression. The tailoring the deformation pattern of the foam cells is a complicated job because of the randomness of their internal architecture. Aluminum foams are in trend for the core material in sandwich structures. Raeisi et al. [5] studied the effect of aluminum pins embedded in the aluminum foam panel to control the deformation pattern of the structure. They found that there is a considerable raise in the elastic modulus, plateau stress, and energy absorption capability of the pinned structure sample. The behavior of the foam core sandwich reinforced with polymer pins under flexural and flat-wise compression has been studied by Abdi et al. [6]. The flat-wise compression and flexural properties of panels have been found to be increased significantly. Further, it has been found that the diameter of polymer pins had also a large influence on the flexural stiffness of the composite sandwich panels.

Polyurethane (PU) foam is also extensively used as the core stuff for sandwich structures, used in aircraft, marine, and automobile body structures. These foams basically have a closed cellular structure implanted in a continuous matrix phase, and this type of closed cell geometry of PU foam is striking for mechanical and insulating properties [7–9]. The integration of reinforcing fillers with the composites upgrades the mechanical properties, low specific weight, and ease of processing [10, 11]. The improvement in mechanical and thermal properties of PU foam can lead it further to the superior bending stiffness and enhanced mechanical properties of the sandwich structures. PU foam is a thermosetting plastic with low-density cellular structure [1, 12], and because of its lightweight as well as high energy absorption capacity and high thermal insulation features, it better suits as a core material for a sandwich structure. On other hand, acrylic is also a lightweight material consisting of higher impact resistance compared to glass and has superior weather resistance compared to other transparent polymer materials. Thus, there is a promising future for the sandwich structure made of PU foam and acrylic sheet with low built-up cost and superior properties. Foams can also be made of metals [13–15].

The present work emphasizes on the numerical modeling and simulation of sandwich composite for the impact with a rigid plate impactor. The mechanical properties of the constituents of the sandwich for the material modeling in ABAQUS have been acquired experimentally. The intended purpose of the study is to ensure the modeled sandwich for its feasibility for the impact in door panel application.

2 Design and Numerical Analysis

In a sandwich composite, a soft core material is confined by two hard outer phases. Further, there can be some reinforcement embedded in the core for property enhancement. In present work, two types of sandwich composites have been modeled. First, Model-1 is a simple sandwich structure with PU foam core and the acrylic as the skin materials. Second, Model-2 is a reinforced sandwich structure composed of PU foam core, acrylic skin, and deformable silicone rubber pins embedded in the core providing the reinforcement. The PU foam core thickness is 25 mm, and the thickness of acrylic skin on both side of the core is 5 mm each. The length and width of the core and skin is of 50 × 50 mm dimension. Four cylindrical silicone rubber pins are vertically embedded in the core and between the skin sheets, all having diameter of 10 mm and length of 25 mm. So, the typical size of both the sandwich samples is 50 × 50 × 35 mm. Figure 1 represents the models of the sandwich composite structures.

Numerical simulation for analyzing the drop weight impact at 4 J of on the sandwich structure was carried out using ABAQUS software. To apply this load, inertia mass of 2 kg with a velocity of 2 m/s at the initial step was provided to the reference point of the rigid plate, and the simulation running time was set to 12 ms. The properties used for material modeling have been obtained experimentally. The fracture strength of the acrylic sheet was found to be about 22.1 MPa. The acrylic skins and silicone pins have been modeled as isotropic linear elastic materials, while the PU foam core has been modeled with elastic plastic properties. The properties used for the constituent material are listed in Table 1. The nonlinear numerical models are consisted of 3D stress family eight-node linear brick finite elements C3D8R with reduced integration and hourglass control for all the parts. A total of 87,500 and 102,825 such elements are used in Model-1 and Model-2, respectively. Figure 2 shows the meshed three-dimensional finite element model of the two structures. A general contact interaction with penalty friction formulation with friction coefficient of 0.3 has been provided between the mating rigid impact plate and the upper skin of the sandwich. Tied constraint has been utilized to model the connection of skin-core, skin-pins, and core-pins with analysis default discretization method. Fully constrained boundary condition was applied at the bottom support of the sandwich.

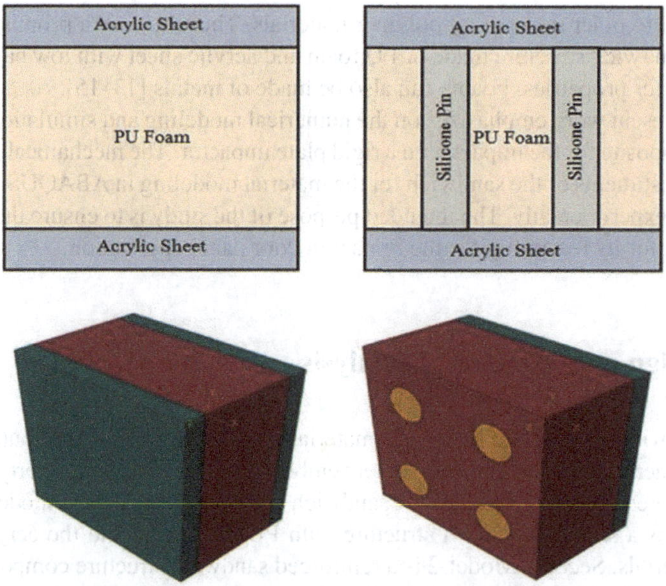

Fig. 1 Sandwich composite structure Model-1 (left) and Model-2 (right)

Table 1 Properties of constituent materials of sandwich composite model

Material	Property		
	Mass density (kg/m^3)	Young's modulus (MPa)	Poisson's ratio
Acrylic	1.2	3000	0.35
Silicone	2	25.5	0.48
PU foam	0.1	2.19	0.40

Fig. 2 Finite element meshed model of Model-1 (left) and Model-2 (right) sandwich structures

Mesh convergence study was conducted to ascertain the accuracy of the numerical approach adopted here.

3 Results and Discussion

Figure 3 shows the deformation behavior and stress distribution in Model-1 sandwich composite structures at different time steps. The deformation at time steps 0.24, 6.24, and 12 ms are shown in the figure. These time steps represent nearly undeformed, maximal deformed, and elastically recovered deformation, respectively. Maximum deformation of 6.98 mm was observed in the top acrylic skin at t = 6.24 ms. Maximum deformation occurs in the top half of the structure which is in direct contact with moving down rigid plates. The barreling in the middle can be attributed to the squeezing-out of the soft PU foam due to compression by relatively much stiffer acrylic face sheets. Stress results show that the maximum stresses are

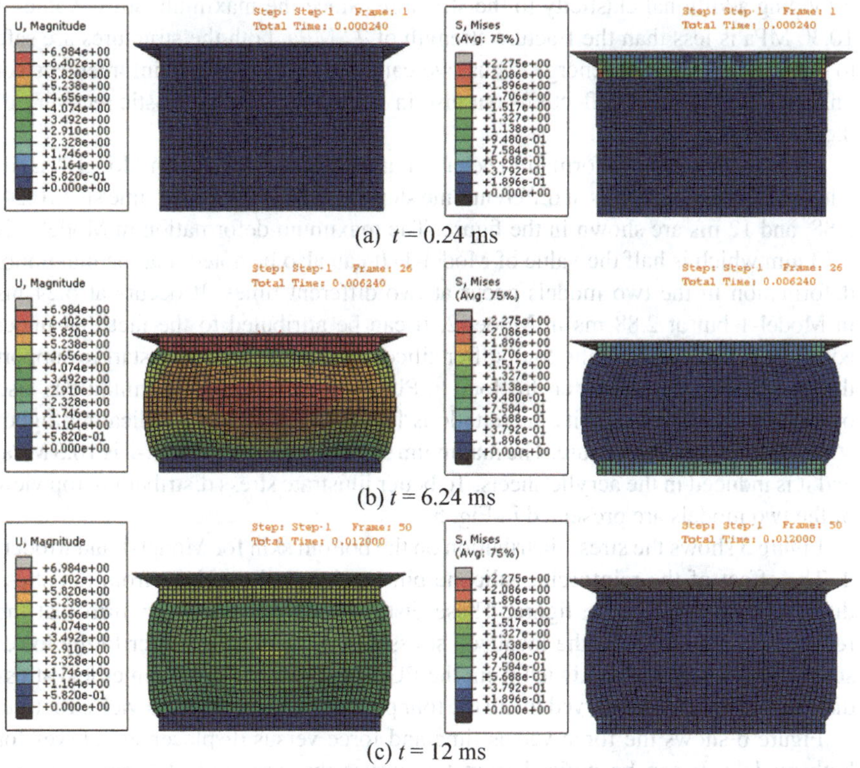

Fig. 3 Deformation behavior and stress distribution in Model-1 sandwich composite structure at different time steps

Table 2 Comparison of two models through simulation results

Structure	Max. deformation (mm)	Max. force (N)	Max. stress (MPa)
Model-1	6.98	825.77	2.28
Model-2	3.53	2091.26	10.30

generated at the beginning of the impact load and then after are spread down with lower intensity in the entire structure. At maximum deformation, acrylic face sheets are exposed to higher stress levels than the soft PU foam core. Table 2 compares the numerical results obtained for Model-1 and Model-2 sandwich structures under an impact of 4 J. Results show that the reinforcing pin helped in reducing the deformation by about half in Model-2 but increased its impact force and maximum von Mises stresses generated. The reduction in deformation can be attributed to the increased compressive stiffness of Model-2 due to addition of reinforcement pins. Higher force and stresses values for Model-2 can be attributed to its higher stiffness which resulted in exerting higher reaction force to the impacting surface. In addition, deformable pins also helped the sandwich structure to recover its original shape and size by providing additional elasticity to the structure. Since the maximum stress value of 10.30 MPa is less than the fracture strength of 22 MPa, both the structures are safe to absorb 4 J of impact energy. Thus, we can conclude that pin reinforcement not only resulted in lower deflections but also in higher post-impact elastic recovery of the structure.

Figure 4 shows the deformation behavior and stress distribution in Model-2 sandwich composite structures at different time steps. The deformation at time steps 0.24, 2.88, and 12 ms are shown in the figure. The maximum deformation in Model-2 is 3.53 mm which is half the value of Model-1. It can also be noted that the maximum deformation in the two models occur at two different times. It occurs at 6.24 ms in Model-1 but at 2.88 ms in Model-2. It can be attributed to the fact that under equi-strain compression, the four stiffer silicone reinforcement pins start to support the load initially and transfer this load to PU foam. Barreling in the middle is also observed in this case, but its magnitude is less than in Model-1, indicating higher stiffness of Model-2 structure. The maximum stress occurred in Model-2 is 10.3 MPa, and it is induced in the acrylic sheets. To better illustrate stress distribution, top view of the two models are presented in Fig. 5.

Figure 5 shows the stress distribution on the bottom skin for Model-1 and Model-2. The effect of the reinforcing silicone pin can be easily noticed from the stress distribution pattern in the figure. These distributions of stresses confirm that the reinforcing pins enhance the localized stress values and then transfer to the lower supporting skin of the sandwich from the PU foam core of it. A symmetrical stress distribution has been observed for all the four pins and also for the sandwich structure.

Figure 6 shows the force versus time and force versus displacement curves for both models. It can be noticed from the curves that the contact force first goes on increasing with time or displacement and obtain a maximum magnitude at the maximum displacement stage. As soon as, the kinetic energy of the impactor plate

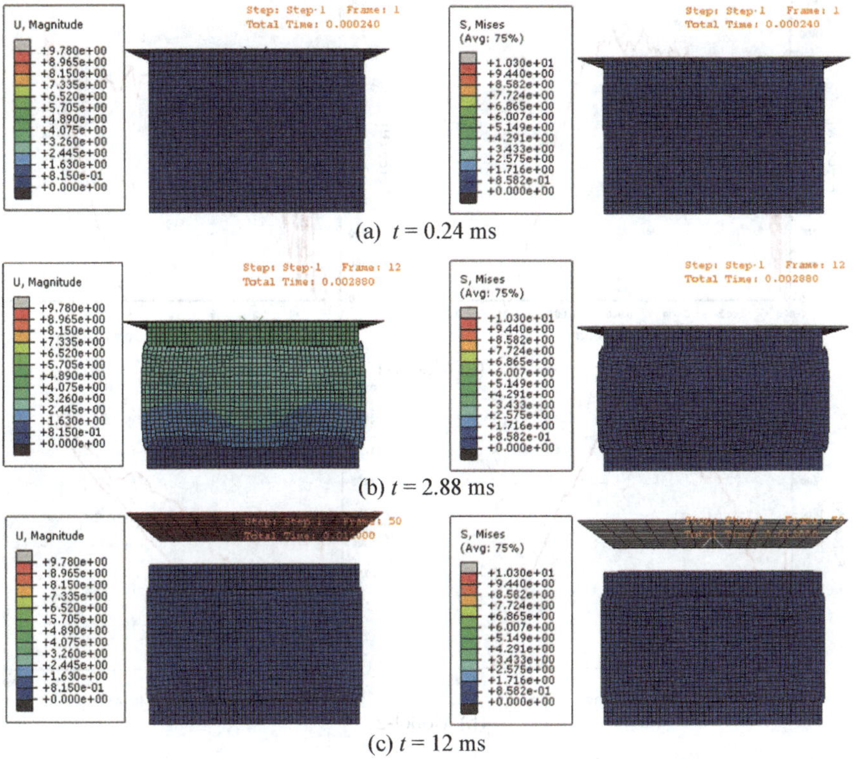

Fig. 4 Deformation behavior and stress distribution in Model-2 sandwich composite structure at different time steps

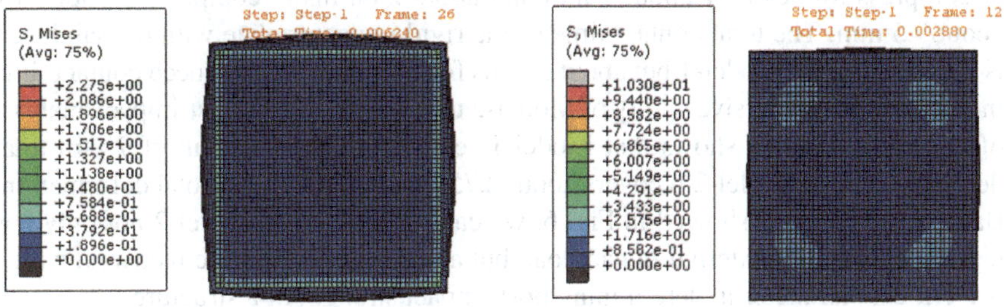

Fig. 5 Stress distribution Model-1 (left) and Model-2 (right) sandwich composite structures

reduces to zero, the sandwich starts regaining its original shape by repulsing the rigid impactor plate in the opposite direction. This repulsive action of the sandwich results in reduced contact force and in the end contact force becomes zero as the rigid plate is now no more in the contact with the sandwich structure. The rigid impactor plate with 4 J of initial kinetic energy has taken about 6.24 of millisecond

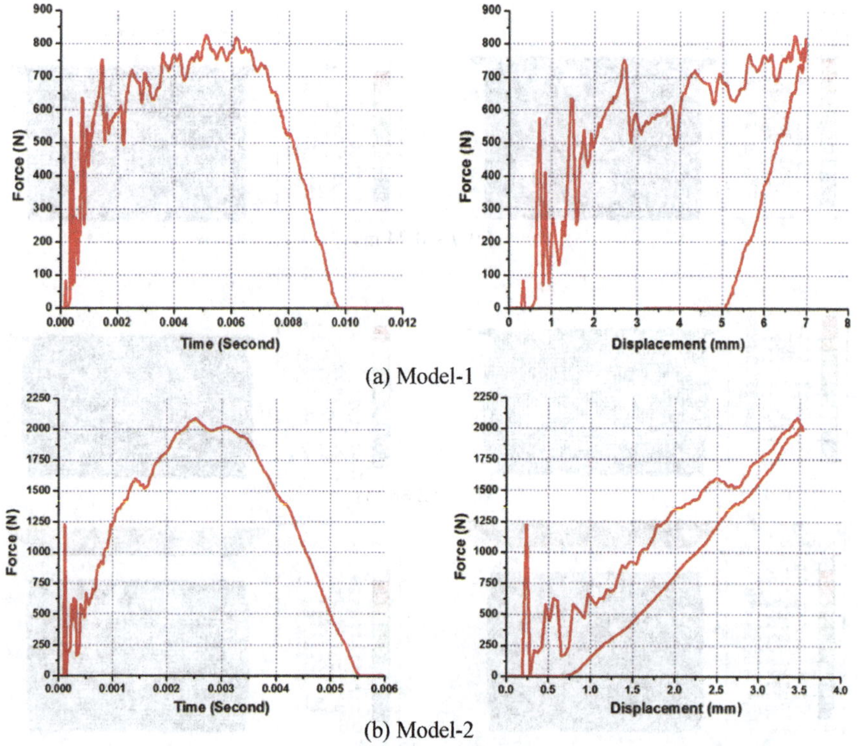

Fig. 6 Force, time, and displacement curves for Model-1 and Model-2 sandwich composite structures

to compress Model-1 by about 7 mm and about 2.88 ms to compress Model-2 by about 3.5 mm. The total contact time of the rigid impactor plate with the sandwich is about 10 ms for Model-1 but about 5.5 ms for Model-2. This reduced contact time indicates higher repulsive ability of Model-2 to impact loads. A post-impact analysis of the two structures shows that Model-1 recovers only 2 mm out of 7 mm total deformation, but Model-2 recovers about 2.75 mm out of 3.5 mm total deformation. Based on the results showed in Fig. 6, we can conclude that Model-2 not only has higher repulsion to external impact load but also has higher elastic recovery, which is a critical parameter in determining post-impact strength of a structure.

4 Conclusion

The present work used finite element analysis to investigate the effect of deformable pins on increasing impact absorption behavior of polymer sandwich composite structures for 4 J of impact energy. Results showed that reinforcing pin helped in reducing maximum deformation from 6.98 to 3.53 mm. It, however, increased the impact force from 825.77 N to 2091.26 N and stresses from 2.28 to 10.30 MPa. Since the maximum stress is less than the fracture strength of 22 MPa for acrylic, the structures is safe to absorb 4 J of impact energy. It was also found that the maximum deformation occurs early in Model-2 than in Model-1. Post-impact analysis of the two structures shows that Model-1 recovers only 2 mm out of 7 mm total deformation, but Model-2 recovers about 2.75 mm out of 3.5 mm total deformation. Thus, based on the study conducted here, we can conclude that Model-2 not only has higher repulsion to external impact load but also has higher elastic recovery, which is a critical parameter in determining post-impact strength of a structure. These results can find applications in designing lightweight, low-cost impact resistant structures with better thermal insulation and mechanical load bearing features.

Acknowledgements This study was supported by IIITDM Jabalpur.

References

1. Gibson J, Ashby F (1999) Cellular solids: structure and properties. Cambridge University Press
2. Kulkarni N, Mahfuz H, Jeelani S, Carlsson L (2003) Fatigue crack growth and life prediction of foam core sandwich composite under flexural loading. J Compos Struct 59(4):499–505
3. Burman M, Zenkert D (1997) Fatigue of foam core sandwich beam-1: undamaged specimen. Int J Fatigue 19(7):551–561
4. Shuchang L, Xiaohu Y, Heran W, Xiaoqing Z (2018) Failure analysis and modeling of foam sandwich laminates under impact loading. Compos Struct 197:10–20
5. Sajjad R, Javad K, Andres T (2019) Mechanical properties and energy absorbing capabilities of Z-pinned aluminum foam sandwich. Compos Struct 214:34–46
6. Abdi B, Azwan S, Abdullah MR, Ayob A, Yahya Y, Xin L (2014) Flatwise compression and flexural behavior of foam core and polymer pin-reinforced foam core composite sandwich panels. Int J Mech Sci 88:138–144
7. Lee Y, Yoon KB (2005) Effect of composition of Polyurethane foam template on the morphology of silicate foam. Microporous Mesoporous Mater 88:176–186
8. Song B, Chen W, Dou S, Winfreeb A, Kang H (2005) Strain-rate effects on elastic and early cell-collapse responses of a polystyrene foam. Int J Impact Eng 31(5):509–521
9. Yu L, Li R, Hu S (2006) Strain-rate effect and micro-structural optimization of cellular metals. Mech Mater 38(1–2):160–170
10. Chauhan S, Bhushan RK (2018) Improvement in mechanical performance due to hybridization of carbon fiber/epoxy composite with carbon black. Adv Compos Hybrid Mater (1):602–611
11. Chauhan S, Bhushan RK (2017) Study of polymer matrix composite with natural particulate/fiber in PMC: a review. Int J Adv Res Ideas Innov Technol 3(3):1168–1179
12. Tu Z, Shim V, Lim C (2001) Plastic deformation modes in rigid polyurethane foam under static loading. Int J Solids Struct 38(50–51):9267–9279
13. Sahu S, Mondal DP, Cho JU, Goel MD, Ansari MZ (2019) Low-velocity impact characteristics of closed cell AA2014-SiCp composite foam. Compos B Eng 160:394–401
14. Sahu S, Ansari MZ (2019) Effect of impactor shapes on low-velocity impact characteristics of Al plate. Ships Offshore Struct 14(1):53–63
15. Sahu S, Mondal DP, Cho C, Ansari MZ (2019) Quasi-static compressive behaviour of aluminium cenosphere syntactic foams. Mater Sci Technol 35(7):856–864

Sensitivity Improvement of Piezoelectric Mass Sensing Cantilevers Through Profile Optimization

Shivanku Chauhan, Mohd. Zahid Ansari, Sonika Sahu, and Afzal Husain

Abstract This work aims to design a piezoelectric cantilever mass sensor in which the cantilever profile was changed to achieve high resonant frequency, so that small mass changes can be measured precisely. A rectangular profile of cantilevers was designed and fabricated using stainless steel with a piezoelectric patch mounted on its top surface. The experimental analysis was executed using an impedance analyzer. To improve the sensitivity, the rectangular profile of the cantilever is made stepped. The first mode of resonant frequencies of the rectangular and stepped-rectangular cantilever profiles was acquired from the experimental analysis and compared with theoretical results. Results showed good consistency between the two results and that stepped-rectangular profile cantilever has shown higher sensitivity than rectangular.

Keywords Cantilever mass sensor · Piezoelectricity · Impedance analyzer · Resonant frequency · Sensitivity

1 Introduction

Mass sensing using cantilever is a principle used in various applications such as explosive and chemical vapor detection [1], picogram protein mass sensor [2], mass detection [3], and biosensor applications [4]. In these applications, the resonant frequency of the cantilever with and without the addition of mass on its tip is compared, and by measuring the frequency shift, the added mass is determined. Cleveland et al. [5] used the mass addition technique on the cantilever tip for the calculation of stiffness

S. Chauhan · Mohd. Z. Ansari (✉)
Mechanical Engineering Discipline, PDPM-Indian Institute of Information Technology, Design and Manufacturing, Airport Road, Dumna, Jabalpur, MP 482005, India
e-mail: zahid@iiitdmj.ac.in

S. Sahu
School of Automation, Banasthali Vidyapith, Niwai, Rajasthan 304022, India

A. Husain
Department of Mechanical and Industrial Engineering, Sultan Qaboos University, PC-123, Muscat, Oman

© The Author(s), under exclusive license to Springer Nature Singapore Pte Ltd. 2021

R. M. Singari et al. (eds.), *Advances in Manufacturing and Industrial Engineering*, Lecture Notes in Mechanical Engineering,
https://doi.org/10.1007/978-981-15-8542-5_89

of the cantilever. Piezoelectric cantilevers are advantageous than others because they provide a self-actuating and self-sensing feature that can be exploited to design ultra-sensitive mass sensors. Cantilever can be actuated at its resonant frequency with the help of piezoelectric layer or patch mounted directly on the cantilever or through the base-excited achieved by placing them on piezo-stack and exciting the whole structure by applying sinusoidal input voltages. Stepped profile cantilevers have shown better sensitivity characteristics then rectangular ones when used as mass sensor [6] or as a biosensor [7–9] or as Peter et al. [10] used a functionalized layer applied on the tip of the cantilever to attract the analyte which was needed to detect. The analyte stacked to the functionalized layer on the cantilever and the mass of the cantilever increased, which resulted in a decrease in the resonant frequency of the cantilever. The shift of resonant frequency can be used to detect the mass change. Gao et al. [11] fabricated cantilever mass sensor with a piezoelectric patch as the actuator on it and studied the effect of geometry of the cantilever on the mass sensitivity of the cantilevers.

In the present work, cantilever made of stainless steel and having rectangular and stepped-rectangular profiled is fabricated using micro-milling and surface grinding operations. A piezoelectric patch is pasted to top cantilever surface and is used to excite as well as measure the frequency shift using impedance analyzer. Square pieces of a polymer adhesive tape of known surface density are used as the precise masses. And, the mass sensing characteristics of the two cantilever types are compared and discussed.

2 Theory and Experiments

Piezoelectric effect is a material property of a substance that results in producing electrical charges on its surface when compressed. This effect can be direct or converse. In direct effect, compressive strains produce electrical field across the material surface, whereas in converse effect the application of electrical field generates strain. We use converse effect in mass sensor applications using piezoelectric layer or patch.

Cantilever sensors have been developed which can be used in variety of applications. The application depends on the mode of operation of the cantilever. In the present work, dynamic mode of operation was used for mass sensing. This mode is based on the resonant frequency of the cantilever. The resonant frequency shift is observed on the increment the effective mass of the system. This fact can be used for mass sensing purpose in dynamic mode. The resonant frequency of a rectangular profile cantilever can be given as [8]:

$$f_0 = \frac{1}{2\pi}\sqrt{\frac{E}{\rho}\frac{t_0}{l_0^2}} \tag{1}$$

where E is Young's modulus of the material, ρ is density of cantilever, t_0 is the thickness, and l_0 is the length of cantilever. The resonant frequency of stepped-rectangular profile cantilever can be estimated as [8]:

$$f_0 = \frac{1}{2\pi}\sqrt{\frac{Et_0^3}{\rho(l_0 t_0 + lt)(l_0 + l)^3}} \tag{2}$$

where t_0 and t are the cantilever thickness at fixed and free ends of the cantilever and l_0 and l are the cantilever length of thick and thin sections. If Δm is the change in mass, k is the stiffness of cantilever, and f_0 and f_1 are the frequencies of vibration in the cantilever before and after the mass change. Then, the mass change at the cantilever tip can be related to its frequency change as [8]:

$$\Delta m = \frac{k}{4\pi^2}\left(\frac{1}{f_1^2} - \frac{1}{f_0^2}\right) \tag{3}$$

The rectangular profile and stepped-rectangular profile have been fabricated by milling followed by grinding operation. Stainless steel having mass density is 7800 kg/m^3, Young's modulus is 200 GPa, and the Poisson's ratio 0.3 was used for fabrication. The total length of the cantilevers was 25 mm. The width and the thickness at the fixed-end were 5 mm and 1 mm, respectively. In stepped-rectangular profile, the length of thick segment was 10 mm, and the length of thin segment was 15 mm. And the corresponding segment thicknesses were 1 mm and 0.6 mm, respectively. A piezoelectric material PZT-5H of size 5 mm $\times$ 10 mm was used for actuation and sensing function. The schematic and fabricated cantilever sensors with piezoelectric connections are shown in Fig. 1.

Figure 2 shows the experimental setup for performing experiments. The cantilevers shown above were placed under a transparent glass cover to eliminate external disturbances. Experiments were performed in normal atmospheric condition. Mass was added to the cantilever tip using precisely measured adhesive polymer film of known area density of 33.65 μg/mm^2. Wayne Kerr 6500B impedance analyzer was used for the experiments to calculate the resonant frequency of the cantilever profiles. The frequency versus impedance graph was plotted using data obtained from by the impedance analyzer. Two wires from the cantilever system inside the experimental chamber is taken out from the base of the chamber and fed to the electrodes of the impedance analyzer. At initial stage of frequency, the impedance starts continuously decreasing, and after some time interval at some particular frequency the impedance starts increasing, goes increasing up to some value, and then again starts decreasing in the same trend. Thus, a peak is observed in the impedance versus frequency graph. The value of frequency at which the impedance stars increasing was marked as the resonant frequency.

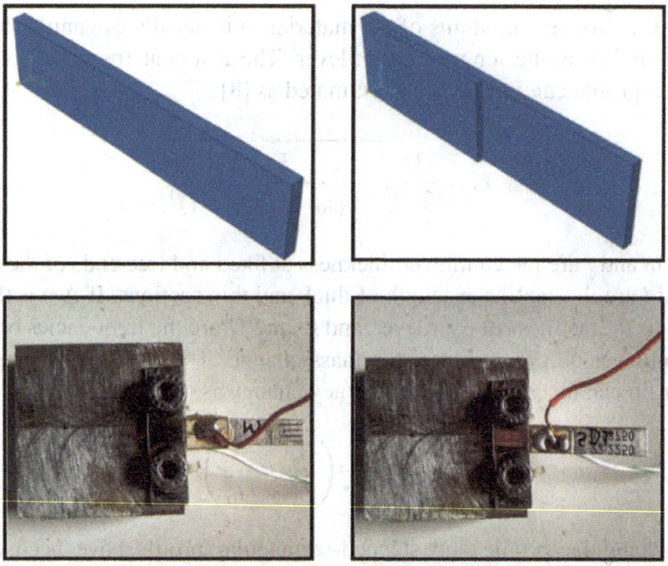

Fig. 1 Schematic design of piezoelectric cantilever profiles (top row) and corresponding fabricated cantilever sensors (bottom row)

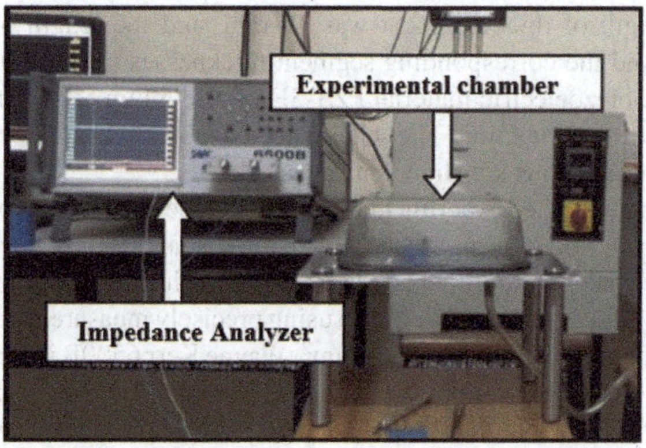

Fig. 2 Experimental arrangement for impedance analysis

3 Results and Discussion

Table 1 presents a comparison between analytical and experimental results for fundamental resonant frequencies of rectangular (R) and stepped-rectangular (SR) profiles of piezoelectric cantilever sensors. Analytical results were obtained using Eqs. (1) and (2) and the experimental from impedance analyzer. This comparison indicates a reasonable conformity between the two results with a deviation of 5.86% and

Table 1 Comparison between fundamental resonant frequencies of cantilevers

Cantilever profile	Fundamental resonant frequency (Hz)		Error
	Analytical	Experimental	
Rectangular	1289.46	1213.91	5.86%
Stepped-rectangular	1479.11	1395.32	5.66%

5.66% for rectangular and stepped-rectangular profile, respectively. Lower frequencies observed in experiments are due to increase in cantilever mass because of piezoelectric patch mounted on its top surface. In addition, the damping offered by ambient air and extra mass of electrical wire connections can also be influential. It can be noticed from the results that the shape of cantilever profile affects its frequency significantly. Stepped-rectangular cantilever shows 14.7% higher resonant frequency than that of rectangular profile of cantilever. The higher the resonant frequency, the higher its mass detection sensitivity will be.

Frequency versus impedance graphs was plotted for the rectangular profile cantilever with different mass on tip, normal atmospheric condition, as shown in Fig. 3. The frequency versus impedance plots are for the rectangular cantilever for

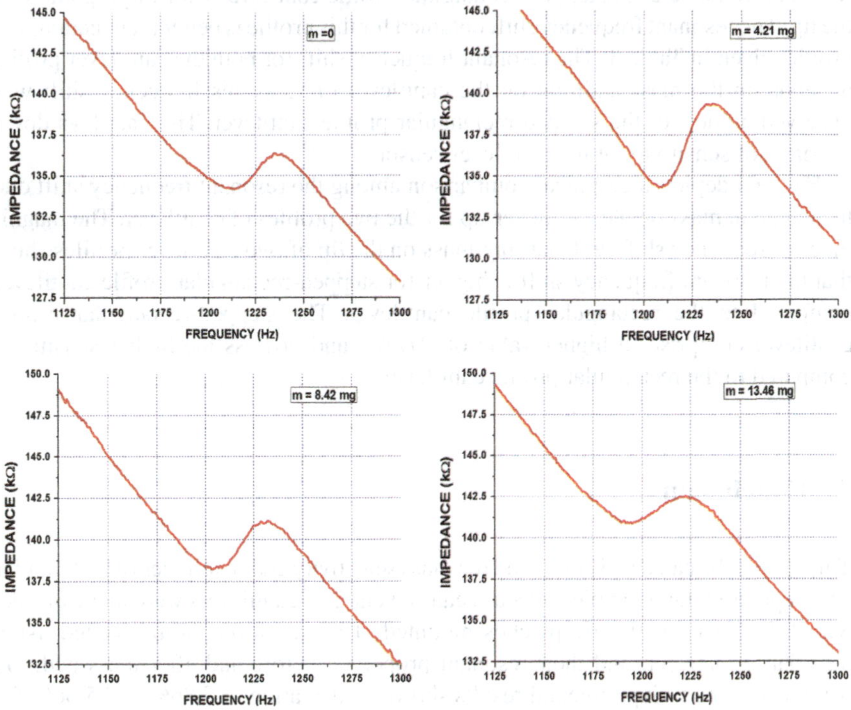

Fig. 3 Frequency versus impedance plots for rectangular profile piezoelectric cantilever for different mass

Table 2 Resonant frequency and frequency shift of piezoelectric rectangular profile cantilever for different mass loading conditions

Mass (mg)	Resonant frequency (Hz)	Resonant frequency shift (Hz)
0	1213.91	–
4.21	1208.49	5.42
8.42	1203.09	10.82
13.46	1194.66	19.25

no mass, 4.21, 8.42, and 13.46 mg. The fundamental resonant frequency of the rectangular profile cantilever without any mass on the tip is 1213.91 Hz, and this resonant frequency further shift down to 1208.49, 1203.09, and 1194.66 Hz for the 4.21, 8.42, and 13.46 mg mass on the tip of the cantilever, respectively. The resonant frequency values for rectangular profile with varying mass on the tip of the cantilever are given in Table 2.

Frequency versus impedance graphs was plotted for the stepped-rectangular profile cantilever with different mass on tip, under normal atmospheric condition. The graphs in Fig. 4 depict the frequency versus impedance plots for the stepped-rectangular cantilever for no mass, 4.21, 8.42, and 13.46 mg. Table 3 contains the resonant frequencies of stepped-rectangular profile cantilever with varying mass on the tip. The resonant frequency shift obtained for this profile is better than rectangular profile, given in Table 3. The resonant frequency shift for both the cantilever profile depends on the mass addition on the cantilever tip, and this frequency shift was observed higher for the stepped-rectangular profile cantilever. This fact lays down the basis of sensitivity of the cantilever sensor.

Figure 5 depicts a graphical comparison among the resonant frequency shift due to change in mass on the cantilever tip for the two profiles of cantilever. The magnitude of frequency shift with varying mass on the tip of two cantilever profiles show that the resonant frequency shift is higher for stepped-rectangular profile cantilever compared to the rectangular profile cantilever. The stepped-rectangular profile cantilever comprises a higher value of $\Delta f/\Delta m$ and possess the higher sensitivity compared to the rectangular profile cantilever.

4 Conclusions

This study fabricated and compared the mass sensitivity characteristics of rectangular and stepped-rectangular piezoelectric cantilevers. The cantilevers were made of steel with PZT-5H piezoelectric patches mounted. The cantilevers were excited using impedance analyzer, and their resonant properties before and after mass addition were determined. Experimental results showed a deviation of 5.86% and 5.66% for the rectangular profile and stepped-rectangular profile cantilever than analytical. In addition, stepped-rectangular cantilever showed 14.7% higher resonant frequency

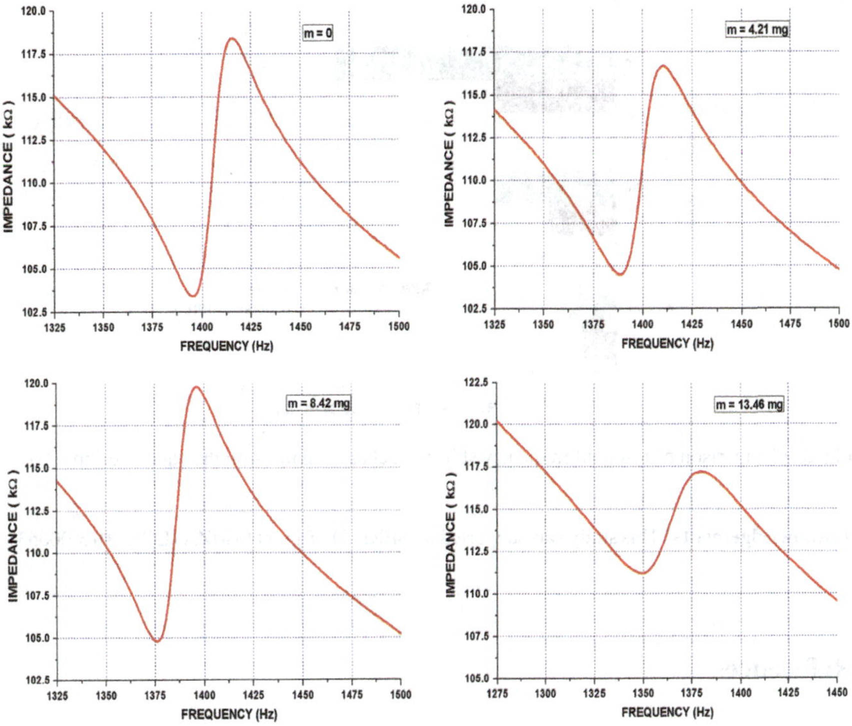

Fig. 4 Frequency versus impedance plot for piezoelectric stepped-rectangular profile cantilever for different mass

Table 3 Resonant frequency and frequency shift of piezoelectric stepped-rectangular profile cantilever under different mass loading conditions

Mass (mg)	Resonant frequency (Hz)	Resonant frequency shift (Hz)
0	1395.32	–
4.21	1388.08	7.24
8.42	1376.09	19.23
13.46	1349.84	45.48

than rectangular profile cantilever under no mass condition. For the three masses of 4.21, 8.42, and 13.46 mg used here, stepped-rectangular cantilevers showed about 33.58%, 77.72%, and 136.26% higher mass sensitivity than rectangular. Thus, based on these results, we can conclude that stepped cantilever has much better sensing characteristics than rectangular.

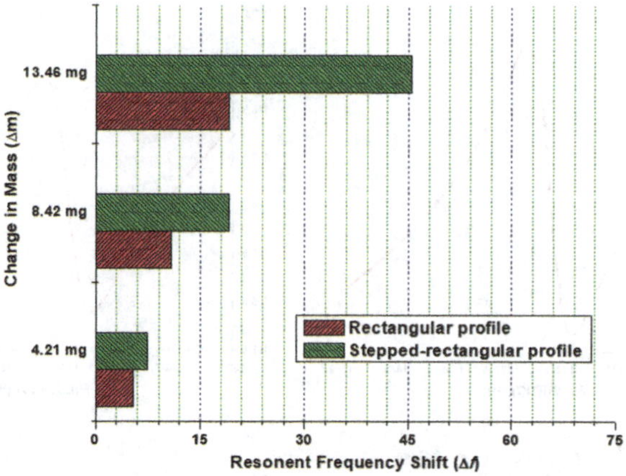

Fig. 5 Comparison of resonant frequency shift with change in mass on the cantilever tip

Acknowledgements This study was supported by SERB-DST grant to MZA (ECR/2015/000531).

References

1. Seena V, Fernandes A, Pant P, Mukherji S, Rao VR (2011) Polymer nanocomposite nanomechanical cantilever sensors: material characterization, device development and application in explosive vapour detection. Nanotechnology 22(29)
2. Lee SM, Hwang KS, Yoon HJ, Yoon DS, Kim SK, Lee YS, Kim TS (2009) Sensitivity enhancement of a dynamic mode microcantilever by stress inducer and mass inducer to detect PSA at low picogram levels. Lab Chip 9:2683–2690
3. Zhao J, Zhang Y, Gao R, Liu S (2015) A new sensitivity improving approach for mass sensors through integrated optimization of both cantilever surface profile and cross section. Sens Actuators B Chem 206:343–350
4. Johnson BN, Mutharasan R (2012) Biosensing using dynamic-mode cantilever sensors: a review. Biosens Bioelectronics 32(1):1–18
5. Cleveland JP, Manne S, Bocek D, Hansma PK (1993) A nondestructive method for determining the spring constant of cantilevers for scanning force microscopy. Rev Sci Instrument 64(2):403–405
6. Arya D, Mondal S, Ansari MZ (2018) A high sensitive mass sensor for robotic applications. In: International Conference on Robotics and Smart Manufacturing, Procedia Computer Science, India, vol 133, pp 799–803
7. Ansari MZ, Cho C, Urban G (2012) Stepped piezoresistive microcantilever designs for biosensors. J Phys D Appl Phys 45(21)
8. Ansari MZ, Cho C (2009) Deflection, frequency, and stress characteristics of rectangular, triangular, and step profile microcantilevers for biosensors. Sensors 9(8):6046–6057
9. Ansari MZ, Cho C (2017) Optimised step profile microcantilevers for biosensors. AIP Conf Proc 1859:1–7
10. Peter, H., Martin, L., Christoph, H.: Microfabricated Cantilever Array Sensors for (Bio-) Chemical Detection. Applied Scanning Probe Methods IV, 183–213 (2006).
11. Gao R, Huang Y, Wen X, Zhao J, Liu S (2017) Method to further improve sensitivity for high order vibration mode mass sensors with stepped cantilevers. IEEE Sens J 17(14):4405–4411

Current Status, Applications, and Factors Affecting Implementation of Additive Manufacturing in Indian Healthcare Sector: A Literature-Based Review

Bhuvnesh Chatwani, Deepanshu Nimesh, Kuldeep Chauhan, Mohd Shuaib, and Abid Haleem

Abstract Additive manufacturing (AM) refers to manufacturing of products or prototypes using layer by layer addition of material. It is quite different from traditional manufacturing which is essentially classified as subtractive manufacturing. AM technology has tremendous versatile applications in the healthcare sector, thus making it a popular choice of the sector. AM technology is not only used for training and simulation products but also for producing medical tools and equipment. This paper discusses the current status of various healthcare applications of AM and the factors affecting implementation of AM in Indian Healthcare Sector.

Keywords Additive manufacturing · Healthcare sector · Prototype

1 Introduction

Additive manufacturing (AM) is the collective term for processes in which materials are added layer by layer to create parts [1]. In AM production methods, not only the technologies and their equivalent manufacturing processes are included, but infrastructures like CAD interface is also present. Its presence can help in making drastic changes in manufacturing thereby making it extremely efficient [2].

Additive manufacturing has been described as the '3rd Industrial Revolution' by the leading scientists of the world. It has a lot of change in its offering. It can completely change the way the industrial planners think.

AM is being adopted as the primary manufacturing technique across various industries in different parts of the world. AM is most favorable for design-sensitive industries, particularly aviation and healthcare sector. Boeing and Airbus being the

B. Chatwani (✉) · D. Nimesh · K. Chauhan · M. Shuaib
Delhi Technological University, Shahbhad Daulatpur, New Delhi 110042, India
e-mail: bchatwani7@gmail.com

A. Haleem
Jamia Millia Islamia, Okhla, New Delhi 110025, India

© The Author(s), under exclusive license to Springer Nature Singapore Pte Ltd. 2021
R. M. Singari et al. (eds.), *Advances in Manufacturing and Industrial Engineering*,
Lecture Notes in Mechanical Engineering,
https://doi.org/10.1007/978-981-15-8542-5_90

major players of the aviation industry use AM technology to a great extent. Hearing aid and dental crown are an example of AM's customization.

Huge number of relevant research papers on the additive manufacturing's application in healthcare sector shows an increased amount of research in this field. But, not much research has been carried out on AM applications in Indian Healthcare System. This review paper will focus on Indian Healthcare Sector. The paper begins with discussing various types of additive manufacturing techniques. It goes on to discuss the benefits of additive manufacturing in healthcare sector. The paper discusses the current status of AM in Indian Healthcare System. Further, steps used in creating 3D healthcare prototype using AM and its applications in Indian Healthcare Sector are identified.

2 Research Questions

Some basic research questions are addressed in this review paper. Those are:

2.1 What Is the Benefit of Using Additive Manufacturing in Healthcare Sector?

The benefits of additive manufacturing in healthcare sector needs to be discussed in order to show the importance of increasing its penetration in Indian Healthcare Sector. Thus, it is the first research question to be answered in this paper.

2.2 What Is the Present Status of Additive Manufacturing in Indian Healthcare Sector?

The present status of AM would tell us the extent of penetration of AM in Indian Healthcare Sector. It is necessary to know the status of additive manufacturing in Indian Healthcare Sector in order to increase the presence of AM in the given sector.

2.3 What Are the Steps to Create a 3D Printed Body Part Used in Healthcare Sector by Additive Manufacturing?

The whole procedure of creating a 3D printed body part used in healthcare sector by additive manufacturing would be discussed stepwise in the paper.

What are the steps to create a 3D printed body part used in healthcare sector by additive What are the applications of Additive Manufacturing in Indian Healthcare Sector?

2.4 *What Are the Applications of Additive Manufacturing in Indian Healthcare Sector?*

As we know great applications of AM in various fields, it is very important to address its applications in the healthcare sector and particularly in Indian Healthcare Sector. This would help us to make a well-defined path to integrate AM in the sector.

2.5 *What Are the Factors Affecting the Implementation of Additive Manufacturing in Indian Healthcare Sector?*

The motive of this query is that many big firms have shown a lot of curiosity in AM, but still they have not been able to implement AM in their manufacturing systems. We would look at various factors which would either act as a barrier or enabler in implementation of AM in Indian Healthcare Sector.

3 Benefits of Additive Manufacturing in Healthcare Sector

AM is influencing medical science in several important ways, most significantly in designing and developing medical devices and instruments. Any domain where it is essential to lower the product preparation time and at the same time provide users with operative work is a perfect outlook for AM [4]. And because our lives depend on the usability and quality of several medical products, there is an added advantage to use AM technologies for their development and production stages. For example, surgical instruments designed with the AM technologies include scalpels, retractors, surgical screws, displaying systems, etc. Following are the salient reasons behind increasing use of AM processes in medical area:

- Complex, free-form, and intricate geometries such as vertebral implants, bone models, prototypes can be easily fabricated which is not possible by conventional methods.
- Design iterations which is held viewed and studied.
- Fewer constraints in part design and sculpture shapes such as bone head and scaffolds can be easily fabricated.
- Perfectly accurate models and implants with no approximations are required. No tooling cost and no fixtures are required.

- Fully customized implants and prosthesis can be fabricated.
- Wide spectrum of materials can be used ranging from titanium alloys, stainless steel mesh to bio-degradable polymer resins to suit different medical requirements.
- Surgeons are increasingly using AM models for planning, diagnosis, and explaining complex operations such as maxillofacial and craniofacial surgeries.
- A transparent or color model fabricated by stereolithography or 3D printing proves very useful in clear viewing of tumors or other abnormalities [5].
- Scientists such as paleontologists, anthropologists, and forensic experts also use AM models to identify skeletons and fossils, develop exact duplicates of infrequent findings, and build the museum exhibition models.
- Latest advances in AM technology are most important to the increasing usage of RP for the renewal of teeth, dentures, etc., and also for creating all hearing aids [6].
- The biologically active implants are being explored for direct manufacturing. It is possible to print hard parts like bones as well as complex soft tissue structures such as kidney liver, heart, etc., using AM technologies [7].

4 Research Status on Additive Manufacturing in Indian Healthcare Sector

In the current situation, AM technology is applied greatly in medical fields, such as printing of medical models and biomaterials [8]. Rapid prototyping (RP) technology is used in several medical practices such as modern orthopedic products, making of customized maxillofacial prosthetic implant, AM also has appropriate usages in odontology [9]. AM also generates good results in the skeleton models [10], and they provide a general idea about the medical work. According to Tuomi [11], AM utilization in the medical field can be classified into these five major areas:

I. Surgical implants [12]
II. External aids [13]
III. Surgical guides [14]
IV. Medical models [15]
V. Bio-manufacturing [16].

There are a few companies pioneering the additive manufacturing in India such as Stratasys India (Bengaluru), Divide by Zero (Mumbai), Novabeans (Gurugram), and think3D (Delhi).

In India, the adoption of additive manufacturing has been largely strewed. Industry analysis from the Wohlers Report, published in 2018, shows that India accounts for roughly 3% of total units installed across the Asia Pacific region when China hits 35% and Japan 30+%. However, according to 6Wresearch [17], additive manufacturing in India is projected to cross $79 million by 2021.

"In India, the adoption did not happen in the expected areas, but there are new avenues and application areas like education and medical pre-surgery where 3D printing is being consumed,"says Swapnil Sansare, chief executive officer, Divide By Zero Technologies, a Mumbai-based 3D printer company [18].

The slow adoption in India can be attributed to the lack of understanding about additive manufacturing. Understanding of additive manufacturing benefits by customers will definitely help increase adoption in growing markets such as India.

Additive manufacturing is being used by engineering students to design prototypes, in medical laboratories to study body organs, and by art students for artwork. IISC's Society of Innovation and Development and WIPRO 3D are working to build India's first industrial-scale 3D printing machine, which is a clear example of this [18].

Equipment and manufacturing costs are some of the other barriers to the adoption of additive manufacturing. According to Swapnil Sansare, additive manufacturing is still very expensive, cost ranges from as low as ₹ 80,000 to ₹ 1.5 crore, depending on technology and build size for polymer printing. Material cost varies from ₹ 2 per gram to ₹ 15 per gram depending on the choice of material. One-time deployment cost with all the infrastructure will range from ₹ 150,000 to ₹ 750,000, depending on the printer technology and size.

According to Rajiv Bajaj, the managing director of Stratasys India, Indian Healthcare Sector has a high potential growth for additive manufacturing as they have seen a continuous rise in the adoption of it for the past few years with increased general market awareness.

5 Steps Used to Create a 3D Printed Body Part Used in Healthcare Sector by Additive Manufacturing [22]

See Table 1.

5.1 Applications of Additive Manufacturing in Healthcare Sector

Major areas in medical applications are [24] (Table 2).

6 Factors Affecting Implementation of Additive Manufacturing in Indian Healthcare Sector

Research status of additive manufacturing in Indian Healthcare Sector is not very encouraging. There are various factors affecting AM implementation in Indian Healthcare Sector. Broadly those are (Table 3).

Table 1 Steps to create 3D printed part by AM

Steps	Description	Benefits
Image acquisition [19]	It is helpful in creating internal parts that are not visible as they are hidden under the skin. E.g.: Heart, bone, etc. It helps in analysis of tissues and organs using techniques like MRI and CT	It helps in locating any extra and intra-vascular parts of human bodies
Segmentation [20]	It involves dividing the image into sections that collectively represent the body part	Analysis of human structure, intra-surgery navigation, surgical planning and virtual surgery simulation
Computer-aided design [21]	With CAD software 3D model is designed and printed using suitable AM technology	It accelerated the process of creation of implants, tools, etc.
Rapid prototyping (3D printing) [22]	It involves the usage of different processes or technologies to create a 3D model that is fused with tissues and made up of biomaterials, living cells	Flexibility in technology to make changes without any requirement of additional input. It is also less time consuming which gives the room for exploring all possible scenarios
Clinical translation [23]	It helps in understanding the disease on molecular level. In long run, it improves the treatment process	Analysis of humans sample with genetic information of the patient

7 Discussion

Additive manufacturing is the biggest innovation of twenty-first century. Its applications and benefits in the healthcare sector have been examined. Several body implants and dental devices can be manufactured in very customized way using AM technology, and thus, it is a boon for this industry requiring highly customized products. Although AM is taking over the traditional manufacturing methods in design-sensitive industries like healthcare sector all over the world rapidly, its pace of penetration in Indian Healthcare Sector is comparatively slow.

Various factors affecting AM's implementation in India are studied and assessed. Most of the factors act as barriers in AM's implementation while only a few act as enablers. These barriers need to be analyzed thoroughly, and solutions need to be formulated. A proper long-term planning is required in order to merge AM with the Indian Healthcare Sector. Certain steps have been taken in this direction by the government which is a positive indication. With such a huge market for healthcare sector and rapidly increasing awareness about AM, India is expected to successfully integrate this technology in its design sensitive industries in the near future.

Table 2 Applications of additive manufacturing in healthcare sector

Area of medical application	Objectives	Major benefits	References
Surgical planning	It acts as a visual aid for doctors and helps in surgery tactics and subsequently reduces the operation time, cost, and risk through beforehand study	It helps in obtaining diagnostic quality by predicting problem cause, and therefore, the AM models benefits during difficult procedure and compound anatomy	Giannatsis et al. [3], Ching et al. [25], Quaresma et al. [26], Leu et al. [27], Weber et al. [28], Bond et al. [29], Bezerra et al. [30], Fukuoka et al. [31], Beerens et al. [32], Manios et al. [33], Rios et al. [34], Miner et al. [35], AAJ et al. [36]
Medical education and training	Its prime focus is to provide better indication of interior and exterior of human body structure	It provides young doctors and students practical experience free of any patient distress	Leep et al. [37], Schmitt et al. [27], Nakao et al. [38]
Designing and developing of products and instrumentation utilized in medical profession	The objective here is to design and develop products, devices and instrumentation that are utilized in medical profession	It designs and develops model which can then be effectively used as medical equipment	Noort [39]
Customized implant design	One of the best technologies for customized implants and devices	CAD and AM technology provides flexibility in manufacturing and designing process for surgical implants at reasonable cost	Noorani [40], MC et al. [27], Kopac et al. [41], Trajanovic [42]
Scaffoldings and tissue engineering	Construction of implants with their special geometric features like scaffolds for the recovery of tissues	The scaffold acts as a guidance system and support structure for patient's defective bone or damaged growing tissue	Mooney [43], Risbud et al. [44]
Prosthesis and orthotics	This process is useful in prosthesis and orthotics area which is unique for every patient	For biomechanically correct geometry, development accuracy in orientation as a feature is required and also for creating custom prosthesis with high accuracy in size	Noorani [40], Liu et al. [27]

(continued)

Table 2 (continued)

Area of medical application	Objectives	Major benefits	References
Mechanical bone and replicas	Bones are created mechanically and model fabrication is also achieved with the AM	SLA creates compound structure having properties similar as in bones	Noorani [40]
Forensics	Helpful tool for crime scene investigator (CSI) where crime scene reconstruction is required	The models help in recreating evidence from fragments such as reconstruction of pieces of crime scenes accurately for assistance during cases	Liu et al. [27]
Drug dosage	It is used to create the drug dosage forms which are difficult if not possible to make using conventional manufacturing	It is possible to fabricate tablets which dissolve and almost instantly deliver the dosage	Tiwari [5]

8 Future Scope

There is a need for further study on the barriers in the implementation of the AM in the Indian Healthcare Sector in future so as to formulate new solutions for this challenge.

9 Conclusion

The paper started with a systematic analysis of the existing literature on AM in the industry. Current research status of its penetration in Indian Healthcare Sector, benefits and applications of AM in the industry are discussed. Steps to create a healthcare prototype using AM technology are further discussed. This paper then discusses the factors affecting the implementation of AM in Indian Healthcare Sector. These factors are acting as barriers in AM's implementation and need to be addressed in order to increase AM's penetration. Finally, the future scope of AM in Indian Healthcare Sector is discussed which seems to be promising. Thus, AM is expected to charge from a disruptive technology used by only innovators to a common method for core production in Indian Healthcare Sector. And by doing this, additive manufacturing is ready to deliver to humanity.

Table 3 Factors affecting implementation of AM in Indian Healthcare Sector

Factor	Description	References
Diffusion of innovation theory	One of the most important factors is that rate of adoption of any new technology is little less in India. On the diffusion curve, India would be present on one of the highest points of late majority section. Thus, doctors in India are not getting such technology for use in treatment. This slow pace of the AM adoption can be eliminated by conducting AM classes for colleges and professional engineers	Anna Aimar et al. [49]
Lack of awareness	The lack of awareness related to AM technologies is also an important barrier for AM applications. There is a dearth of awareness about AM in healthcare industry, and very few people understand different AM methodologies. There is lack of AM software, lack of knowledge about AM materials, guidelines of complete AM process. Thus, the companies would need to hire experts in this field in order to incorporate AM technology	Christopher Gläßner et al. [1]

(continued)

Table 3 (continued)

Factor	Description	References
High investment cost	High investment cost is a big challenge for AM applications by companies. AM machines and equipment are very costly and not appropriate for every healthcare institution. Also AM technology would require highly skilled workers which means additional wages to pay. All this increases investment cost to a much higher level which everyone cannot afford. Combined with lack of awareness, the risk of incorporating AM becomes double for Indian companies. AM's very high investment costs are the biggest problem in countries like India where AM is still in its initial phase	Yi et al. [1]
Organizational transformation	Complete existing organization would need to be adapted toward AM. Major changes would be required to accommodate AM. A complete reorganization of supply chain would be required. New AM quality features would be defined, and its standard would be established. Various safety precautions would be incorporated for safety of workers. This seems to be quite difficult in Indian Healthcare Sector as whole infrastructure of hospitals and medical centers need to be changed drastically in order to accommodate AM	Christopher Gläßner et al. [1]

(continued)

Table 3 (continued)

Factor	Description	References
Unpredictable value and risk including legal security concerns	Many healthcare sector's companies cannot exactly predict the benefits of AM for them, thus creating unpredictable value and risk. Hence, they try to stay with conventional technologies. The risk associated with AM includes product piracy and intellectual property protection. Also AM may help competitors to imitate their products fraudulently. As healthcare sector's AM usage involves high level of customization, legal security is not a direct concern but an indirect concern, as it may reveal patient's medical history to others which can be misused	Aurich et al. [1]
Infrastructure unavailability	Hospitals all over the country are not much ready to incorporate such modern and advanced technologies such as AM machines and equipment. Fewer R&D centers for 3D printing are present in India. Thus, infrastructure unavailability plays a major barrier toward implementation of AM technology	Christopher Gläßner et al. [1]
Capability challenges	A successful transition to AM will require new engineering and management skills to exploit its full benefits. But the problem is that we are facing a significant skills gap. It is difficult to find a well-trained and skilled workforce that are capable of applying 3D printing to real-world production. Even though current engineering graduates may have learned about the technology, it is unusual to find potential recruits who understand the holistic capabilities of the technology [45]	Deloitte [45]

(continued)

Table 3 (continued)

Factor	Description	References
Governmental policies	The current government has taken several steps for improvement in manufacturing. Project like 'Make in India' is a step ahead in this direction. However, bureaucratic procedures are still a big obstacle to faster adoption and implementation. Lenient tax system would help adopt AM faster. Also schemes like Ayushman Bharat which covers over 100 million poor and vulnerable families do not cover AM products and its services. Hence, doctors do not want to transfer high costs to patients as they would not be able to afford it. Ayushman Bharat should also cover this technology as well for its faster adoption	Prakash Panneerselvam [48]
Ecosystem requirement	A certain kind of ecosystem is required for any technology to propel. Indian hospitals have always been overloaded by patients creating an environment not suitable for adopting AM technologies. If we want to adopt AM at a very much faster rate, it is necessary to bring together the fragmented pieces of ecosystems. A proper ecosystem would help integration of AM with Indian medical industry very efficiently	Justin Scott et al. [50]
Engineering design	In AM, the image of the object gets scanned, and its coordinates are noted. Remaining 85% of the design work requires a lot of skills and hard work. It is a limitation as it is very difficult to make design changes in the object	Thomas-Seale et al. [46]

(continued)

Table 3 (continued)

Factor	Description	References
Raw material	There is a huge difference in the number of materials available for AM and that for conventional methodologies. Dearth of materials for AM in the healthcare sector is a big challenge. Titanium alloy although expensive is a popular material in AM's healthcare products	Akerfeldt et al. [47]

References

1. Yi L, Gläßner C, Aurich JC (2019) How to integrate additive manufacturing technologies into manufacturing systems successfully: a perspective from the commercial vehicle industry. J Manuf Syst 53:195–211
2. Pour MA, Zanardini M, Bachhetti A, Zanomi S (2016) Additive manufacturing impacts on productions and logistics systems. In: International Federation of Automatic Control, pp 1679–1684
3. Giannatsis J, Dedoussis V (2009) Additive fabrication technologies applied to medicine and health care: a review. Int J Adv Manuf Technol 40:116–127
4. O'Brien FJ, Taylor D, Lee C (2007) Bone as a composite material: the role of osteons as barriers to crack growth in compact bone. Int J Fatigue 29:1051–1056
5. Sanadhya S, Vij N, Chaturvedi P, Tiwari S, Arora B, Modi YK (2015) Medical applications of additive manufacturing. Int J Sci Prog Res 12:12–13
6. Cohen A, Laviv A, Berman P, Nashef R, Abu TJ (2009) Mandibular reconstruction using stereolithographic 3-dimensional printing modeling technology. Oral Surg Oral Med Oral Pathol Oral Radiol Endodontol 108(5):661–666
7. Gibson I, Cheung LK, Chow SP, Cheung WL, Beh SL, Savalani M, Lee SH (2006) The use of rapid prototyping to assist medical applications. Rapid Prototyping J 12(1):53–58
8. Shuaib M, Kumar L, Javaid M, Haleem A, Khan MI (2016) A comparison of additive manufacturing technologies. In: International conference on advance production and industrial engineering held at DTU Delhi in Dec 2016
9. Javaid M, Kumar L, Kumar V, Haleem A (2015) Product design and development using polyjet rapid prototyping technology. Int J Control Theory Inf (2015)
10. Cheng X, Peng Q et al (2017) Models partition for 3D printing objects using skeleton. Rapid Prototype J
11. Tuomi J, Paloheimo KS, Vehviläinen J et al (2014) A novel classification and online platform for planning and documentation of medical applications of additive manufacturing. Surg Innov
12. Hieu LC, Bohez E, Vander Sloten J et al (2003) Design for medical rapid prototyping of cranioplasty implants. Rapid Prototype J 9:175–186
13. Singare S, Dichen L, Bingheng L, Zhenyu G, Yaxiong L (2005) Customized design and manufacturing of chin implant based on rapid prototyping. Rapid Prototype J 11:113–118
14. Malley OF (2014) Research to inform the improved accuracy of zygomatic implants placed using computer design and additive manufactured surgical guides. In: ADT 5th triennial congress Beijing, China 6th–8th Sept 2014
15. Hieu LC, Zlatov N, VanderSloten J et al (2005) Medical rapid prototyping applications and methods. Assembly Autom 25:284–292
16. Ozbolat IT, Yu Y (2013) Bioprinting toward organ fabrication: challenges and future trends. IEEE Trans Biomed Eng 60:691–699
17. Wresearch. Indian 3D Printer Market
18. Livemint. Here's what holding back the application, adoption of 3D printing in India
19. Valverde I, Gomez-Ciriza G, Hussain T, Suarez-Mejias C, Velasco-Forte MN, Byrne N (2017) Three-dimensional printed models for surgical planning of complex congenital heart defects: an international multicentre study. Eur J Cardio-thorac Surg 52:1139–48
20. Sutherland M, Firek L, Kałmucki P (2014) 3D heart model printing for preparation of percutaneous structural interventions: description of the technology and case report. ISSN 0022–9032
21. Haleem A, Javaid M (2018) Role of CT and MRI in the design and development of orthopaedic model using additive manufacturing. J Clin Orthopaedics Trauma
22. Kanwal M, Farooqi MD, Feroze Mahmood MD (2017) Innovations in preoperative planning: insights into another dimension using 3D printing for cardiac disease. J Cardiothoracic Vascular Anesthesia
23. Haleem A, Javaid M (2019) 3D printed medical parts with different materials using additive manufacturing. Clin Epidemiol Global Health

24. Javaid M, Haleem A (2017) Additive manufacturing applications in medical cases: a literature based review. Alexandria J Med
25. Kai CC, Meng CS, Ching LS, Hoe EK, Fah LK (1998) Rapid prototyping assisted surgery planning. Int J Adv Manuf Technol 14:624–630
26. Faber J, Berto PM, Quaresma M (2016) Rapid prototyping as a tool for diagnosis and treatment planning for maxillary canine impaction. Am J Orthod Dentofac Orthop 129:583–589
27. Liu Q, Leu MC, Schmitt SM (2006) Rapid prototyping in dentistry: technology and application. Int J Adv Manuf Technol 293:317–335
28. Sodian R, Weber S, Markert M et al (2007) Stereolithographic models for surgical planning in congenital heart surgery. Ann Thoracic Surg 83:1854–1857
29. Guarino J, Tennyson S, McCain G, Bond L, Shea K, King H (2007) Rapid prototyping technology for surgeries of the pediatric spine and pelvis. J Pediatric Orthopaedics 27:955–960
30. Paiva WS, Amorim R, Bezerra DAF, Masini M (2007) Application of the stereolithography technique in complex spine surgery. Arq Neuropsiquiatr 65:443–445
31. Mizutani J, Matsubara T, Fukuoka M et al (2008) Application of full-scale three dimensional models in patients with rheumatoid cervical spine. Eur Spine J 17:644–649
32. Poukens J, Laeven P, Beerens M et al (2008) A classification of cranial implants based on the degree of difficulty in computer design and manufacture. Int J Med Robot Comput Assisted Surg 4:46–50
33. Maravelakis E, David K, Antoniadis A, Manios A, Bilalis N, Papaharilaou Y (2009) Reverse engineering techniques for cranioplasty: a case study. J Med Eng Technol 32:115–121
34. Zenha H, Azevedo L, Rios L et al (2010) The application of 3-D bio-modelling technology in complex mandibular reconstruction experience of 47 clinical cases. Eur J Plast Surg 34:257–265
35. Mehra P, Miner J, D'Innocenzo R, Nadershah M (2011) Use of 3-D stereolithographic models in oral and maxillofacial surgery. J Maxillofac Oral Surg 10:6–13
36. Peltola MJ, Vallittu PK, Vuorinen V, Aho AAJ, Puntala A, Aitasalo KMJ (2011) Novel composite implant in craniofacial bone reconstruction. Eur Arch Oto-rhino-Laryngology 269:623–628
37. Nyaluke AP, An D, Leep HR, Parsaei HR (1995) Rapid prototyping: applications in academic institutions and industry. Comput Ind Eng 29:345–349
38. Mori K, Yamamoto T, Oyama K, Nakao Y (2008) Modification of three-dimensional prototype temporal bone model for training in skull-base surgery. Neurosurg Rev 32:233–239
39. Van Noort R (2012) The future of dental devices is digital. Dent Mater 28:3–12
40. Noorani R (2006) Rapid prototyping: principles and applications. John Wiley and Sons Inc, Hoboken (NJ) ISBN: 978–0-471-73001-9
41. Balazic M, Kopac J (2007) Improvements of medical implants based on modern materials and new technologies. J Achieve Mater Manuf Eng 25:31–34
42. Milovanovic J, Trajanovic M (2007) Medical applications of rapid prototyping. Mech Eng 5:79–85
43. Kim BS, Mooney DJ (1998) Development of biocompatible synthetic extracellular matrices for tissue engineering. Trends Biotechnol 16:224–230
44. Hutmacher DW, Sittinger M, Risbud MV (2004) Scaffold-based tissue engineering rationale for computer-aided design and solid free-form fabrication systems. Trends Biotechnol 22:354–362
45. Deloitte. Challenges of additive manufacturing. Why companies don't use additive manufacturing in serial production
46. Thomas-Seale LEJ, Kirkman-Brown JC, Attallah MM, Espino DM, Shepherd DET (2018) The barriers to the progression of additive manufacture: perspectives from UK industry. Int J Prod Econ 112–113
47. Akerfeldt P, Antti ML, Pederson Influence of microstructure on mechanical properties of laser metal wire-deposited Ti-6Al-4V. Mater Sci Eng A-Structure 674:428–437
48. Panneerselvam P (2018) Additive manufacturing in aerospace and defence sector: strategy of India. J Defence Stud 39–60

49. Aimar A, Palermo A, Innocenti B (2019) The role of 3D printing in medical applications: a state of art. J Healthcare Eng
50. Scott J, Gupta N, Weber C, Newsome S (2012) Additive manufacturing: status and opportunities, pp 11–12

System Optimization for Economic and Sustainable Production and Utilization of Compressed Air (A Case Study in Asbestos Sheet Manufacturing Plant)

Debashis Pramanik and Dinesh Kumar Singh

Abstract This paper presents the findings of "performance optimization of a compressed air system—a case study in a large size asbestos sheet plant. The study was aimed at assessing the large compressed air network optimization, evaluating improvements, and suggests energy conservation measures (ECMs). Free air delivery (FAD) was evaluated to be 48–95% and specific energy consumption (SEC) as 0.08–0.12 kW/m^3/min. The pneumatically controlled units were consuming electricity also at different loads. Total air consumed in the main machines, associated systems in sheet processing units (SPUs), receiver was estimated about 25% of 520 cfm generated (de-rated minimum) and equated to total FAD. The lines leakage was also estimated to be 20.3–65.1%. The study revealed that the strategic operation of compressors (include replacing or overhauling), ensuring merit-order-rating/ECMs and realize the saving of about 225 kWh/day (8–12 months payback). The housekeeping could plug the leakage in SPUs up to 10.3–55.1% (large part about 400 cfm leaked; allowable 10%) for energy saving of 30–700 kWh/day. Additional optimizing systems installations were also in progress. The sheet processing (kg asbestos per m^3 compressed air or kWh) rate would be improved by 11–27 kg/m^3 and 232–295 kg/kWh. Asbestos units throughout the world have been undergoing considerable pneumatic system modifications and technological advanced designs. The overall efforts on multiple factors provided a solution for the progressive capabilities, economic production, and utilization of compressed air, without which it could not be fully sustained with poor performing facilities and not maintaining its norms and processes of the upgraded systems.

Keywords System optimization · Sustainable production · Compressed air · Asbestos sheet

D. Pramanik · D. K. Singh (✉)
Mechanical Engineering Department, Netaji Subhas University of Technology, Sector 3, Dwarka, New Delhi 110078, India
e-mail: dks662002@yahoo.com

D. Pramanik
e-mail: debashisteri@gmail.com

© The Author(s), under exclusive license to Springer Nature Singapore Pte Ltd. 2021
R. M. Singari et al. (eds.), *Advances in Manufacturing and Industrial Engineering*, Lecture Notes in Mechanical Engineering,
https://doi.org/10.1007/978-981-15-8542-5_91

1 Introduction

An asbestos processing medium scale industry in India produces item in three groups, viz. (1) asbestos sheet, (2) asbestos fibre rope, and (3) asbestos fibre blanket. The asbestos industry mainly has three basic types of units; those producing sheet only, those producing other asbestos items only, and those produces three items. Process technology is well established for these processed products. Basically, the 'raw asbestos' are fibrous variety of distinct mineral species and chemically they are hydrous silicate of calcium, magnesium, and iron only. The material does not contain any chemical/toxic or harmful elements. The finished products that we are using in daily life for working conditions cannot be regarded as unsafe because they are used for various purposes.

All asbestos products manufacturing units used two forms of energy, viz. thermal (heat) energy and electrical energy [1, 3]. The power to heat ratio vary from 3:1 to 8:1 and its magnitude depended upon the type of finished products (form of sheet, pipe, boards, fireproofing, insulating materials, sewer pipes, corrugated and flat roofing sheets and wall lining, etc., in standard batches) and equipment used. The pneumatically controlled plant consumed a large quantity of compressed air and electricity at different loads and used as an utility from transporting asbestos powder and slurry materials, to operate machinery, pneumatic conveying, cylinders for machine actuation, power tools, controls/instrumentation, product cleansing, and blow-offs to produce different products. Therefore, the conservation of electrical energy was more important than thermal energy in asbestos industries. The electrical energy could be conserved in significant quantity by the network optimization of compressed air (non-electrical form considered as thermal energy category). The conservation schemes (with an improving focus on monitoring/maintenance) in compressed air system (realized in thermal energy stream) could be implemented broadly in three areas: compressed air generation (using auto-governor fitted compressors), its distribution, consumption, and utilization.

The total installed production capacity of the plant (two sections/sheet processing unit—SPU and bulk material handling unit/BHU) was 90,000 tonnes of asbestos sheet per annum. The specific consumption of compressed air and electrical energy was found to be about 90 m^3/T asbestos sheet and 4 kWh/T sheet, respectively. In general, the method of energy audit may be regarded as the translation of conservation ideas into realities by blending technically feasible solutions with economic, environmental, and other organizational considerations within specified time frame. It is difficult to carry out energy conservation programme (ECP) in the continuously running process, which would interfere plant yield and the product quality. With this practical condition and optimization, attempt for economic production possibility in view the energy conservation study (with major focus on compressed air generation to utilization) was taken in the utilities of an asbestos sheet manufacturing plant. Two sections were operating between 4–24 h a day for 334–350 days/year. The actual total production in the previous year was about 82, 000 tonnes of asbestos sheet. The cost of electricity was Rs. 7.00 per unit (kWh). In the present paper, a diagnostic approach of energy audit is focussed based on the extensive field measurements conducted on most energy-consuming system, major equipments/devices/components, instruments, controls and detailed analysis of the compressed air system in one of the two units in asbestos sheet manufacturing plant.

The main part of this research is consists of analysing the effects of various parameters on actual free air delivery (FAD; available quantity, pressure, and temperature/humidity) of air compressors or efficiency, critical parameters such as high suction air temperature/energy consumption, its fluctuations, etc. The operating status was also reviewed for compressed air storing header's inlet/exit, distribution ducts for minimum pressure fluctuation/reduction, end use system's operating incoming air pressure, abnormality in air moisture/temperature level, air pressure losing ducts, temperature differences of cooling medium, quality consistency with fluctuating air pressure for sustainable production, and smooth changeover in power supply [3].

2 Design and Objectives of the Study

The major thrust of the project was on the performance (test of FAD) study of compressed air systems/network. The main objectives of the network study were listed below:

- Scheduling/ensuring the peak/maximum production and associated process load in the plant;
- Ensure maximum load condition of all operating systems in the area of compressed air generation, distribution, and utilization in three shifts;
- Observation on operating parameters comparison and deviation from design values;
- Determination of FAD or efficiency of the individual air compressor with highest accuracy level;
- Estimation and balancing of compressed air demand and supply at steady load;
- Analysis of various loss making pneumatic areas/segments and identification of short and medium term energy conservation measures (ECMs) for improving the performance of compressed air generation, distribution, and utilization;
- Identification of long-term measures and study the feasibility of retrofitting/replacing the existing inefficient compressors and pneumatic areas/segments with energy-efficient systems;
- Techno-economic feasibility of implementing the recommended energy conservation measures; and
- Preparation of a set of guidelines for better operation, maintenance (O&M), and more reliability of the system.

The overall goal of the project report and tips preparation was to provide technical information and best practice guidance with regards to compressors at two sections and associated equipment installation, O&M, and instruments calibration. The plant provided the necessary data to evaluate the technology's performance and allowed for accurate comparison with available upgraded technologies/techniques.

3 Details of Process Systems and Major Equipment of Compressed Air Network Studied

The asbestos powder and slurry are used to prepare the asbestos products of different quality. The typical asbestos sheet manufacturing process primarily consists of various steps, including material and energy flow streams. The plant had two processing units (SPU I and SPU II) with similar equipment. The third unit was for the material handling (BHU). All the raw materials were properly handled in BHU and then processed in SPU I & II for the final product.

The compressed air consumption was calculated by computing the equipment/devices/tools piston displacement and operating frequency. Second method could not be tried out using existing three compressed air totalizers due to the unavailability of week-wise data. The compressed air was used for the large number of pneumatic operations, controls of production machines, and conveying equipment.

3.1 Operating Systems and Energy Use

The existing compressed air network for the plant's process, instrument, and equipment was radial type. There was no air purification equipment (dryer) installed in the network. Moreover, the overall cost of radial type system is always less than the loop/ring type. The exiting header sizing (oversized to some extent) and pressure levels were adequate to meet unusual air supply demand.

Major processes were mixing, heating, cooling, drying, and storing. The major fuels being used by the plant were LDO and HSD. There were two sources of power supply to the plant—SEB supply and the plant's own generation through D.G. sets. The energy conservation opportunities were more promising in operating compressed air systems (generation, distribution, and utilization) than any thermal systems, because the pneumatic system is at the heart of plant [3–5].

The suction air in few compressors was carrying certain amount of moisture in the vapour form. Air was compressed (volume reduces) after the piston stroke, but absolute amount of moisture remained the same. Hence, the volumetric percentage of moisture content increased (eightfold in a 7 kg/cm^2,g compressor) all along the pipeline. When the moisture concentration became more than its holding capacity, the excess moisture would condense out. Compressed air system could be optimized by improving the compressors performance, installing appropriate device, maintaining the system properly, and plugging the variety of leaks. It gives the major saving of compressed air and electrical energy.

3.2 Technical Information

There were nine reciprocating air compressors (RCs) installed in the BHU and SPU. One compressor was continuously running in the BHU and two compressors continuously running in SPU. Each of these units had three standby RCs. The annual operating days (days in a year) of the BHU, SPU were 334 and 350, respectively. The details of five RCs are given in Table 1. These RCs were double acting, water cooled, and lubricating type.

In BHU and SPU, two and three RCs with associated components were located within two separate compressor rooms. In BHU, two RCs were connected with two separate receivers and other two RCs were directly connected with the pneumatically operated equipment. There were three receivers in the SPU. They were interconnected through valves. This arrangement was made to connect any compressor to any receiver and to supply the compressed air to the desired equipment. The operating pressure at the receiver of compressor #1 and #5 in BHU was 3.9 kg/cm^2(g) and 3.4 kg/cm^2(g), respectively. The operating pressure at the receiver of compressor #1, #2, and #3 in SPU was 7.6 kg/cm^2(g), 7.2 kg/cm^2(g), and 7.1 kg/cm^2(g), respectively. The RCs were switched-on and off automatically as the lower and upper pressure limits are reached. One compressor was running in BHU and two RCs were in SPU. The number of stand by working compressors in BHU and SPU was one in each. Four RCs were undertaken for the maintenance work.

4 Methodology

Capacity of an air compressor is the full rated volume of flow of air compressed and delivered at conditions of total temperature, total pressure, and composition prevailing at the compressor inlet. It sometimes means actual flow rate (per unit time), rather than rated volume of flow. This also termed as FAD, i.e. air at atmospheric conditions at any specific location [2, 3], because the altitude, local ambient pressure and temperature may vary at different localities and at different times. The FAD of air compressor can be evaluated by two methods. The methodology adopted for carrying

Table 1 Details of five reciprocating compressors (RCs: 2 in BHU and 3 in SPU)

Parameter	BHU, Comp. #1	BHU, Comp. #5	SPU, Comp. #1	SPU, Comp. #2	SPU, Comp. #3
Capacity (cfm/m^3 per min)	1000/28.5	1200/34.0	350/9.91	350/9.91	225/6.37
No. of stages	1	1	2	2	2

Note Cooling water is routed through the intercooler or aftercooler to reduce the thermal stresses and the chance of water condensation inside the compressor out the energy audit of air compressors and its network included the measurements using portable instruments, making necessary observation, which are described in the following sections.

4.1 FAD Test of RCs by Pump-Up Method

The pump-up test of a compressor needs the isolation of the air receiver and compressor from rest of the plant for 45–60 min period and repeat same trials in other shifts also. This could be carried out for all the operating compressors because the individual receiver could be isolated from the individual operating machine. Most consistent sets data/readings were considered for the FAD calculation.

Compressors are designed to deliver the fix quantity of air at certain pressure. But due to ageing, wear, and tear or poor maintenance, compressor may not deliver the specified quantity of air. By performing this test, actual output of a compressor can be assessed. This pump-up test determines the pumping capacity of the compressors in terms of FAD, i.e. air pumped at atmospheric conditions. The capacity of the compressor was calculated using the following formula (taken from "Energy Audit Handbook") [2, 3]:

$$\text{FAD} = \{\text{V} \times (\text{P}_2 - \text{P}_1)\}/\text{P}_\text{o} \times \text{T}$$

FAD = actual pumping capacity of the compressor (m^3/min).
V = total volume (m^3) = $V_r + v$.
V_r = volume of the receiver (m^3).
V = volume of the pipeline connected from air compressor to air receiver (m^3).
P_o = atmospheric pressure (1.013 bar absolute).
P_1 = initial pressure of the receiver (bar absolute).
1 bar = 1.019 kg/cm^2.
P_2 = final pressure of the receiver (bar absolute).
T = average time taken (min) = $(t_1 + t_2 + t_3)/3$.
t_1, t_2, t_3 = time taken to fill the receiver at working pressure of the system.
kW = input power monitored at full load condition.

The actual SEC (kW/m^3/min) of the compressors was calculated. It is the actual shaft power to generate one m^3/min of compressed air, when the compressor is running at full load.

Specific power consumption (kW/m^3/min) = Actual electrical power input/FAD (m^3/min).

4.2 FAD Test by Suction Velocity Method

The lesser accurate suction velocity tests (5–6 sets data) would have been carried out during continuous, heavy load operating condition in case pump-up test attempt failed due to non-availability of compressor and receiver isolation. But the suction velocity tests were not required here. The method involves measuring the following parameters:

- Suction air temperature (at 4 quadrants);
- Air velocities (at 4 quadrants) at the suction filters;
- Cross section of filters;
- Maximum operating pressure; and
- Input power at full load and no load condition.

5 Observations on Operating Practices

The compressed air from all the operating RCs was supplied to the header in the respective section/area through the different intermediate receivers and it further went to the branch pipes in the plant. The operating practices, observations for the existing continuous duty RCs and associated components are given in the following sections:

5.1 Operating Pressure

The pressure gauge was installed with the individual air compressor and air receiver. The actual pressure was monitored on the installed pressure gauge on the air receiver at the time of RCs loading–unloading operation during pump-up test of the following compressors given in Tables 2 and 3, respectively. Generally, the difference between the two pressure levels in the compressor happens to be due to ageing, wear, and tear, or due to modifications (in piston) carried out.

The loading and unloading pressure was adjusted in the RCs. But in actual practice, most of the RCs were running almost continuously and no frequent stoppage was observed during the visit of energy audit team. However, the atmospheric temperature would be low during peak winter and the efficiency of the RCs would remain at the high level. It would then lead to the lower running hours of the RCs for the same duty. That is why the frequent stoppages might have also been noticed in peak winter only. Each RC had the automatic type governor to monitor the switch-on and off of the compressor at desired pressure settings.

Table 2 Operating parameters of receiver, 2 RCs, and FAD in BHU

Sl. No	Parameter	Compressor #1	Compressor #5
1	Capacity of the receiver (m^3)	2.08	3.19
2	Capacity of the interconnecting pipe (m^3)	0.2774	0.3332
3	Initial pressure (bar abs.)	1.013	1.013
4	Final pressure (bar abs.)	3.824	3.138
5	Pump-up time (minutes)	0.47	0.39
6	Actual FAD (m^3/min)	13.87	18.84
7	Design FAD (m^3/min)	28.5	34.0
8	Actual power cons. (kW)*	69.0	90.0

*Input power monitored at full load condition

Table 3 Operating parameters of receiver, 3 RCs, and FAD in SPU

Sl. No	Parameter	Compressor #1	Compressor #2	Compressor #3
1	Capacity of the receiver (m^3)	1.65	1.63	2.06
2	Capacity of the interconnecting pipe (m^3)	0.4109	0.396	0.5053
3	Initial pressure (bar abs.)	1.013	1.013	1.013
4	Final pressure (bar abs.)	7.453	7.061	6.963
5	Pump-up time (minutes)	1.4	1.4	2.5
6	Actual FAD (m^3/min)	9.36	8.64	6.03
7	Design FAD (m^3/min)	9.91	9.91	6.37
8	Actual power cons. (kW)*	51.0	51.0	33.15

*Input power monitored at full load condition

5.2 *Inlet Air Temperature, Air Filter Position, Suction Side Piping/Fittings*

For RCs, the short sized individual air intake upward vertical/horizontal duct portion, filter, and long interconnecting pipes (at delivery side) were located within the compressor room/house. One small sized air intake/suction pipe, filter at the compressor #2 in SPU was located adjacent to the compression cylinder of compressor #1. The air intake filter for the individual compressor was located at the level of accessible range. The inlet/suction duct end should not give any chance of air temperature to rise and impurities to come inside. The air inlet was protected against the entry of rain and wind blow dust. It helped to avoid the clogging of filters and prevented energy loss.

Temperature of air in the ambient, compressor room, and the individual suction (around periphery) air were measured 6–8 times in each shift using thermocouple, and it was found that the average temperature of the inlet air is 4.7 °C higher than the ambient temperature for the compressor #2. The normal ventilation in compressors room was of moderate level. Most of the RCs provided inlet was allowing the suction air as ambient. However, the RC #2's air intake pipe of was towards the surrounding/next RC installation, which did not to allow the suction air as ambient. Thus, lesser the temperature of air at compressor inlet or close to ambient temperature, lesser would be the energy consumption of compressor.

5.3 *Drain Valves for Moisture Removal*

The compressed air receivers were also checked for the type of water drainage facility. Three air receivers at SPU were individually fitted with automatically operated drain valve to facilitate the water drainage. The air receivers at BHU were in the manual mode during inspection for moisture drainage, although these were fitted with the automatically operated drain valve. By using the automatic drain valve, water would only be allowed to pass intermittently, thereby reducing the wastage of compressed air. The manual mode of valve was used when the automatic mode fails or in case repair of automatic adjustment or water drainage during emergency.

5.4 *Aftercoolers and Intercoolers*

The actual difference (ΔT) of inlet/outlet water temperature was good for RCs based on the measurement of actual inlet/outlet water temperatures to/from intercoolers. The actual ΔT for the compressors should be compared with the designed ΔT level and if it is well below the designed level (considering design flow rate of water) then cleaning of the intercoolers should also be done.

5.5 *Compressed Air Delivery System (up to Distribution Headers/branch Pipes)*

Operating practices pertaining to the suction air pipes, air filter position, bends and valves, governors, air delivery to the header and receiver, and discharge in the distribution pipelines/branch pipes were noted in each compressor of the plant. These factors influence the quantity and pressure of compressed air delivered.

6 Analysis of Test/Performance Results

6.1 FAD by Pump-Up Tests

FAD test was carried out (12 times in 2 shifts for each compressor) by the pump-up method for the two compressors in BHU and three compressors in the SPU. The conducted testings were high in number for having better data averaging and obtain results/recommendations and ECMs implementation for best round-the-year compatible solution in the plant. The detail results of actual FAD tests (estimation) of RCs using the pump-up test are given in Tables 2 and 3, respectively. The pressure switches fitted with the RCs are adjusted and the actual pressure levels recorded.

The actual FAD of compressors in BHU is 48.6%, 55.4% and SPU was 94.4%, 87.1%, 94.6%, respectively. The actual FAD of lower generating compressor #2, #3 in SPU was calculated to be 8.64 m^3/min (305 cfm) and 6.03 m^3/min (213 cfm), respectively. Therefore, the minimum total generation of compressed air at any time was 518 cfm. Due to ageing of some of the compressors and inherent inefficiencies in the internal components (comprised with poor materials), the FAD was significantly less than the design values and specific energy consumption (SEC) was comparatively on higher side in SPU. The SEC required for compressing air in BHU compressor #1, #5 and SPU compressor #1, #2, #3 was estimated to be 0.083 kW/m^3/min, 0.080 kW/m^3/min and 0.091 kW/m^3/min, 0.098 kW/m^3/min, and 0.092 kW/m^3/min, respectively. Normally, the SEC range of the operating RCs installation is between 0.08 and 0.12 kW/m^3/min with the power consumption of 33–90 kW. However, the design or manufacturers recommended SEC between 0.06–0.08 kW/m^3/min could also be achieved by best operating large size piston type RCs. The annual consumption (last year's) of compressed air, electrical energy was reported to be 7,393,680 m^3 and 352,800 kWh. Therefore, the asbestos sheet was processed at the rate of 11 kg/m^3 and 232 kg/kWh in SPU.

6.2 Compressed Air Leakage Tests/Estimation and Its Supply and Demand Balancing

Normally, most of the energy losses in the total compressed air system were as a result of the operating compressors improper selection and inefficiencies in air distribution/utilization system. The overall leakage test, to quantify the leakage in the systems was carried out only when the machines in the two units using compressed air were stopped. During the study, considerable leakage of compressed air was noticed from the pipe, joints, etc., in both units. The pressure of compressed air at the application point of different pneumatically operated equipments/components was not available. As distribution lines and fitted components/accessories age, corrosion and contamination collect in low areas and at joints. The demand was estimated based on the details data on compressed air consumption in main machines, associated systems in SPU I & II, other units (III & IV), receiver and interconnecting pipes provided by the plant personnel.

6.2.1 Leak Estimation Test

Compressed air leakage forms a major source of energy wastage [6, 7]. The leakage in the compressed air system was quantified by running the compressor with all air using equipment shut off. The leakage rate in the compressed air line of BHU and SPU is given in Table 4. Even leaving a margin of 10% leakage for untraceable points in the compressed air lines, the reduction in electricity consumption was achievable in the units.

The standard method of the overall leakage test [2, 3] in the compressed air distribution line was conducted as described below. Noted down the time when system attained desired pressure or compressor unloads. This pressure reduces because of leakage in the system and the compressor loading again. Noted this time also, the period (4–5 times) was to be recorded for which the compressor was on load or off load and calculated an average value.

The overall leakage in percentage was to be estimated as

$$L = (100 \times t_1)/(t_1 + t_2)$$

The gross leakage is to be estimated as

$$L_g = (FAD \times t_1)/(t_1 + t_2)$$

L_g = gross leakage (m^3/min).
FAD = actual free air delivery of the compressor (m^3/min).
t_1 = average on load time of compressor (sec or min).
t_2 = average off load time of compressor (sec or min).

The leakage in the compressed air lines of BHU, SPU was estimated to be 20.3% and 65.1% (very high), respectively. The 5–6 trials/sets testing were done for detailed leakage calculation and 10% margin leaving for leakage through non-traceable points in both the lines. The simple housekeeping measure could reduce the leakage in BHU, SPU by 10.3% and 55.1%, respectively. The annual electrical energy saving

Table 4 Leakage rate related parameters in air lines with BHU and SPU (using compressor #1 in both units)

Sl. No	Parameters	BHU	SPU
1	Initial pressure (Kg/cm^2)	2.5	6.9
2	Final pressure (Kg/cm^2)	3.0	7.4
3	Loading time (seconds)	12	193.8
4	Unloading time (seconds)	51	103.8
5	Leakage rate (%)	20.3	65.1

Table 5 Details of compressed air demand estimation

Sl. No	Compressed air using area/system	Unit	Level
1	Pneumatic control systems in SPU I & II	m^3/hr	53.38
2	Other systems of SPU I & II	m^3/hr	18.88
3	Unit III	m^3/hr	0.010623
4	Unit IV	m^3/hr	0.102807
	Sub total	m^3/hr	72.37
		cfm	42.6
5	Volume of air receiver + interconnecting pipe	m^3	2.1274
		cfm	75.14
	Total compressed air available for work = (42.6 + 75.14)	cfm	117.76

was estimated to be 9950 kWh and 236,100 kWh, and the corresponding monetary saving would be Rs. 0.7 lakhs and Rs. 16.0 lakhs, respectively, for the two units.

6.2.2 Demand Estimation in SPU I and II

The demand estimation of compressed air in the main machines, associated systems in SPU I and SPU II, other units (III & IV), receiver and interconnecting pipes was computed per shift (8 h) basis. Compressed air consumption of pneumatic machines/equipment/ components, cylinders for machine actuation, power tools, controls, and instrumentation (about 2500 nos.) estimated by computing their piston displacement and operating frequency. The compressed air available for work in one minute (cfm) was derived for all units. The detail of compressed air demand estimation is given in Table 5.

The consumption of compressed air for all units worked out to be 42.6 cfm. The volume of receiver and interconnecting (R and IP) pipes fitted with compressor was 75.14 cu. ft. As discussed in the previous section, the allowable leakage in the compressed air distribution system should be about 10% [7, 8]. Therefore, the total quantity of compressed air required in the main machines, associated systems in SPU I and SPU II, receiver and interconnecting pipes was estimated to be 129.53 cfm (=117.76 + 11.77), which is about 25.1% of the minimum total compressed air generated (518 cfm). It is the minor part of the compressed air generated. Hence, the large part [=518-(117.76 + 11.77) cfm = 388.47 cfm or 74.9%] of the compressed air generation is lost through the major leakage points in the distribution network. However, the estimated 65.1% (as indicated by loading and unloading time method in Table 4) leakage was 9.8% lower than the 74.9%. This analysis clearly establishes that the smallest compressor (#3) in SPU could be operated to fulfil the total demand of compressed air in that unit [7, 14] and another properly sized small compressor could be installed as standby.

7 Use of Automatic/Manual Drain Valve in Air Receiver

It was observed that the most of the air receivers drain valve was manually operated to facilitate the water drainage. Moisture is drained from the receivers once a day. This provision not only passes the water but also allow the compressed air to pass to the atmosphere intermittently. By using the automatic mode drain valve, water would only be allowed to pass intermittently, thereby reducing the wastage of compressed air.

8 Measures to Improve Performance of Compressed Air System

The study was aimed at assessing the compressed air generation, distribution network optimization by observing the deviations of parameters, evaluating compressors performance improvements and suggests ECMs. The feasibility of replacement of inefficient compressor with lower size efficient compressor or overhaul was explored.

The effective O&M is one of the most cost-effective methods for ensuring energy performance/efficiency, reliability, and safety [9, 10]. Inadequate maintenance of energy using systems is a major cause of energy wastage and deteriorating reliability at individual compressed air system in plant [10, 11]. Energy losses from air leaks, maladjusted/uncalibrated or inoperable controls, and other losses from poor maintenance are often significant/considerable [12, 13].

Following measures were recommended for the performance improvement of air compressors, rest of the network, energy consumption, and sheet production system.

8.1 Long-Term Measures

1. It is therefore recommended to change/revise the operating strategy of the compressors in the plant based on their merit-order-rating (FAD performance and SEC) and get compressors overhauled in few stages and one set after the other) with the appropriate priority and immediately to have the better productivity and lower energy consumption.
2. The immediate overhauling of 2 compressors in BHU can give the average annual energy saving of about 2527 kWh, 3020 kWh and annual monetary saving of about Rs. 0.18 lakhs and 0.21 lakhs, respectively. The compressor manufacturer's recommended SEC norm is used as target for the energy-saving calculation [13, 14]. Investment required on account of above suggestion would be moderate.

3. The study revealed the inefficient compressors stopping/overhauling, immediate plugging, running 01 efficient compressor (in place 02 nos.), ensuring merit-order-rating and ECMs to be implemented to make the energy-efficient SP unit and BH unit. Moreover, the SEC of RCs #1, #2, #3 in SPU is much higher than the recommended/design level. It leads to average annual energy loss of about 2440 kWh, 2760 kWh, 1600 kWh, and annual monetary loss of these compressors works out to be about Rs. 0.17 lakhs, Rs. 0.19 lakhs and Rs. 0.11 lakhs, respectively. The immediate improvement in motors performance would give the same amount of saving of energy and monetary units.
4. Plug the leakage in the compressed air distribution system in SPU I & II and BHU immediately. Most of the threaded joints in the overhead/underground/inaccessible pipes/segments should be made leak proof by welding at the pipe connections/joints/ends. It is recommended to get the entire compressed air distribution system tested with soap solution thrice in a year.
5. Operate the compressor #3 in SPU and install a lower compressor as standby. The achieved savings was reported and it was due to second compressor's (small one of 4.5 m^3/min or 160 cfm) full load working in SPU replacing earlier large second at part load. Stop the very inefficient compressor #1 in BHU and undertake for major overhauling immediately, which will improve the actual FAD. It is also recommended to get the compressor #2 in BHU overhauled in phased manner to have the lower energy consumption. Plant personnel should undertake the overhauling work under the guidance of manufacturer. Proper maintenance has also to be ensured for the FAD of all the compressors in SPU and BHU above the satisfactory level.
6. Relocate the air inlet (suction pipe with filter) of the compressor #2 in SPU away from the compression cylinder of compressor #1 to allow the suction air (temperature and flow) as ambient. The reduction in electricity consumption is achievable in the compressor.
7. Temperature of the air supplied to the suction of the compressors should be monitored periodically to ensure that it is, at least, close or equal to the ambient temperature throughout the compressor operation in all seasons.
8. Calibrate and operate automatic drain valves with receivers fitted with the compressors at BHU. Ensure proper maintenance of the automatic drain valve with SPU compressors.
9. Install the air dryer to prevent moisture from condensing in the compressed air distribution system and to have a lower dew point.
10. Replace old and redundant pipes in the existing compressed air distribution pipelines.
11. The automatic drain valves should not allow the compressed air to pass to the atmosphere. Ensure proper maintenance of the automatic drain valves. Water should only be allowed to pass intermittently.
12. Additional optimizing systems installation was also in progress: (i) Savair managers (switch off when no air demand) to generation points. (ii) Variable feed control (VFC)—tracks demand side and receiver and meets demand) to

control flow. (iii) Ring main distribution (replacing existing radial network) to avoid unidirectional flow and farthest user.

8.2 *Short Term and Housekeeping Measures*

1. Operate one compressor (in SPU) out of two at a time and full load condition.
2. Ensure periodic calibration of the pressure gauges/switches fitted with compressors, receivers to avoid zero error and automatic system so that the differential pressure setting for the compressor on–off operation does not go beyond 0.5 kg/cm^2 (g).
3. Most of the leakage points should be plugged before the in-house trial for overall leakage test on some other day during the next shut down, which would reduce the leakage about 5%. It was informed that the plant personnel attended detection, plugging of leakage points and achieved better utilization of compressed air in the entire distribution line in last 3 months before the follow-up visit of team.

9 Conclusions

Asbestos units throughout the world have been undergoing considerable pneumatic system modifications and technological advancement compared to older conventional designs. Economic production and utilization of compressed air required close attention to multiple factors, overall stringent control and investment in O&M/ECMs which would enable the system to operate at highest efficiency [13, 14]. The efforts provided a solution for the progressive capabilities (over existing utility system), without which it could not be fully sustained with poor performing facilities and not maintaining its norms and processes of the upgraded systems.

The major saving of compressed air (non-electrical form and as thermal energy) in distribution network could be obtained by using proper item/component for pipe connection and sealing/welding technique at right locations, like overhead/non-accessible pipe segments. In the proposed (after adopting the schemes) scenario, the annual consumption of compressed air, electrical energy was estimated to be 3,039,120 m^3 and 277,200 kWh in SPU. The achieved saving was due to use of second compressor's (small one of capacity 4.5 m^3/min or 160 cfm) full load working replacing earlier large second compressor working at part load. During the energy audit follow-up visit, plant personnel reported about the major savings achieved by full load SPU operation using two compressors, one compressor of 6.03 m^3/min (213 cfm) capacity and another compressor of smaller capacity (4.5 m^3/min or 160 cfm).

Therefore, the asbestos sheet would be processed at the rate of 27 kg/m^3 and 296 kg/kWh. The specific consumption of compressed air, electrical energy in SPU would be 37 m^3/tonne asbestos sheet and 3.4 kWh/tonne asbestos sheet. It also indicated that the production would improve by 2.4 times per m^3 compressed air and

1.3 time per kWh electricity. Stringent control of air compressors, related parameters and leakage in distribution enable the compressed air system operation at highest efficiency. It would certainly improve the overall cost of asbestos sheet production by significant amount.

(The following bibliography provides a reference list with entries for journal articles [1], an LNCS chapter [2], a book [3], proceedings without editors [4], as well as a URL [5] as prefix to my References sequence no.)

References

1. Bureau of Energy Efficiency (2005) Energy efficiency in electrical utilities. New Delhi, Book 3 Second Edition Guide book for National Certification Examination for Energy Managers and Energy Auditors, pp 45–68
2. The Energy Research Institute (2007) Handbook on energy audits and management, New Delhi
3. DOE, USA (1998) Compressed Air Systems Fact Sheet #6 (www.compressedairchallege.org/library/factsheets/factsheet09.pdf), Compressed Air System Controls. Improving Compressed Air System Performance A Source Book for Industry Rev. 0, pp F6.1–F6.4
4. Expert Training, UNIDO (2013) Compressed air system optimization
5. Clair Hessmer PE, James Olmsted PE, Samantha Meserve, ANTARES Ram K. Kondapi, CPE, CEA, National Grid (2015) Compressed air system energy efficiency upgrades Implemented vs. Underutilized Measures. ACEEE Summer Study on Energy Efficiency in Industry
6. Murphy W, McCkay G (1995) Energy Management. Butterworth, Heinemann
7. CII (Confederation of Indian Industry), Manual on Compressors and Compressed Air Systems, pp. 1–61. www.greenbuisnesscentre.com/documents/Manual_compressors.pdf
8. U.S. Department of Energy (2003) Improving compressed air system performance. A Source Book for Industry Energy Efficiency and Renewable Energy, Washington D.C., pp 35–45
9. Subir Raha (2004) Energy options for India. A supplement to IEI News, pp 4–5
10. Scot FR (2005) Optimizing the compressed air system. Energy Eng 102(1):49–60
11. Scherff R (2005) The benefits of variable speed drive for air compressors. In: Conference Proceedings 4th International Conference on Energy Efficiency in Motor Driven Systems Heidelberg, Germany, vol 2, pp 320–328
12. Chaturvedi P (2004) Energy sector development and planned economy. A Supplement to IEI News, pp 13–18
13. O'Connor M (2003) Managing compressed air systems. Plant Eng 57(12):58–60
14. Sathaye J, Price L, de la Rue du Can S, Fridley D (2005) Assessment of energy use and energy savings potential in selected industrial sectors in India. Energy Analysis Department Environmental Energy Technologies Division, Lawrence Berkeley National Laboratory Berkeley LBNL-57293, pp 1.1–7.4

Investigating the Prospects of E-waste and Plastic Waste as a Material for Partial Replacement of Aggregates in Concrete

Abhishek Singh, Ahmad Sahibzada, Deepak Saini, and Susheel Kumar

Abstract With the advent in technology, large amount of E-waste is produced together with plastic waste. The most fruitful way of decomposing E-waste is landfills which require large area; whereas, plastic being a non-biodegradable material takes approximately thousand years to decompose. Together these types of wastes possess a great threat to the environment and their incorporation in making of concrete may provide us with the best way for their disposal. In this paper, we are partially replacing coarse aggregates with E-waste (2.5%, 5%, 7.5%, 10%) and fine aggregates with plastic waste (2%, 4%, 6%, 8%) to achieve an optimal strength. Specimen has been cured for 28 days and tested for compression, flexure, workability and post-heat compression test. Results showed that the maximum compressive strength is achieved at the replacement of 2% and 5% of plastic waste and E-waste, respectively, and further increase in concentration of E-waste and plastic waste results in decreased compressive strength. A standard mix of grade M-25 is prepared for the test purpose, and the compressive strength achieved at the combination of 2% and 5% replacement yielded an increase of 9.6% in compressive strength compared to standard mix.

Keywords Plastic waste · E-waste · Compressive strength · Flexural tensile strength · Post-heat compressive test

1 Introduction

With the onset in urbanization and industrialization along with rise in population, there has been an increased demand of resources like electronic gadgets and plastic consumables. Without any realization, we tend to use enormous amounts of plastic daily such as plastic bags and water bottles. In the development and advancement in technology, we have become dependent heavily on electronic gadgets. According to a survey, almost 40% of Indians want to change or upgrade their smartphones within

A. Singh (✉) · A. Sahibzada · D. Saini · S. Kumar
Delhi Technological University, Delhi 110042, India
e-mail: abs15.dtu@gmail.com

© The Author(s), under exclusive license to Springer Nature Singapore Pte Ltd. 2021
R. M. Singari et al. (eds.), *Advances in Manufacturing and Industrial Engineering*, Lecture Notes in Mechanical Engineering,
https://doi.org/10.1007/978-981-15-8542-5_92

a year [1]. With such heavy consumption of these materials, it is but obvious that there will be a large amount of waste to be disposed. For the disposal of E-waste, the most effective way known is dumping the waste into landfills which utilizes a large extent of area which is a scarcity in over populated countries. On the other hand, plastic being a non-biodegradable substance takes an enormous amount of time to decompose completely approximated around 1000 years.

Ankit Kumar (RAETCS-2018) conducted a research by partially replacing coarse aggregates by E-waste in concrete notably with 5%, 7.5% and 12.5% of coarse aggregates. The concrete strength is found optimal at 7.5% replacement of E-waste which is found to be 17.8% greater than the standard mix [2]. Shagata Das (CUET-2016) conducted a research on partial replacement of fine aggregates with plastic waste notably with 2%, 4%, 6% and 8% of fine aggregates. It was found that there was a decrease in compressive strength with increase in amount of plastic waste. The optimum compressive strength is obtained at a replacement of 2% which is found to be 12% greater than the standard mix [3].

Recently, Bangalore International Airport Pvt Ltd (BIAL) has launched a campaign called #PlasticBeku asking people to donate plastic. These used plastics would then be utilized for construction of roads using a mix of plastic and bitumen. In the hindsight of this project, we got the inspiration of incorporating E-waste and plastic waste in the making of concrete. In this paper, we are trying to replace a portion of coarse aggregates and fine aggregates with E-waste and plastic waste, respectively, in concrete and test the same for compressive strength, flexural tensile strength, workability and post-heat compression test. The variation of strength at ideal level of replacement of aggregates with E-waste and plastic wastes to achieve maximum strength and to reduce the cost of construction along with providing a safe way of disposal of these wastes is the main objective of our research.

2 Methodology

2.1 Experimental Investigation

The purpose of this research is to compare the properties of concrete made with partial replacement of E-waste and plastic waste to that of standard mix prepared.

2.2 Experimental Programme

Standard specimens have been designed according to standard mix for M25. Thereafter, coarse aggregates are replaced by E-waste in ratios of 2.5%, 5%, 7.5%, 10% while fine aggregates are replaced with plastic waste in ratios of 2%, 4%, 6%, 8%

by proportion of weight. These specimens are then tested after 28 days of curing for compressive strength test and flexural tensile strength test.

2.3 *Material Properties*

Cement. In this experiment, we have used Ordinary Portland Cement (OPC) of grade 43 of brand UltraTech Cement. The properties of cement are as follows (Table 1):

Aggregates. Standard coarse aggregates were used which is available in the market, and for fine aggregates we have used Ennore Sand which is obtained from a place in Tamil Nadu (Table 2).

Waste. We obtained plastic waste from an industrial facility in Narela while the E-waste is obtained from a recycling facility in Nehru Place, New Delhi. E-waste used in our research consisted of PCBs, used wires, motherboards of smartphones and computers and LEDs (Table 3).

Table 1 Properties of cement

S. No	Properties	Values
1	Normal consistency %	34%
2	Initial setting time	31 min
3	Final setting time	600 min
4	Soundness of cement (Le chatelier expansion)	2.65 mm
5	Fineness of cement (% retained on 90 μm IS sieve)	3.81%

Table 2 Properties of aggregates

Properties	Fine aggregates	Coarse aggregates
Fineness modulus	2.7	6.90
Specific gravity	2.5	2.7
Unit weight	1560 kg/m^3	1620 kg/m^3
Absorption capacity (%)	0.32	0.2
Moisture content (%)	4.5	2.02

Table 3 Properties of waste

Properties	Plastic waste	E-Waste
Fineness modulus	3.68	2.40
Melting temperature	75–80 °C	120–130 °C
Specific gravity	1.36	1.68
Shape	Angular	Granular
Absorption	<0.2%	<0.1%

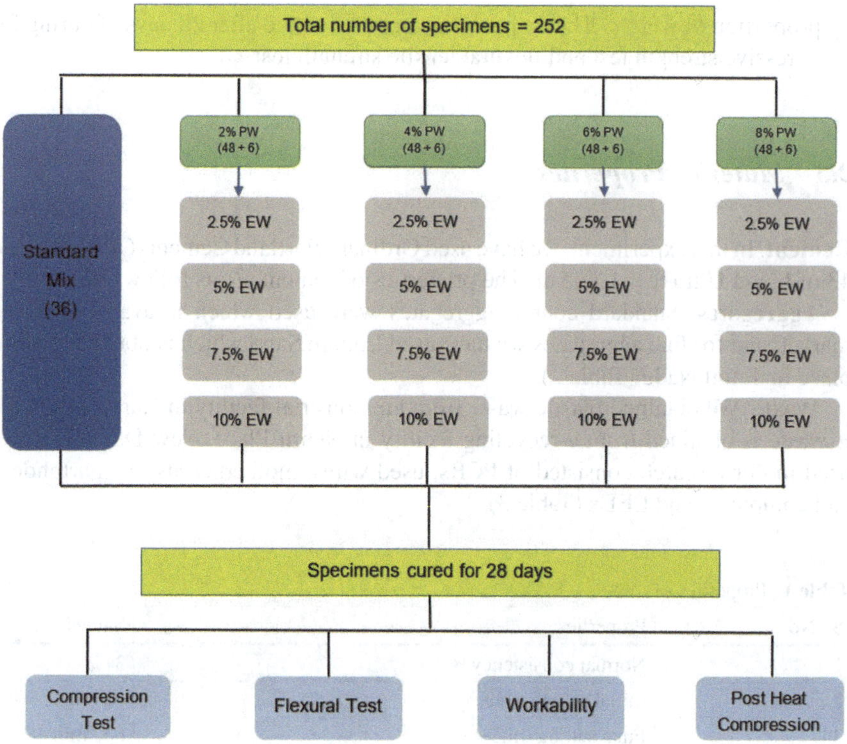

Fig. 1 Flowchart of experiments conducted

Similarly, plastic waste consisted of used water bottles (single use was preferred) and used PVC pipes. All these wastes were crushed in a pulverizing machine and added in concrete (Fig. 1).

2.4 Flowchart of Experiment

2.5 Tests Conducted

Compressive strength test. Cubes of 150 × 150 × 150 mm are casted and cured for a period of 28 days, and destructive testing is done in a compression testing machine. For each combination of waste, six specimens of cubes are tested [4].

$$\text{Compressive Strength} = \frac{\text{Crushing Load}}{\text{Area}} \quad (1)$$

Flexural tensile strength test. Specimens of 150 × 150 × 700 mm are casted and cured for a period of 28 days and are tested in moist condition. The test is conducted on flexural testing machine. The specimens are placed in testing machine on two 38 mm diameter rollers with a center to center spacing of 600 mm. The load is applied without shock increasing continuously with a rate of 0.7 N/mm^2/min until the specimen fails [4].

$$\text{Modulus of Rupture} = \frac{pl}{bd^2} \tag{2}$$

where,
p = maximum load applied at failure,
l = length of span on which specimen is supported,
b & d = width and depth of specimen.

Post-Heat Compression test. Specimens of 150 × 150 × 150 mm are casted and cured for a duration of at least 7 days after which the samples are removed and burned in a fire crucible to a temperature of 200 C for 2–3 hours and tested in compression testing machine.

Workability. It is defined as the ease with which concrete flows and is calculated by conducting slump test on freshly prepared concrete [5].

2.6 *Standard Design Mix*

The design of standard mix was done for grade M-25 according to IS 10262 (2009) [6]. Further different trial mixes were made with varying water/cement ratio as to achieve optimum target strength. Three different samples consisting of six specimens each for compressive and flexural tests were tested in varying level of water/cement ratio to obtain a concrete of desired target mean strength of 32 MPa. Compression strength testing and flexure strength testing were done after 28 days of curing. The required strength was obtained at water/cement ratio of 0.44 (Table 4).

Table 4 Trial mix strength

Water/Cement ratio	Specimen used	Average compressive applied load (kN)	Average compressive strength (MPa)	Average flexural strength (MPa)	Workability (slump in mm)
0.42	6 + 6	789.75	35.10	3.80	30
0.44	6 + 6	733.50	32.60	3.62	33
0.46	6 + 6	693	30.80	3.46	38

3 Results

See Figs. 2, 3, 4 and 5.

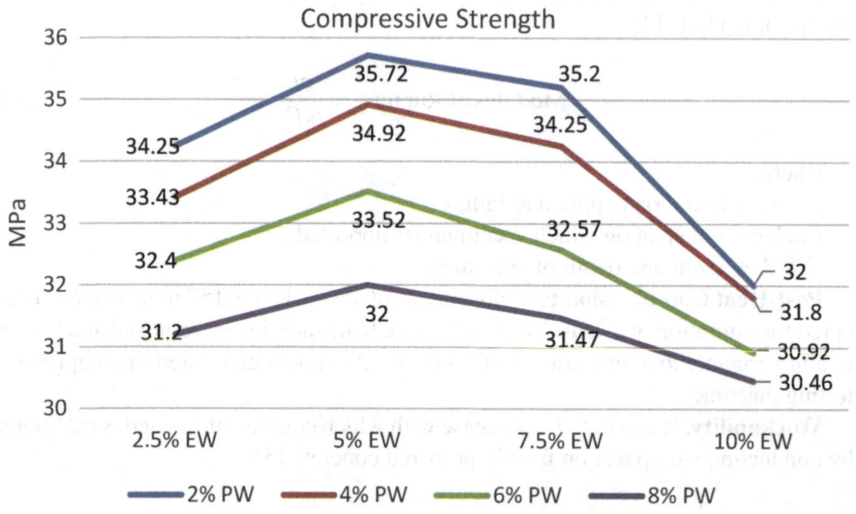

Fig. 2 Variation of compressive strength with different combination of wastes

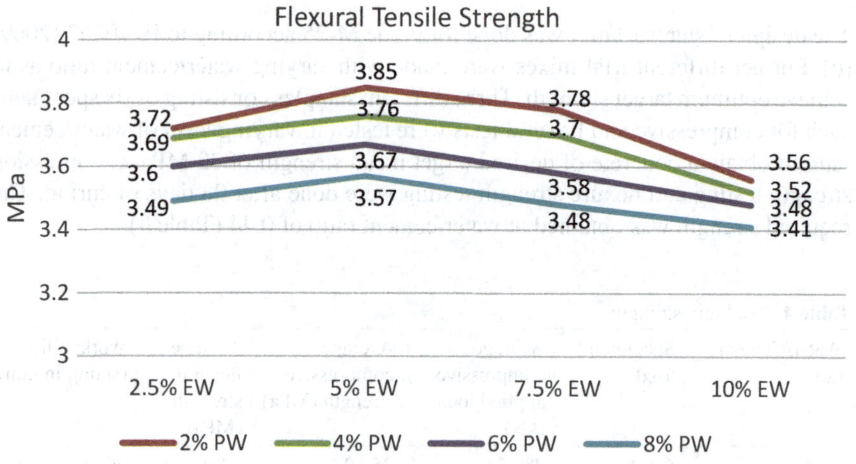

Fig. 3 Variation of flexural strength with different combination of wastes

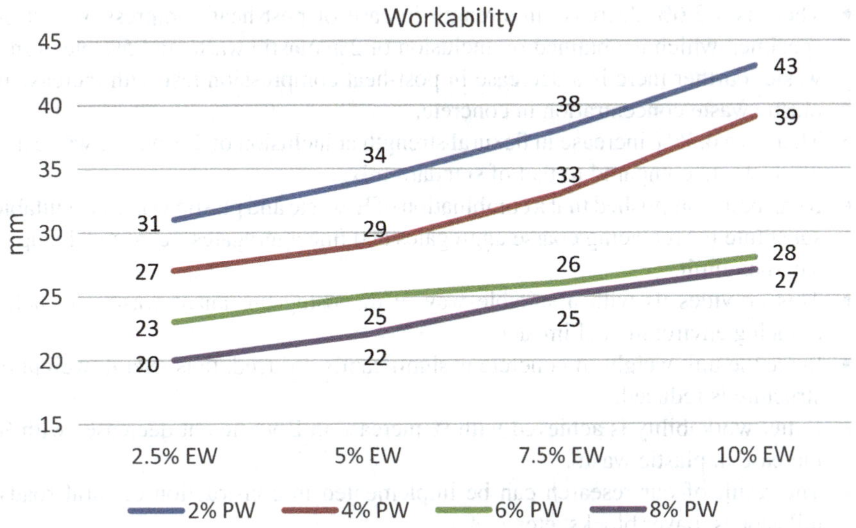

Fig. 4 Variation of workability with different combination of wastes

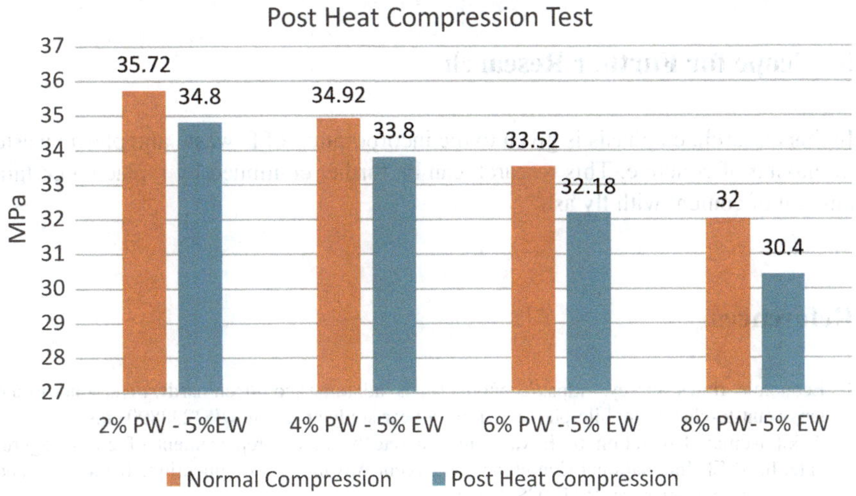

Fig. 5 Variation of post-heat compression test with normal compression test for different combination of wastes

4 Conclusion

- The optimum compressive strength of concrete is found to be 9.6% greater than the standard mix which is obtained at the inclusion of combination of 2% plastic waste and 5% electronic waste.

- There is a 2.6% decrease in strength in case of post-heat compression test on specimen which is obtained on inclusion of 2% plastic waste and 5% electronic waste. Further there is a decrease in post-heat compression test with increase in plastic waste concentration in concrete.
- There is a 6.35% increase in flexural strength at inclusion of 2% plastic waste and 5% E-waste compared to that of standard mix.
- It can be accomplished that a combination of E-waste and plastic waste is a suitable substitute for replacing coarse aggregates and fine aggregates, respectively, up to a certain limit.
- This provides us with a suitable way of disposing unwanted waste and helps reducing environmental impact.
- Since the unit weight of concrete is significantly reduced, thus overall weight of structure is reduced.
- Better workability is achieved with % increase in E-waste but decreases with % increase in plastic waste.
- The result of our research can be implemented in construction of rural roads, milestones, paver blocks, etc.
- Areas which otherwise would have been used for landfills can now be used more efficiently.

5 Scope for Further Research

In this research, emphasis is given to the incorporation of E-waste and plastic waste in making of concrete. This research can be further continued by replacing certain amount of cement with fly ash.

References

1. Economic times survey. https://economictimes.indiatimes.com/tech/hardware/around-40-indians-want-to-change-mobile-phones-within-a-year-study/articleshow/63238890.cms
2. Ankit Kumar. Utilization of E-waste in concrete by partial replacement of coarse aggregate. In: IEEE International Conference on Recent Advances in Engineering, Technology and Computational Sciences (RAETCS-2018)
3. Das S, Alam MT, Chowdhury I (2016) Utilization of plastic waste in concrete as a partial replacement of fine aggregate. In: Proceedings of 3rd International Conference on Advances in Civil Engineering, 21–23 December 2016, CUET, Chittagong, Bangladesh
4. IS 516 (1959) Method of tests for strength of concrete
5. IS 1199 (1959) Methods of sampling and analysis of concrete
6. IS 10262 (2009) Guidelines for concrete mix design proportioning

Investigation of Combustion, Performance and Emission of Aluminium Oxide Nanoparticles as Additives in CI Engine Fuels: A Review

Manish Kumar, Naushad A. Ansari, and Samsher

Abstract Over recent years, it has been observed that the excessive demand for conventional fuels has been aggravating due to the exploitative and unsustainable consumption by human beings. Thus, it is a rising ecological concern across the globe for the need to utilize alternatives in the existing fuels to reduce harmful emissions like CO, CO_2, UHC, NO_x, etc. The present study analyses and summarizes the effect of the addition of only alumina-based nanoparticles in the fuels of diesel–biodiesel blend and considering the research work conducted by different researchers. Also, the paper investigates various performance and combustion parameters like BTE, BSFC, HRR, BP, and EGT, etc. This work discusses the advantages, disadvantages, and challenges associated when nanoparticle additions to the fuel are being done to enhance the performance of CI engines.

Keywords Aluminium oxide · Nanoparticles · Emission · Combustion · Performance

Nomenclature

Al_2O_3	Aluminium oxide
CO	Carbon monoxide
CO_2	Carbon dioxide
NO_x	Oxide of nitrogen
UHC	Unburned hydrocarbon
CI	Compression ignition
BTE	Brake thermal efficiency
BSFC	Brake-specific fuel consumption
BSEC	Brake-specific energy consumption

M. Kumar · N. A. Ansari (✉) · Samsher
Department of Mechanical Engineering, Delhi Technological University, New Delhi, Delhi 110042, India
e-mail: naushad@dtu.ac.in

© The Author(s), under exclusive license to Springer Nature Singapore Pte Ltd. 2021
R. M. Singari et al. (eds.), *Advances in Manufacturing and Industrial Engineering*, Lecture Notes in Mechanical Engineering,
https://doi.org/10.1007/978-981-15-8542-5_93

HRR	Heat release rate
EGT	Exhaust gas temperature
ppm	Parts per million
bmep	Brake mean effective pressure
GSOME	Grape seed methyl ester
D80GSOME20	Diesel 80% and grape seed methyl ester 20%
D80GSOME20$Al_2O_3$50	Diesel 80%, grape seed methyl ester 20% and 50 ppm of aluminium oxide
D80GSOME20$Al_2O_3$100	Diesel 80%, grape seed methyl ester 20% and 100 ppm of aluminium oxide
D80J20	Diesel 80% and Jatropha biodiesel 20%
D80J2$Al_2O_3$20	Diesel 80%, Jatropha biodiesel 20% and 20 ppm of aluminium oxide
D80J20$Al_2O_3$30	Diesel 80%, Jatropha biodiesel 20% and 30 ppm of aluminium oxide
D80J20$Al_2O_3$40	Diesel 80%, Jatropha biodiesel 20% and 40 ppm of aluminium oxide
PSME	Palm stearin methyl ester
PSME + $Al_2O_3$50	Palm stearin methyl ester 100% and 50 ppm of aluminium oxide
PSME + $Al_2O_3$150	Palm stearin methyl ester 100% and 150 ppm of aluminium oxide
PSME + $Al_2O_3$200	Palm stearin methyl ester 100% and 200 ppm of aluminium oxide
PLOME	Poultry litter methyl ester
D80PLOME20	Diesel 80% and poultry litter methyl ester 20%
D80PLOME20A30	Diesel 80%, poultry litter methyl ester 20% and 30 ppm of aluminium oxide
J100	Jatropha biodiesel
J100A100	Jatropha biodiesel 100% and 100 ppm of aluminium oxide

1 Introduction

The diesel engine has a significant role in the day-to-day energy-related applications. Today, it is being most capable engine used in transportation, industrial and agricultural sector and also fulfils the basic need of human. For the development of industrial and transportation sector, petroleum fuel became a key resource. Diesel engine has some advantages such as its low fuel consumption, high torque capability, good fuel conversion efficiency, durability, reliability and high efficiency. Since diesel engines have a variety of uses, therefore it causes a high demand for diesel which in turn raises the concern for its depletion [1, 2].

The recent global energy scenario demands various actions to be implemented for solving the issues regarding an increase in the level of pollution, a continuous rise in the price of petroleum fuel and simultaneously, looking into various alternative of fuel which is more efficient and less polluting. Among various alternative fuels, biodiesel is a topmost choice by researchers as it does not need much modification in design and has several other technical and environmental benefits [3–8]. Extensive experimental work is being carried out by using various blends of biodiesel feedstock for its engine performance test [9]. Authors observed that the engine performance of biodiesel fuelled engines is less than pure diesel-fuelled engines; whereas, the pollutants like smoke, CO and UHC were significantly decreased but NO_x slightly increased [10–13]. The foremost aim of this research is to achieve substantially less emission without alteration in engine design. Earlier experimental works prioritize improving fuel quality while using different emulsification and additives. While studying a range of nanoscale, there exist problems like conglomeration, inhomogeneous distribution in size and sedimentation [14–19].

But with recent developments in nanoscience, now it has been possible to prepare particle size which is less than 100 nm easily and thus used in the engine as additives, thus solving the problem of conglomeration, sedimentation and non-uniform size distribution [20]. Researchers also concluded that on the addition of metallic particles of nanoscale, the overall properties of fuel improved. Using nanoparticles as additives improves the chemical and physical properties of base fuel. It may improve viscosity, thermal conductivity and diffusivity which results in complete fuel combustion. The high-energy density of metallic particles increases the gross calorific value which improves the HRR. Nano fuel has a good impact on ignition temperature and ignition delay which can increase the combustion rate and improves engine performance. The nanoparticle has a high ratio of surface area to the volume which provides a larger contact area between fuel and oxygen which improves the atomization of fuel and provides effective combustion. It causes lesser emission either by reacting with water that produces hydroxyl radical which in turn increases the oxidation of soot or reacts with carbon to decrease the temperature of oxidation [21].

The size of particles mainly used with base fuel lies between 5 and 100 nm. There are many nanoparticles which are used as additives with a blend of diesel–biodiesel with different concentration such as an oxide of aluminium, magnesium, cerium, zinc, titanium, copper, cobalt and silver. In this literature review paper, mainly oxides of aluminium have been studied and focussed on their effect on combustion, performance and emission characteristics of a diesel engine. The experimental perusal carried out by the researchers has been presented in this paper along with the requisite research gap that is to be examined further.

1.1 Alumina (Aluminium Oxide Al_2O_3) Nanoparticles

Karthikeyan et al. [22] examined the combustion, emission and performance parameters of diesel engine run by using aluminium oxide (Al_2O_3) as additives with a blend of GSOME biodiesel. The author used 50 and 100 ppm concentrations of alumina dosing with 80% diesel and 20% GSOME. Tests were conducted at D80GSOME20, D80GSOME20$Al_2O_3$50 and D80GSOME20$Al_2O_3$100. The experiment is conducted in a diesel engine single-cylinder vertical air-cooled with varying speed. On the addition of alumina nanoparticles, improvement in calorific value flashpoint was observed compared with D80GSOME20 leads to improvement in BTE and BSFC. The BTE increased with bmep on increasing the dosing level of nano additives and BSFC reduces accordingly. It was also observed that EGT of alumina blend is lower than D80GSOME20 results reduction in average cylinder temperature. On addition of alumina oxide nanoparticles to the blend emission of UHC and CO was significantly reduced. There was a reduction in NO_x emission on the addition of nano alumina due to low EGT of the blend. The smoke emissions are lower in D80GSOME20 with nanoparticles due to a reduction in the ignition delay period. At peak bmep, BTE was 4–5 and 13% for D80GSOME20$Al_2O_3$50 and D80GSOME20$Al_2O_3$100 higher than D80GSOME20, respectively. Also, BSFC reduced by 6% and 15–16% for D80GSOME20$Al_2O_3$50 and D80GSOME20$Al_2O_3$100, respectively, compared to D80GSOME20. Compared to D80GSOME20, UHC and CO emissions were decreased by 8–9 and 9% for D80GSOME20$Al_2O_3$50, and for D80GSOME20$Al_2O_3$100 it was reduced by 17 and 13–14%, respectively, especially at peak bmep. Smoke and NO_x opacity are reduced by 11–12 and 4% for D80GSOME20$Al_2O_3$100, respectively, compared to D80GSOME20.

Razek et al. [23] showed the effect of alumina (Al_2O_3) nanoparticles as additives mixed with diesel and Jatropha biodiesel blend on diesel engine having single-cylinder direct ignition water-cooled with constant compression ratio run with 1500 rpm. Tests were conducted at pure diesel, D80J20, D80J20$Al_2O_3$20, D80J20$Al_2O_3$30 and D80J20$Al_2O_3$40 to study combustion, performance and emission parameters of biodiesel blend and results compared with neat diesel and pure D80B20. For D80J20$Al_2O_3$40, the BSFC was decreased to a maximum value of 12.5% compared to D80B20; whereas, it was increased to a maximum value of 5% when compared to fossil diesel. At full-load condition, a maximum increase of 12% in the BTE was obtained for D80J20$Al_2O_3$40 compared to D80B20 and 3.8% decreases for D80J20$Al_2O_3$40 when compared to diesel. The EGT for D80J20$Al_2O_3$20, D80J20$Al_2O_3$30 and D80B20$Al_2O_3$40 was decreased by 5.5, 9.5 and 13%, respectively, when compared with D80B20 at peak load. The percentage reductions of NO_x emission for D80J20$Al_2O_3$20, D80J20$Al_2O_3$30 and D80J20$Al_2O_3$40 were 7.5, 10 and 13%, respectively, compared to D80B20. The smoke opacity was reduced by 12, 15, 17 and 21% compared with pure diesel fuel, respectively, at full load. UHC was reduced by 17, 22, 27 and 32% for D80J20$Al_2O_3$20, D80J20$Al_2O_3$30 and D80J20$Al_2O_3$40 compared with neat diesel fuel on average at all loads while CO emission was reduced by 5, 7.5, 10 and 12% when compared with neat diesel fuel, respectively (Table 1).

Krishna et al. [24] conducted an experiment to investigate the effect of alumina nanoparticle used as additives with palm stearin methyl ester (PSME) biodiesel as a base fuel on a four-stroke single-cylinder water-cooled Kirsloskar diesel engine

Table 1 Value of various parameters of diesel and different blends are shown below [23]

Parameters	Diesel	D80B20	D80J20$Al_2O_3$20	D80J20$Al_2O_3$30	D80J20$Al_2O_3$40
BSFC (kg/kWh)	0.24	0.288	0.277	0.26	0.252
CO (%)	0.04	0.038	0.037	0.036	0.034
UHC (ppm)	22	18	17	15	13
NO_x (ppm)	217	272	250	240	230
Smoke opacity (%)	63	55	52	49	44
EGT (°C)	210	255	240	230	220
HRR (kJ/°)	39.1	37.03	37.3	38.1	38.8
Cylinder peak pressure (bar)	71.22	67.2	68.5	69.2	70.5

at a constant speed of 1500 rpm. The authors studied the effect of combustion, performance and emission of biodiesel dispersed with different compositions of alumina particles. Tests were conducted at PSME + $Al_2O_3$50, PSME + $Al_2O_3$150 and PSME + $Al_2O_3$200, and results were compared with diesel and PSME biodiesel. There is a reduction in BTE for PSME + $Al_2O_3$150 and PSME + $Al_2O_3$200 compared to fossil diesel, due to the enrichment of oxygen in the alumina nanoparticles. BSFC is a minimum for diesel. It was observed that the CO_2 emissions steadily increase as the load increases for all conditions. By using PSME + $Al_2O_3$50, the CO emission was decreased by 39.21% when compared with diesel. The NO_x emissions were decreased by 9.70% for PSME + $Al_2O_3$50 ester compared to diesel. PSME has higher NO_x emissions.

Ramesh et al. [25] used nano alumina as an additive in a biodiesel blend, which was used as base fuel in a computerized single-cylinder four-stroke diesel engines run at 1500 rpm. By transesterification process of poultry litter oil with methanol, biodiesel is obtained in which H_2SO_4 and KOH are used as a catalyst. They used PLOME biodiesel D80PLOME20 and D80PLOME20 biodiesel blended with 30 mg/l additives alumina and examined their effect on performance, emission and combustion characteristics and compared them with neat diesel. The experiment was conducted at D80PLOME20 and D80PLOME20A30 by varying the load. The experimental analysis showed that the BTE of D80PLOME20A30 is higher than D80PLOME20 by 3–4% and diesel by 8–9% due to proper atomization and quick evaporation. Slight increase in HRR was observed with D80PLOME20 and D80PLOME20A30 compared to diesel at full-load conditions. Also, D80PLOME20A30 showed lower HRR than D80PLOME20. The peak pressure was rising steadily with the load, and it is greater than diesel when the engine runs with D80PLOME20 and D80PLOME20A30 at all load conditions. Results indicated that the emission of NO_x in the D80PLOME20 blend is more than by 1% due to its higher HRR while D80PLOME20A30A30 release NO_x slightly less than diesel by 4–5%. On the addition of alumina nanoparticles, D80PLOME20A30 showed drastical decrease in the emission of UHC due to the complete combustion of fuel. Exhaust gas analyser showed that the quantity of CO is decreasing in part load condition for all tested blends and increasing in full condition, and it is lower for D80PLOME20 and D80PLOME20A30 than diesel. Smoke opacity of exhaust gas is more than diesel for all blends due to its low volatility and high viscosity.

Hossain and Hussain [26] conducted an experiment on a model LPWSBio3 three-cylinder CI engine of rated speed 1500 rpm with varying load using a Jatropha biodiesel blend with additives Al_2O_3 and CeO_2 (cerium oxide) nanoparticles to study the combustion, performance and emission parameters. The fuel was prepared through a chemical process in which nanoparticles are mixed with neat Jatropha and a surfactant Triton-X100 of 1000 ppm concentration. It was observed that CeO_2 did not dissolve properly and leaves some sediment at the bottom of the container for both the dosing level. This makes unsuitable for use in the engine; therefore, Al_2O_3 nanoparticles are used as additives with neat Jatropha oil. The tests were examined by using diesel, J100 (Jatropha biodiesel) and Jatropha blend with alumina nanoparticles with 100 ppm concentration J100A100. It was seen that BSFC of J100A100 is 13% higher than neat diesel and J100 4.5% is lower than J100A100 at peak load condition. On the other hand, it was observed that BSEC of J100A100 was lower than neat diesel at all load range. It decreases by 6% at a condition of full load. On average, it was noticed that BTE was increased by 3% for J100A100 fuel compared with fossil diesel. Low CO emission was observed for J100A100 when compared with diesel fuel at low load conditions. On the other hand, CO emission of J100A100 was higher than neat diesel due to high BSFC and more oxygen content in the fuel. The emission of CO_2 of J100A100 was higher than diesel and J100 due to high BSFC. It was found that the release of UHC is lower for J100A100 when compared with diesel and J100 because of complete combustion by the catalytic effect. Higher emission of NO_x from J100 and J100A100 was found than seen for pure diesel. It was seen that the biodiesel nanoparticles blends release much less smoke opacity than fossil diesel due to more oxygen content. The biodiesel blend with nanoparticles generates the lowest HRR in comparison with neat diesel and J100 at 60% load due to longer ignition delay time, and at 100% peak load condition the HRR of pure diesel and J100A100 was found to be approximately same.

2 Summary

Though very little literature is available on nanoparticles as additives, authors have concluded that reduction in emission, as well as enhancement in performance, was basically dependent on the type, size and volume fraction of nanoparticle used. It also depends on the used main fuel quality.

The literature review shows that the experiment works were carried out by altering the engine speed and load while the other parameters were kept constant. The influence of nanoparticles on parameters such as variable injection pressure, compression ratio and injection timing is still under study. However, there are contradictions by researchers concerning variation in pressure rise rate and pollutants like CO, UHC and NO_x. The fuel properties like calorific value, density and viscosity are dependent on the type of nanoparticles used, its size and type along with the type of biodiesel used. Hence, much research is awaited to rule out the relation of the varying volume fraction, operating parameters with various nanoparticles and fuel origin.

The distribution of fuel and its mixing with air in the cylinder also depend on the size of nanoparticles. However, the variation in size of additives and their dependency on the fuel emission and performance characteristics are yet to be reported.

When alumina nanoparticles are used with fuel, they oxidize it and might result in a reduction of UHC and CO emissions. But these nanoparticles will eventually be thrown out causing pollution. Ferro-based nano additives have an advantage as these can be gathered near the exhaust that causes no air pollution. Hence, the effects of nanoparticles that are oxides of iron, in emission characteristics and engine combustion, are highly needed to be researched. When nanoparticles were added to the base fuel, it was found that the rate of heat transfer of the mixture increases causing ebbing in ignition delay which leads to the rise in pressure and thus a better thermal efficiency has been achieved. But when the concentration of nanoparticle increases beyond the limit knocking and unusual combustion have been observed. Thus, in order to reduce the emission and improvement in performance, a little more research has to be carried out regarding the optimum concentration value of nanoparticles.

Before considering the addition of nanoparticles to diesel or biodiesel, some aspects have still to be studied throughout. When nanoparticles dispersed in base fuel, the injection properties such as atomization, diffusion and penetration vary. The clustering of nanoparticles in base fuel is a field of extended study that should also focus on their stability. Therefore, detailed experimental data should aid the usability of nanoparticles in diesel/biodiesel blends.

3 Conclusion

The CI engines that run on biodiesel produce lesser performance than those running on pure diesel. The emission of NO_x also increases substantially with the use of biodiesel. This paper presents the work of different researchers that have used nanoparticles as an additive to form biodiesel blend in an effort to reduce emissions and enhance engine performance characteristics. These results should also be augmented with suitable experimental research to validate their findings. The improvement in engine performance parameters depends upon the origin of biodiesel used and also upon the size of nanoparticle additives. The data from this paper can be used to conclude that biodiesel blends with nanoparticle additives possess the potential of reduced emissions and enhanced performance from the engine that will reduce our dependence on fossil fuels and will also help in curbing environmental pollution.

References

1. Ansari NA, Kumar J, Amitkumar, Trivedi D (2013) Emission characteristics of a diesel engine using soyabean oil and diesel blends. Int J Res Eng Technol 2. ISSN: 2319-1163
2. Jindal S, Salvi BL (2012) Sustainability aspects and optimization of linseed biodiesel blends for compression ignition engine. J Renew Sustain Energy 4:1–10. https://doi.org/10.1063/1.4737922
3. Knothe G, Razon LF, Madulid DA et al (2018) Methyl esters (biodiesel) from *Pachyrhizus erosus* seed oil. Biofuels 9:449–454. https://doi.org/10.1080/17597269.2016.1275493
4. Yatish KV, Lalithamba HS, Suresh R et al (2016) Synthesis of biodiesel from *Garcinia gummi-gutta, Terminalia belerica* and *Aegle marmelos* seed oil and investigation of fuel properties. Biofuels. https://doi.org/10.1080/17597269.2016.1259524
5. Kanakraj S, Dixit S (2016) A comprehensive review on degummed biodiesel. Biofuels 7:537–548. https://doi.org/10.1080/17597269.2016.1168021
6. Bahadur S, Goyal P, Sudhakar K, Bijarniya JP (2015) A comparative study of ultrasonic and conventional methods of biodiesel production from mahua oil. Biofuels 6:107–113. https://doi.org/10.1080/17597269.2015.1057790
7. Ansari NA, Sharma A, Singh Y (2018) Performance and emission analysis of a diesel engine implementing Polanga biodiesel and optimization using Taguchi method. Process Saf Environ Prot. https://doi.org/10.1016/j.psep.2018.09.009
8. Suresh S, Sinha D, Murugavelh S (2016) Biodiesel production from waste cotton seed oil: engine performance and emission characteristics. Biofuels. https://doi.org/10.1080/17597269.2016.1192442
9. Mofijur M, Masjuki H, Kalam M et al (2013) Effect of biodiesel from various feedstocks on combustion characteristics, engine durability and materials compatibility: a review. Renew Sustain Energy Rev 28:441–455. https://doi.org/10.1016/j.rser.2013.07.051
10. Igbokwe JO, Nwufo OC, Nwaiwu CF (2015) Effects of blend on the properties, performance and emission of palm kernel oil biodiesel. Biofuels 6:1–8. https://doi.org/10.1080/17597269.2015.1030719
11. Senthil R, Silambarasan R (2015) Annona: a new biodiesel for diesel engine: a comparative experimental investigation. J Energy Inst 88:459–469. https://doi.org/10.1016/j.joei.2014.09.011
12. Das D, Pathak V, Yadav AS, Upadhyaya R (2017) Evaluation of performance, emission and combustion characteristics of diesel engine fueled with castor biodiesel. Biofuels 8:225–233. https://doi.org/10.1080/17597269.2016.1221298
13. Rajesh S, Kulkarni BM, Banapurmath NR, Kumarappa S (2018) Effect of injection parameters on performance and emission characteristics of a CRDi diesel engine fuelled with acid oil biodiesel–ethanol blended fuels. Biofuels 9:353–367. https://doi.org/10.1080/17597269.2016.1271628
14. Dinesha P, Jagannath K, Mohanan P (2018) Effect of varying 9-Octadecenoic acid (oleic fatty acid) content in biofuel on the performance and emission of a compression ignition engine at varying compression ratio. Biofuels 9:441–448. https://doi.org/10.1080/17597269.2016.1275491
15. Kanakraj S, Rehman A, Dixit S (2017) CI engine performance characteristics and exhaust emissions with enzymatic degummed linseed methyl esters and their diesel blends. Biofuels 8(3):347–357
16. Talamala V, Kancherla PR, Basava VAR et al (2017) Experimental investigation on combustion, emissions, performance and cylinder vibration analysis of an IDI engine with RBME along with isopropanol as an additive. Biofuels 8(3):307–321
17. Pochareddy YK, Ganeshram AK, Pyarelal HM et al (2017) Performance and emission characteristics of a stationary direct injection compression ignition engine fuelled with diethyl ether–sapote seed oil methyl ester–diesel blends. Biofuels 8:297–305. https://doi.org/10.1080/17597269.2016.1225646

18. Azahari SR, Salahuddin BB, Noh NAM et al (2016) Physico-chemical and emission characterization of emulsified biodiesel/diesel blends. Biofuels 7:337–343. https://doi.org/10.1080/17597269.2015.1135374
19. Dinesha P, Mohanan P (2018) Combined effect of oxygen enrichment and exhaust gas recirculation on the performance and emissions of a diesel engine fueled with biofuel blends. Biofuels 9:45–51. https://doi.org/10.1080/17597269.2016.1256551
20. Singh G, Sharma S (2015) Performance, combustion and emission characteristics of compression ignition engine using nano-fuel: a review. Int J Eng Sci Res Technol 4(6):1034–1039
21. Sani FM, Abdul Malik IO, Rufai IA (2013) Performance and emission characteristics of compression ignition engines using biodiesel as a fuel: a review. Asian J Nat Appl Sci 2(4):65–72
22. Karthikeyan S, Elango A, Silaimani SM, Prathima A (2014) Role of Al_2O_3 nano additive in GSOME biodiesel on the working characteristics of a CI engine 2014. Indian J Chem Technol 21:285–289
23. Razek SMA, Gad MS, Thabet OM (2017) Effect of aluminum oxide nano-particle in Jatropha biodiesel on performance, emissions and combustion characteristics of DI diesel engine. IJRASET 5:358–372
24. Krishna K, Prem Kumar BS, Kumar Reddy KV et al (2017) Effects of alumina nano metal oxide blended palm stearin methyl ester bio-diesel on direct injection diesel engine performance and emissions. IOP Conf Ser Mater Sci Eng 225. https://doi.org/10.1088/1757-899X/225/1/012212
25. Ramesh DK, Dhananjaya Kumar JL, Hemanth Kumar SG et al (2018) Study on effects of alumina nanoparticles as additive with poultry litter biodiesel on performance, combustion and emission characteristic of diesel engine. Mater Today Proc 5:1114–1120. https://doi.org/10.1016/j.matpr.2017.11.190
26. Hossain AK, Hussain A (2019) Impact of nanoadditives on the performance and combustion characteristics of neat Jatropha biodiesel. Energies 12. https://doi.org/10.3390/en12050921

A Review of CI Engine Performance and Emissions with Graphene Nanoparticle Additive in Diesel and Biodiesel Blends

Varun Kr Singh, Naushad A. Ansari, and Akhilesh Arora

Abstract Nowadays, very advance and experimental researches are going on in the field of nanoparticles blending in conventional diesel and biodiesel fuel. Because of deteriorating impacts on the environment by its combustible emissions like CO, CO_2, NO_x, UHC, etc., of conventional diesel fuel and to some extent biodiesel blended fuel, the current analysis studied the effects of blending of GNPs in diesel and biodiesel blends by reviewing various studies conducted in this field. In this article, observation has been also carried out on how the specific performance and combustion parameters such as BTE, BSFC, EGT and HRR are influenced by blending of GNPs along with its concentration in diesel and biodiesel blend. This work also discusses advantage, disadvantage and challenges associated with GNPs addition to the fuel in CI engines.

Keywords Nanoparticles · Performance · Emission · Graphene

Nomenclature

CI	Compression ignition
BSFC	Brake-specific fuel consumption
BSEC	Brake-specific energy consumption
BTE	Brake thermal efficiency
BMEP	Brake mean effective pressure
CO	Carbon monoxides
UHC	Unburned hydrocarbon
NO_x	Nitrogen oxides
HOME	Honge oil methyl ester
ppm	Parts per million
GNPs	Graphene nanoparticle

V. K. Singh · N. A. Ansari (✉) · A. Arora
Department of Mechanical Engineering, Delhi Technological University, New Delhi, Delhi 110042, India
e-mail: naushad@dtu.ac.in

© The Author(s), under exclusive license to Springer Nature Singapore Pte Ltd. 2021
R. M. Singari et al. (eds.), *Advances in Manufacturing and Industrial Engineering*, Lecture Notes in Mechanical Engineering,
https://doi.org/10.1007/978-981-15-8542-5_94

HOME100GNPs25	Honge oil methyl ester, 25 ppm of GNPs
HOME100GNPs50	Honge oil methyl ester, 50 ppm of GNPs
WCOME	Waste cooking oil methyl ester biodiesel
WCOME100GNPs20	Waste cooking oil methyl Ester, 20 ppm of GNPs
WCOME100GNPs40	Waste cooking oil methyl ester, 40 ppm of GNPs
WCOME100GNPs60	Waste cooking oil methyl ester, 60 ppm of graphene nanoparticles
B0	Diesel 100%, biodiesel 0%
B10	Diesel 90%, biodiesel 10%
B20	Diesel 80%, biodiesel 20%
GO	Graphene oxide
B20GO60	Diesel 20%, biodiesel 20% and graphene oxides nanoparticles 60%
VIT	Variable injection timing
CR	Compression ratio

1 Introduction

Nowadays, most of the industries like as energy sector, automobiles sector and transportation sector, etc., use a diesel engine because of its high efficiency, heavy load carrying capacity and low BSFC. Diesel engines require a continuous supply of fuel, which leads to fuel reserve and its vanishing rate.

In this era, we have various concerns regarding the use of petroleum fuel because of pollutant emission, increasing price of petroleum and its limited reserve so we have to think about the alternatives of diesel fuel which causes less pollution, low price and more efficient. Biodiesel fuel is fulfilling all the desirable requirements of a diesel engine without modification of the design in diesel engine [1–5]. Many experiments have been done on diesel engines and their performance by using various types of biodiesel and its blends [6]. On the basis of these experimental results, researchers claimed that BSFC and BTE were slightly reduced but emission, like CO, UHC and smoke, was decreased but NO_x emission is increased for some biodiesel [7–10]. Research, which are conducted nowadays, have the aim to enhance the engine performance characteristics and decrease the pollutant emission simultaneously. To improve these parameters, it is mandatory to improve the quality of fuel. Many researchers in this field claimed that the blending of nanoparticles as additives improved the engine performance as well as decrease the pollutant emission from the engine [11–16].

The GNPs improve the calorific value, decrease ignition delay, reduce the flashpoint, reduce BSFC, increase BTE, improved combustion characteristics and reduced the CO and UHC emission, and it does not affect the NO_x emission [17].

2 GNPs Nanoparticles

Bhagwat et al. [18] carried out an experiment in which they had taken GNPs as an additive in Honge oil derivate biodiesel called (HOME). The test was conducted on HOME100GNPs25 and HOME100GNPs50. The test was run at on Kirloskar four-stroke diesel engine with water-cooled cooling the system direct injection, the engine run at 1500 rpm. It was observed from the test that if we use graphene as an additive in HOME biodiesel, then brake thermal efficiency (BTE) at a load 80% of the full load is increased for HOME100GNPs25 is 4.16% and for HOME100GNPs50 is 8.33% as compared to HOME biodiesel but it is reduced from 10.71% to 7.142%, respectively, from diesel fuel. Harmful emissions from the engine, the exhaust, are decreased when GNPs are used as additives in HOME biodiesel, for HOME100GNPs25 CO quantity released was reduced by 42.85%, NO_x emission reduced by 8.88%, UHC reduced by 22.72% and smoke opacity reduced by 6.66% as compared to neat HOME biodiesel but as compared to diesel fuel the CO emission increased by 50%, NO_x emission was increased by 18%, UHC emission was increased by 54.54%, and smoke and opacity increased by 12.90%. For HOME100GNPs50, CO released was reduced by 57.14% and NO_x produce was reduced by 22.22% UHC emission reduced by 27.27% and smoke opacity is reduced by 23.52% as compare to HOME biodiesel but as compared to diesel fuel CO emission is increased by 33.33%, NOx emission is reduced by 42.85%, UHC emission increased by 31.25%, and smoke opacity is increased by 3.07%.

Sunilkumar et al. [19] carried out an experimental analysis in which they had taken GNPs as additives in WCOME biodiesel fuel GNPs. The experiment was performed on a single-cylinder water-cooled four-stroke CI engine, and it developed 5.2 kW of power at speed of 1500 rpm. The static injection period was 23° BTDC. It was conducted on three blends WCOME100GNPs20, WCOME100GNPs40 and WCOME100GNPs60. BSFC was decreased as an increase in BP for all blends; it also decreases as they increase the dosing level of additives in the blends but it is still more than neat diesel fuel. BTE was increased with an increase in the dosing level of GNPs in biodiesel but it decreases compared to diesel fuel. The CO emission decreased by (13–14)%, (9–10)% and (9–10)% for WCOME100GNPs60, WCOME100GNPs40 and WCOME100GNPs20, respectively, compared to neat WCOME but it was still more than neat diesel. The UHC emission decreased by (8–9)%, (7–8)% and (4–5)% for WCOME100GNPs60, WCOME100GNPs40 and WCOME100GNPs, respectively, as compared to neat WCOME biodiesel, but it was increased compared to diesel. NO_x emission increased with an increase in the mass fraction of GNPs additives but it was lesser than neat diesel fuel. Smoke opacity was decreased by (11–13)%, (5–7)% and (2–4)% for WCOME100GNPs60, WCOME100GNPs40 and WCOME100GNPs20, respectively, as compared to neat WCOME biodiesel.

El-Seesy et al. [20] studied the impact of nanoparticles used as additives with biodiesel–diesel blend fuel. They obtained the fuel by mixing Jatropha oil with neat diesel and adding GNPs as an additive. The composition of fuel was prepared by mixing 20% of Jatropha oil and 80% of pure diesel with different concentrations of

GNPs 25, 50, 75 and 100 mg/l. The tests were conducted at D80J20, D80J20GNPs25, D80J20GNPs50, D80J20GNPs75 and D80J20GNPs100 on an engine test rig with constant speed 2500 rpm with different load conditions. The addition of GNPs into the combustion chamber results in the collection of mass in the cylinder that has a higher surface to volume ratio, appreciating fuel-efficient combustion. This proper combustion leads to higher peak pressure inside the engine cylinder. It was found that the mixing of 75 and 100 ppm GNPs particles produces lower peak pressure as compared to 50 ppm GNPs. During the ignition delay period, it was found that the HRR of fuel became negative. The gross HRR was found to be decreased on the addition of GNPs with the D80J20 blend when compared with D80J20. On the addition of nanoparticles with blends, it was found that BSFC was reduced by 20% and a hike in BTE by 25%. On the addition of GNPs with D80J20, the EGT is decreased for all loading conditions due to the reduction of rich mixture zone in diffusion mechanism.

Hoseini et al. [21] studied the impact of graphene oxide (GO) nanoparticles mixed in biodiesel blend, namely evening primrose (*Oenothera lamarckian*), the fruit of tree of heaven (*Ailanthus altissimo*) and camelina (*Camelina sativa*) and compared its physicochemical, performance and emissions characterization with diesel. The oil content of evening primrose, camelina and the fruit of tree of heaven are 26%, 29%, and 38%, respectively. The test was performed in a single-cylinder CI engine air cooled at a rated speed of 2100 rpm on peak load condition. The fuel was prepared by using biodiesel feedstocks mentioned above at different concentrations B0, B10 and B20. The GO nanoparticles are added with B20 with 60 ppm concentration. The experiment was studied on B10, B20 and B20GO60 (Table 1).

Table 1 Physicochemical characterization of different is shown in the table [21]

Properties	Limits	Test method	Camelina oil	Tree of heaven	Evening primrose
Density at 15 °C (g/cm^3)	0.86–0.90	ASTM D4052	0.883	0.873	0.878
Kinematic viscosity @ 40 °C (mm^2/s)	1.9–6.0	ASTM D446	4.01	4.74	5.68
Flash point, closed cup (°C)	Min 130	D93	154	169	196
Cloud point (°C)	−3 to 12	ASTM D6751	−2	2	7
Pour point (°C)	−15 to 10	ASTM D6751	−6	−4	4
Acid number (mg KOH/g)	Max 0.50	ASTM D664	0.27	0.37	0.39
Cetane number	–	ASTM D613	46	49	51
High heating value (MJ/kg)			41.3	40.5	39.6

On comparing the BP among B10, B20 and B20GO60 of three mentioned biodiesels, it was found that camelina oil has high BP in all tested blend. There is a significant hike in BP when GO is added with three feedstocks, and it has maximum power for camelina oil. The BSFC when using different biodiesel blends like B10 and B20 was found the minimum for camelina oil and maximum for the tree of heaven in comparison with pure diesel. The composition B20GO60 for camelina oil has the lowest BSFC because nanographene has a high surface area which enhances the reaction rate for proper combustion leads to lower fuel consumption. On studying the CO exhaust, it was found that with biodiesel the emission of CO is decreased when compared by using diesel as a fuel. The evening primrose oil has a minimum generation of CO with composition B10 due to its high cetane value. The emission of CO is also reduced in the case of GO nanoparticles blend, and it is lowest for camelina oil with concentration B20GO60. This might be attributed to the GO particles having high surface area volume ratio which increases the chemical reaction rate and shortened the ignition delay results incomplete combustion of fuel and reduces the generation of CO, since there is a high content of oxygen in biodiesel due to which oil gets completely ignite and leaves no residue in emission. Therefore, UHC pollutant was decreased in the case of biodiesel blend mentioned above, and it is the minimum for camelina oil with composition B10 and B20 when compared with neat diesel. On the addition of nanoparticles of GO, the generation of UHC is also reduced, and it is lowest for camelina oil with composition B20GO60. The amount of NO_x emission is increased when fuelled with three different feedstocks compared to pure diesel due to the high temperature of combustion. The evening primrose generates minimum NO_x among others two oils with the composition of B10 and B20. By the addition of nanographene GO, the emission of NO_x also increased due to an increase in combustion temperature, and it is the maximum for camelina oil with a concentration of B20GO60. On the overall study, it was concluded that the performance, physicochemical and emission characteristics of camelina oil are better than the other two oils.

3 Summary

Last but not least, now I want to summarize this with nanographene diffused in base fuel (diesel/biodiesel blend). So, as per many experiments after done, it was noted that nanoparticle as an additive to biodiesel can improve the performance, also depends upon that which type of fuel we are using and also which type of nanoparticle used. Experiments also claim that by selecting the size of the particle and having a specific range, then by doing this, we can decrease the exhaust content and increase the engine performance. Researchers also claim that most of the experiments are done on the specific conditions, for instance, the experiments done by changing load or varying load and the other parametric quantity keeping the constant. By doing this, we can analyse the behaviour of GNPs on combustion and emission, now, here I am using GNPs as an additive which has played a vital role. Like that graphene has its unique structure of excellent properties; it is a thinnest and strongest material and has good electrical conductivity.

Moreover, graphene has a honeycomb structure of carbon atoms. In 1 mm of graphite, there are about 3 million layers of graphene. Also, it is harder than diamond, more elastic than rubber and lighter than aluminium, though many things are still there which is not studied yet, for instance, the performance parameter like VIT, the pressure at injection, and CR. Besides that, some more useful information regarding this was contradicting experiments by researchers like that variation of pressure rise rate and emission like NO_x, UHC and CO. So these experiments need yet to do, so more research needs to be done on this behaviour of nanoparticles. However, one more thing needs to be an experiment for best fit to nanoparticles in diesel or biodiesel, some like that behaviour of fuel injection, for instance, dispersion, atomization and penetration of nanoparticles mixed with base fuel still not observed. Therefore, to get a good result of blending nanoparticle with diesel or biodiesel, we need to examine the fuel injection properties.

4 Conclusion

When different types of biodiesel are using as fuel in diesel engines, it was concluded that engine performance, like BSFC and BTE, is slightly reduced when compared to fossil diesel. The exhaust quantity of NO_x was increased when neat biodiesel is used as a fuel. Therefore, enhancement in the performance parameter and reduction in exhaust pollutant from diesel engines has been observed, when GNPs are used as an additive in base fuel. This paper concludes the result of many types of research about the impact of GNPs in biodiesel and its blends. The use of GNPs in biodiesel increased in-cylinder pressure and accelerates the combustion. Addition of GNPs in biodiesel decreased the ignition delay of engine and BSFC, and it also increased the BTE of the engine. Most pollutant emissions like CO, UHC and the smoke opacity of diesel engines decreased as compared to neat biodiesel by using GNPs as additives but it was still more than the neat diesel fuel. It was concluded that smoke opacity decreases by raising the concentration of GNPs in the biodiesel blend. It is required that very advanced level of experimental research is to study the impacts of GNPs on diesel engine under variable condition. Detailed analysis is required considering GNPs and their size, types of biodiesel and injection characteristics, etc., on addressing the above challenge we could consider the main ingredient of fuel like types biodiesel and mass fraction of GNPs as a variable for the decrease in engine emission and for enhancing the engine performance parameters.

References

1. Knothe G, Razon LF, Madulid DA et al (2018) Methyl esters (biodiesel) from *Pachyrhizus erosus* seed oil. Biofuels 9:449–454. https://doi.org/10.1080/17597269.2016.1275493
2. Yatish KV, Lalithamba HS, Suresh R et al (2016) Synthesis of biodiesel from *Garcinia gummi-gutta, Terminalia belerica* and *Aegle marmelos* seed oil and investigation of fuel properties. Biofuels. https://doi.org/10.1080/17597269.2016.1259524
3. Kanakraj S, Dixit S (2016) A comprehensive review on degummed biodiesel. Biofuels 7:537–548. https://doi.org/10.1080/17597269.2016.1168021
4. Bahadur S, Goyal P, Sudhakar K, Bijarniya JP (2015) A comparative study of ultrasonic and conventional methods of biodiesel production from mahua oil. Biofuels 6:107–113. https://doi.org/10.1080/17597269.2015.1057790
5. Suresh S, Sinha D, Murugavelh S (2016) Biodiesel production from waste cotton seed oil: engine performance and emission characteristics. Biofuels. https://doi.org/10.1080/17597269.2016.1192442
6. Mofijur M, Masjuki H, Kalam M et al (2013) Effect of biodiesel from various feedstocks on combustion characteristics, engine durability and materials compatibility: a review. Renew Sustain Energy Rev 28:441–455. https://doi.org/10.1016/j.rser.2013.07.051
7. Igbokwe JO, Nwufo OC, Nwaiwu CF (2015) Effects of blend on the properties, performance and emission of palm kernel oil biodiesel. Biofuels 6:1–8. https://doi.org/10.1080/17597269.2015.1030719
8. Senthil R, Silambarasan R (2015) Annona: a new biodiesel for diesel engine: a comparative experimental investigation. J Energy Inst 88:459–469. https://doi.org/10.1016/j.joei.2014.09.011
9. Das D, Pathak V, Yadav AS, Upadhyaya R (2017) Evaluation of performance, emission and combustion characteristics of diesel engine fueled with castor biodiesel. Biofuels 8:225–233. https://doi.org/10.1080/17597269.2016.1221298
10. Rajesh S, Kulkarni BM, Banapurmath NR, Kumarappa S (2018) Effect of injection parameters on performance and emission characteristics of a CRDi diesel engine fuelled with acid oil biodiesel–ethanol blended fuels. Biofuels 9:353–367. https://doi.org/10.1080/17597269.2016.1271628
11. Dinesha P, Jagannath K, Mohanan P (2018) Effect of varying 9-octadecenoic acid (oleic fatty acid) content in biofuel on the performance and emission of a compression ignition engine at varying compression ratio. Biofuels 9:441–448. https://doi.org/10.1080/17597269.2016.1275491
12. Kanakraj S, Rehman A, Dixit S (2017) CI engine performance characteristics and exhaust emissions with enzymatic degummed linseed methyl esters and their diesel blends. Biofuels 8(3):347–357
13. Talamala V, Kancherla PR, Basava VAR et al (2017) Experimental investigation on combustion, emissions, performance and cylinder vibration analysis of an IDI engine with RBME along with isopropanol as an additive. Biofuels 8(3):307–321
14. Pochareddy YK, Ganeshram AK, Pyarelal HM et al (2017) Performance and emission characteristics of a stationary direct injection compression ignition engine fuelled with diethyl ether–sapote seed oil methyl ester–diesel blends. Biofuels 8:297–305. https://doi.org/10.1080/17597269.2016.1225646
15. Azahari SR, Salahuddin BB, Noh NAM et al (2016) Physico-chemical and emission characterization of emulsified biodiesel/diesel blends. Biofuels 7:337–343. https://doi.org/10.1080/17597269.2015.1135374
16. Dinesha P, Mohanan P (2018) Combined effect of oxygen enrichment and exhaust gas recirculation on the performance and emissions of a diesel engine fueled with biofuel blends. Biofuels 9:45–51. https://doi.org/10.1080/17597269.2016.1256551
17. Singh G, Sharma S (2015) Performance, combustion and emission characteristics of compression ignition engine using nano-fuel: a review. Int J Eng Sci Res Technol 4(6):1034–1039

18. Bhagwat VA, Pawar C, Bamapurmath NR (2015) GNPs biodiesel blended diesel engine. Int J Eng Res Technol 4:2278–0181
19. Sunilkumar T, Manavendra G, Banapurmath NR, Guruchethan AM (2017) Performance and emission characteristics of graphene nano particle-biodiesel blends fuelled diesel engine. Int Res J Eng Technol 4:552–557
20. El-Seesy AI, Attia AMA (2018) The effect of nanoparticles addition with biodiesel-diesel fuel blend on a diesel engine performance. In: International conference on renewable energy: generation and applications
21. Hoseini SS, Naja G, Ghobadian B, Ebadi MT, Mamat R, Yusaf T (2020) Biodiesels from three feedstock: the effect of graphene oxide (GO) nanoparticles diesel engine parameters fuelled with biodiesel. Renew Energy 145:190–201. https://doi.org/10.1016/j.renene.2019.06.020

Synthesis and Study of a Novel Carboxymethyl Guar Gum/Polyacrylate Polymeric Structured Hydrogel for Agricultural Application

Khushbu, Ashank Upadhyay, and Sudhir G. Warkar

Abstract The objective was to synthesize an eco-friendly hydrogel that can sustain the crops for longer period even with lesser irrigation, supporting the cause of water conservation. To assist in identifying the utility of efficient materials in water delivery for agricultural applications, this study investigates the use of carboxymethyl guar gum (CMG), a modified form of a natural biopolymer of guar gum and sodium polyacrylate in the synthesis of hydrogels for efficient water delivery systems. A series of hydrogel samples were prepared by varying concentration of CMG, acrylic acid and cross-linker. The swelling and water retention behaviour of synthesized hydrogels were investigated with respect to time, temperature and pH. Results from swelling data showed it was highly dependent on acrylic acid concentration. The FTIR data showed that the copolymerization and the hydrogel formation reaction were successfully performed. In addition to that, the hydrogel showed response in relation to pH, temperature and more importantly with respect to the growth of plants.

Keywords Carboxymethyl guar gum · Growth promoter · Superabsorbent hydrogels

1 Introduction

In desert areas, water scarcity is a major factor influencing the agriculture. In such terrains, crops rely majorly on rainfall, which is occasional and scanty. It can be very costly and ambiguous to irrigate the fields only depending on the rainfall. Hence, there is a need of an alternative irrigation system that is reliable and requires less investment, low water supply and low maintenance.

In the recent years, superabsorbent hydrogels have become very promising material to be used in industrial as well as agricultural application [1–3]. They can hold up to several hundred times the amount of water than their weight and sustain it even under pressure. Hydrogels used in agriculture are generally non-ionic or ionic

Khushbu · A. Upadhyay · S. G. Warkar (✉)
Department of Applied Chemistry, Delhi Technological University, New Delhi, Delhi, India
e-mail: sudhirwarkar@gmail.com

© The Author(s), under exclusive license to Springer Nature Singapore Pte Ltd. 2021
R. M. Singari et al. (eds.), *Advances in Manufacturing and Industrial Engineering*, Lecture Notes in Mechanical Engineering,
https://doi.org/10.1007/978-981-15-8542-5_95

derivative of polyacrylamide or polyacrylates to increase water retention capacity. Due to the same reason, they have been modified to enhance the physical properties of soil by increasing water holding capacity, increasing efficiency in use of water, reducing irrigation frequency, avoiding erosions, enhancing plant performance and increasing sorption capacity as favouring the uptake of some nutrient elements by the plants [4–12].

Guar gum (GG) is a non-ionic galactomannan polysaccharide seed gum derived from Cymaposis tetragonolobus. It is one such biopolymer of interest in super-absorbing polymer (SAP) chemistry that has been exploited in various areas, for example, thickening agent, ion exchange, suspending agent and controlled release devices for biomedical applications as well as agriculture application [13, 14]. Carboxymethylation of guar gum enhances properties of crude guar gum by increasing its efficacy to be polymerized and easy solution processability [15]. Guar gum–polyacrylic acid interpenetrating network hydrogel has previously been prepared by Li et al. [16]. Poorna Chandrika et al. [14] studied hydrogels formed from the guar gum with polyacrylate for agricultural application. Carboxymethyl guar gum hydrogels with polyacrylamide have been previously studied for its swelling properties and found to have maximum swelling to be 7000% [17]. But combination of CMG with poly (sodium acrylate) has been tried out for the first time through our research. Studies have also shown that the polarity in a molecule affects water absorbency [18]. Hence, sodium polyacrylate has been taken to synthesize the hydrogels. MBA being polar in nature shows the highest absorbency among other cross-linkers [2]. Various polyacrylate-based commercial hydrogels have been employed in agricultural use like Stokosorb, Agrosoak, Luquasorb, etc. Guar gum polyacrylate-based hydrogels have also shown maximum swelling upto 500 g/g [14].

In the present study, we have synthesized carboxymethyl guar gum–polyacrylate hydrogel using free radical polymerization, and the effect of composition of various constituents is reported. Also, its performance as a function of external stimuli, namely temperature, pH salts as well as water absorption, retention in sandy loam soil has been presented.

2 Experimental

2.1 Material

CMG was procured from Hindustan Gum Ltd. India, N,N methylene bis acrylamide (NNMBA) was brought from Merck, Germany, potassium persulphate (KPS) from CDH, New Delhi. NaOH and acrylic acid from Sigma-Aldrich. All the reagents were used as received, and all the solutions were prepared using double distilled water.

2.2 Synthesis of the Hydrogel

A weighed amount of CMG was mixed in 10 ml of double distilled water and constantly stirred until it is fully soluble. Now varying amounts of NNMBA and sodium acrylate solution were added while stirring the solution in composition described in Table 1. Sodium acrylate was prepared by neutralization of acrylic acid with NaOH in a specified ratio. After bubbling the nitrogen gas in the reaction beaker for 15 min, KPS was added to initiate the free radical solution polymerization. After polymerization was carried out for 10 min, the mixture was transferred to the test tubes and kept in water bath for 2 h at 60 °C. The hydrogels obtained were cut into pieces and immersed in deionized water for 3 days with refreshing the water every day. The hydrogels then were dried to xerogel under vacuum at 40 °C and stored in desiccators for further use. The optimum value of acrylic acid was derived from a previous study [19] and has been taken from [20]; whereas, the schematic mechanism of the synthesized (CMG/PAA) hydrogel is shown in Fig. 1.

2.3 Swelling Studies

After obtaining the dried hydrogel, the weighed hydrogel was kept in distilled water at 30 °C and weights of swelled hydrogel were recorded at regular intervals. Finally, the water absorption capacity of hydrogel (S) (g water uptake/g hydrogel) and equilibrium swelling S_{eq} were calculated by using the following Eqs. 1 and 2, respectively.

$$S = \left(W_s - W_g\right) / W_g \quad (1)$$

$$S_{eq} = \left(W_e - W_g\right) / W_g \quad (2)$$

where

W_s Weight of swelled hydrogel.
W_g Weight of dried hydrogel.
W_e Weight of hydrogel at equilibrium.

The swelling profile of hydrogels was studied with respect to variation in composition of its constituents. Swelling studies were done at varying temperature, pH and in 0.1 molar NaCl solution.

Table 1 Swelling, gel content and composition of constituents for preparation of semi-IPN hydrogels

Hydrogel	CMG (g/l)	AA (ml)	NaOH (8 M)	NNMBA (mg)	KPS (mg)	Gel content (%)	Max swelling (w/w)		
							Distilled water	pH 12	1 M NaCl
I	0	7.2	2.68 g	1	50	62.67	423.4	202.8	101.1
II	4	7.2	2.68 g	1	50	72.61	511.8	256.8	109.1
III	8	7.2	2.68 g	1	50	69.58	507.6	353.2	113.6
IV	16	7.2	2.68 g	1	50	60.38	348.7	312.9	75.4
V	20	7.2	2.68 g	1	50	61.73	408.7	269.7	73.6
VI	8	10.8	4.08 g	1	50	53.63	359.8	298.5	86.5
VII	8	14.4	5.44 g	1	50	51.94	390.8	262.0	70.4
VIII	8	3.6	1.36 g	1	50	46.83	264.6	205.1	82.3
IX	8	7.2	2.68 g	5	50	93.7	396.9	168.9	65.5
X	8	7.2	2.68 g	15	50	83.3	258.7	130.8	49.7

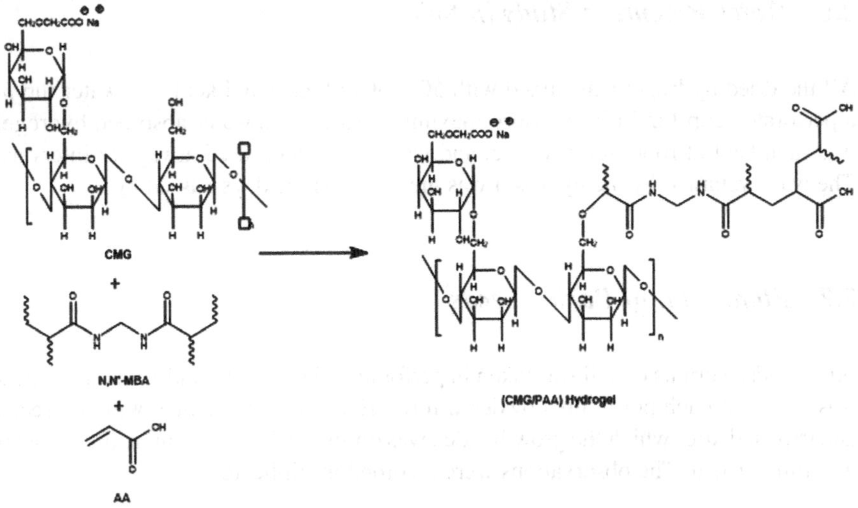

Fig. 1 Schematic diagram representing the formation of (CMG/PAA) hydrogel

2.4 Characterization

The synthesized hydrogels were characterized by using FT-IR spectroscopy using KBr pellets on Thermo Scientific Nicolet 380 Spectrometer in transmittance mode to study the evidence of formation of CMG-poly(acrylate) hydrogel. For thermal studies, Perkin Elmer Thermogravimetric Analyser TGA 4000 was used in nitrogen atmosphere for the temperature sweep from 0 to 900 °C.

2.5 Gel Content of Hydrogels

Gel content was calculated after the dissolution of the soluble part of hydrogel in water. This was done (shown in Table 1) as follows: the obtained gel after cross-linking was dried for 48 h at 30 °C and weighed (W_d) after which it was kept in distilled water for 72 h with constantly changing water every day and stirring to dissolve the soluble content. After this it was thoroughly dried and then again weighed which was termed as W_g. The gel content was then obtained by the following Eq. 3.

$$G = (W_d - W_g)/W_d \quad (3)$$

2.6 Water Retention Study in Soil

All the dried hydrogel was mixed with 50 g of soil each and kept in a water tub in a perforated cup for 24 h to absorb maximum water. The water absorbed hydrogel was then kept at room temperature, and weight was measured at regular intervals. The water retained by the hydrogel was measured as the days passed by.

2.7 Plantation of Barley Seeds

Five hundred grams of soil was taken in perforated plastic pots, and 1 g of hydrogels was mixed in each pot in the powdered form. Each pot was planted with ten seeds and irrigated after which the growth index was observed for each sample till the plant was fully grown. The observations were recorded in triplicate.

3 Result and Discussion

3.1 Swelling Studies

Swelling was studied in distilled water, at pH 7, pH 12 and in the 0.1 molar brine solution. In distilled water, the highest swelling was observed for the sample with CMG concentration 4 g/l and acrylic acid volume 7.2 ml by keeping the minimum concentration of MBA to be 1 mg (Fig. 2a). Table 1 briefly summarizes the absorption of hydrogels at equilibrium, and it is clear from the table that the water absorption capacity of the hydrogel increases with increase in AA concentration (Fig. 2c) which can be attributed to the increase in ionic groups but the absorbency decreases on further increasing the AA concentration which can be attributed to the increase in homopolymer content [2]. The swelling index for the synthesized hydrogel increases (Fig. 2b) as the CMG concentration increases because of the COO^- groups, and the number of counter ions (Na^+) along the polymeric chains also got increased within the gel phase, resulting an amplification in the chain relaxation because of the repulsion among like charged COO^- groups. Increasing concentration of NNMBA (Fig. 2d) resulted in decrease in value of absorption suggesting the increase in the degree of cross-linking between the polymer chains leading to the reduction in size of the intermolecular spaces.

Table 1 also compared the swelling studies obtained in pH 12 basic solution and 1 M brine solution. In basic media despite increase in the number of OH, which is responsible for hydrogen bonding with water molecule, the effect of repulsion between ions due to increase in total number of ions was more predominant and almost 50% decrease in total absorbency was observed. Similar observations were found in 1 M NaCl solution where the water absorbance capacity reduced to 25% of

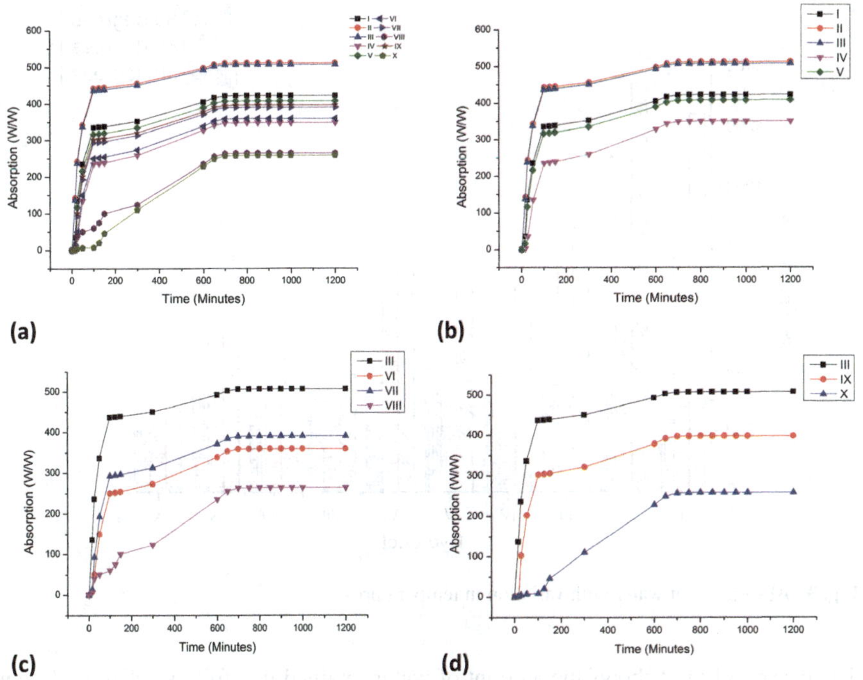

Fig. 2 Swelling studies of hydrogels **a** in distilled water, **b** w.r.t CMG concentration w.r.t time, **c** w.r.t AA concentration w.r.t time, **d** w.r.t NNMBA concentration w.r.t time

that in distilled water, and this can be attributed to the replacement of water molecules by anions in the hydrogel causing repulsion between ions leading to decrease in absorbency.

3.1.1 Swelling Studies with Respect to Temperature

On raising temperature, the maximum swelling almost doubled for most hydrogel samples. This can be attributed to the breakage of H bonding on heating, increase in repulsion of like ions and thus increasing the space to accommodate more water molecules in the hydrogel (Fig. 3).

3.2 Maximum Water Absorption Capacity

This study reveals how effective the hydrogels are when mixed with soil and kept at ambient temperature after complete absorption of water. Almost all hydrogels containing soil showed 150% higher water absorption than the hydrogel less soil.

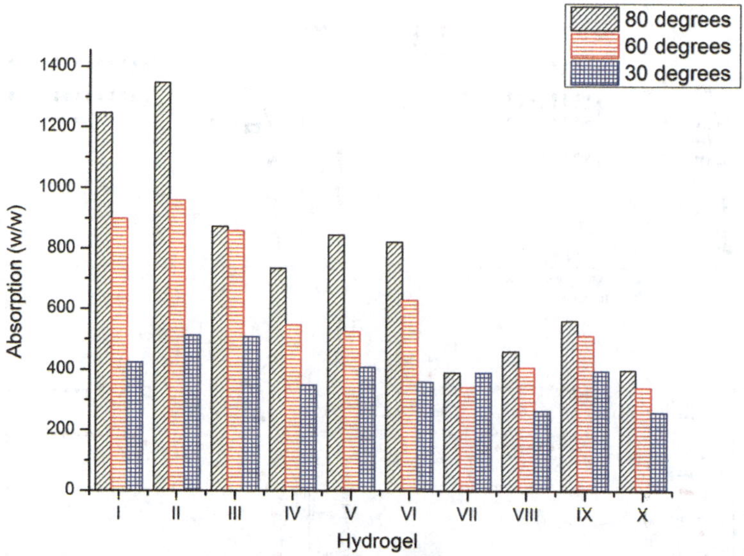

Fig. 3 Absorption of water with variation in temperature

The curve in Fig. 2 shows the amount of water retained in a fully swelled hydrogel with respect to number of days when kept in ambient temperature. It suggests that the hydrogels with curve on the outer side are better at retaining water than hydrogels having lower curve. Hydrogel V was found to be best in retaining water with respect to other hydrogels showing larger water retention in soil. Table 2 showed the maximum water absorbed by the hydrogels in soil.

Table 2 Maximum water absorbed by hydrogels in soil

Sample	Maximum water absorbed
I	34.38
II	32.96
III	31.76
IV	31.3
V	31.58
VI	30.56
VII	30.08
VIII	31.29
IX	29.47
X	28.06
Control (without hydrogel)	23.52

Table 3 Effect of hydrogel on barley plant growth

Sr. No.	Germination/Growth	Barley seeds	Barley + Hydrogel polymer
1.	Germination (%)	79%	88%
2.	Shoot Height (cm)	9.38 ± 0.44	13.12 ± 0.47
3.	Root Height (cm)	3.13 ± 0.25	5.21 ± 0.32

3.3 Plantation

3.3.1 Plant Growth Indices

Plants were irrigated as discussed in Sect. 2.7 in a sandy loamy soil containing 13.03% gravel, 85.95% sand and 12% clay and silt found using sieve analysis. Plant length was measured using a ruler, and it was found that the comparatively higher seeds were germinated and length of plant was found to be higher for the soil containing hydrogels. El-Rehim [1] attributed the increase in yield of plant to the function of alginate group to increase soil fertility. In this study, the yield directly depends on the water absorption capacity of the hydrogel as shown in Table 3.

3.4 Characterization

3.4.1 FTIR

Figure 4 shows the FT-IR curve of prepared hydrogel sample, at 3417 cm^{-1} broad peak shows –OH stretching of alcohol. Peak at 1429 cm^{-1} can be attributed to carboxymethyl group, and peak at 1188 and 1384 cm^{-1} can be attributed to CH_2–O–CH_2 bending and –CH_2 group bending. The peak attributing to alkene groups was missing suggesting the successful polymerization of acrylic acid.

3.4.2 TGA

TGA curve in Fig. 5 shows the decomposition profile of the CMG as well as CMG-PAA hydrogel. The first slope between 100 and 300 °C shows the loss of water. Second slope represents the degradation of polymer backbone after 200–450 °C which is gradual in hydrogel and abrupt in CMG, and the gradual decomposition of hydrogel can be attributed to the thermal stability of polyacrylate chains in the IPN structure. Slope between 450 and 550 °C attributes to the degradation of carboxymethyl group shown by both the curves.

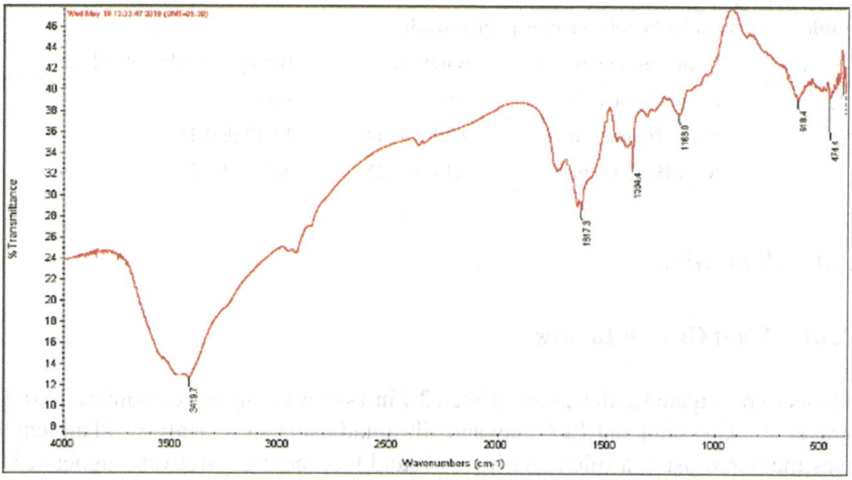

Fig. 4 FTIR spectra of CMG polyacrylate hydrogel

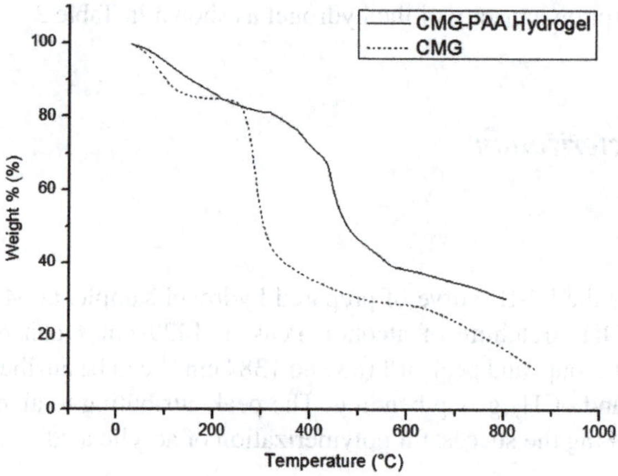

Fig. 5 TGA of CMG and its polyacrylate-based hydrogel

4 Conclusion

This study focused on synthesizing and finding a novel IPN hydrogel by varying constituents so that it can be used in agricultural applications, mainly desert areas. The hydrogel was synthesized keeping into consideration its non-toxicity varied by using cross-linker concentration. The swelling and water retention behaviour showed consistency, and most swelled sample showed most water retention capacity. The hydrogel also showed variation to different stimulus like pH, salt solution and heat.

The hydrogel was then employed irrigation and found to be better than the soil without hydrogel in terms of number of seed germinated and shoot length.

References

1. El-Rehim HAA (2006) Characterization and possible agricultural application of polyacrylamide/sodium alginate crosslinked hydrogels prepared by ionizing radiation. J Appl Polym Sci 101:3572–3580
2. Mohana Raju K, Padmanabha Raju M (2001) Synthesis of novel superabsorbing copolymers for agricultural and horticultural applications. Poly Int 50:946–951
3. Kashyapa PL, Xiang X, Heiden P (2015) Chitosan nanoparticle based delivery systems for sustainable agriculture. Int J Biol Macromol 77:36–51
4. Johnson MS, Leah RT (1990) Effects of superabsorbent polyacrylamides on efficiency of water use by crop seedlings. J Sci Food Agric 52:431
5. Sojka RE, Westermann DT, Lentz RD (1998) Water and erosion management with multiple applications of polyacrylamide in furrow irrigation. Soil Sci Soc Am J 62:1672
6. Sojka RE, Lentz RD (1997) Reducing furrow irrigation erosion with polyacrylamide (PAM). J Prod Agric 10(1):47
7. Fonteno WC, Bilderback TE (1993) Impact of hydrogel on physical properties of coarse-structured horticultural substrates. J Am Soc Horticultural Sci 118:217
8. Rubio HO, Wood MK, Cardenas M, Buchanan BA (1990) Seedling emergence and root elongation of four grass species and evaporation from bare soil as affected by polyacrylamide. J Arid Environ 18:33
9. Ben-Hur M, Keren R (1997) Ploymer effects on water infiltration and soil aggregation. Soil Sci Soc Am J 61:565
10. Ingram DL, Yeager TH (1987) Effects of irrigation frequency and a water-absorbing polymer amendment on ligustrum growth and moisture retention by a container medium. J Environ Horticulture 5:19
11. Hedrick RM, Mowry DT (1952) Effect of synthetic polyelectrolytes on aggregation, aeration, and water relationships of soil. Soil Sci 73:427
12. Wang YT (1989) Medium and hydrogel affect production and wilting of tropical ornamental plants. HortScience 24:941
13. Wang W, Zhai N, Wang A (2010) Preparation and swelling characteristics of a superabsorbent nanocomposite based on natural guar gum and cation-modified vermiculite. J Appl Polym Sci 119:3675
14. Poorna Chandrika KSV, Singh A, Sarkar DJ, Rathore A, Kumar A (2014) pH-Sensitive crosslinked guar gum-based superabsorbent hydrogels: swelling response in simulated environments and water retention behavior in plant growth media. J Appl Polym Sci 41060
15. Dodi G, Hritcu D, Popa MI (2011) Carboxymethylation of guar gum: synthesis and characterization. Cellulose Chem Technol 45(3–4):171–176
16. Li X, Wenhui Wu, Wang J, Duan Y (2006) The swelling behavior and network parameters of guar gum/poly(acrylic acid) semi-interpenetrating polymer network hydrogels. Carbohyd Polym 66:473–479
17. Gupta AP, Warkar SG (2015) Synthesis, characterization and swelling properties of poly (acrylamide-Cl carboxymethyl guargum) hydrogels. Int J Pharm Bio Sci 6(1):(P)516–529
18. El-Rehim HAA, Hegazy E-SA, El-Mohdy HLA (2004) Radiation synthesis of hydrogels to enhance sandy soils water retention and increase plant performance. J Appl Polym Sci 93:1360–1371
19. Wang WB, Xu1 JX, Wang AQ (2011) A pH-, salt- and solvent-responsive carboxymethylcellulose-g-poly(sodium acrylate)/medical stone superabsorbent composite with enhanced swelling and responsive properties. eXPRESS Poly Lett 5(5):385–400

20. Khanlari S, Dube MA (2015) Effect of pH on poly(acrylic acid) solution polymerization. J Macromol Sci Part A Pure Appl Chem 52:587–592

Artificial Neural Network (ANN) for Forecasting of Flood at Kasol in Satluj River, India

Abhinav Sharma and Anshu Sharma

Abstract Every year there occurs a huge loss to mankind and property due to flood. Therefore, forecasting and warning are considered to be the most significant, credible and economical measures for prevention of the losses from flood. One of the easy approaches to forecasting of flood is the modeling of statistical connections and relations among the hydrologic inputs and outputs. In this method, the mathematical and physical relationships of the variables and parameters are not required to be known. Artificial neural network (ANN) is the similar tool that offers a fast and adaptive approach for model development. In this work, the development of an ANN model for flood forecasting for Satluj River at Kasol station has been performed. The statistical parameters such as autocorrelation (ACF), partial correlation (PACF) and cross-correlation (CCF) have been computed, and with its use, the input vector (IV) of the model has been selected. The lead times considered in the model development for flood forecasting are 1–7 days lead. The results of the ANN models are analyzed using statistical indices such as correlation coefficient, root mean squared error (RMSE), percentage error in peak flow estimation. The results of ANN models indicate that the developed ANN model structures simulate the nonlinearity in the data with reasonable accuracy. There is deterioration in the model performance with increment in lead time from 1 to 7 days.

Keywords Flood forecasting · ANN · Backpropagation algorithm

A. Sharma (✉)
Graphic Era University, Dehradun, Uttarakhand 248001, India
e-mail: abhinav_sharma2008@yahoo.com

A. Sharma
Graphic Era Hill University, Dehradun, Uttarakhand 248001, India

© The Author(s), under exclusive license to Springer Nature Singapore Pte Ltd. 2021

R. M. Singari et al. (eds.), *Advances in Manufacturing and Industrial Engineering*, Lecture Notes in Mechanical Engineering,
https://doi.org/10.1007/978-981-15-8542-5_96

1 Introduction

India has suffered the damage of agricultural crops, transport and communication networks, infrastructure and also to human lives due to floods over the years. An area of more than 40 million in India has been identified as flood prone [1]. India is a land where a large area is cultivated by various river systems. Most of these rivers are seasonal and are prone to flood. The country experiences severe floods in perennial rivers of northern and middle India at the time of rainy especially the monsoon season (June–September) due to high intensity of rainfall in the catchment. Therefore, there is utmost need to develop real time flood forecasting system to prevent the severe loss due to the natural disaster of flood. There can be some conventional rainfall–runoff models developed to target the problem. But development and calibration of these require to optimize a number of physical parameters interacting in a complex way, and therefore, a large amount of historic data is needed. Instead, black box models are proven to produce good result from the input–output mapping when a detailed physical description of the process is not required. In recent years, ANN is implemented successfully for forecasting of flood due to its competency to map any nonlinear function of given sufficient complexity. ANNs are proven to produce improved performance over other black box models in numerous hydrological studies. The ANN models are advantageous as for development of these models the mathematical form of the information about the complex interaction among the physical variables is not required.

2 The Study Area and Data Collection

In this work, the Sutlej River basin up to Kasol had been used. The location of the basin for study is on the upriver of Bhakra reservoir. The total area of basin up to Kasol is 56,980 km^2. Figure 1 represents the location of the reservoir. For this work, the per day rainfall data for Namagia, Bhakra, Suni, Kalpa, Rackchham, Berthin, Rampur, Kahu, Kaza, and Kasol was gathered among the year 1987–2000.

3 Artificial Neural Networks

It is a collection of linked input output units where every link contains a weights. It assists to create models that predict the output from massive databases. This model is replica of the human brains. It assists to implement the tasks of pattern or image recognition, speech recognition, etc. Artificial neural networks can be understood as weighted graphs with directions that contains artificial neurons as nodes and edges with direction and weights as links among outputs of neurons and inputs of neurons and have been elaborated in various hydrologic papers [1–4].

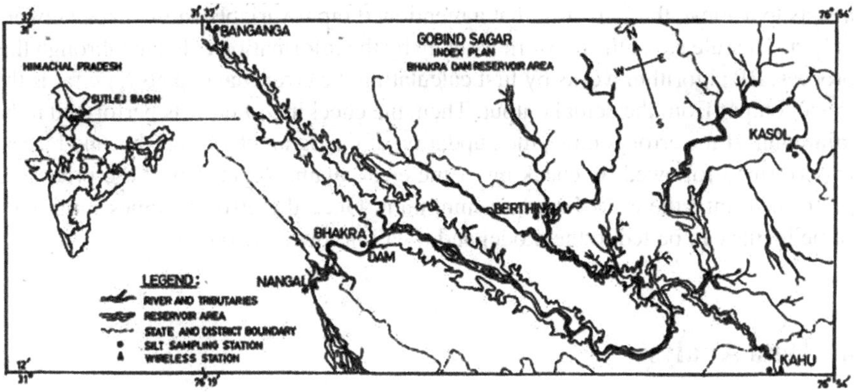

Fig. 1 Index map—Bhakra reservoir

For designing an ANN, initially, weights are initialized with some values. Whatever values of the weights are picked initially may not be correct, or might not fit the model the best. Output of the model may be different than the actual output, i.e., the error value can be significant. A typical three-layered feedforward ANN is shown in Fig. 2.

In the ANN technique, an algorithm is chosen such that it forces the model to alter the parameters (weights), to minimize the error.

It is a popularly practiced algorithm that is well proven for being accurate as it permits itself to improve by itself; therefore, it achieves high degree of accuracy [5]. It is a standardized technique of training ANN. This technique assists to compute the rate change parameter of a loss function w.r.t. all the network weights. Backpropagation is the main tool of neural network training.

In this algorithm, network is simplified by connections with weight that contain very less or no influence on the trained network. A collection of input and activation values are applied to discover the connection among the input and hidden layers. It

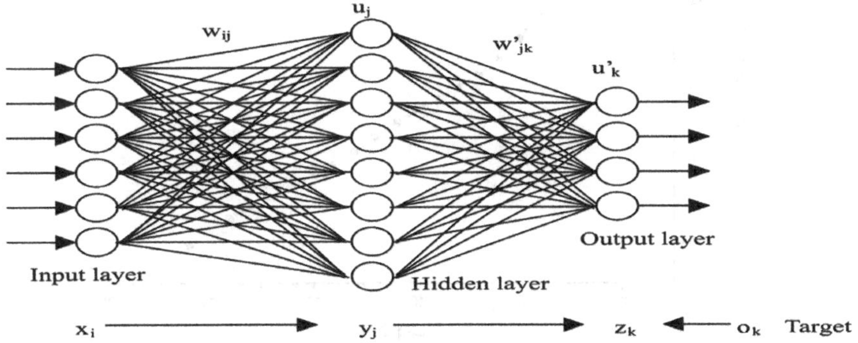

Fig. 2 ANN with three layers

assists to achieve the influence that a mentioned input variable causes on a network output. The rule set of the network is set up by the information collected through this process. BP algorithm works by first calculating the error that depicts how far is the model output from the actual output. Then, the checking of error is performed to be minimum. If the error is huge, then updation of the parameters (weights and biases) is performed followed by checking of the error again. Repetition of the process is performed until the error becomes minimum. Once the error becomes minimum, some inputs can be fed to the model and it will produce the output.

4 Data Analysis

For the development of model, daily rainfall data for Kahu, Kasol, Rampur, Suni, Berthin, and Bhakra was available for the period from 01-01-1987 to 31-12-2004. The daily discharge data for Kasol was also available for the same period. These rain gauge and river gauge stations are marked in Fig. 1. In this work, the ANN models have been developed to forecast the river flow at Kasol at 1–7 day in advance.

To set up the ANN model, the IVs are taken as rainfall and discharge values of stations which are to the upstream of the target station. By determining the rainfall lags and values of discharge that significantly affect the forecasted flow, the count of future rainfall and values of discharge are determined. These values that correspond to different lags can be determined accurately by performing statistical analysis of data sequence. The IV is elected commonly by hit-and-trial iteration method; however, a statistical procedure has been presented that avoids the hit-and-trial iteration process [6]. The ACF, PACF and CCF may be applied for the current application. It is evident from Fig. 3, which presents the ACF plot of river flow at Kasol, that it is

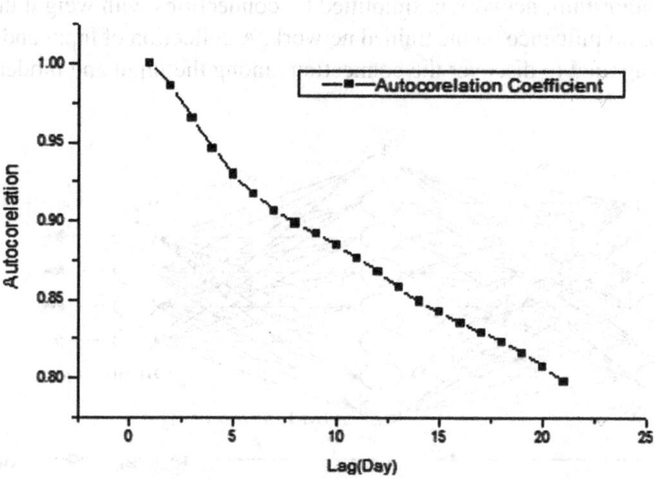

Fig. 3 Autocorrelation of the discharge series—Kasol

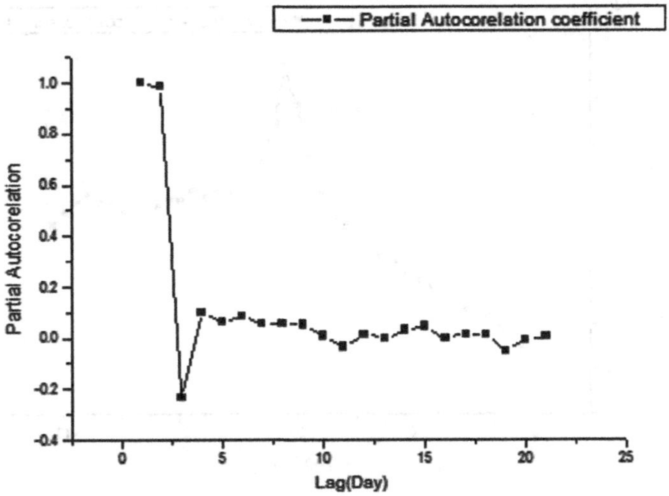

Fig. 4 Partial autocorrelation of the discharge series–Kasol

autoregressive.

Figure 4 showing the PACF of flow sequence of Kasol gives potential antecedent runoff values that influence the runoff value at the current period. It can be seen from Fig. 4 that the runoff series up to 2 lags must be the parts of in the IV. The PACF of the flow series at Kasol and CCF of flow series at Kasol between rainfall series at Rampur, Bhakra, Kahu, Berthin, Kasol and Suni indicate the IV to the ANN model. The PACF is always a helpful tool in the analysis of the large set of data.

PACF along with ACF and CCF is a tool used in the current research also to find which set of data series is most influencing.

Cross correlations of all other stations have also been taken into consideration to decide the influencing lags.

It has always helped researchers for making the decision of the influencing lags of the target class.

Figure 5 suggests the rainfall at 1 lag influences the runoff. In the same way, the influencing lags of runoff series at Rampur, Suni, Berthin, Kahu and Bhakra have been decided.

After observing Figs. 3, 4 and 5, the training or calibration IV chosen is as follows

$$Q_{\text{Kasol},t} = f(Q_{\text{Kasol},t-1}, Q_{\text{Kasol},t-2}, R_{\text{Rampur},t}, R_{\text{Berthin},t}, R_{\text{Bhakra},t}, R_{\text{Kahu},t}, R_{\text{Kasol},t}, R_{\text{Suni},t}) \quad (1)$$

in which Q and R are discharged and spatially averaged rainfall values, respectively.

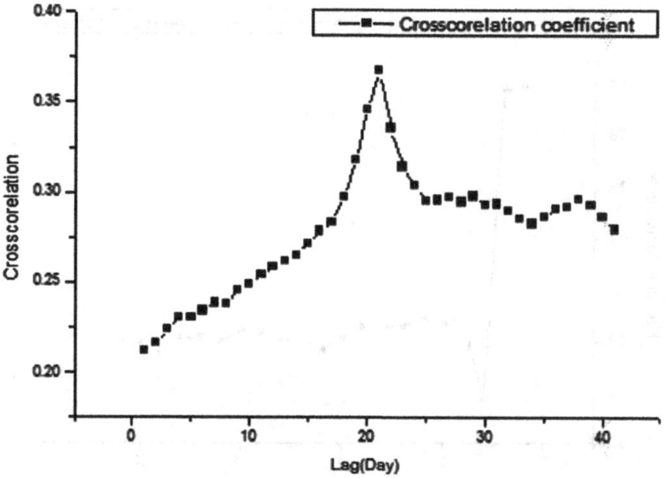

Fig. 5 Cross-correlation of rainfall—Kasol with discharge series—Kasol

4.1 Assessment of Performance

For assessment of performance of ANN, all of the data is bifurcated, one is used for the purpose of training (calibration) and other one for testing (validation). The following performance indices are used to evaluate the calibration and validation performance:

$$\text{Root Mean Square Error (RMSE)} = \sqrt{\frac{\sum_{k=1}^{K}(t-y)^2}{K}} \tag{2}$$

$$\text{Efficiency} = 1 - \frac{\sum(t-y)^2}{\sum(t-\bar{t})^2} \tag{3}$$

$$\text{Correlation Coefficient} = \frac{\sum TY}{\sqrt{\sum T^2 \sum Y^2}} \tag{4}$$

where K is the number of observations; t is the observed data; y is computed data; in which is the mean of the observed data, and in which is the mean of the computed data.

4.2 Model Set-Up

Backpropagation algorithm has been used for setting up the model. 1 day lead to 7 day lead is taken as the lead times for development of models. The dataset from the year 1987 to year 2000 is considered for calibration and that from year 2001 to year 2004 is used for validation of the model. MATLAB (The Mathworks, Inc., 2013) has been used as the software for training the models.

5 Results and Discussions

The performances of the forecasts have been summarized in the form of tables. Calibration results and validation results forecasted have been presented in the form of hydrographs for 1-day lead-time in Figs. 6 and 7, respectively, together with the observed flow series corresponding to it. The viability of the developed ANN model in forecasting the flow at Kasol is clearly depicted in the hydrographs. The results are analyzed using statistical indices too. Tables 1 and 2 present the results of calibration and validation for all the lead-times in terms of various statistical indices (Tables 3 and 4).

As the tables suggest that the highest coefficient of correlation for both calibration and validation has been achieved for Table 1 (1-day lead time), it is indicated that developed ANN model is remarkable in estimating the forecasts with reduced error (Tables 5, 6 and 7).

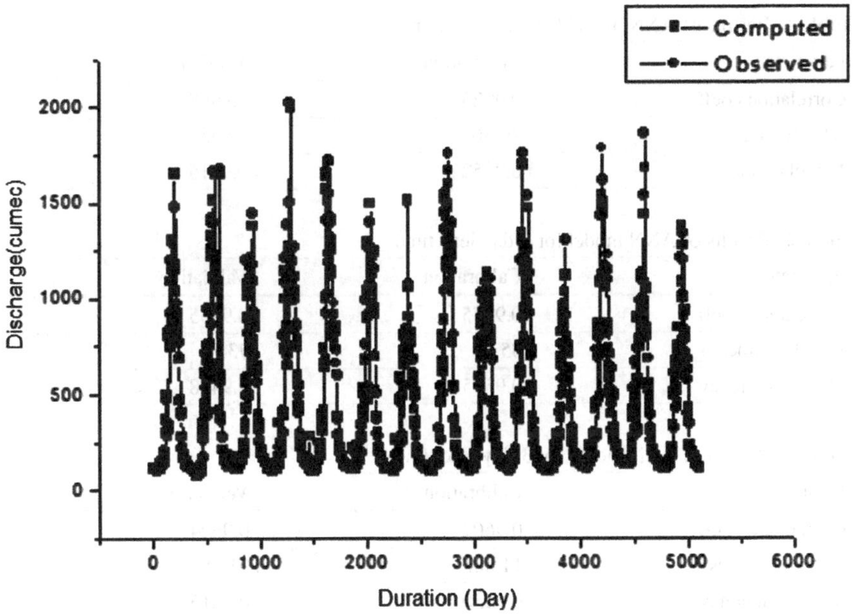

Fig. 6 Calibration result of data (1-day lead)

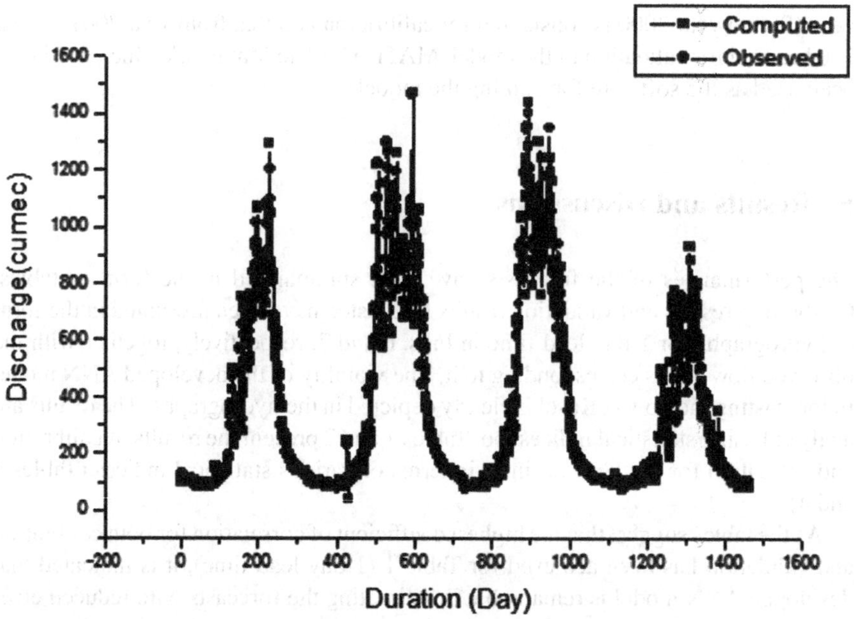

Fig. 7 Validation result of data (1-day lead)

Table 1 Results of ANN model for 1-day lead time

Parameter	Calibration	Validation
Correlation coeff	0.9925	0.9908
RMSE (cumecs)	49.56	44.55
Model efficiency	0.9852	0.9815

Table 2 Results of ANN model for 2-day lead time

Parameter	Calibration	Validation
Correlation coeff	0.9775	0.9755
RMSE (cumecs)	85.94	73.37
Model efficiency	0.9555	0.9498

Table 3 Results of ANN model for 3-day lead time

Parameter	Calibration	Validation
Correlation coeff	0.9607	0.9624
RMSE (cumecs)	113.06	91.78
Model efficiency	0.9230	0.9215

Table 4 Results of ANN model for 4-day lead time

Parameter	Calibration	Validation
Correlation coeff	0.9469	0.9506
RMSE (cumecs)	130.91	104.98
Model efficiency	0.8967	0.8974

Table 5 Results of ANN model for 5-day lead time

Parameter	Calibration	Validation
Correlation coeff	0.9364	0.9407
RMSE (cumecs)	142.98	114.67
Model efficiency	0.8768	0.8776

Table 6 Results of ANN model for 6-day lead time

Parameter	Calibration	Validation
Correlation coeff	0.9264	0.9316
RMSE (cumecs)	153.42	122.74
Model efficiency	0.8582	0.8598

Table 7 Results of ANN model for 7-day lead time

Parameter	Calibration	Validation
Correlation coeff	0.9194	0.9239
RMSE (cumecs)	160.17	128.87
Model efficiency	0.8454	0.8456

With the increment in time lead between one day to seven days, ANN model-performance deteriorates. The ANN model with lead time as 1 day can be used for the computation of flood forecast at Kasol.

The forecasted calibration and validation results have been presented in the form of hydrographs for 1-day lead time in Figs. 6 and 7, respectively, together with the observed flow series corresponding to it.

6 Conclusion

In this work, the development of an ANN model for flood forecasting for Satluj River at Kasol station was performed using the average daily rainfall data for Kahu, Kasol, Rampur, Suni, Berthin and Bhakra and daily discharge data at Kasol from 1987 to 2004. The IVs of the model were selected using the ACF, PACF and CCF of the sequence. The lead times considered in the model development for flood forecasting are 1–7 days lead. The results of the ANN models are analyzed using statistical indices as ACF, RMSE and percentage of error in peak flow estimation. It is concluded from the validation and calibration results that the ANN model for 1-day lead time can be used for issuing flood warnings at Kasol. There is deterioration in the performance with increment in lead time between 1 and 7 days.

References

1. Mitra P, Ray R, Chatterjee R, Basu R, Saha P (2016) Flood forecasting using internet of things and artificial neural networks. In: IEMCON, IEEE. https://doi.org/10.1109/IEMCON.2016.7746363
2. Senthil Kumar RA, Goyal MK, Ojha CSP, Singh RD, Swamee PK (2013) Application of artificial neural network, fuzzy logic and decision tree algorithms for modeling of streamflow at Kasol in India. Water Sci Technol 68(12):2521–2526
3. Elsafi SH (2014) Artificial neural networks (ANNs) for flood forecasting at Dongola Station in the River Nile, Sudan. AED 53(3):655–662
4. Vora K, Yagnik S. A survey on backpropagation algorithms for feedforward neural networks. IJEDR. ISSN: 2321-9939
5. See L, Dougherty M, Openshaw S (1997) Some initial experiments with neural network models of flood forecasting on the river Ouse. In: Proceedings of geocomputation '97 and SIRC '97, University of Otago, New Zealand, pp 15–22
6. Bruen M, Yang J (2005) Functional networks in real-time flood forecasting—a novel application. Adv Water Resour 28(9):899–909

Sewage Treatment Using Alum with Chitosan: A Comparative Study

Jaya Maitra, Athar Hussain, Mayank Tripathi, and Mridul Sharma

Abstract Nowadays, economic and ecologically sound treatment systems are highly encouraged to treat large-scale wastewater emanating from different sources. In the present study, an effort has been made using chitosan as a natural coagulant and alum as a chemical coagulant for the treatment of raw sewage. The coagulation process has been carried out to evaluate the optimum dose of coagulant under different pH values. Also, the alum and chitosan dosages in different mix proportions have been used to observe the turbidity, conductivity, salinity, TDS, COD, and TSS removal from raw sewage. The optimum pH value of 7.1 has been observed under different coagulants dosages. Also, at alum:chitosan ratio of 1:1 the maximum turbidity removal of 64% has been observed at turbidity concentration of 88 NTU. The synergistic effect is found to be more effective in COD. The results indicated that the synergistic effect shows maximum percent removal with a minimum dose of alum and chitosan, which would yield significant environmental and economic benefits.

Keywords Coagulation · Chitosan · Wastewater

J. Maitra
Department of Applied Chemistry, School of Applied Sciences, Gautam Buddha University, Greater Noida 201310, India

A. Hussain (✉) · M. Sharma
Civil Engineering Department, Ch. Brahm Prakash Government Engineering College, Jaffarpur, New Delhi 110073, India
e-mail: athar.hussain@gov.in; athariitr@gmail.com

M. Tripathi
Environmental Engineering Section, Civil Engineering Department, School of Engineering, Gautam Buddha University, Greater Noida, UP 201312, India

© The Author(s), under exclusive license to Springer Nature Singapore Pte Ltd. 2021
R. M. Singari et al. (eds.), *Advances in Manufacturing and Industrial Engineering*, Lecture Notes in Mechanical Engineering,
https://doi.org/10.1007/978-981-15-8542-5_97

1 Introduction

Coagulation and flocculation is a simple and rapid technique for the treatment of wastewater [1]. Coagulation is normally used for the treatment of water to be supplied for drinking usage after removing the turbidity and natural organic matter present in it [2-4]. In addition, this at the final stage of treatment, disinfection process is normally used in order to make water safe against microbiological species such as bacteria and viruses including traces of organic matter present in the distribution network [3].

The coagulation process efficiently increases the particles size in water as flocs which can be abolished by the sedimentation process very effectively. The process, however, produces high sludge yield consisting metal hydroxides formed during the coagulation and sedimentation process while treatment of water containing high amount of turbidity. Also, the usage of high aluminum dose in the process ultimately results in high aluminum concentration in treated water. It possibly raises the threat of Alzheimer's disease in human beings [5]. Due to economical point of view, the salts of aluminum and iron are most commonly used coagulants around the globe. However, in recent advancements and developments, new products such as synthetic polymers have gained a lot of popularity [6-8]. Use of natural coagulants may save chemicals and reduce the cost of sludge treatment [9, 10]. Chitosan is a natural, eco-friendly, non-hazardous, linear biopolyaminosaccharide prepared in the laboratory by the process of alkaline deacetylation of chitin and very effective as a coagulant. It is a natural eco-friendly linear cationic polymer that can be used as a coagulant. It can be effectively used for the removal of colloidal particles, COD, metal ions, cryptosporidium cysts, hemic acids, and another bio-microsphere from wastewater. The properties of chitosan are alike as synthetic polymers but are comparatively quite less toxic to humans and aquatic species as compared to polymers. However, the presence of hydroxyl and amino functional groups significantly increases its adsorption capacity toward numerous pollutants [11]. The important property of chitosan is its non-solubility in water under ordinary circumstances; however, it can be easily dissolved in carboxylic acid solutions including acetic acid. However, the use of hydrochloric acid as an alternative proves to be effective to assess the coagulation capability of HCl-prepared chitosan. The chitosan/AA coagulant proves to be quite effective the ratio of TOC/AA after coagulation considerably increases [12].

The studies carried out by various researchers elucidated that the use of aluminum and iron salts proves to be ineffective coagulants in removal of nano-sized particles from the wastewater. Furthermore, the overdose of such chemical coagulants produces toxic chemical sludge with elevated chemical costs, which further makes its disposal in the category of scheduled waste. This high dose of chemical coagulants ultimately results in bulky usages of such chemicals with high cost of treatment. This also makes the system less efficient and high increment in toxicity in the treated wastewater [13, 14] in a study have used chitosan effectively and efficiently for the removal of turbidity from water. The maximum turbidity removal of greater than

90% in the raw water has been reported which was found to be lower than the target study.

A study carried out by [15] reported the use of chitosan as a natural coagulant for the treatment the turbid river water. The study results showed that at turbidly concentrations below 1000 NTU the turbidity removal efficiency was quite high. However, at turbidity concentration greater than 1000 NTU, the coagulation process failed completely. Therefore, considering the above-mentioned facts, in view the present study has been carried out with the objective to find the combined effect so alum and chitosan for the treatment of raw sewage.

2 Materials and Methods

Chitosan preparation was carried out by dissolving chitosan powder in 1% chloride acid solution to make 1% (w/w) stock solution by stirring at 150 rpm during overnight. The wastewater samples with the required turbidity were prepared by mixing known amounts of clay or bentonite with distilled water. Sodium bicarbonate ($NaHCO_3$) was added to produce the alkalinity as calcium carbonate ($CaCO_3$). Jar tests were performed using Phipps and Byrd programmable jar test apparatus. Before and after treatment, the samples were measured for pH, turbidity, conductivity, salinity, total dissolved solids (TDS), chemical oxygen demand (COD), and total suspended solids (TSS) [16, 17]. All the experiments were carried out as per the standard methods of [18] manual.

3 Results and Discussion

The optimum pH values and dose of alum and chitosan was being determined using jar test apparatus. The determined values at different alum and chitosan concentrations are summarized in Table 1. The reduction in pH value was seen on increasing alum dose from 1 to 20 mg/L. The optimum pH value of 7.08 was observed at a concentration of 20 mg/L. The study can be well compared with the study being carried out by [17]. Similarly, the chitosan concentration was being varied from 1 to 25 mg/L. The optimum pH value of 7.09 has been obtained at chitosan concentration of 20 mg/L as shown in Table 1. Also, the alum:chitosan ratio was varied in order of 1:1, 1:10, and 10:1, respectively.

The optimum pH value of 7.09 was being obtained at alum: chitosan ratio of 10:1 as shown through Fig. 1. Alum:chitosan ratio of 1:1 the highest pH value of 7.61 has been obtained. Figure 1 shows that there was considerable variation in pH value as the ratio of alum and chitosan was increased. Therefore, it is to infer that the pH value of around 7.1 is found to be optimum at different alum and Chitosan ratio of 10:1 as compared to individual concentration of 20 mg/L of each. The dose of alum and chitosan has been varied under different turbidity concentrations. The turbidity

Table 1 pH values of different samples after dissolving of coagulants

Initial pH—6.98						
S. No.	Alum (mg/L)	pH	Chitosan (mg/L)	pH	Alum and chitosan mix proportions	
					Alum:Chitosan (mg/L)	pH
1	1	7.78	1	7.15	1:1	7.28
2	5	7.30	5	7.21	1:10	7.41
3	10	7.21	10	7.20	10:1	7.09
4	15	7.24	15	7.18	10:10	7.39
5	20	6.99	20	7.09	1:1	7.61
6	25	7.08	25	7.17	5:5	7.37

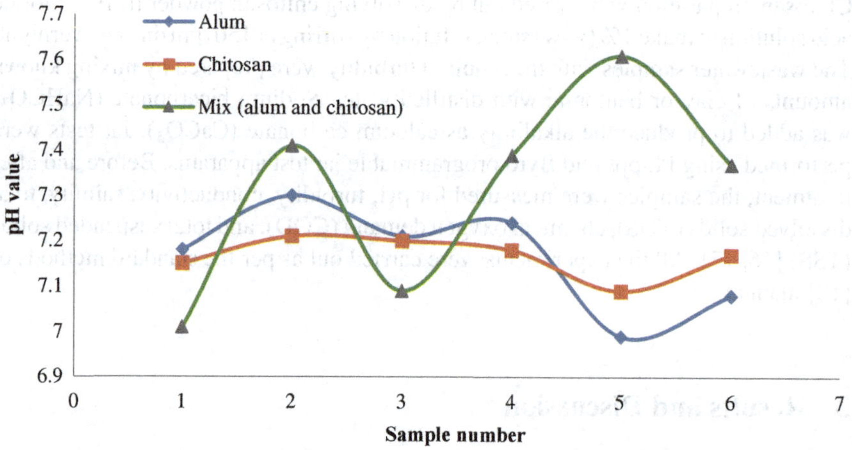

Fig. 1 Determination of optimum pH for different coagulants dosages

removal percent under different alum and chitosan concentrations has been obtained and is summarized in Table 1. Perusal of the data as summarized in Table 1 indicates the maximum turbidity removal of 83% at turbidity concentration of 38 NTU under alum concentration of 15 mg/L.

In case of chitosan of 10 mg/L concentration, the maximum turbidity removal of 78% has been obtained at turbidity concentration of 53 NTU. The lowest turbidity removal of 40% has been observed at chitosan concentration of 1 mg/L and turbidity concentration of 144 mg/L. Also, at alum:chitosan ratio of 1:1, the maximum turbidity removal of 64% has been observed at turbidity concentration of 88 NTU. The reduced dosages of alum and chitosan were found to be more effective in removing turbidity, depicting a synergism between the two coagulants. Turbidity results obtained in the present study are comparable with the results obtained by [18] for treating the raw industrial wastewater and sludge with high organic load. The same is depicted

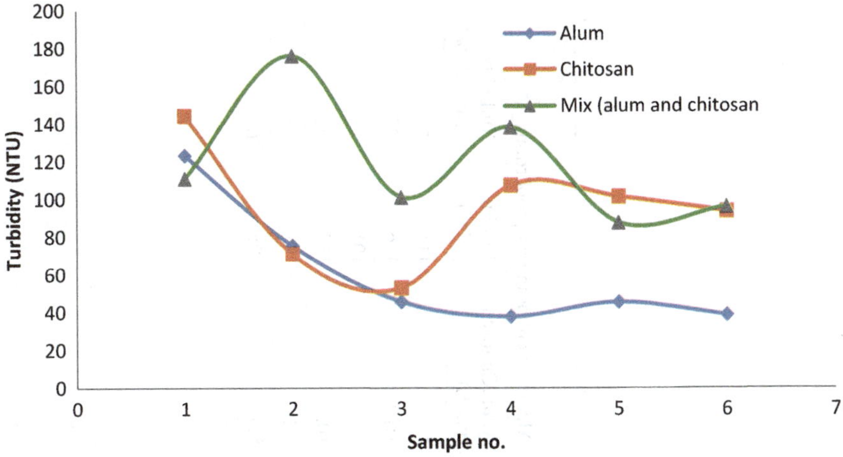

Fig. 2 Maximum turbidity removal for different coagulants dosages

through Fig. 2 graphically showing maximum percentage removal at alum:chitosan of 1:1 mg/L dosage (Table 2).

The initial electrical conductivity of the sewage sample was high indicating high dissolved solids concentration. The variation in conductivity at various coagulants dosages is summarized in Table 3. The addition of alum as a coagulant led to maximum decrease in conductivity value of 331 μS/cm at alum dose of 20 mg/L while maximum conductivity of 345 μS/cm has been obtained at an alum dose of 1 mg/L. Similarly, for chitosan dose of 10, the minimum conductivity of 338 μS/cm was seen and maximum conductivity of 349 μS/cm has been observed at chitosan dose of 15 mg/L. The individual consequence of alum and chitosan in decreasing conductivity was not more pronounced. When alum and chitosan have been used in combination, the results have been found to be promising. At alum; chitosan ration of 1:10, the minimum conductivity of 314 μS/cm was obtained. In the present study, the results for conductivity indicate that the synergistic effect is more effective. Results obtained can be comparable to the study conducted by [19] on raw industrial sewage. The data indicates decrease in conductivity at a alum and chitosan dose in ratio of 1:10 mg/L. High concentrations of salts affect the microbial degradation processes in the wastewater treatment systems.

Results from Table 4 indicates that the alum at 20 mg/L dose remove maximum salinity of 5.40% and a minimum salinity removal of 2.54% at alum dose of 5 mg/L. Similarly, for chitosan, maximum salinity removal of 5.16% has been observed at a chitosan dosage of 20 mg/L which is same as in case of alum. Combination of alum and chitosan yielded the best results compared to individual alum and chitosan dosage. The maximum salinity removal 15.96% has been obtained alum:chitosan ratio of at 10:10.

The obtained results can be well compared with the study being conducted by [20] on raw industrial wastewater. Their synergistic effect of coagulants dose is

Table 2 Turbidity values of different samples after dissolving of coagulants

Initial turbidity—240 NTU									
S. No.	Alum (mg/L)	Turbidity (NTU)	Turbidity removal (%)	Chitosan (mg/L)	Turbidity (NTU)	Turbidity removal (%)	Alum and chitosan mix proportions		
							Alum:Chitosan (mg/L)	Turbidity (NTU)	Turbidity removal (%)
1	1	123	49	1	144	40	1:1	111	54
2	5	75	69	5	71	70	1:10	176	27
3	10	45	81	10	53	78	10:1	101	58
4	15	38	84	15	107	55	10:10	138	43
5	20	45	81	20	101	58	1:1	88	64
6	25	39	83	25	93	61	5:5	96	60

Table 3 Conductivity values of different samples after dissolving of coagulants

Initial conductivity—343 μS/cm						
S. No.	Alum (mg/L)	Conductivity (μS/cm)	Chitosan (mg/L)	Conductivity (μS/cm)	Alum and chitosan in mix proportions	
					Alum:Chitosan (mg/L)	Conductivity (μS/cm)
1	1	345	1	343	1:1	334
2	5	336	5	342	1:10	314
3	10	338	10	338	1:1	322
4	15	343	15	349	10:10	323
5	20	331	20	340	1:1	335
6	25	333	25	347	5:5	315

Table 4 Salinity values of different samples after dissolving of coagulants

Initial salinity—168.5 mg/L									
S. No.	Alum (mg/L)	Salinity (mg/L)	Salinity removal (%)	Chitosan (mg/L)	Salinity (mg/L)	Salinity removal (%)	Alum and chitosan in mix proportions		
							Alum:Chitosan (mg/L)	Salinity (mg/L)	Salinity removal (%)
1	1	161	4.33	1	163	3.20	1:1	159	5.40
2	5	162	2.54	5	160	4.92	1:10	151	10.26
3	10	161	4.21	10	162	3.85	10:1	151	10.14
4	15	162	3.73	15	163	2.04	10:10	141	15.96
5	20	159	5.40	20	160	5.16	1:1	154	8.24
6	25	160	5.34	25	162	3.67	5:5	147	12.58

depicted through Fig. 2 indicating the curve trending downward. The maximum TDS removal of 9.91% was observed at alum dose 5 mg/L and minimum TDS removal of 2.83% has been obtained at alum dose of 1 mg/L. In case of chitosan, the maximum TDS removal of 7.02% has been observed at chitosan dose of 20 mg/L and minimum TDS removal of 2.83% has been obtained at chitosan dose of 25 mg/L. Total dissolved solids (TDS) comprise inorganic salts (mainly calcium, magnesium, potassium, sodium, bicarbonates, chlorides, and sulfates) and some small amounts of organic matter that are dissolved in water [21].

Table 5 presents a reduction in TDS at different coagulant dosages. Also, the maximum TDS removal of 12% has been obtained at mix alum:chitosan dose ratio of 5:5 mg/L. In comparison, it can be understood that the combined dose was more pronounced giving maximum TDS removal of 12%. The obtained results can be well compared with the study conducted on raw industrial sewage. Notably, the coagulant dose used in combination was less in comparison with individual dosages thus indicating a lower risk of inorganic release in the system. It is evident from the present results that the synergistic effects of these coagulants are more effective in removing TDS.

The coagulants dosages have been varied and COD removal values being observed are summarized in Table 6. In case of alum, the maximum COD removal of 46% has been observed at alum dose of 5 mg/L and the minimum COD removal of 18% has been found at alum dose of 5 mg/L. While in case of chitosan, the maximum COD removal of 48.5% has been obtained at a chitosan dose of 20 mg/L and minimum COD removal of 41.3% a chitosan dose of 15 mg/L dosage.

However, the COD removal of 83.2% has been observed at alum:chitosan ratio of at 5:5 mg/L. The synergistic effect is found to be more effective in COD. The results of present study can be well compared to study being conducted by [22] on raw industrial wastewater. The dose of alum and chitosan is less when used in combination as compared with the individual dosage of alum, and chitosan used in the present study indicating the optimum and effective usage of both the coagulants. As shown in Fig. 3, the mix proportions of alum and chitosan result in better COD percentage removal producing treated water of better quality.

The effect of TSS removal using alum and chitosan as coagulants has also been studied and the obtained results are shown in Table 7. In case of alum, the maximum TSS removal of 79% has been observed at alum dose of 15 mg/L and minimum TSS removal of 15% at alum dose of 10 mg/L. In case of chitosan, the maximum TSS removal of 85% has been attained at a chitosan dose of 15 mg/L while the minimum TSS removal of 25% at chitosan dose of 1 mg/L. However, the maximum TSS removal efficiency of 70% has been observed at alum and chitosan dose ratio of 10:1 mg/L. The combined dose of alum and chitosan was less than their individual dose proving that the synergistic effect is found to be more effective in eliminating TSS from wastewater. Bazrafshan et al. [23] carried out similar study using the industrial sewage and obtained the similar results indicating the present study can be well compared. The same can be shown through Fig. 3 indicating maximum TSS percentage removal using coagulants in combination.

Table 5 TDS values of different samples after dissolving of coagulants

Initial TDS—212 mg/L

S. No.	Alum (mg/L)	TDS (mg/L)	TDS removal (%)	Chitosan (mg/L)	TDS (mg/L)	TDS removal (%)	Mix (alum and chitosan)		
							Alum:Chitosan (mg/L)	TDS (mg/L)	TDS removal (%)
1	1	206	2.83	1	203	4.24	1:1	191	9.76
2	5	200	9.91	5	204	3.77	1:10	195	7.87
3	10	198	6.41	10	201	5.18	10:1	187	11.6
4	15	198	6.60	15	203	4.24	10:10	190	10.0
5	20	199	6.13	20	197	7.02	1:1	195	8.01
6	25	200	5.84	25	206	2.83	5:5	186	11.98

Table 6 COD values of different samples after dissolving of coagulants

Initial COD—537 mg/L

S. No.	Alum (mg/L)	COD (mg/L)	COD removal (%)	Chitosan (mg/L)	COD (mg/L)	COD removal (%)	Alum and chitosan in mix proportions		
							Alum:Chitosan (mg/L)	COD (mg/L)	COD removal (%)
1	1	289	46.11	1	293	45.44	1:1	281	47.67
2	5	289	46.23	5	286	46.71	1:10	123	77.02
3	10	365	32.06	10	298	44.45	10:1	111	79.24
4	15	440	18.10	15	315	41.31	10:10	203	62.15
5	20	313	41.74	20	276	48.53	1:1	135	74.87
6	25	382	28.80	25	303	43.49	5:5	90	83.26

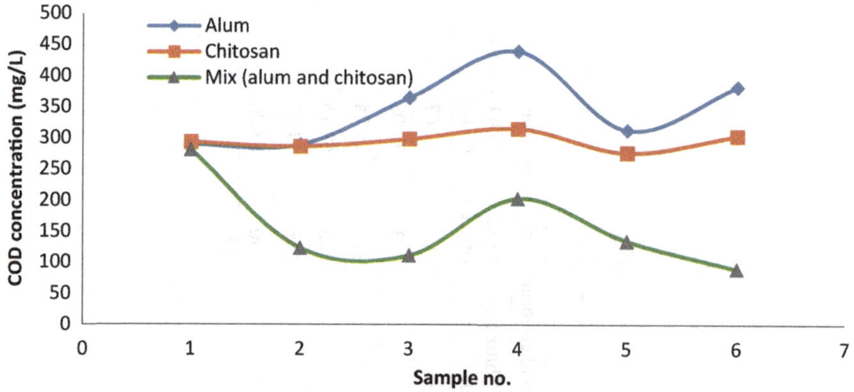

Fig. 3 Maximum COD removal at different coagulants dosages

It has been observed that the combined usage of these coagulants is quite effective in terms of pollutants removal and as well as minimum quantity of these coagulants can be best utilized at maximum efficiency. Therefore, from the present study, it can be inferred that the same can be implemented on real scale systems for providing maximum pollutants removal, and secondly cost on coagulants usage can be reduced considerably as the results pertaining to synergistic effects of these coagulants are found to be very compromising.

4 Conclusion

In the present study, it has been observed that alum and chitosan as coagulants can be effectively used to treat raw sewage. However, using alum and chitosan as coagulant can effectively remove 78–84% of turbidity, 5% of salinity, 6–7% of total dissolved solids, 46–48% of chemical oxygen demand and 78–85% of total suspended solids from raw sewage. The synergistic effect of alum and chitosan has also been determined in the present study. Results from the present study show that it can be assessed that 58% of turbidity, 16% of salinity, 12% of TDS, 83% of COD and 70% of TSS can be removed effectively. Therefore, the present study confirms that alum and chitosan have a pronounced effect in coagulation process and the raw sewage can be effectively treated at low cost.

Table 7 TSS values of different samples after dissolving of coagulants

Initial TSS—83 mg/L									
S.No.	Alum (mg/L)	TSS (mg/L)	TSS removal (%)	Chitosan conc. (mg/L)	TSS (mg/L)	TSS removal (%)	Alum and chitosan in mix proportions		
							Alum:Chitosan (mg/L)	TSS (mg/L)	TSS removal (%)
1	1	55	34	1	62	25	1:1	65	21
2	5	66	20	5	23	72	1:10	43	48
3	10	71	15	10	30	63	10:1	25	70
4	15	18	78	15	12	85	1:10	27	68
5	20	62	25	20	34	59	1:1	40	51
6	25	63	25	25	28	67	5:5	30	64

Acknowledgements The research was done under the premises of Gautam Buddha University, Greater Noida, U.P., India. The authors want to acknowledge everyone who supported his project directly or indirectly.

References

1. Prakash NB, Sockan V, Jayakaran P (2014) Waste water treatment by coagulation and flocculation. Int J Eng Sci Innov Technol (IJESIT) 3(2):479–484
2. Volk C, Bell K, Ibrahim E, Verges D, Amy G, LeChevallier M (2000) Impact of enhanced and optimized coagulation on removal of organic matter and its biodegradable fraction in drinking water. Water Res 34(12):3247–3257
3. Rizzo L, Belgiorno V, Gallo M, Meric S (2005) Removal of THM precursors from a high-alkaline surface water by enhanced coagulation and behaviour of THMFP toxicity on *D. magna*. Desalination 176(1–3):177–188
4. Uyak V, Toroz I (2007) Disinfection by-product precursors reduction by various coagulation techniques in Istanbul water supplies. J Hazard Mater 141(1):320–328
5. Rondeau V, Commenges D, Jacqmin-Gadda H, Dartigues JF (2000) Relation between aluminum concentrations in drinking water and Alzheimer's disease: an 8-year follow-up study. Am J Epidemiol 152(1):59–66
6. Gao B, Yue Q, Miao J (2003) Evaluation of polyaluminium ferric chloride (PAFC) as a composite coagulant for water and wastewater treatment. Water Sci Technol 47(1):127–132
7. Shi Y, Fan M, Brown RC, Sung S, Van Leeuwen JH (2004) Comparison of corrosivity of polymeric sulfate ferric and ferric chloride as coagulants in water treatment. Chem Eng Process 43(8):955–964
8. Rizzo L, Belgiorno V, Meriç S (2005) Organic THMs precursors removal from surface water with low TOC and high alkalinity by enhanced coagulation. Water Sci Technol Water Supply 4(5–6):103–111
9. Brostow W, Lobland HH, Pal S, Singh RP (2009) Polymeric flocculants for wastewater and industrial effluent treatment. J Mater Educ 31(3–4):157–166
10. Šćiban MB, Klašnja MT, Stojimirović JL (2005) Investigation of coagulation activity of natural coagulants from seeds of different leguminose species. Acta Periodica Technol 36:81–90
11. Hu CY, Lo SL, Chang CL, Chen FL, Wu YD, Ma JL (2013) Treatment of highly turbid water using chitosan and aluminum salts. Sep Purif Technol 104:322–326
12. Huang C, Chen S, Pan JR (2000) Optimal condition for modification of chitosan: a biopolymer for coagulation of colloidal particles. Water Res 34(3):1057–1062
13. Yatish E, Gopinath R, Kumar A (2016) Feasibility analysis for chitosan as an effective coagulant in public health engineering. Int J Adv Res Eng Appl Sci 2:12–21
14. Divakaran R, Pillai VS (2002) Flocculation of river silt using chitosan. Water Res 36(9):2414–2418
15. Sekine M, Takeshita A, Oda N, Ukita M, Imai T, Higuchi T (2006) On-site treatment of turbid river water using chitosan, a natural organic polymer coagulant. Water Sci Technol 53(2):155–161
16. American Society of Testing Materials (1995) Standard practice for coagulation/Flocculation jar test of water EI-1994 R.D 2035-80. Annual book of ASTM Standards 11(2)
17. Kawamura S (2000) Integrated design and operation of water treatment facilities. Wiley, Hoboken
18. APHA (2012) Standard methods for the examination of water and wastewater, 22 edn. American Public Health Association, American Water Works Association, Water Environment Federation, Washington, DC

19. Patel H, Vashi RT (2013) Comparison of naturally prepared coagulants for removal of COD and color from textile wastewater. Global NEST J 15(4):522–528
20. Carmen Z, Daniela S (2012) Textile organic dyes—characteristics, polluting effects and separation/elimination procedures from industrial effluents—a critical overview. In: Organic pollutants ten years after the Stockholm convention—environmental and analytical update. InTech
21. Lakherwal D (2014) Adsorption of heavy metals: a review. Int J Environ Res Dev 4(1):41–48
22. Asif MB, Majeed N, Iftekhar S, Habib R, Fida S, Tabraiz S (2016) Chemically enhanced primary treatment of textile effluent using alum sludge and chitosan. Desalin Water Treat 57(16):7280–7286
23. Bazrafshan E, Alipour MR, Mahvi AH (2016) Textile wastewater treatment by application of combined chemical coagulation, electrocoagulation, and adsorption processes. Desalin Water Treat 57(20):9203–9215

Automatic Plastic Sorting Machine Using Audio Wave Signal

S. M. Devendra Kumar, S. Prashanth, and Rani Medhashree

Abstract In this modern society due to rapid urbanization and developing technologies, the detection and segregation of plastic play a major role in human life. Currently, this task of segregating the plastic is manual and that requires higher effort in hazardous condition and it is dangerous too in some places (Gopinath et al. in International conference on technological advancements in power and energy, 2017) [1]. Due to the high rate of plastic consumption, it is important to develop a machine that can separate the plastic from normal waste material in regions wherein manual sorting is difficult to be followed and the segregation process should be efficient compared to the primitive techniques (Hussain et al. in 6th International colloquium on signal processing & its applications (CSPA), 2010) [2]. Currently, there are different types of methods implemented to segregate plastic such as manual sorting, post grinding waste sorting, optical waste sorting and floating waste sorting. In this project, a mechanical device is developed which contains a sensing unit that segregates the plastic based on audio wave signals that are produced during the crushing operation of the waste materials. The sensing unit will be pretrained to detect only plastic out of different element such as wood, steel, metal and plastic using machine learning technique. Mel frequency cepstral coefficients are used for plastic segregation, ultimately the proposed design of plastic segregation.

Keywords Machine learning · Mel frequency cepstral coefficient · Feature extraction · Audio feature extraction · Sensor unit servomotor · Actuation unit

S. M. Devendra Kumar (✉) · S. Prashanth
Govt. SKSJ Technological Institute, Bangalore 560001, India
e-mail: smdevendrakumar@gmail.com

R. Medhashree
Delhi Technological University, New Delhi, Delhi, India

© The Author(s), under exclusive license to Springer Nature Singapore Pte Ltd. 2021
R. M. Singari et al. (eds.), *Advances in Manufacturing and Industrial Engineering*, Lecture Notes in Mechanical Engineering,
https://doi.org/10.1007/978-981-15-8542-5_98

1 Introduction

Nowadays, one of the most used material is plastic due to its easy availability and flexibility but since they are non-biodegradable, they have to be segregated in order to recycle and reuse them efficiently [3]. According to the Indian reports obtained from the Central Pollution Control Board (CPCB) in the year 2017–18, it was mentioned that nearly 6 lakh tonnes of plastic were generated. In general, there are around 25 thousand tonnes of plastic are generating in a day, on a global scale in the year 2015 the total plastic generated was 381 million tonnes. China is the highest contributor for plastic waste followed by Philippines and Vietnam. Hence, proper methods should be followed in order to segregate these plastics from biodegradable wastes. Manual and automatic segregation are the two approaches that can be used [4]. Manual sorting of plastic requires worker who individually identifies it and it is very time consuming and difficult in most of hazardous location. Automatic sorting machine automatically detects the plastic using some feature and separates the plastic automatically without any hectic process. In the modern days, there are two different methods used for the detection of plastic: One is by using image processing technique and the other by using Fourier transform infrared rays method. In the image processing technique based on the polymer color, the classification is carried on but this method fails for black polymer materials [5].

1.1 Plastic Detection Using Fourier Transform

In this method, hyper spectral imaging (HSI) sensing device is used for detection. Using these HIS images, the spectrum of each pixel is obtained which helps in the detection process [6]. Sorting is used for identifying different kind of polymer at different levels. Polymer recognition takes place by near infrared detectors and using color in case of polyethylene terephthalate (PET) containers. Density and thickness of various polymers are measured, and the spectral images of different material are obtained. Based on the wavelength obtained from these spectral images, the material is classified as plastic if its wavelength ranges between 780 and 2500 mm. For accurate measurement, it should vary between 1000 and 1700 mm.

1.2 Image Processing Technique

This is another field of research for the object detection and sorting process where in the object image is captured using webcam. Along with the detection, a mechanical device is developed to detect the plastic using computer vision. The RGB images obtained from the webcam are applied with contours which give the information of size and shape that can be used for the plastic separation. The process is carried in

Table 1 Percentage of plastic production in major countries

Countries	Plastic production (%)
China	28
Indonesia	10
Philippines	6
Vietnam	6

two steps: One is extraction of feature and the other is classification based on these extracted features [7]. The features may be size, shape and color based on which the plastic is separated. In general, human eyes are sensitive to different shape and size so these features can be extracted very easily using the image processing technique by applying the contours and filtering of thc image [8]. The efficiency of this technique is quite acceptable (Table 1).

2 Methodology

The sound which is produced by the materials during the crushing and grinding process is unique in nature. So generally, when the waste materials are crushed, they produce some sort of audio waves which can be used for sorting operation. The audio signals that are produced by crushing a soft plastic or trapping sound wave generated by hard plastic are completely different from one another and also with respect to other materials crushing sound such as glass, wood and paper [9]. These audio signals are captured using sensor which senses the acoustic waves in a real-time fashion, and these sensed data will be sent to the machine learning algorithm which are developed for differentiating plastic audio wave and normal audio waves. The ML algorithm will be trained based on some unique characteristic feature of the plastic waste. Hence, the audio signal classification can be done using audio feature extraction. Based on this acoustic feature, the process of classification whether object is plastic or non-plastic can be done. All this function works inside a sealed mechanical device which consists of pulley which is used to open the container for plastic separation-based real-time decision made by the combination of audio signal obtained at that moment and the machine learning algorithm.

2.1 Extraction of Audio Information

The Mel frequency cepstral coefficient (MFCC) features are obtained for the analysis of the audio signal which are basically a representation of short-term power [10]. This MFCC gives the information of overall shape of the spectral envelope, and it is based on the inverse cosine function and logarithmic power spectrum of nonlinear frequency [11]. The audio signals which are obtained are trimmed to a short period of 18–22 ms segment. These segments are passed through hamming window to minimize the signal distortion, and as a result, the signal will be having slight discontinuity. As the analysis is easier and efficient in spatial domain, the signal is applied to fast Fourier transform later when the magnitude response is obtained which gives the information of power at different region of frequency [12]. Equation (1) is used to get the Mel filter coefficients, and Mel filter is used to increase the quality of information obtained.

$$m = 259 * \log *10(1 + F/100) \quad (1)$$

Later, the logarithmic function is applied for each of the Mel frequency obtained followed by discrete cosine function applied to get the cepstral coefficient. These coefficients are used for training the machine learning model which are different when compared to plastic and other materials. The MFCC values are the amplitude of the resultant spectrum. The artificial neural network (ANN) algorithm is used for classification [13]. Initially, the coefficient data are fed to the input layer of the neural network [14]. Between the input layer and output layer, there are hidden layers which help in classification. Once the input is fed to the primary input layer, it computes and forwards it to the next layer, and this process continues until the output pattern is obtained at the output layer. If there is any error, then in the hidden layer the weights get updated till the process completes. As the number of hidden layer increases, training period increases and the efficiency also increases (Fig. 1).

2.2 Mechanical System Design

Materials required:

1. Conveyor belt
2. Audio sensor
3. Servomotor
4. Freewheeling lid
5. Rectangular sealed box
6. Crusher blades
7. Bins (one to collect plastic waste and other for collecting normal waste).

For proper working, a perfect mechanical device must be present. Conveyor belt, freewheeling roller, segregator and two separate bins for plastic and non-plastic material are used for the prototype designing. The system should be designed such that there is less noise from its own system as the detection is carried through audio signal information. Hence, the number of motors used inside the system should be less (Fig. 2).

The conveyor should be synchronized because for larger object, and it should move slowly as the crushing operation takes more time; whereas, for smaller object it should move fast based on the weight the conveyor is designed to adjust the speed

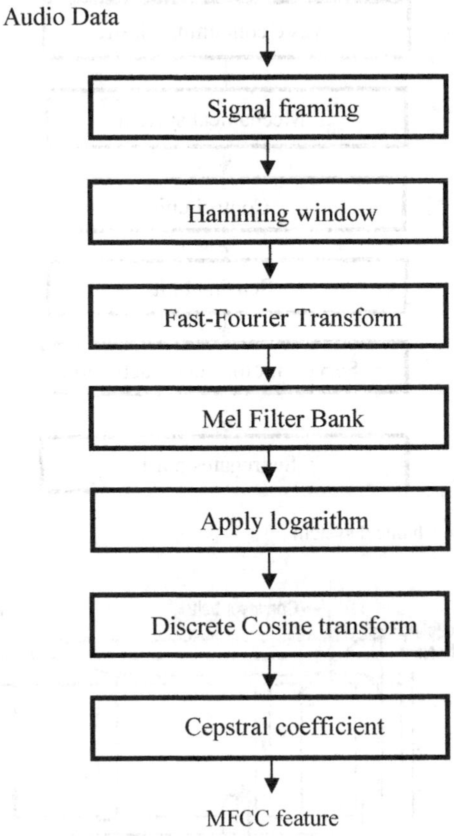

Fig. 1 MFCC feature extraction algorithm

of the motion. The audio signal which is generated is recorded using universal serial bus (USB) microphone and that signal is preprocessed before giving to the pre-trained machine learning algorithm after which the object is segregated to plastic or non-plastic waste. In the preprocessing stage, the noises in the audio signal are removed and a selected portion of the signal is cut and boosted. The preprocessed signal is next compared with the pre-trained machine learning algorithm for the final decision for classification. Till this process, the conveyor should pause moving. All this process takes less time to be executed and will be working on real-time scenario (Fig. 3).

The mechanical system has a conveyor belt of 0.3 m dimension along with free-wheeling roller for the motion. By the combination of control circuit and motor, the speed can be varied accordingly, the waste material is moved through the conveyor belt, the conveyor base collects the waste garbage, next the crushing process takes place which produces the audio signal, and these audio signals are recorded using the USB recorder and fed to the computer which contains the machine learning

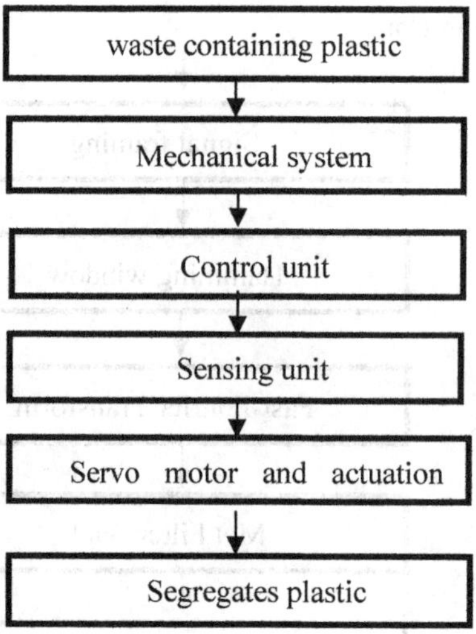

Fig. 2 Block view of mechanical system

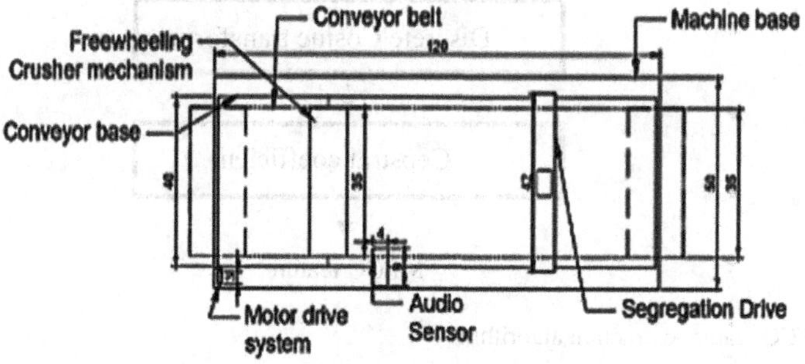

Fig. 3 Mechanical system top view

algorithm for the process of segregation. After the evaluation process, the crushed material is sent to plastic bin or normal bin based on plastic and non-plastic material, respectively.

The machine learning algorithm is trained initially using audio set of crushing different plastic waste and normal waste. The supervised learning process is carried where the dataset which is used for training the model is labeled dataset. These data are obtained during the crushing of different materials in real-time environment, later different predefined model is implemented to get best accuracy, and one of them is convolution neural network (CNN) algorithm. This algorithm is trained for different dataset, and the results are checked for testing audio set which obtained a result of 85% accuracy and 9% loss percentage (Fig. 4).

The sensing unit contains a USB microphone that records the audio of crushing materials when the object passes through the freewheel roller on a conveyor belt.

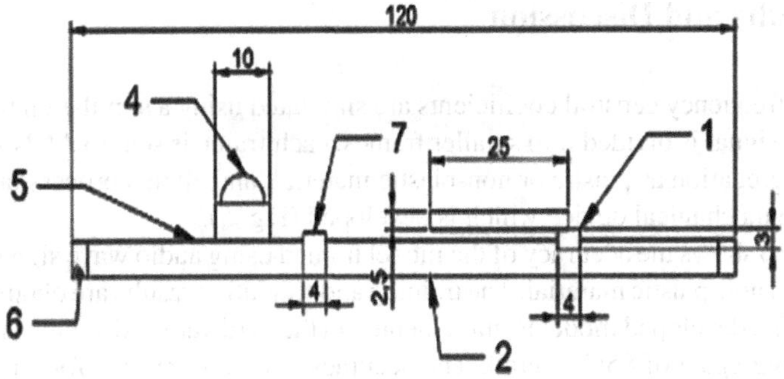

Fig. 4 Side view of the mechanical system

The motion of the conveyor belt varies in accordance with the weight of the waste materials. Controller unit contains a i5 processor with 8 GB RAM which contains the machine learning code for the separation of the waste garbage material. The actuation unit contains a 13 kg/cm motor for the mechanism and 17 cm lever. The system is maintained in a closed loop fashion in order to perform efficiently. The servomotor is used for the mechanical motion of the leverage.

3 Working in Real-Time Scenario

During the process, different waste materials are placed on the moving conveyor belt. This conveyor belt is designed in such a way that it moves slowly for larger substance as it takes more time for crushing process and moves rapidly for lighter substance. The audio waves which are developed are sensed by the audio sensor using USB modem, these sensed signals are passed on to the controller unit wherein it is preprocessed, the MFCC components are obtained, and these components are compared with pre-trained model which is trained to differentiate the MFCC of normal waste signal and MFCC of plastic waste signal. So, based on the decision made by the machine learning algorithm, lever is used to open either of the bin for storing the waste material.

Solid edge software is used for designing the mechanical system, and accurate measurements are considered in this process. Based on the designing, considerable hardwares are used which are mentioned in Sect. 2.2. For the implementation of software module, integrated development environment (Spyder) is used for the classification of normal waste and plastic waste. Supervised learning process is followed along with classification model such as support vector machine.

4 Results and Discussion

The Mel frequency cepstral coefficients are simulated using a simulation tool where the audio signal is divided into smaller frames, each frame is sent to ANN algorithm for the segregation as plastic or non-plastic material, and all this process takes place inside the mechanical device which is developed (Fig. 5).

Figure 6 shows the accuracy of the model trained using audio wave signal for both plastic and non-plastic material. The training and validation results are obtained for 30 epochs. The developed model is implemented in the hardware and validates correctly with an aggregate of 85% overall. The accuracy can be improved for large dataset and good GPU for training the model.

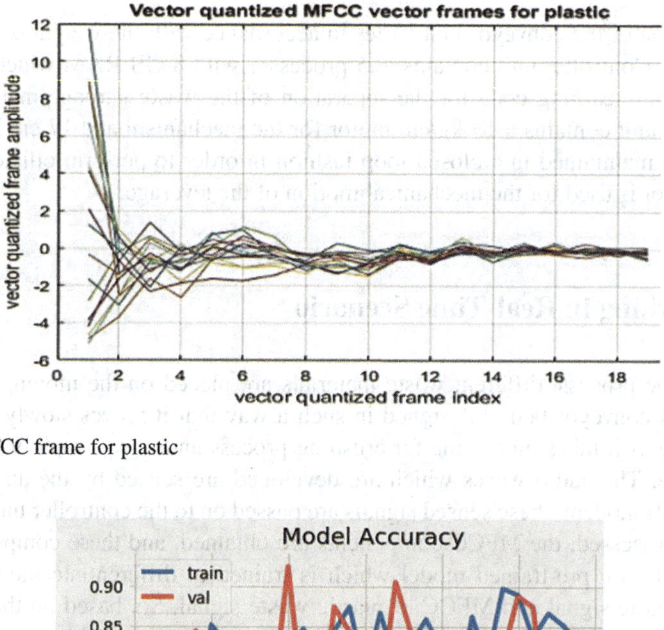

Fig. 5 MFCC frame for plastic

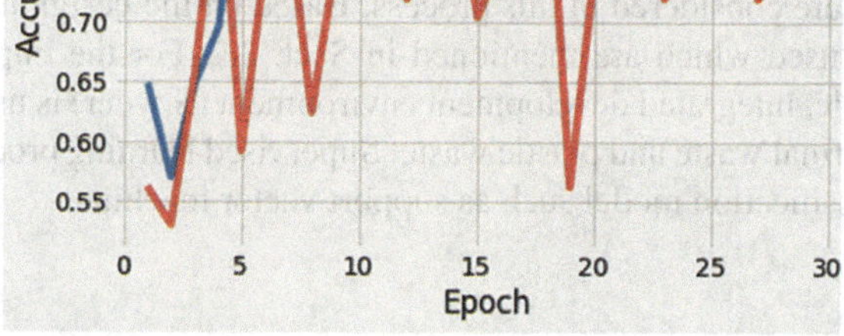

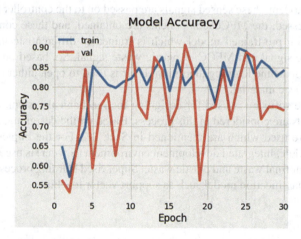

Fig. 6 Audio wave model trained accuracy

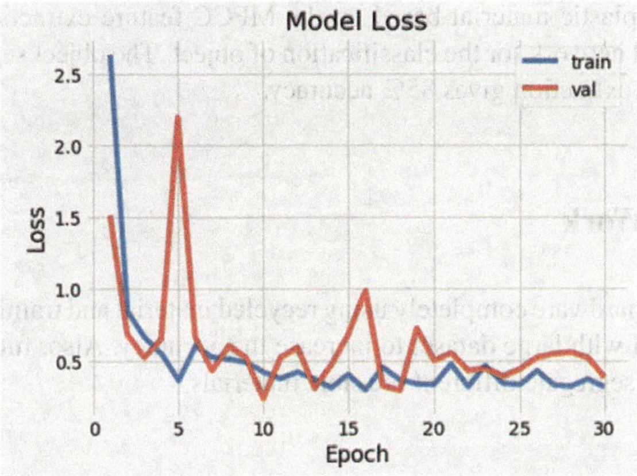

Fig. 7 Loss percent of the trained model

From the graph in Fig. 7, we can conclude that the loss percentage is very less as the training and validation curves are almost similar to one another.

5 Application

Segregating of plastic is very important in this modern world which results in good recycling of plastic for developing new products effectively. It takes only two-third of the actual energy to develop a product from the recycled material [15]; Whereas, developing a product using new plastic materials requires more energy and is a time-consuming process.

6 Conclusion

Plastic is one of the most dangerous wastes in the environment. A single milk jug takes up to 1 million year to decompose. In the meanwhile, of decomposition, many plastic materials get mixed with air which results in harmful environment for human as well as for animals and birds. The toxic released from plastic joins our blood and tissues, and this causes birth defects immunity problems. But the segregated plastic can also be used for beneficial works such as road construction. Hence, by adding the plastic content on to the road construction material, it strengthens the road and increases the durability of the road. Some kinds of plastic are also used for absorbing smoke. So, a low-cost automatic plastic segregation machine is developed. The plastic is segregated based on their audio signal obtained during the crushing process. Audio signals are recorded and fed to the computer which contains the ML algorithm for the sorting of plastic material based on the MFCC feature extraction process and artificial neural network for the classification of object. The object segregation using MFCC feature extraction gives 85% accuracy.

7 Future Work

Designing the hardware completely using recycled material and training the machine learning model with large dataset to increase the accuracy. Also, future technology can be used to segregate different reusable materials.

References

1. Gopinath H, Indu V, Dharman MM (2017) Development of autonomous underwater inspection robot under disturbance. In: International conference on technological advancements in power and energy
2. Hussain A et al (2010) Support vector machines for automated classification of plastic bottles. In: 6th International colloquium on signal processing & its applications (CSPA)
3. Wahab DA, Hussain A, Scavino E, Mustafa MM, Basri H (2006) Development of a prototype automated sorting system for plastic recycling. Am J Appl Sci 3(7):1924–1928
4. Shilpa B, Indu V, Rajasree SR (2017) Design of an underactuated self balancing robot using linear quadratic regulator and integral sliding mode controller. In: IEEE International conference on circuit, power and computing technologies, ICCPCT
5. Saiter JM et al. Different ways for re-using polymer-based wastes. The examples of works done in European countries. Recent developments in polymer recycling, pp 261–291
6. Bonifazi G, Di Maio F, Potenza F, Serranti S (2016) FT-IR Analysis and hyperspectral imaging applied to postconsumer plastics packaging characterization and sorting. IEEE Sens J 16
7. Scavino E, Wahab DA, Hussain A, Basri H, Mustafa M (2009) Application of automated image analysis to the identification and extraction of recyclable plastic bottles. J Zhejiang Univ Sci A 10(6):794–799
8. Tharuvana D, Sundaresh A, Sreelakshmi VJ, Das A, Nair B, Madhavan A, Pal S (2018) Sand and charcoal as immobilization matrices of phage's and probiotics for wastewater treatment. Pollut Res
9. Sadeghi M, Marvi H (2017) Optimal MFCC features extraction by differential evolution algorithm for speaker recognition. In: 3rd Iranian conference on signal rocessing and intelligent systems (ICSPIS)
10. Smitha, Shetty S, Hegde S, Dodderi T (2018) Classification of healthy and pathological voices using MFCC and ANN. In: 2nd International conference on advances in electronics, computers and communications
11. Ramli S, Mustafa MM, Wahab DA, Hussain A (2010) Plastic bottle shape classification using partial erosion-based approach. In: 6th International colloquium on signal processing & its applications (CSPA)
12. Vázquez-Guardado A, Money M, McKinney N, Chanda D (2015) Multi-spectral infrared spectroscopy for robust plastic identification. Appl Opt 54
13. Wahyuni ES (2017) Arabic speech recognition using MFCC feature extraction and ANN classification. In: 2nd International conferences on information technology information systems and electrical engineering

14. Shijith N, Poornachandran P, Sujadevi VG, Dharman MM (2017) Breach detection and mitigation of UAVs using deep neural network. In: Recent advancements in control, automation & power engineering
15. Schlesinger ME (2017) Aluminium recycling, 2nd edn. CRC Press, Boca Raton

GSM Constructed Adaptable Locker Safety Scheme by Means of RFID, PIN Besides Finger Print Expertise

S. M. Devendra Kumar, B. Manjula, and Rani Medhashree

Abstract The goal of this paper is to plan an ease framework that gives high security. It is an independent framework and furthermore minimal in size. It has low power utilization. Unique mark check is one of the most dependable individual recognizable proof strategies in biometrics. This framework comprises of microcontroller, RFID pursuer, GSM modem, fingerprint module, and LCD. Security is of essential concern and in this occupied aggressive world. In this modern era, human automates new technique to provide security without any manual work.

Keywords Arduino interface through RFID · GSM · Fingerprint segment · LCD · Keypad · Buzzer · PIR sensor

1 Introduction

Because of increment in burglary and robbery step by step, security at certain spots is significant. So, the principle point of our task is to give high security to the bank storage spaces, ATM, verified workplaces, diamond setter showroom, explore center, and so on. In the current society, people can access their data whenever and wherever without any hectic process, this flexibility is also more prone to hackers. Right now, personal identification numbers, passwords, or recognizable proof cards are utilized for individual distinguishing proof [1]. Be that as it may, cards can be taken, and passwords and numbers can be speculated or overlooked. To take care of these issues, biometric confirmation innovation which distinguishes individuals by their remarkable natural data is standing out. Biometrics can be characterized as perceiving

S. M. Devendra Kumar (✉)
ECE, GOVT SKSJ Technological Institute, Bengaluru, Karnataka, India
e-mail: smdevendrakumar@gmail.com

B. Manjula
ECE, Government Engineering College (Guest Faculty), Chamarajanagar, Karnataka, India

R. Medhashree
Delhi Technological University, New Delhi, Delhi, India

© The Author(s), under exclusive license to Springer Nature Singapore Pte Ltd. 2021
R. M. Singari et al. (eds.), *Advances in Manufacturing and Industrial Engineering*, Lecture Notes in Mechanical Engineering,
https://doi.org/10.1007/978-981-15-8542-5_99

and distinguishing an individual dependent on physiological or conduct attributes. In biometric verification, a record holder's body qualities or practices (propensities) are enrolled in a database and afterward contrasted with other people who may attempt with different database to get in-to that record to check whether the endeavor is real. To plan and execute a profoundly verified and solid keen storage security framework utilizing RFID, biometric unique finger impression, secret key, and GSM innovation. In reality, people are more worried about their security for their significant things like gems, cash, significant records, and so on. So, the bank storage spaces are the most secure spot to store them. The advancement in security enables clients to have high security software with distinguishable electronic proof choices. These recognizable proof innovations incorporate bank lockers and ATM just as other smart cards [2], client IDs, and secret key-based frameworks. Be that as it may, sadly, these are unbound as robberies, programmer assaults, and overlooked passwords [3]. In any case, the biometric or unique finger impression confirmation and RFID based recognizable proof are the most effective and dependable answer for stringent security.

2 Methodology

The model works on Arduino, GSM module, RFID, fingerprint module, and PIR sensor.

2.1 Basic Approach

The two strategies are unique finger impression method and RFID procedure. At whatever point, we need to open the entryway initially, the approved individual composed the secret word, and at that point they spot unique mark. On the off chance that it is coordinated, at that point, entryway will open, and this is one of the systems that we utilized. At that point, another procedure is RFID method [4]. On the off chance that we need to open the entryway, on the off chance that approved individual is exhibited, then chosen one individual ought to have RFID card, and afterward put the RFID in to RFID pursuer; at that point, OTP will be sent to their cell phone, and you need to type the received OTP, in the event that it is coordinated, at that point the entryway will open; generally, entryway will not open [5]. Here, we utilize the change to turn on exchanging mode, and in this mode PIR sensor is enacted in evening time. In the event that any gatecrasher is recognized, it will send the SMS to cell phone, and the ringer will begin to caution the framework.

2.2 Architecture

The security framework comprises of microcontroller, RFID, GSM, fingerprint module, PIR sensor, LCD show in Fig. 1 [6], keypad, solenoid valve, relay, buzzer, and power supply. At the point when security framework begins its procedure, we enter the secret key through keypad. On the off chance that the secret phrase will be right, at that point place your unique mark module. In the event that the secret key will not be right, at that point return the secret key through keypad. In the event that the set unique mark coordinated with the enlisted finger impression, at that point the entryway will be open, else it does not open the entryway. When RFID label interacts with the RFID pursuer, it creates the ID on the off chance that it is legitimate. At that point, it will send the OTP to cell phone through GSM [7]. The got OTP is composed through keypad; in the event that it is coordinated, at that point the entryway will be open, else it does not open.

Here in Fig. 1, we utilize the change to turn on exchanging mode, this mode PIR sensor is enacted in evening time [8]. In the event that any gatecrasher is identified, it will send the SMS to cell phone, and the signal will begin to caution the framework.

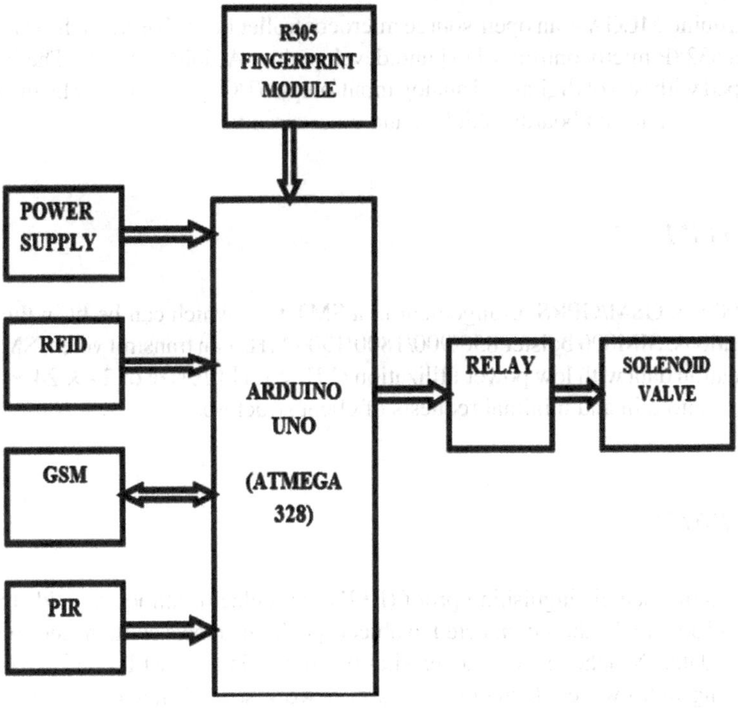

Fig. 1 Block view of locker security system

3 Implementation and Results

In the propelled framework, the security framework is intended for securing the significant information and even cash. Here, the Arduino is interfaced with GSM, RFID, fingerprint module, keypad, and PIR sensor. By joining more than one lock framework [9] to give high security, PIR sensor distinguishes the human movement, and if people go into the storage space, Arduino sends the message to the approved individual cell phone. Arduino checks the twofold secret key through GSM module. On the off chance that the secret key is right, the entryway lock will be open, and generally the entryway lock stays closed. On the off chance that the approved individual misfortune their RFID card [10], they use unique mark innovation to ensure and keep up the secrecy of information. By consolidating these innovations, we can give more security than other innovation. Here, they are giving two alternatives that is secret key, unique mark innovation, and another is RFID innovation.

3.1 Arduino

The Arduino MEGA is an open-source microcontroller board based on the microchip ATmega328P microcontroller [11] and developed by Arduino.cc [12]. The board is equipped with sets of digital and analog input/output (I/O) pins that may be interfaced to various expansion boards (shields) and other circuits.

3.2 GSM

SIM800 is a GSM/GPRS arrangement in a SMT type which can be fit in the client applications. SIM800 bolster 850/900/1800/1900 MHz can transmit voice, SMS, and information data with low power utilization [13]. With little size of 24 × 24 × 3 mm, it can fit into thin and minimal requests of client structure.

3.3 RFID

Radio-recurrence distinguishing proof (RFID) uses electro-attractive fields to naturally find and track labels connected to objects [14]. The labels contain electronically cast out data. Detached labels gather vitality from a close by RFID pursuer's cross-examining radio waves. Dynamic labels may work several meters from the RFID pursuer.

3.4 Unique Mark Module

Fingerprint handling consists of two sections, listing of finger impression and coordination of unique finger impression (the coordinating can be 1:1 or 1:N). During selection process, client needs to enter the finger multiple times [15]. The software will process the double cross-finger pictures, produce a format of finger dependent on forming results, and load the layout. At the time of coordination, client's finger is sensed through optical sensor, and framework will develop a format of the finger and compares it with layouts of the finger library.

3.5 Arduino IDE

This is the Arduino IDE, and once it is been opened, it opens into a clear sketch where you can begin programming right away. In the first place, we ought to design the board and port settings to enable us to transfer code [16]. Associate your Arduino board to the PC by means of the USB link. Also connect C/C++ Code Completion, Multi-venture, Theme support, Cross-stage support, Serial port screen, Library and the board framework, together for Continuous coordination.

3.6 Storage Mode

When we type the secret phrase through keypad, in the event that it is coordinated with the put away secret word, at that point it shows "ENTER FINGERPRINT" within 15 s as appeared in underneath Fig. 2.

Subsequent to putting unique finger impression, in the event that it is coordinated with enlist unique mark, at that point entryway will be opened for 3 s, and within 3 s, we need to open the entryway; generally, entryway will be secured as appeared in Fig. 2.

Fig. 2 Fingerprint is matched

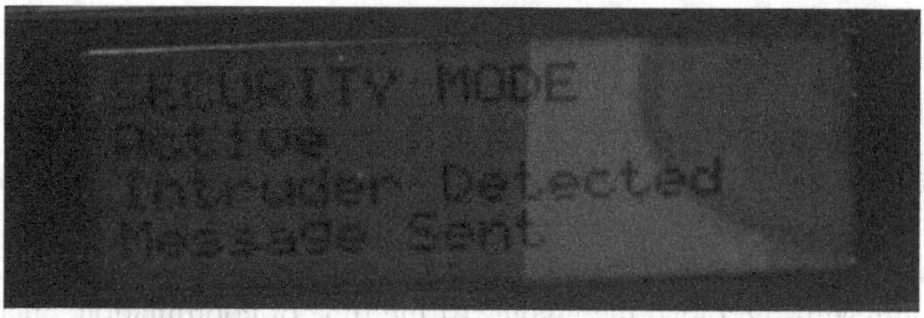

Fig. 3 When intruder detected

3.7 Security Mode

In evening, we need to turn on the security mode for intruder alert.

In the security modem we use PIR sensor for recognizing the gatecrasher as shown in Fig. 3. In the event that any gatecrasher recognized, at that point GSM [17] sends message to regarded versatile number and just indicates as single signal, which will be ON that means blare sound will have landed for alarming reason.

4 Application

Residential reason: It is utilized for security in home, structures, and so on.

Business reason: It is utilized for security in bank, ATM, workplaces, and so on. Also provide security in Modern and restorative field Jeweler Shops and in Research focuses.

5 Conclusion

In this venture, we have executed a storage security framework utilizing RFID, fingerprint, password, and GSM. It is a minimal effort, low in control origination, reduced in size, and independent framework. In this task, we have first audited the as of late proposed utilizing storage key for banking; however, they are verified, and there are a few disservices [18]. It might be giving erroneous individual access the record. So in this, we are actualizing security framework dependent on biometric. This framework is secure and less cost, and it will be a best financial framework. Biometric and GSM security are given right and quick client confirmation. Since biometric cannot be overlooked, they are hard for assailants to manufacture and for client to revoke. Finger impression is an exceptional ID for everybody. He discovered if somebody is attempting to open his storage. The framework has effectively beaten a portion of the perspectives existing with the present advances, by the utilization of unique finger impression biometric as the validation innovation.

6 Future Enhancements

A propelled entryway locking security framework ought to be created by consolidating the current security systems or by presenting another method [19]. We will create security framework dependent on iris scanner for visual recognizable proof of the individual. In future, Arduino is supplanted by Raspberry Pi, lessening in cost. Utilize all the more bolting framework for high security.

Acknowledgements It is so far creating and totally capably. Any necessities, this endeavor are done at this point they have to revive and adjust a couple of modules. We are persistently contemplating affiliation necessities moreover creating bit by bit. We for the most part need execute something more.

References

1. Vaid V (2014) Comparison of different attributes in modeling a FSM based vending machine in 2 different style. In: 2014 ICES international conference on embedded system, pp 18–21
2. Sarce J, Gaikwad AN (2016) Fingerprint and iris biometric controlled smart banking machine embedded with GSM technology for OTP
3. Turner M, Naber J (2010) The design of a bi-directional, RFID-based ASIC for interfacing with SPI bus peripherals. In: 2010 53rd IEEE international Midwest symposium on circuits and systems, Seattle, WA, pp 554–557
4. Parvathy A, Gudivada VRR, Venumadhav Reddy M, Manikanta Chaitanya G (2011) RFID based exam hall maintenance system. IJCA Spec Issue Artif Intell Techn Novel Approach Pract Appl AIT
5. Rane KP, Nehete PR, Chaudhari JP, Pachpande SR (2016) Literature survey on lock security systems. Int J Comput Appl 153(2):13–18
6. Sumalatha C, Viyayamanasa A, Ramasrujana K, Hema Rani K (2016) Bank locker security system using RFID and GSM technology. IJRASET 4(April)
7. Hasan R, Khan MM, Ashek A, Rumpa IJ (2015) Microcontroller based home security system with GSM technology, June 2015
8. Leens F (2009) An introduction to I^2C and SPI protocols. IEEE Instrum Measur Mag, Feb 2009
9. Nehete PR, Chaudhari JP, Pachpande SR, Rane KP (2007) Literature survey on lock security systems. Int J Comput Appl 153(2):13–18
10. Farooq U, Hasan M, Amar M, Hanif A, Usman Asad M (2014) RFID based security and access control system. IACSIT Int J Eng Technol 6(4):309–314
11. Shafiul Islam M (2014) Home security system based on PIC18F452 microcontrollers, vol 978. IEEE
12. Verma GK, Tripathi P (2010) A digital security system with door lock system using RFID technology. Int J Comput Appl (0975-8887) 5(11)
13. Nelligani BM, Uma Reddy NV, Awasti N (2016) Smart ATM security system using FPR, GSM, GPS
14. Choudhury S, Singh GK, Mehra RM (2014) Design and verification serial peripheral interface (SPI) protocol for low power applications. IJIRSET Int J Innov Res Sci Eng Technol 1675016758
15. Ghanekar A, Kishor B, Bandewar S (2016) Design and implementation of SPI bus protocol. IJAREEIE Int J Adv Res Electr Electron Instrum Eng 4155–4157

16. Ackland K, Lotya M, Finn DJ, Stamenov P (2016) Use of slits of defined width in metal layers within ID-1 cards, as reactive couplers for near-field passive RFID at 13.56 MHz. In: 2016 IEEE international conference on RFID (RFID), Orlando, FL, pp 1–4
17. Vidhya Sagar K, Balaji G, Narendra Reddy K (2015) RFID-GSM imparted school children security system 2(2)
18. Good T, Benaissa M (2009) A low-frequency RFID to challenge security and privacy concerns. In: 2009 IEEE 6th international conference on mobile adhoc and sensor systems, Macau, pp 856–863
19. Roy B, Mukherjee B (2010) Design of a coffee vending machine using single electron devices: (an example of sequential circuit design). In: 2010 International symposium on electronic system design, Bhubaneswar, pp 38–43

Low-Voltage Squarer–Divider Circuit Using Level Shifted Flipped Voltage Follower

Swati Yadav and Bhawna Aggarwal

Abstract In this paper, a low-voltage one-quadrant squarer–divider (SD) circuit based on level shifted flipped voltage follower (LSFVF) has been proposed. The proposed squarer–divider circuit utilizes low-voltage SD based on MOS translinear (MTL) principle. In the proposed design, flipped voltage follower (FVF) configuration used for biasing of MTL transistors has been replaced by LSFVF configuration. This proposed LSFVF-based squarer–divider circuit not only maintains the low supply voltage requirement of the FVF-based squarer–divider but it reduces the error present in output current significantly. To validate the working of proposed SD circuit, it has been simulated in LTspice using 0.18 μm CMOS technology. These simulations show that result shows, that a significant improvement is achieved at low values of input current in the proposed LSFVF SD circuit as compared to FVF-based SD circuit.

Keywords Current mode · MTL loop · FVF · Low voltage · Body effect · Stacking

1 Introduction

The historical growth in integrated circuit (IC) industry has profoundly changed the way in which the data is processed and information is stored and this evolved due to many technological developments from small-scale integration to ultra-large-scale integration. The reason for this phenomenal growth is due to exponential growth in the ability to shrink transistor dimensions, every few years. The ability to scale transistor supply voltage is determined by the lowest voltage required to switch a transistor between ON and OFF states.

For designing low power devices, the crucial parameters are threshold voltage, channel length, oxide thickness, and doping concentration of channel. When a device is scaled variations in process parameters become critical. Circuits designed using these scaled down devices need to be operated at low supply voltages. Reducing

S. Yadav · B. Aggarwal (✉)
ECE Division, NSIT, Delhi, India
e-mail: kbhawnagarg@yahoo.co.in

© The Author(s), under exclusive license to Springer Nature Singapore Pte Ltd. 2021
R. M. Singari et al. (eds.), *Advances in Manufacturing and Industrial Engineering*, Lecture Notes in Mechanical Engineering,
https://doi.org/10.1007/978-981-15-8542-5_100

voltage in power supply further results in lower power consumption as dynamic power has a quadratic relationship with power supply voltage.

There are many methods defined in the literature to reduce the required supply voltage in analog and mixed-signals circuits. A few among them are: folded-cascode, triode-mode operation, subthreshold operations mode of MOS, floating-gate MOS, and current-mode processing circuits.

In recent years of development, engineers have shifted their interest more toward current-mode processing circuits, as they have gained high importance due to the following advantages [1]: (a) improved linearity; (b) more accuracy; (c) simple circuitry; (d) increase in bandwidth; (e) low electrostatic discharge due to low impedance at input side.

Flipped voltage follower is one of the naïve techniques employed to reduce supply voltage and power consumption. It possesses the advantages offered by current-mode circuits.

The objective of this work is to design a low-voltage FVF-based squarer–divider circuit. The paper is structured as: Sect. 2 explains the various configurations of voltage follower with their advantages and disadvantages. Section 3 describes MOS translinear principle, the working of squarer–divider with conventional biasing and with different FVF configurations. Section 4 discusses the simulation result. Section 5 concludes the paper.

2 Voltage Follower

2.1 Voltage Follower

The voltage follower (shown in Fig. 1) is an amplifier formed by a MOSFET connected in common drain configuration. This circuit is popularly recognized as voltage buffer and source follower and is characterized by the following properties [2]: (a) high impedance at input terminal (b) low impedance at output terminal (R_{out}

Fig. 1 Common drain configuration of MOS

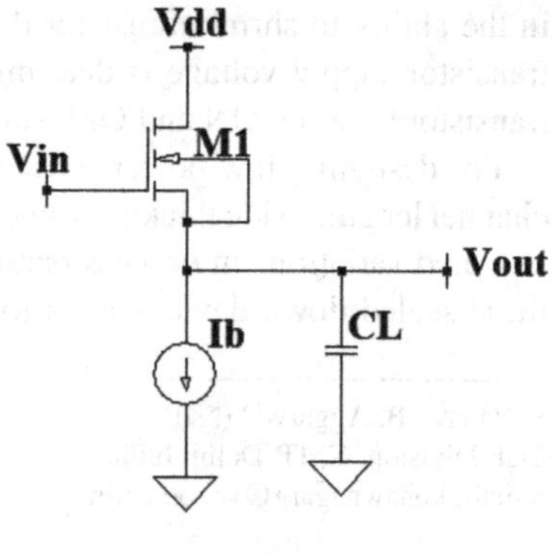

$= 1/g_m$) (few kilo ohms) (c) high bandwidth, ($BW = g_m/(2\pi.C_L)$) (d) unity gain (e) $R_{in} \approx$ infinite (f) large input and output swing.

Common drain configurations have following limitations in practical [2]:

- Output resistance is not very low, to achieve low output resistance, high values of transconductance (g_m) has to be used. This can be accomplished by providing, a biasing through large current source and large aspect ratio.
- Output current is given by $I_{out} = V_{out}/\omega.X_L$ where X_L is load impedance; thus, gate-source voltage varies with respect to the input and results in distortion.
- Slew rate is not symmetrical as current sourcing capability is high, and current-sinking capability is limited by biasing current source I_b.

2.2 *Flipped Voltage Follower (FVF)*

FVF corresponds to a voltage follower with enhanced properties [2]. It is basically a common drain amplifier cascaded with common source amplifier. In this circuit, common source amplifier is connected in negative feedback with common drain amplifier to improve its characteristics. NMOS configuration of FVF is shown in Fig. 2. Here MOSFET M1 is in common drain configuration and MOSFET M2 is

Fig. 2 NMOS FVF

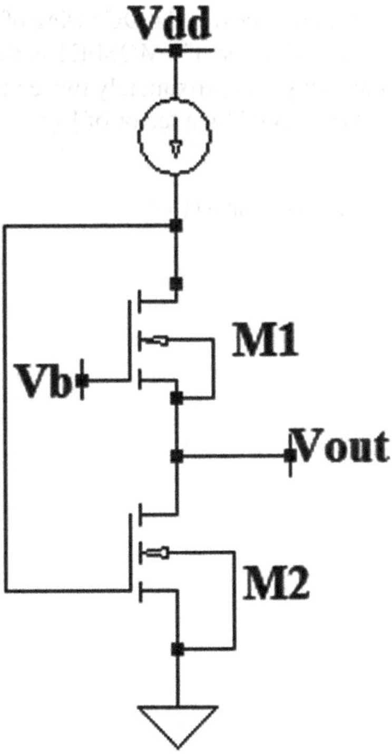

connected in common source configuration. The gate terminal of M2 is connected with drain terminal of M1 resulting in negative feedback. Gate of M1 is used to apply input and output is taken from its source terminal as shown in NMOS-based FVF of Fig. 2.

In FVF, variations of current at output terminal node are absorbed by M2 (current sinking transistor), while current in M1 stays practically constant due to dc current source I_1. This ensures constant gate-source voltage of M1 and maintains low values of distortion even at high frequencies, which was one of the major drawbacks of the source follower.

Flipped voltage follower shown in Fig. 2 has the following characteristics [2]: (a) a very low resistance at output terminal ($R_{\text{out}} = 1/(g_{\text{m}_1} g_{\text{m}_2} r_{\text{o}_1})$); (10's of Ωs), due to negative feedback, enabled by the presence of M2 (b) quite low supply requirements ($V_{\text{DDmin}} = V_{\text{GS2}} + V_{\text{DSsat}}$), approximately equal to threshold voltage (c) high current-sinking capability (d) high gain bandwidth product ($\text{GBW} = g_{\text{m}_2}/C_{\text{C}}$).

Problems faced by Flipped Voltage Follower:

Practically, it allows very small input to output signal swing $V_{\text{in}} = V_{\text{T}} - V_{\text{DSsat}}$. Swing limitation is due to the fact that the swing of M1 (Fig. 2) is "strangled" by the V_{GS2} of the current sensing transistor M2. Swing does not increase with the supply voltage (V_{DD}) as it is the case for most circuits.

This problem of FVF is minimized using different configurations of FVF and one of them is flipped voltage follower with level shifter (LSFVF) [2].

LSFVF came one of the modifications of conventional FVF [3] and is shown in Fig. 3. It improves the DC swing of output. This increase in swing is possible due to the introduction of a MOSFET in the feedback loop of M1 and M2. Mathematically, the swing is approximately twice of the threshold voltage. LSFVF's operating range gets extended by a factor of [4]:

Fig. 3 Level shifted FVF

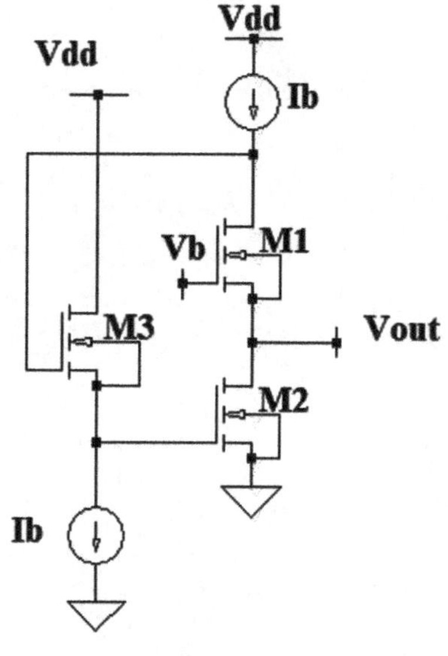

$$V_{TM3} + \sqrt{\frac{2I_B}{K_N(W/L)_{M3}}} \quad (1)$$

However, introducing the third MOSFET consumes more quiescent power and deteriorates bandwidth.

3 Squarer–Divider Circuit

In analog and mixed signal processing circuits, analog multiplier circuits such as geometric mean, squarer–divider, and other circuits are of critical importance. Squaring or squarer–divider circuit is one of the most practically used circuitry and considered as a basic building block in analog and mixed signal processing circuits like [5] rms-dc converter, fuzzy logic circuits, defuzzification, artificial neural network, phase synchronizer, vector sectors, phase synthesizers, biomedical industry, and instrumentation industry.

In analog domain, SD circuit can be designed using methods or circuitry like delta-data converter, MOS translinear loop (MTL) principle, or class AB transconductance.

3.1 MOS Translinear (MTL)

MTL principle [6–10] is analogous to bipolar translinear principle given by Gilbert in [6], and this applicable on "even number of MOSFET operating in strong inversion or saturation mode are taken in closed loop such that gate-source voltage of half of these MOSFET appears in clockwise direction and for rest half it appears in anticlockwise direction.".

In CMOS technology, squarer–divider circuit has been implemented using MOS translinear principle with three different configurations: (a) stacked loop technology configuration (b) up-down topology configuration (c) electronically simulated loop configuration.

Stacked loop configuration operates similar to class AB transconductance, but has a disadvantage of body effect as substrate and body are biased at different voltage while completing the loop. Up-down topology alleviates the problem of mismatch in V_{th} due to virtual ground. Electronically simulated loop can be employed using basic electronic components such as resistors and opamps.

Figure 4 shows the conventional circuit of squarer–divider with up-down topology [3]. However, suffers from body effect due to the stacking of transistor M11 and M16. MTL loop is applied on transistors M11, M12, M14, and M17 in conventional circuit of squarer–divider shown in Fig. 5.

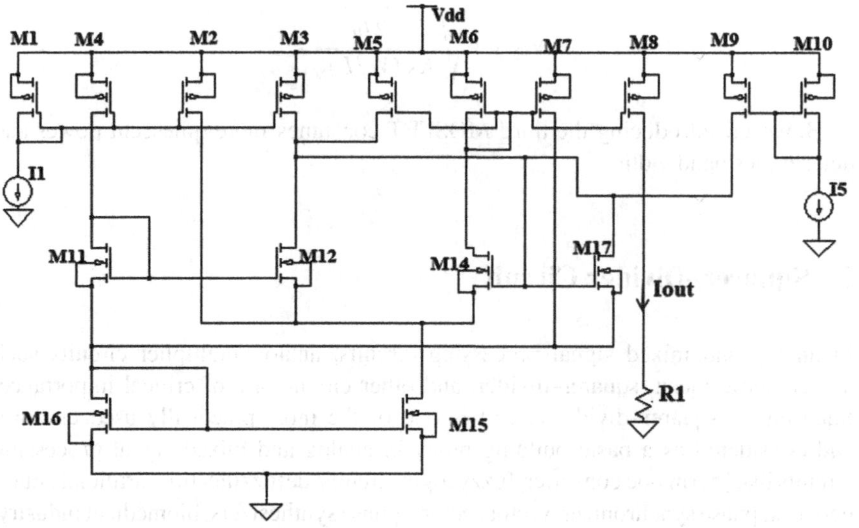

Fig. 4 Conventional squarer–divider circuit [3]

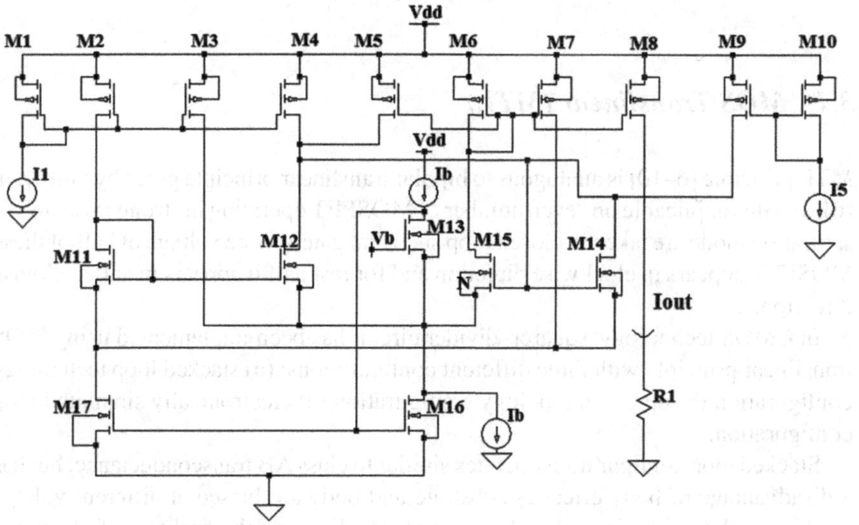

Fig. 5 Low-voltage squarer–divider with FVF [3]

Application of MTL principle on these MOSFETS gives:

$$V_{GS11} + V_{GS14} = V_{GS12} + V_{GS17} \quad (2)$$

Considering these MOSFETs to be matched and operating in saturation mode with equal process parameters and threshold voltage, and their gate to source voltage can be written in terms of current as follows:

$$\sqrt{I_{11}} + \sqrt{I_{14}} = \sqrt{I_{12}} + \sqrt{I_{17}} \tag{3}$$

For proper functioning of circuits, current of MOSFET M12 and M17 is forced to be equal source I_5 to satisfy the following relation [3]:

$$I_{12} = I_{17} = (I_{11} + I_{14} + 2I_5)/4 \tag{4}$$

The output of squarer–divider circuit is shown in Fig. 4 and can be derived by analyzing the circuit and using above equations. This current can be obtained as:

$$I_{\mathrm{OUT}} = I_{\mathrm{R1}} = I_5^2/I_1 \tag{5}$$

A modified version of conventional squarer–divider circuit that operates at low voltage has been proposed by Fig. 5 shows the squarer–divider circuit with FVF [3] for Carvajal et al. in 2005 and is shown in Fig. 5. This low-voltage squarer–divider uses FVF configuration in the MTL loop formed by M11, M12, M14, and M15. Here MOSFETs M13 and M16 along with a small biasing current I_b. This FVF improves the biasing of MTL transistors and is very helpful in achieving very low-voltage squarer–divider circuit.

In conventional squarer–divider (Fig. 4), M16 and M11 cause stacking of transistor leading to increase in supply voltage requirement.

By inserting FVF the node, voltages of loop are maintained at adequate value. The current mirror requirement is met by high swing current mirror formed by M13 and M16 working at low voltage. Moreover, low impedance output node of FVF does not allow frequent changes in voltage irrespective of variation in input and output current.

In this paper, Fig. 6 depicts a new low-voltage squarer–divider circuit that has been designed using level shifted FVF configuration for biasing the MTL transistor of conventional squarer–divider is shown in Fig. 6. In this circuit MOSFETs, M13, M15, and M18 along with current source biasing configurations I_b, form the desired LSFVF transistor for biasing of MTL transistors. Here, M18 is the MOSFET in feedback loop of M13 and M15.

This proposed LSFVF-based squarer–divider operates at low supply voltage, as LSFVF configuration reduces the supply voltage headroom caused by stacking of biasing transistors in conventional circuit. The proposed design operates at low voltage similar to FVF-based squarer–divider circuit. However, as shown in the next section, this circuit reduces the error observed in DC characteristics of FVF-based squarer–divider. Thereby, the proposed design proves to be useful for circuit designers operating at low supply voltage.

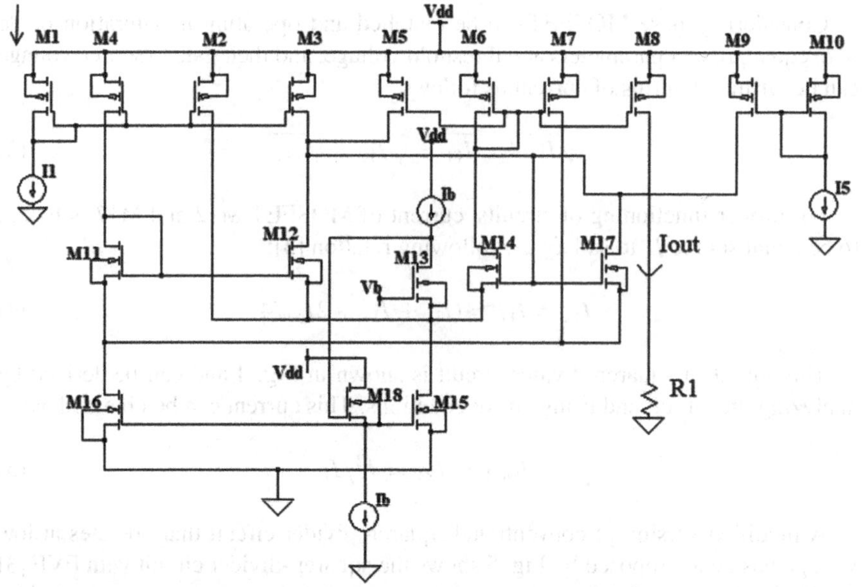

Fig. 6 Proposed low-voltage squarer–divider with LSFVF

4 Simulation Result

The simulations have been performed in LTSpice with 0.18 μm technology. A supply voltage of 1.8 V has been used to simulate conventional squarer–divider circuit, whereas a voltage of 1.5 V has been used to simulate FVF-based squarer–divider circuit. For biasing of FVF and LSFVF configuration, $I_b = 2$ μA and $V_b = 1.3$ V have been used. Aspect ratio of 80 μm/4.8 μm has been chosen for all the MOSFETs used in different circuits.

For simulations, a DC sweep in the range of 5–20 μA has been taken for current source I_1.

Figure 7 shows the DC analysis, of all three SD circuits discussed in the paper. The analysis has been carried out for input range from 5–20 μA. This figure shows that for higher currents (greater than 10 μA), behavior of all three circuits is approximately similar. However, for low values of I_1 output current of proposed LSFVF SD show significant deviation compared to the values of output current obtained in rest two circuits. To verify the authenticity of proposed design, percentage error in output current has been plotted in Fig. 8. This figure clearly shows the reliability of the proposed design compared to rest of the circuits. From Fig. 8, it can be observed that proposed design offers minimal error at low currents.

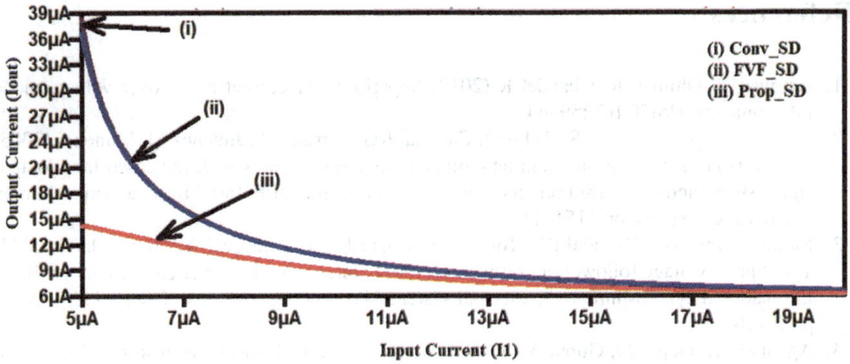

Fig. 7 Output current of SD circuit (i) conventional SD (ii) FVF SD (iii) proposed LSFVF SD

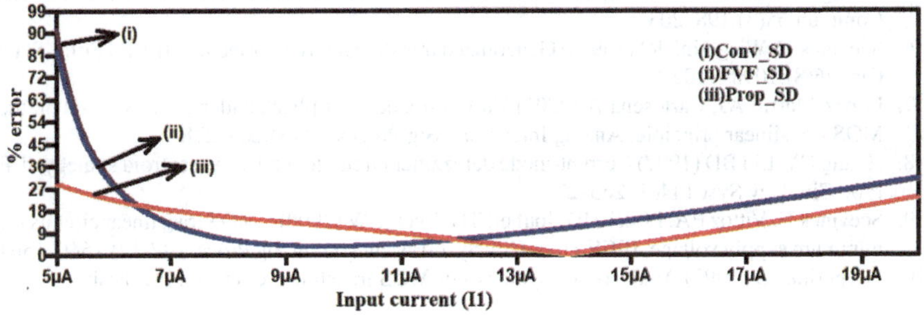

Fig. 8 Percentage error in output current (i) conventional SD (ii) FVF SD (iii) proposed LSFVF SD

5 Conclusion

In this paper, a LSFVF-based low-voltage squarer–divider circuit has been designed and proposed. Carvajal et al. have given FVF-based squarer–divider circuit, at lower voltage. In the present work, the FVF configuration has been replaced with LSFVF configuration. The simulations of all the circuits have been carried out in LTspice using 0.18 μm technology. It has been examined that the proposed circuit operates at low voltage of 1.5 V, similar to FVF-based SD in [3] and the proposed circuit functions with much lower error as compared to conventional SD or FVF-based SD, especially for low values of input current. Thereby, it can be inferred that proposed LSFVF-based circuit is a good choice for circuit designers working on low-voltage analog-based mixed signal processing circuits.

References

1. Agrawal Y, Dhiman R, Chandel R (2012) Superiority of current mode over voltage mode interconnects. IJSET 1(2):59–64
2. Ramírez-Angulo J, Gupta S, Padilla I, Carvajal RG, Torralba A, Jimenez M, Munoz F (2005) Comparison of conventional and new flipped voltage structures with increased input/output signal swing and current sourcing/sinking capabilities. In: IEEE 48th Midwest symposium on circuits and systems, pp 1151–1154
3. Ramirez-Angulo J, Carvajal RG, Torralba A, Galan JAGJ, Vega-Leal AP, Tombs JATJ (2002) The flipped voltage follower: a useful cell for low-voltage low-power circuit design. In: IEEE International symposium on circuits and systems. Proceedings (Cat. No. 02CH37353), vol 3, pp III–III
4. Aggarwal B, Gupta M, Gupta AK (2013) A low voltage wide swing level shifted FVF based current mirror. In: IEEE International conference on advances in computing, communications and informatics (ICACCI), pp 880–885
5. Tangsrirat W, Pukkalanun T, Mongkolwai P, Surakampontorn W (2011) Simple current-mode analog multiplier, divider, square-rooter and squarer based on CDTAs. AEU-Int J Electron Commun 65(3):198–203
6. Seevinck E, Wiegerink RJ (1991a) Generalized translinear circuit principle. IEEE J Solid-State Circ 26(8):1098–1102
7. López-Martín AJ, Carlosena A (2001) Current-mode multiplier/divider circuits based on the MOS translinear principle. Analog Integr Circ Sig Process 28(3):265–278
8. Huang CY, Liu BD (1997) Current-mode defuzzifier circuit to realise the centroid strategy. IEE Proc Circ Dev Syst 144(5):265–271
9. Seevinck E, Vittoz EA, Plessi MD, Joubert TH, Beetge W (2000) CMOS translinear circuits for minimum supply voltage. IEEE Trans Circ Syst II Analog Digit Sig Process 47(12):1560–1564
10. Wiegerink RJ (1992) Analysis and synthesis of MOS translinear circuits. PhD thesis

Memristor-Based Electronically Tunable Unity-Gain Sallen–Key Filters

Bhawna Aggarwal, Manshul Arora, Marsheneil Koul, and Maneesha Gupta

Abstract Memristor is considered to be the fourth fundamental circuit element, along with resistor, capacitor and inductor. It is assumed to occupy very less chip area, consume low power and provides good manufacturability at low nanometer technologies. In this paper, memristor-based unity-gain Sallen–Key filters have been designed and proposed. These filters provide the flexibility of electronic adjustment of parameters, something that was not possible in conventional Sallen–Key filters. Moreover, the proposed filters provide better performance at low nanometer technologies due to its better characteristics at these technologies. To justify the stated results, simulations of the proposed circuits have been carried out using EldoSpice by Mentor Graphics and using linear ion drift model of memristor. Simulated results depicted that memristor behaves as a simple resistor at frequencies greater than 100 Hz as well as at initial resistance (R_{INIT}) greater than 11 kΩ. The cut-off frequency of simulated filters has been observed to be 14.46863 Hz which is well in agreement with theoretically calculated values. This value and thereby bandwidth and Q-factor can be varied by changing R_{INIT}.

Keywords Memristor · Sallen–Key filter · Electronic tuning · Nanometer technologies · EldoSpice · Linear ion drift model

1 Introduction

In 1971, Prof. Leon Chua proposed memristor as fourth fundamental circuit element, along with resistor, capacitor and inductor [1]. Through Maxwell equations, it was presented that this element fills the gap in the relation between charge and flux ($d\phi = Mdq$) [1]. Since then, memristor is considered to be a basic element for unique applications that are not feasible to be designed using networks based on RLC circuits only. The term 'memristor' refers to 'memory resistor' and its electrical characteristic is known as memristance (M). It is a two-terminal device whose memristance can

B. Aggarwal (✉) · M. Arora · M. Koul · M. Gupta
ECE Division, NSIT, Delhi, India
e-mail: kbhawnagarg@yahoo.co.in

© The Author(s), under exclusive license to Springer Nature Singapore Pte Ltd. 2021

R. M. Singari et al. (eds.), *Advances in Manufacturing and Industrial Engineering*, Lecture Notes in Mechanical Engineering,
https://doi.org/10.1007/978-981-15-8542-5_101

be controlled by appropriately varying the value, polarity and time duration of the voltage signal applied across its terminals [2].

Many devices with the characteristics similar to that of the memristor were developed time-to-time [3]. However, in 2008, Hewlett Packard (HP) scientists successfully linked the memristor postulated by Chua with the device developed by them using titanium dioxide (TiO_2). The group of Stanley Williams fabricated the first memristor and reported the similarity in switching behavior of their device and Chua's memristor [4]. This announcement leads to breakthrough in the field of analysis of elementary memristor characteristics and its applications in wide areas of circuit designing. Various papers demonstrating the use of memristor in band-pass and notch memristor filters, programmable analog ICs, reactance-less oscillators, circuits for performing basic arithmetic operations, etc., have been published in literature [5–9]. In low nanometer technologies, memristive devices are thought off to be devices having improved characteristics with simple structure and device geometry [4, 9]. The ability of memristor to remember its most recent memristance for indefinite time makes it a useful element to be used in non-volatile memory designing and can be helpful in improving the performance characteristics of digital circuits without scaling down transistor sizes. It is observed that for high frequencies (>20 Hz) hysteresis loop of memristor starts shrinking and it behaves almost as a linear resistor beyond 100 Hz [4, 10]. It is being predicted that future memristive analog circuits may work similar to brain.

In this paper, a memristance-based Sallen–Key filter has been designed and analyzed. The paper is presentation is given as: Modeling of memristor using linear ion drift model is explained in Sect. 2. Operation and importance of Sallen–Key filters is briefed in Sect. 3. Section 4 presents the implementation and designing of the proposed memristor-based unity-gain Sallen–Key filters. Results obtained through simulations have been described in Sect. 5. Finally, the paper is summarized in conclusion.

2 Modeling of Memristor

Memristor has been defined as fourth circuit element where total electric charge (q) passing through the device determines the flux (ϕ) between its two terminals [1]. The symbol of memristor and its internal structure as given by HP is shown in Fig. 1 [11]. Pinched hysteresis loop is formed in its V–I characteristic. Excitation of memristor by a periodic signal shows voltage ($v(t)$) = 0, if the current ($i(t)$) = 0 and vice versa, thereby giving identical zero-crossing for current and voltage. It is assumed that a device having V–I hysteresis curve is a memristive device or memritor itself [4, 6].

A memristor is supposed to be controlled by its charge, if $\phi = f(q)$ and if $q = f(\phi)$, it is supposed to be flux controlled. Memristance ($M(q) = \mathrm{d}f(q)/\mathrm{d}(q)$) is a property of the memristor which is measured in ohms [4, 12]. The memristance of HP memristor as per linear ion drift model is given as [4]:

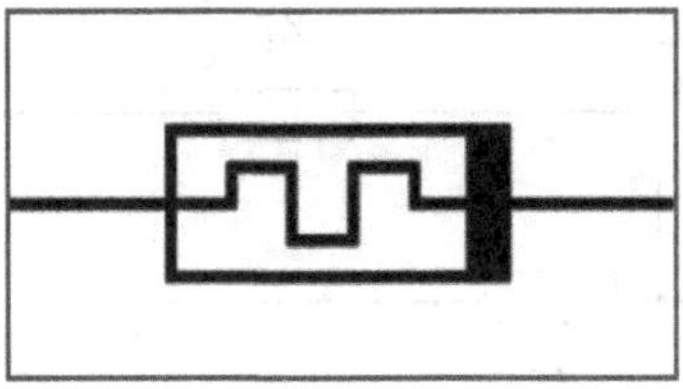

(a)

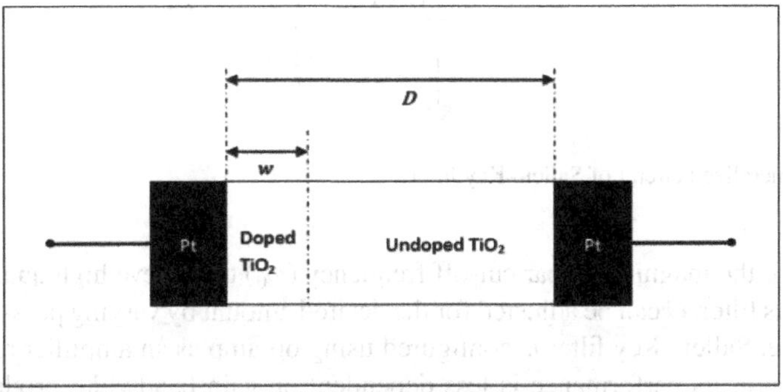

(b)

Fig. 1 **a** Symbol of memristor, **b** internal structure of memristor

$$M(q) = R_{\text{OFF}}\left(1 - \frac{R_{\text{ON}}}{\beta}q(t)\right) \tag{1}$$

where $\beta = \frac{D^2}{\mu_D}$ has the dimensions of magnetic flux, μ_D is dopant mobility having unit $cm^2/V\,s$, D is TiO_2 film thickness, R_{OFF} and R_{ON} are 'off-state' and 'on-state' memristances, respectively, and $q(t)$ is the total charge passing through the memristive device.

Memristor is sensitive to the polarity of applied signal. The application of positive voltage across a memristor decreases its memristance and vice-versa. It also behaves as a memory device, as it remembers the last memristance that it had, when the voltage applied across its terminals was turned off. Since, memristor shows improved characteristics with smaller values of 'D', hence memristor is supposed to be an important element for future nanometric devices [4, 11].

3 Sallen–Key Filter

Sallen–Key configuration is one of the most widely used filter topology which was introduced by R. P. Sallen and E. L. Key of MIT's Lincoln Laboratory in 1955. It is a single amplifier-based filter where controlled amount of positive feedback is provided

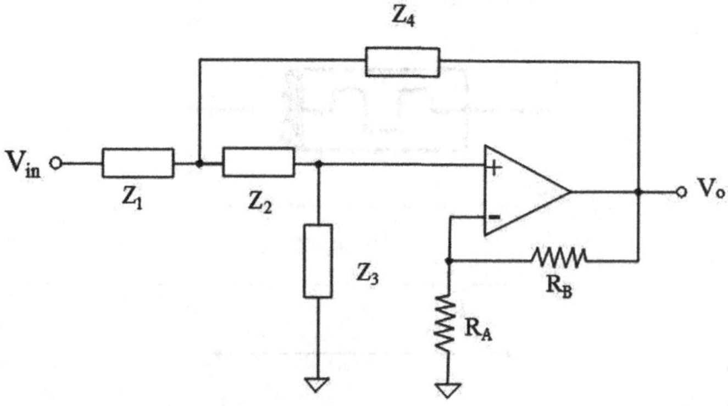

Fig. 2 Generalized circuit of Sallen–Key filter

to increase the magnitude near cut-off frequency (ω_0) to achieve high quality factor (Q). In this filter, Q can be adjusted for the desired amount by varying pass-band gain (K). Since, Sallen–Key filter is configured using op-amp as an amplifier rather than an integrator, its performance is less dependent on gain bandwidth product of the op-amp used. Moreover, in Sallen–Key filter, the component spread of resistors and capacitors being low and it is considered to be a good circuit for manufacturability [13]. The generalized circuit of Sallen–Key filter is shown in Fig. 2.

Here, Z_1, Z_2, Z_3 and Z_4 represent the generalized impedance terms and resistors R_A and R_B set the pass-band gain of the filter. For Sallen–Key low-pass filter configuration, it is assumed that $Z_1 = Z_2 = R$ and $Z_3 = Z_4 = \frac{1}{sC}$. The ideal transfer function of Sallen–Key unity-gain low-pass filter (LPF) is given as:

$$T_{LP}(s) = \frac{\frac{1}{C^2R^2}}{s^2 + \left(\frac{2}{RC}\right)s + \frac{1}{C^2R^2}} \tag{2}$$

High-pass filter (HPF) is obtained from LPF by switching the positions of resistors and capacitors in the circuit and its ideal transfer function is given as:

$$T_{HP}(s) = \frac{s^2}{s^2 + \left(\frac{2}{RC}\right)s + \frac{1}{C^2R^2}} \tag{3}$$

The characteristic parameters of these filters are defined as:

$$\omega_0 = 1/\mathrm{CR};\ \mathrm{Q} = 1/2;\ \mathrm{K} = 1 \tag{4}$$

Both low-pass and high-pass Sallen–Key filters have two poles located at $s = -1/RC$.

The circuit of unity-gain Sallen–Key band-pass filter (BPF) is shown in Fig. 3. The ideal transfer function of Sallen–Key BPF is given as:

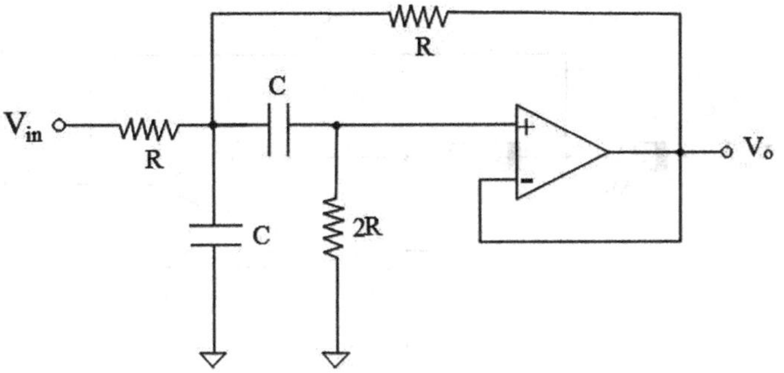

Fig. 3 Sallen–Key band-pass filter

$$T_{BP}(s) = \frac{\left(\frac{1}{RC}\right)s}{s^2 + \left(\frac{2}{RC}\right)s + \frac{1}{C^2R^2}} \tag{5}$$

For Sallen–Key BPF (Fig. 3), all the filter parameters remain unchanged (as defined by Eq. 4), except the pass-band gain at the cut-off frequency, which is given as $K = 1/2$.

4 Memristor-Based Unity-Gain Sallen–Key Filters

Memory-based Sallen–Key filters have been proposed in literature [14]. However, these filters were designed just by replacing pass-band gain resistors (R_A and R_B of Fig. 2) with memristors. They possess very restricted flexibility of electronically tuning the filter parameters. Moreover, the filters proposed in [14] were not developed in resistor-free environment, hence, they do not present all the advantages provided by memristor-based devices at low nanometer technologies. In this paper, memristor-based unity-gain Sallen–Key filters have been designed and proposed. They have been implemented by replacing all the resistors by memristors with $R_{init} = 11$ kΩ. The circuits of proposed memristor-based Sallen–Key low-pass, high-pass and band-pass filters are shown in Figs. 4, 5 and 6, respectively.

In these circuits, memristor exhibits an equivalent resistance (R_{eq}), given as:

$$R_{eq}(t) = M(t) = R_{ON}\frac{w(t)}{D} + R_{OFF}\left[1 - \frac{w(t)}{D}\right] \tag{6}$$

where $R_{ON} = 100\ \Omega$ is the memristance of the memristor in closed state, $R_{OFF} = 16$ kΩ is memristance in fully open state. Normalized width (x) of the doped layer of memristor is denoted as [4]:

$$x(t) = \frac{w(t)}{D} \tag{7}$$

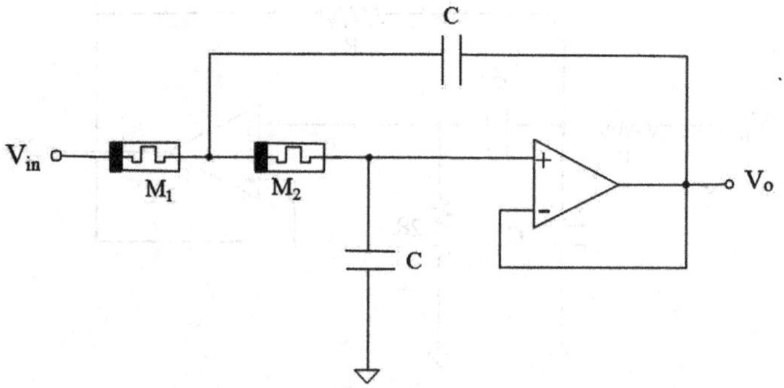

Fig. 4 Memristor-based unity-gain Sallen–Key low-pass filter

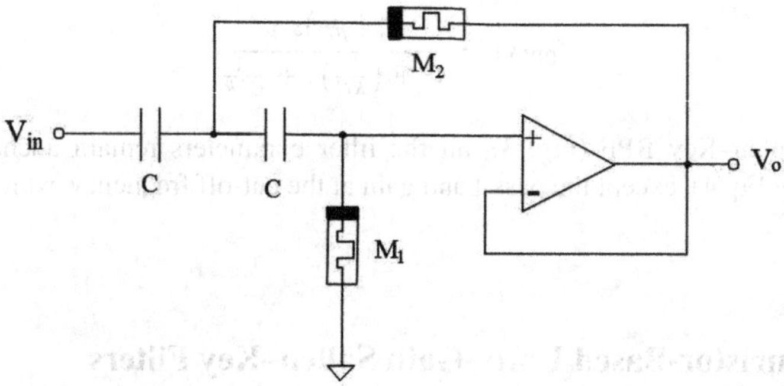

Fig. 5 Memristor-based unity-gain Sallen–Key high-pass filter

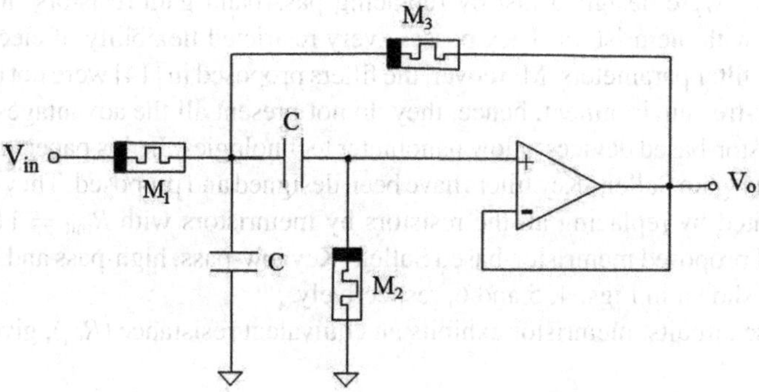

Fig. 6 Memristor-based unity-gain Sallen–Key band-pass filter

The initial resistance of memristor (R_{INIT}) determines the initial state of normalized width of the doped layer (x_0) and is given as:

$$x_0 = \frac{R_{OFF} - R_{INIT}}{\Delta R} \tag{8}$$

Memristor behavior resembles a resistor at high frequencies with memristance equal to its initial resistance (R_{INIT}). Since, its R_{INIT} can be adjusted by external voltage or current pulses, memristance can be tuned electronically. This feature of tenability allows electronic adjustment of parameters for the proposed filters, something that was not possible in conventional Sallen–Key filters. Moreover, the proposed memristor-based Sallen–Key filters occupy much less chip area, consume very less power and provide good manufacturability at low nanometer technologies, thereby providing the better performance.

The 3-dB cut-off frequency or center frequency of these filters is given as:

$$f_0 = \frac{1}{2\pi R_{\mathrm{INIT}} C} \tag{9}$$

5 Simulation Results

The behavior of proposed memristor-based unity-gain Sallen–Key filters has been studied in EldoSpice using TiO_2-based linear ion drift model. Joglekar window with p (control parameter) $= 10$ has been used to satisfy the boundary conditions of the memristor model. In all the circuits discussed above, the values of different parameters for linear ion drift model have been tabulated in Table 1.

Magnitude and phase response of memristor-based Sallen–Key unity-gain LPF are plotted in Fig. 7. It is observed that 3-dB cut-off frequency is equal to 14.46863 Hz, which is much in agreement with theoretically calculated values. Magnitude and phase response of this filter with varying R_{INIT} are plotted in Fig. 8. The curves in this figure show that corner frequency of the filter can be regulated by varying

Table 1 Design parameters for linear ion drift model for memristor

S. No.	Design parameter	Value
1.	Dopant mobility (μ_D)	10^{-10} cm^2 s^{-1} V^{-1}
2.	Thickness of TiO_2 film (D)	10 nm
3.	Off-state memristance (R_{OFF})	16 kΩ
4.	On-state (R_{ON})	100 Ω
5.	Initial resistance (R_{INIT})	11 kΩ
6.	Control parameter (p)	10

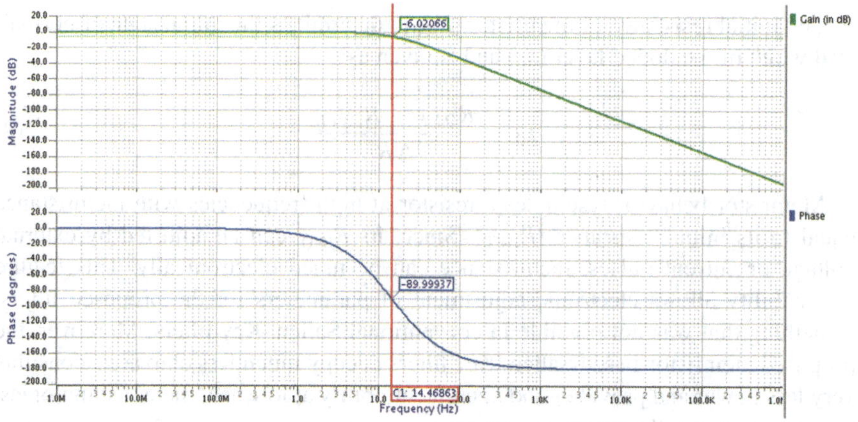

Fig. 7 Magnitude and phase response of memristor-based unity-gain Sallen–Key low-pass filter

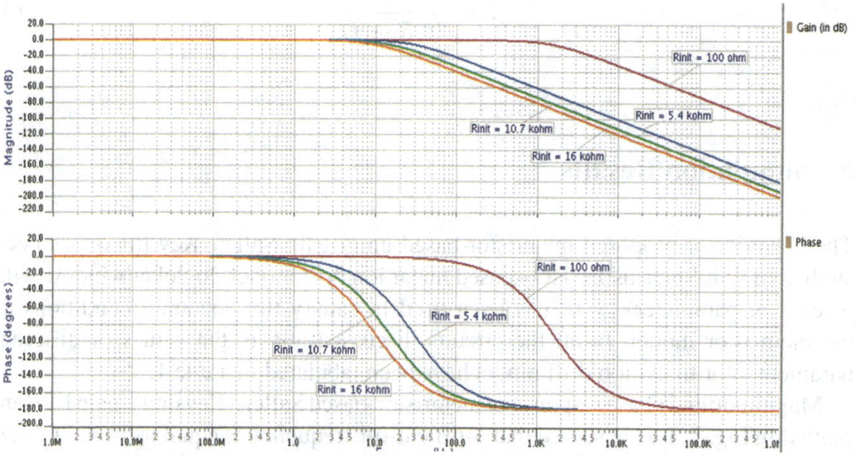

Fig. 8 Magnitude and phase response of memristor-based unity-gain Sallen–Key low-pass filter with varying R_{INIT}

R_{INIT}. Deviation in central frequency with variations in R_{INIT} is shown in Fig. 9. Magnitude and phase response of memristor-based Sallen–Key unity-gain HPF are plotted in Fig. 10 and deviation in these responses with variation in R_{INIT} is shown in Fig. 11. These responses for memristor-based unity-gain Sallen–Key BPF are shown in Figs. 12 and 13, respectively. In band-pass filter also, central frequency of 14.46863 Hz has been observed with $R_{INIT} = 11\ \text{k}\Omega$. This can be varied by changing R_{INIT}. The variation in central frequency and bandwidth of memristor-based unity-gain Sallen–Key BPF with R_{INIT} has been illustrated in Figs. 14 and 15, respectively. All these results validate the effective functionality of proposed memristor-based

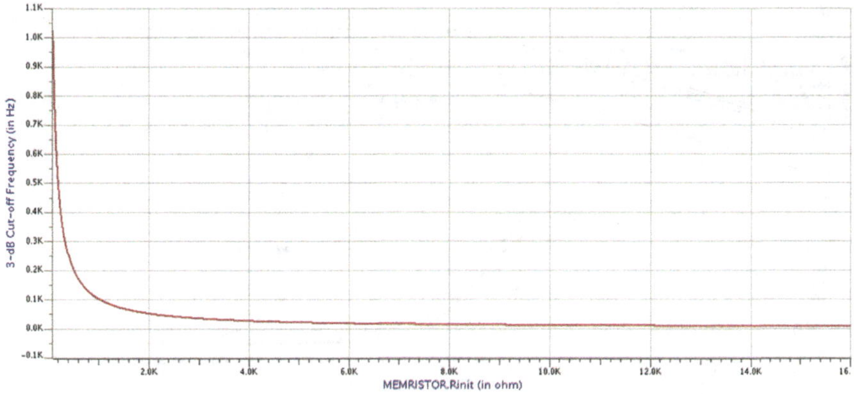

Fig. 9 Variation of 3-dB cut-off frequency of a memristor-based unity-gain Sallen–Key low-pass filter with R_{INIT}

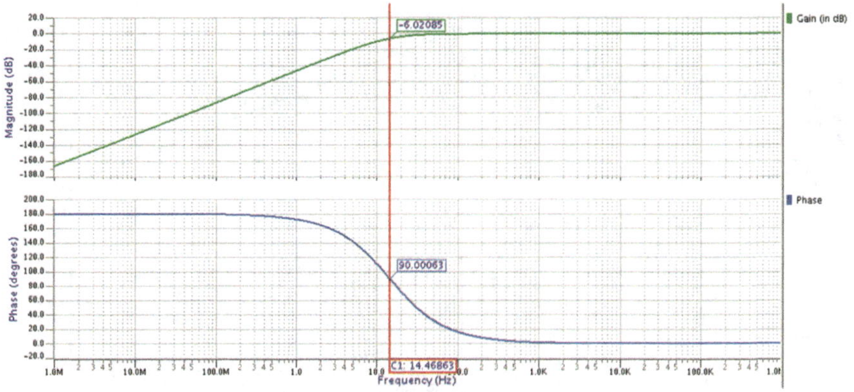

Fig. 10 Magnitude and phase response of memristor-based unity-gain Sallen–Key high-pass filter

Sallen–Key filters. Moreover, they confirm the electronic tunability of the proposed filters.

6 Conclusion

In this paper, unity-gain Sallen–Key filters (LPF, HPF and BPF) have been designed and proposed. These filters have been implemented by replacing all the resistors of Sallen–Key filters by memristors. Thereby, they can be designed in resistor-free environment and occupy very small area. Moreover, they offer high flexibility for electronic tuning of filter parameters like gain and bandwidth. The proposed filters

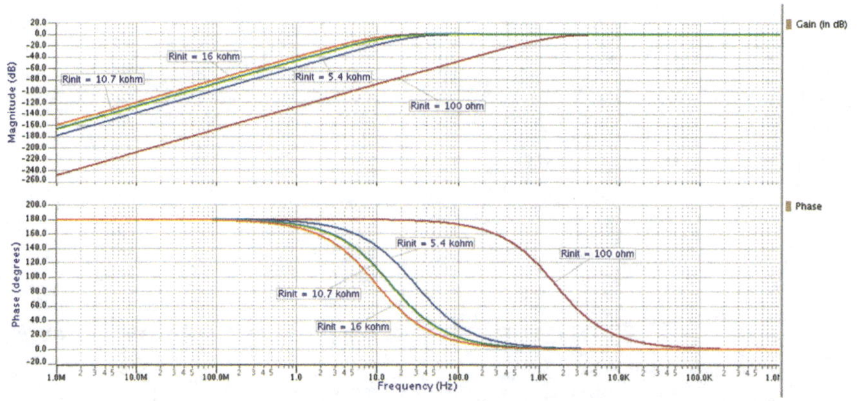

Fig. 11 Magnitude and phase response of memristor-based unity-gain Sallen–Key high-pass filter with varying R_{INIT}

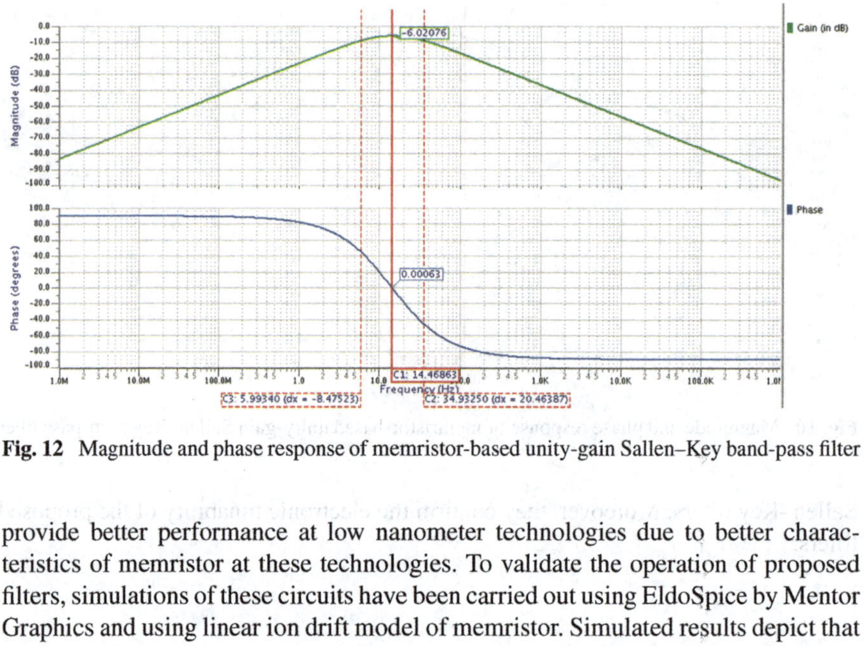

Fig. 12 Magnitude and phase response of memristor-based unity-gain Sallen–Key band-pass filter

provide better performance at low nanometer technologies due to better characteristics of memristor at these technologies. To validate the operation of proposed filters, simulations of these circuits have been carried out using EldoSpice by Mentor Graphics and using linear ion drift model of memristor. Simulated results depict that memristor behaves as a simple resistor at frequencies greater than 100 Hz. The cut-off frequency of simulated filters has been observed to be 14.46863 Hz. This value and thereby bandwidth and Q-factor can be electronically regulated by changing R_{INIT}.

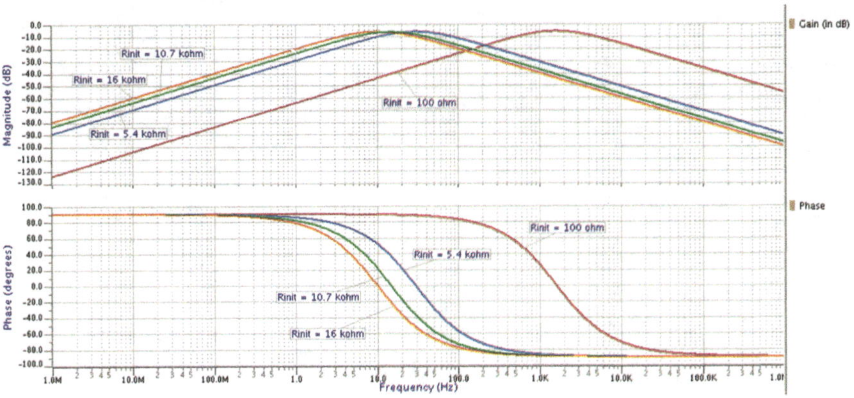

Fig. 13 Magnitude and phase response of memristor-based unity-gain Sallen–Key band-pass filter with varying R_{INIT}

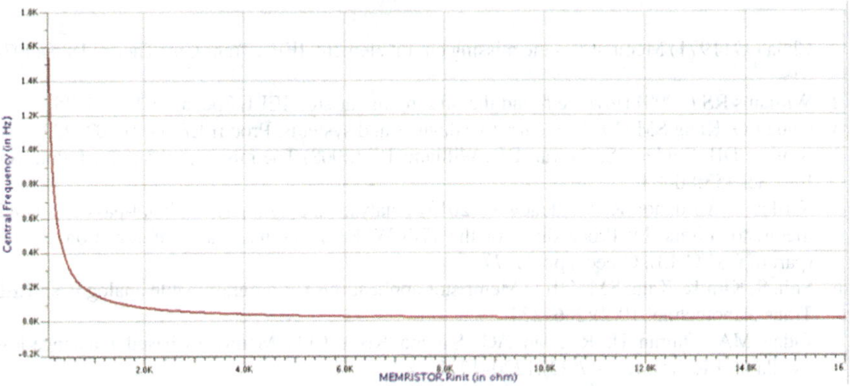

Fig. 14 Variation in central frequency of a memristor-based unity-gain Sallen–Key band-pass filter with R_{INIT}

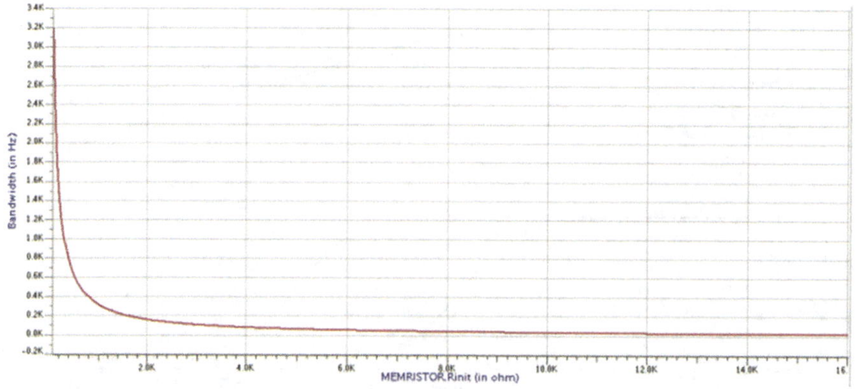

Fig. 15 Variation in bandwidth of a memristor-based Sallen–Key band-pass filter with R_{INIT}

References

1. Chua LO (1971) Memristor—the missing circuit element. IEEE Trans Circ Theory 18(5):507–519
2. Williams RS (2008) How we found the missing memristor. IEEE Spectr 45(12):28–35
3. Chua LO, Kang SM (1976) Memristive devices and systems. Proc IEEE 64(2):209–223
4. Strukov DB, Snider GS, Stewart DR, Williams RS (2008) The missing memristor found. Nat Int J Sci 453:80–83
5. Kirilov S, Yordanov R, Mladenov V (2013) Analysis and synthesis of band-pass and notch memristor filters. In: Proceedings of the 17th WSEAS international conference on circuits (part of CSCC'13), Greece, pp 74–77
6. Shin S, Kim K, Kang SM (2011) Memristor applications for programmable analog ICs. IEEE Trans Nanotechnol 10(2):266–274
7. Zidan MA, Omran H, Radwan AG, Salama KN (2011) Memristor-based reactance-less oscillator. Electron Lett 47(22):1220–1221
8. Bayat FM, Shouraki B (2011) Memristor-based circuits for performing basic arithmetic operations. Procedia Comput Sci 3:128–132
9. ElSamman AH, Radwan AG, Madian AH (2012) Memristor-based oscillator using Deboo integrator. In: 7th International conference on computer engineering & systems (ICCES), Cairo, Egypt, pp 103–107
10. Biolek Z, Biolek D, Biolkova V (2009) SPICE model of memristor with nonlinear dopant drift. Radioeng J 18(2):210–214
11. Kim H, Sah MP, Yang C, Cho S, Chua LO (2012) Memristor emulator for memristor circuit applications. IEEE Trans Circ Syst I Regul Pap 59(10):2422–2431
12. Batas D, Fiedler H (2011) A memristor SPICE implementation and a new approach for magnetic flux-controlled memristor modeling. IEEE Trans Nanotechnol 10(2):250–255
13. Schaumann R, Xiao H, Valkenbur MEV (2010) Analog filter design, 2nd edn. Oxford University Press, USA
14. Yener SC, Muylu R, Kuntman HH (2015) Memory-based Sallen-key filters. J Fac Eng Archit Gazi Univ 30(2):173–184

Influence of Target Fields on Impact Stresses and Its Deformations in Aerial Bombs

Prahlad Srinivas Joshi and S. K. Panigrahi

Abstract A typical aerial bomb has high explosive packed in a metallic cylindrical shell. The bombs explode due to the detonation of high explosive either on impact or in proximity to the target. The metallic shell on explosion disintegrates into various small fragments which strike the various target fields contributing to the fragmentation effect. The present work mainly focuses on finite element (FE) simulation of aerial bomb impact on target fields like sand, concrete and steel using Ansys explicit dynamics for imported three-dimensional model of aerial bomb using solid edge. FE simulation was carried out to find stresses and corresponding deformations within the bombshell for three different target fields. Further, the impact simulation results were compared for the deformation and stress limits of the bombshell material.

Keywords Drop impact · Shell deformation · Target filed · Aerial bomb · FE simulation

1 Introduction

The bombs dropped from aerial platforms are termed as aerial bombs. A typical aerial bomb has huge explosive filled in a steel cylindrical shell. The bombs explode due to the detonation of high explosive either on impact or in proximity to the target. Hence, the bomb has a blast effect. The metallic shell on explosion disintegrates into various small fragments which strike the target contributing to fragmentation effect.

The explosion in a contained atmosphere produces more devastating effect. The bombshell provides this atmosphere till explosion. The material of the bombshell has to withstand the high-velocity impact after drop and also the internal pressure generated due to explosion. The bombshells are usually made either cast or forged or carbon steel [1]. The blast effect is characterised by a wave which comprises of shock wave and blast wind [2]. The blast wave is quantified by its peak pressure and its duration. The peak pressure is dependent on net weight of explosive inside the

P. S. Joshi (✉) · S. K. Panigrahi
Department of Mechanical Engineering, DIAT, Girinagar, Pune 411025, India
e-mail: prahjo20904@gmail.com

© The Author(s), under exclusive license to Springer Nature Singapore Pte Ltd. 2021
R. M. Singari et al. (eds.), *Advances in Manufacturing and Industrial Engineering*, Lecture Notes in Mechanical Engineering,
https://doi.org/10.1007/978-981-15-8542-5_102

bombshell and velocity of detonation of that explosive. A peak pressure of 138 kPa is sufficient to ensure heavy structural damages on concrete buildings and fatalities to humans [3]. Srivastha and Ramakrishnan [4] presented ballistic performance of various metallic materials but there is scope for extending his work by choosing different target fields. Moxley et al. [5] experimentally evaluated response of thin-walled steel projectiles on concrete targets. Xu et al. [6] experimentally calculated peak air and ground pressure for composition B (RDX + TNT trinitrotoluene), research development explosive (RDX) and aluminised explosive (RDX + TNT + Aluminium powder) weighing 2 kg each. It was observed that RDX has a peak ground pressure of 372.91 kPa at a distance of three metres and 0.030 s post-detonation. This value was the highest among all the three explosives tested which is extrapolated for numerical studies in the paper.

This work aims at finding impact stresses and its corresponding deformations induced in aerial bombshell during drop impact on various target fields using Ansys explicit dynamics FE simulation software. The simulation is done on legacy Mk-82 bomb [7]. The FE simulation is presented for three different types of target fields like sand, concrete and steel for the bombshell. The simulation results are compared with the ultimate limits of corresponding bombshell material.

2 3-D Modelling of Aerial Bomb

By knowing the approximate dimensions of the aerial bomb as given in Table 1, a three-dimensional model was developed by the aid solid edge software shown in Fig. 1.

2.1 Finite Element Modelling/Meshing

To generate finite element mesh for aerial bombshell, a 3D model was imported from solid edge and choosing Tetrahedron element form Ansys element library for meshing in the Ansys software. The details of the finite element mesh for aerial bomb and target field are displayed in Fig. 2.

Table 1 Weight and dimensions of Mk-82 bomb for FE simulation

Bomb	Total weight of bomb	Explosive weight	Length of bomb	Diameter of bomb
Bomb 1, 2, 3	500 lb (227 kg)	192 lb (87 kg)	87.4 in. (2200 mm)	10.75 in. (273 mm)

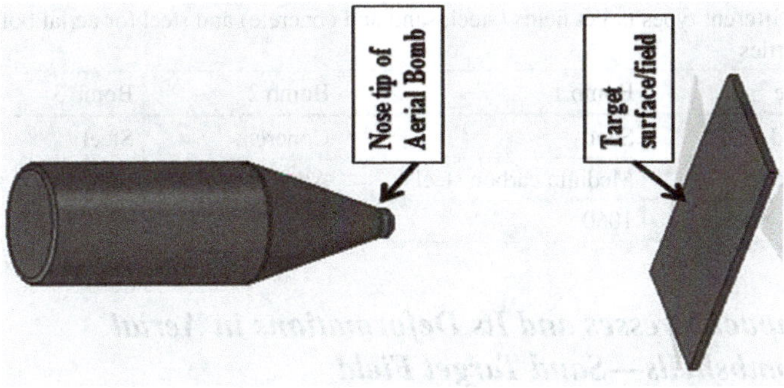

Fig. 1 3D model of aerial bomb and target filed

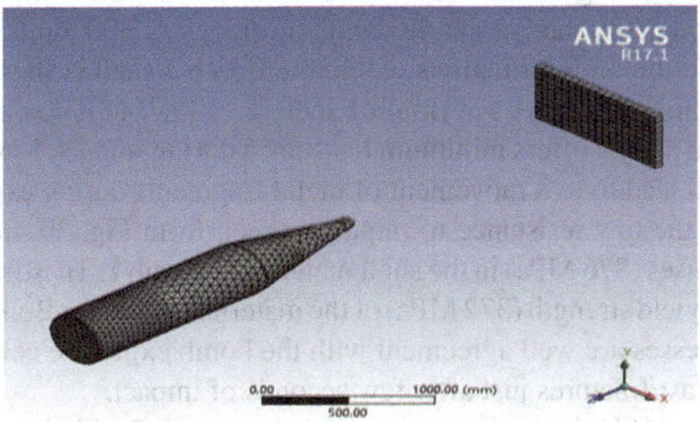

Fig. 2 Finite element (FE) meshing for aerial bomb of a fighter aircraft

3 FE Simulation of Aerial Bomb with Different Target Fields—Results and Discussions

The FE simulation of aerial bomb was done using Ansys explicit dynamics [8] with the following material properties, and there self-weight is listed in Table 2 as inputs to the software. The various constraints used for impact simulation are illustrated in Table 3. Three types of target fields which are given in Table 3 were considered as the primary constraints for FE analysis of the aerials bombs.

Table 2 Material properties and weight of the empty bombshell for FE simulation

Bomb type	Weight of the bomb in N	Density (kg/mm^3)	Young's modulus (MPa)	Poisson's ratio	Bulk modulus (MPa)	Shear modulus (MPa)
Bomb 1, 2, 3	2227	7.85×10^{-6}	2.016×10^5	0.26	1.4×10^5	80,000

Table 3 Different types target fields (steel, sand and concrete) and steel for aerial bombshell and their properties

Bomb type	Bomb 1	Bomb 2	Bomb 3
Target filed	Sand	Concrete	Steel
Type of steel	Medium carbon steel	Alloy steel	Low carbon steel
SAE AISI grade	1060	4140	1215

3.1 Impact Stresses and Its Deformations in Aerial Bombshells—Sand Target Field

The FE models had pre-defined drop height with respective time duration and well-defined sand has the target fields are referred to Bomb 1. Figure 3 indicates the impact of Bomb 1 through the sand target filed with a velocity of 300 mm/s at a distance of 10 m. The variation of deformations within the Bomb 1 shell is shown in Fig. 3a. It was seen that the deformations in Bomb 1 shell were low but it was huge in the sand target filed, and also, it offers minimum resistance during impact. Low deformations within the shell lead to less movement of metal fragments during explosions.

Because of the low resistance to impact as seen from Fig. 3b, that there are no significant stresses (876 MPa) in the shell material of Bomb 1. These stress values are well above the yield strength (372 MPa) of the material chosen for Bomb 1 simulation. The impact stresses are well agreement with the bomb explosive concepts in which bomb has to blast/fractures just after few seconds of impact.

The same set of FE simulation was extended to Bomb 2 with drop height of 10 m but here it was impacted on to a concrete target field. The maximum deformations and impact stresses during impact of Bomb 2 are displayed in Fig. 4.

FE simulation results showing similar trend as sand target filed of Bomb 1 but little lower deformations (Fig. 4a) are found in concrete target fields as in case of

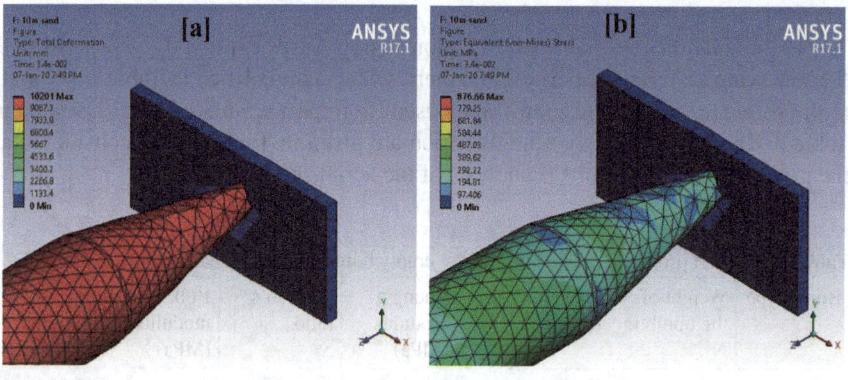

Fig. 3 FE results showing **a** deformations in shell of Bomb 1 and **b** stress distribution for Bomb 1 after impact to sand as the target filed

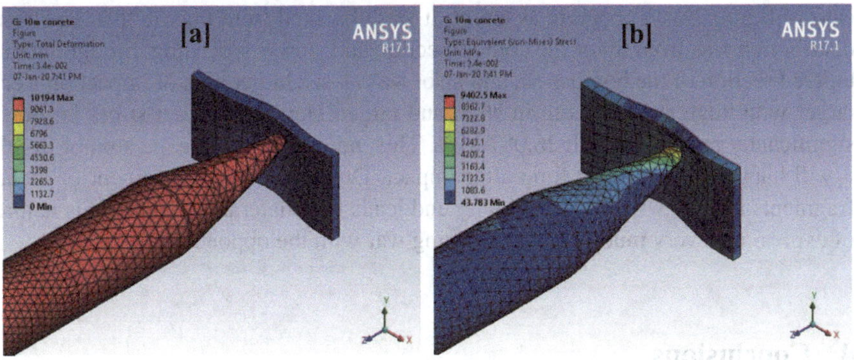

Fig. 4 FE results showing **a** deformation and **b** stress distribution for Bomb 2 impact on concrete as the target filed

Bomb 2. However, due to huge resistance from concrete target leads to large amount of stresses within the conical section of the Bomb 2 were noticed (Fig. 4b); this is due to the influence of higher stiffened concrete target. This large deformation and stress on the conical shell of Bomb 2 are due to gained potential energy leading metal fragments moving at higher velocity resulting into collateral damages. This nature of explosion is sometimes well-suited for attacks in enemy's territory but not suited in case of deployment within the land for Guerilla warfare/anti-naxal operations.

Finally, same FE simulation was done using steel has the target filed as in case of Bomb 3 instead of sand and concrete in Bomb 1 and Bomb 2. Impact simulation results of stresses and deformation induced in shell and target of the Bomb 3 are depicted in Fig. 5.

An attempt was made to know the deformations and its stresses within the source and target filed using Bomb 3 (steel bomb and steel target) impact is shown in Fig. 5a and b. From Fig. 5a, it is evident that the deformation is all most nearly same as in

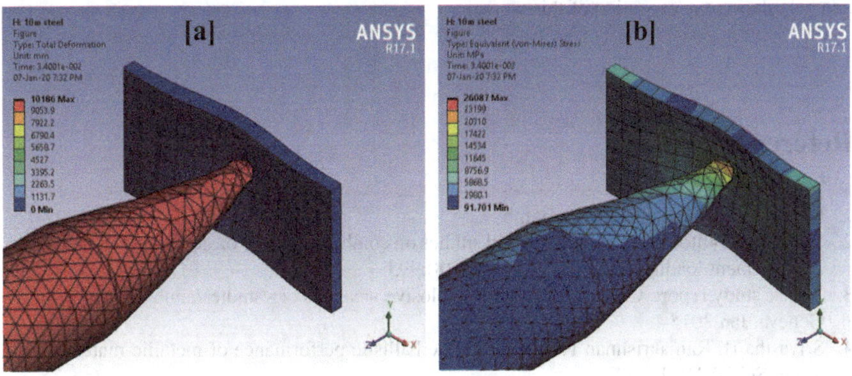

Fig. 5 Impact FE simulation results showing **a** deformation and **b** impact. Stress distribution for Bomb 3 impact on steel as the target filed

case of Bomb 2 with concrete as target. It was observed from Fig. 5b that the stress distribution was from tip of the conical section and slowly spreading throughout the tapered section of the bombshell, and also, we can see larger area of impact on steel target with huge stresses both in shell and target. Here the impact stress is found significantly maximum with 26,087 MPa. This indicates that the explosion of Bomb 3 will happen within short time after impact. Due to this, the movement of metal fragments is huge with higher velocity and leads to collateral damages. This nature of explosion is very much benefited during war with the opponents.

4 Conclusions

Finite element simulation of aerial bomb impact was successfully done using Ansys explicit dynamics R17 by considering three different target fields like sand, concrete and steel with constant drop height and velocity of impact; the following conclusions were drawn from the above FE simulation.

For sand target as in case of Bomb 1, the deformations induced are maximum in target field but the stresses were seen minimal with the value 876 MPa as compared to Bomb 2 and Bomb 3 with concrete and steel target fields. However, these stress well above the yield strength of the materials chosen for the Bomb 1.

Due to stiffened target fields, the impact stresses induced are significantly high with Bomb 2 and Bomb 3 with respect to Bomb 1. But the deformations were nearly same in both the cases (Bomb 2 and Bomb 3).

Finally, it was concluded that as we increase the stiffness of the target a significant enhancement in deformations and associated stress values within the bombshell and target fields caused higher damage, which indicates the large movements of the metal fragments spreads at higher velocity during impact/explosions.

Acknowledgements The authors would like to thank Dr. P. S. Shivakumar Gouda Department of Mechanical Engineering, SDM College of Engineering and Technology, Dharwad, for their support and help during the execution of this work.

Reference

1. https://en.wikipedia.org>wiki>bomb
2. Nystro U, Gylltoft K (2009) Numerical studies on combined effects on combined effects of blast and fragment loading. Int J Impact Eng 36(8):995
3. A case study report. Characterisation of explosive weapons.org/studies/annex-emk82-GICHD, Geneva, Jan 2015
4. Srivastha B, Ramakrishnan N (1997) On the ballistic performance of metallic materials. Bull Mater Sci 20(1):111–123
5. Moxley RE, Adley MD, Rohani B (1995) Impact of thin walled projectiles with concrete targets. Shock Vibr 2(5)355–364

6. Xu W, Wang C, Yuan J, Goh W, Li T (2019) Investigation on energy output structure of explosives near ground explosion. Defence Technol. Available online 10 Aug 2019
7. https://en.wikipedia.org>wiki>Military.wikia.org/wiki/Mark-82-bomb
8. Ansys 17 explicit dynamics user manual

A Review of Vortex Tube Device for Cooling Applications

Sudhanshu Sharma, Kshitiz Yadav, Gautam Gupta, Deepak Aggrawal, and Kulvindra Singh

Abstract Vortex tube devices are mechanical devices. The vortex tube works on the phenomenon referred to as temperature (energy) separation and is a non-conventional cooling system. The absence of moving components increases reliability, reduction in maintenance, and an increase of system life, with air as a working fluid, avoids dangerous environmental leakages. This article begins with a brief overview of vortex tube devices, construction, and operation. A review of recent developments in research, commercial developments, and typical applications of vortex tubes has been done. The article concludes the growth and potential use of vortex tube devices suitable for vortex tube cooling.

Keywords Vortex tube · Vortex tube cooling · Ranque Hilsch vortex tube

1 Introduction

A vortex tube is a mechanical device that generates streams of hot air on one side and cold air on another side. The vortex tube was invented in 1932 by a student named G. J. Ranque. Ranque was able to create two streams of different temperatures successfully. In 1945, Rudolf Hilsch, a German physicist, improvised on the designs provided by Ranque. Hilsch was able to measure the temperatures of inlet and outlet streams but could not explain the process behind it. The device was later named Ranque–Hilsch vortex tube. The vortex tube works on the phenomenon referred to as temperature (energy) separation.

S. Sharma (✉) · K. Yadav · G. Gupta · D. Aggrawal · K. Singh
Department of Mechanical Engineering, Galgotias College of Engineering and Technology, Greater Noida, Uttar Pradesh 201306, India
e-mail: sudhanshu.shr@gmail.com

© The Author(s), under exclusive license to Springer Nature Singapore Pte Ltd. 2021
R. M. Singari et al. (eds.), *Advances in Manufacturing and Industrial Engineering*, Lecture Notes in Mechanical Engineering,
https://doi.org/10.1007/978-981-15-8542-5_103

2 How It Works

A vortex tube is a device that generates cold gas at one end and hot gas at another end by using compressed air as input. It consists of a tangential inlet nozzle, an axial tube with a vortex chamber, a cold gas outlet, a hot gas outlet, and a control valve. Highly compressed air is injected through the inlet into the vortex generator chamber tangentially through inlet nozzles. The accelerating air through the inlet nozzles exits into a cylindrical section that is the vortex generator chamber and generates an intense vortex. The vortex proceeds toward the hot end side of the tube to the conical back pressure control valve. The valve does not eject all the gas striking it, a portion of the gas reverses its direction and moves axially along the center of the tube toward the cold side of the against the oncoming hot flow. As the cold flow moves toward the cold side, it looses its thermal energy to the outer flow, eventually raising its temperature. The temperature of the peripheral flow exceeds the inlet pressure. The phenomenon of energy separation is known as Ranque effect. The hot gas exits the tube from the area between the conical control valve and the tube wall. The gas of the core flow exits from the orifice and the cold end side tube. The conical control valve controls the relative flow rates of the hot and cold tube.

3 Thermodynamic Analysis of Vortex Tube

Figure 1 shows an actual vortex tube. It gets high-pressure air from a compressor through a tangential nozzle. Suffixes c, h, and i stand for cold end, hot end, and inlet to the nozzle, respectively.

The mass and energy conservation for the control volume gives

$$\dot{m}_i = \dot{m}_c + \dot{m}_h \tag{1}$$

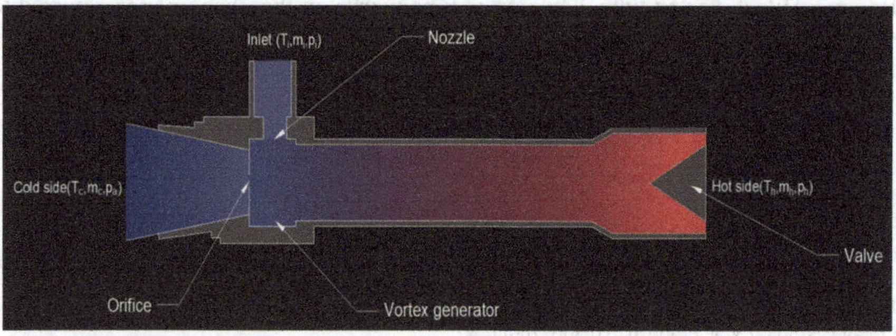

Fig. 1 Schematic diagram of a vortex tube

$$\dot{m}_\mathrm{i} h_\mathrm{i} = \dot{m}_\mathrm{c} h_\mathrm{c} + \dot{m}_\mathrm{h} h_\mathrm{h} \tag{2}$$

where the net change in kinetic energy has been ignored.

Energy balance between hot side and cold side

$$\dot{m}_\mathrm{c}(T_\mathrm{i} - T_\mathrm{c}) = \dot{m}_\mathrm{h}(T_\mathrm{h} - T_\mathrm{i}) = (m_\mathrm{i} - \dot{m}_\mathrm{c})(T_\mathrm{h} - T_\mathrm{i}) \tag{3}$$

If $\mu = \dot{m}_\mathrm{c}/\dot{m}_\mathrm{i}$ is the cold air fraction, then from Eq. (3),

$$\mu = (T_\mathrm{h} - T_\mathrm{i})/[(T_\mathrm{h} - T_\mathrm{i}) + (T_\mathrm{i} - T_\mathrm{c})] = \Delta T_\mathrm{h}/(\Delta T_\mathrm{h} + \Delta T_\mathrm{c}) \tag{4}$$

Isentropic expansion from inlet of the nozzle to atmospheric pressure, p_a, the temperature is obtained to be

$$T'_\mathrm{c} = T_\mathrm{i}(p_\mathrm{a}/p_\mathrm{i})^{(\gamma-1)/\gamma} \tag{5}$$

Then, maximum possible temperature drop

$$\Delta T'_\mathrm{c} = T_\mathrm{i} - T'_\mathrm{c} = T_\mathrm{i}\left[1 - (p_\mathrm{a}/p_\mathrm{i})^{(\gamma-1)/\gamma}\right] \tag{6}$$

The ratio is defined as

$$\Delta T_\mathrm{rel} = \Delta T_\mathrm{c}/\Delta T'_\mathrm{c} = \left(\frac{\Delta T_\mathrm{c}}{T_\mathrm{i}}\right)/\left[1 - (p_\mathrm{a}/p_\mathrm{i})^{(\gamma-1)/\gamma}\right] \tag{7}$$

Vortex tube isentropic efficiency:

$$\eta = \frac{\text{Actual cooling}}{\text{Ideal cooling}} = \frac{\dot{m}_\mathrm{c} c_\mathrm{p} \Delta T_\mathrm{c}}{\dot{m}_\mathrm{i} c_\mathrm{p} \Delta T'_\mathrm{c}} = \mu \Delta T_\mathrm{rel}$$

$$\eta = \mu \Delta T_\mathrm{c}/\left[T_\mathrm{i}\left\{1 - (p_\mathrm{a}/p_\mathrm{i})^{\gamma-1/\gamma}\right\}\right] \tag{8}$$

Energy supplied to the compressor

$$\dot{W} = \dot{W}_\mathrm{ideal}/\eta_\mathrm{c} = \dot{m}_\mathrm{i} c_\mathrm{p} T_\mathrm{a}\left[(p_\mathrm{i}/p_\mathrm{a})^{\gamma-1/\gamma} - 1\right]/\eta_\mathrm{c} \tag{9}$$

COP of vortex tube refrigeration system

$$\mathrm{COP} = \frac{\text{Actual cooling}}{\text{Energy supplied}}$$

$$\mathrm{COP} = \frac{\dot{m}_\mathrm{c} c_\mathrm{p} T_\mathrm{c}}{\dot{m}_\mathrm{c} c_\mathrm{p} T_\mathrm{a}\left[(p_\mathrm{i}/p_\mathrm{a})^{(\gamma-1)/\gamma} - 1\right]/\eta_\mathrm{c}} \tag{10}$$

Using Eq. (8) into Eq. (10), we get

$$\mathrm{COP} = (T_\mathrm{i}/T_\mathrm{a})\eta\eta_\mathrm{c}(p_\mathrm{a}/p_\mathrm{i})^{(\gamma-1)/\gamma} \tag{11}$$

4 Nomenclature

m_i	Mass flow rate of air at the inlet
m_c	Mass flow rate of air at cold side
m_h	Mass flow rate of air hot side
c_p	Specific heat capacity of air
η_c	Compressor efficiency
ΔT_c	Cold temperature difference
T_h	Hot side temperature
T_i	Inlet temperature
T_c	Cooling side temperature
T_a	Atmospheric temperature
P_a	Atmospheric pressure
P_i	Inlet pressure of compressed air
η	Vortex tube isentropic efficiency
COP	Coefficient of performance
γ	Heat capacity ratio
μ	Cold air fraction
h_h	Enthalpy of hot air
h_c	Enthalpy of cold air
h_i	Enthalpy of inlet air
T_c	Ideal temperature of air at cold side.

5 Benefits of Vortex Tube

The vortex tube is a mechanical device that uses air as a working medium. Conventional cooling devices use condensers, compressors, expansion while vortex tube cooling device uses the only compressor. This fundamental difference gives the vortex tube the following advantages over conventional devices.

1. It is environment-friendly as it uses air as a working medium.
2. It can be manufactured in a compact size using less space.
3. Lightweight, which is advantageous where lightweight is more important than COP.
4. The rate of refrigeration and COP can be easily controlled by varying the mass flow rate and inlet pressure.

6 State of the Art Review

6.1 *Performance Investigation*

There have been countless studies done over the past to get a better picture of working and enhance the COP of the vortex tube. In a study, CFD model or computational fluid dynamics model was used to investigate the thermal process and energy separation phenomenon occurring inside a vortex tube. The phenomenon of energy separation is due to the shear force acting on a rotating control surface, which distinguishes the two thermal regions. The transfer of heat in a vortex tube is from cold to hot regions, whereas the transfer of work is vice versa; this in turn reduces the temperature separation effect [1]. An experimental study shows that there is a significant energy separation effect for some inert gases below 80 K [2]. The COP of a vortex tube may increase by using fins (parallelogram and rectangular fins) due to the increase in the temperature difference of the cold side of the tube [3]. Tangential inlet nozzle is used in a study for studying the effect of rotating flow, performance and efficiency of the vortex tube [4]. Studies are being performed exploring the temperature separation in a counter-flow vortex tube. The vortex transforms from a forced vortex at the inlet region to a free vortex at the hot end, which shows vortex transformation [5]. The flow property is analyzed in a study, and the energy distribution density along with the vortex tube velocity was calculated from which the reason for separation was identified [5]. There is a specific and regular flow structure in the intense turbulence of the vortex tube. The flow structure has a significant role in the energy separation process and performance. It reduces the temperature of the air, which revolves opposite hot tube air; that is why the temperature of the cold air decreases, and COP of the vortex tube increases [6]. Between two vortex tubes operating at different pressures, the vortex tube with the higher pressure will also have better efficiency [7]. Depending upon the inlet pressure and cold mass ratio, the curvature has a different effect on the performance of the vortex tube. The maximum temperature difference is of the straight vortex tube, and the maximum refrigeration capacity is of the curved vortex tube. Non-dimensional and cold temperature differences and non-dimensional refrigeration capacity are independent of the curvature of the vortex tube [8]. The energy separation phenomenon is the separation of a single air stream into hot and cold streams. Pressure energy is converted into thermal energy by energy transfers due to shear in different directions [9]. Hot and cold streams are generated by a vortex tube when a fluid is injected from the inlet. The hot region of the tube has no energy transfers [10]. The vortex tube has two concentric vortexes with opposite flow directions. The primary vortexes extract energy from the secondary backflow vortex and transfer it to the outer region [11].

6.2 Design Modification

In a study, the shape and design of the vortex chamber have been modified, and the numerical analysis of the vortex tube with cyclonic type extension of the vortex chamber was performed. By carefully designing the vortex chamber, a high capacity vortex tube was generated [12]. The diameter and the number of inlets are also factors that determine the coefficient of performance of a vortex tube. The outlet rate of flow increases with the increase in the diameter and length of the vortex tube. With the increase in the number of inlets, the temperature at both exits decreases [13]. Annular vortex tube (AVT) and regular vortex tube are compared in a study, and the result shows that the annular vortex tube enhances cooling by about 24% compared to regular vortex tube [14]. In a study, a vortex tube is connected to a natural gas pipeline with a constant pressure of 4 bar. To improve vortex tube efficiency, six generators with different cold orifice angles, five generators with different cold orifice diameter, and three generators with varying areas of nozzle are studied. The result shows that variation of the nozzle area does not affect optimum cold mass fraction while cold mass angle and cold mass diameter move this point. As the angle of the optimum cold increases while the mass flow increases and temperature decreases due to this, the COP of the vortex tube increases [15]. If two sets of vortex generators are tested under different inlet pressure with dry air as working fluid, then the nozzle aspect ratio affects the energy separation phenomenon. Larger the aspect ratio better will be the performance of the vortex tube [16]. Compressed carbon dioxide is used as a working fluid and nozzles which have 2, 4, 6 orifices, which are made up of polyamide and brass material with pressure 150–550 kPa. By using the following design, the coefficient of performance is enhanced [17]. The instability of the finite length vortex tube is shown.

The core of the vortex tube develops axial flow directed inwards from both ends of the vortex tube. This disrupts the vortex structure [18]. The nozzle material also determines the performance of the vortex tube. Studies show that for air as working fluid, steel nozzle proves to be better [19]. Maximum cooling was at the lowest *do*/*D* ratio and at the smallest mass fraction and maximum heating effect at the highest *do*/*D* ratio at the highest possible. Tube fins can be added to increase the cooling effect [20]. In a convergent type vortex tube, cold end temperature drop and cold end mass fraction can be increased, the temperature drop was 3%, and COP was 102% compared to regular vortex tube [21].

Performance of the vortex tube energy separation under different geometrical parameters, mainly tubes length, diameter, and internal tapering angle, concludes that the greater the inlet pressure, the higher the temperature difference [22]. The vortex tube possesses higher cold air temperature and efficiency when hot tubes if cooled by water [23]. Vortex tube based on the principle of energy separation. From the vortex tube, CO_2 can be separated from the air at a fixed pressure and cold mass fraction ratio. In this paper, it is shown that the percentage of mass fraction of CO_2 at cold exit is 7.4% at 3 bar and a mass fraction of 0.6 [24]. By using a truncated cone throttle valve, the efficiency will improve. From this paper, we find that the optimum cone length is

6 mm [25]. In this paper, the thermal and flow characteristics of a plane and curved vortex generator with and without punched holes on their surface are investigated, and the corresponding mechanism of heat transfer is analyzed. The study suggests that the ratio of hole area to the vortex generator area should be optimized to achieve better heat transfer. And punching hole at a lower position and close to the leading edge gives a better thermal-hydraulic performance [26]. By experimental evaluation, it was found that the maximum performance of the vortex tube was obtained with aluminum nozzles with nozzle number 6 at 550 kPa inlet pressure [27]. High inlet pressure results to lower temperature distribution. At the same inlet–outlet pressure ratio, larger vortex tube size leads to high cold temperature difference but has very little influence on hot temperature difference [28]. Vortex tubes of finite length are unstable. They have an instability which disrupts the velocity structure quickly. Axial flow is generated within the core of vortex directed inwards from both ends of the tube. This disrupts the vortex structure [18]. By the experiment and study about the vortex tube, it depends on the injection pressure in the vortex tube, and the flow generation into the vortex tube due to this more temperature difference can be generated to give better efficiency [10]. The performance of the vortex tube depends upon several factors such as inlet pressure (P_i), conical valve angle (Q), length to diameter ratio (L/D), and mass fraction if accurately used these parameters give better performance in the efficiency of the vortex tube [29].

6.3 Altering Parameters

The higher the inlet pressure, the maximum is the temperature difference, and energy separation decreases when the inlet nozzle flow reaches choked condition [22]. The efficiency of a vortex tube is enhanced when the hot side is cooled [23]. When the inlet parameters like temperature, pressure, velocity, and fluid material are chosen correctly, the performance of the vortex tube can be enhanced [30]. In a detailed study, the maximum cooling effect and the maximum heating effect were obtained when the cold mass fraction was 0.3 and 0.8, respectively [31]. The cold end temperature drop increases with the increase in cold mass fraction and decreases for insulated and non-insulated cases. Hot end temperature drop increases for a whole range of cold mass faction for non-insulated one [32]. The vortex tube performs better when CO_2 is used as a working fluid instead of air or nitrogen [33]. A vortex tube is a simple device with an industrial application and is used for spot cooling devices in the industry. The fluid enters the vortex tube through one or more nozzles and produces a cold vortex flow. From the previous result, we come to know that the number of nozzles has a vital role in the performance of the vortex tube. But in the present study, a circular model is introduced and examined mathematically for reducing the effect of several nozzles. From this, it can be seen that the impact of nozzle number can be reduced [34]. The hydrogen as a working medium lowers the temperature difference by 7–8 K than when air is used [35]. During the study in this experiment, dry air is used as a working fluid, with a single nozzle, gives better efficiency than two and multiple nozzles [16]. In this experiment it was observed that certain geometric parameters, including nozzle type, hot tube and cold tube with proper diameter, setting of the length of the vortex tube and the flow inside the tube give better efficiency [5]. A study done on vortex tube found that its performance depends upon three different parameter inlet pressure and mass flow ratio of the cold stream to the mass flow ratio of the hot stream [36].

6.4 Using Devices and Attachments

The double circuited vortex tube is used instead of a regular vortex tube to create a refrigeration device with high thermodynamic efficiency [37]. Analysis of the thermo-hydraulic characteristics of the heat exchanger fitted tube. If the dimensions are chosen under appropriate length ratio and the vortex tube pitch ratio, the heat transfer rate increases with a moderate increase in the pressure drop. The result is indicated that the COP of vortex tube increases due to an increase in heat transfer rate [38]. A Peltier device is used to extract heat from the hot tube and transfer it to the surrounding. It reduces the temperature of the hot vortex tube side; that is why its temperature reduced than the cooling effect of refrigeration increases due to this coefficient of performance of the vortex tube increases [39].

7 Conclusion

Vortex tube is a promising and emerging technology with a simple design and working. The only downside is its low coefficient of performance. Many studies have been done, and different methods have been adopted to enhance the COP of the device. The coefficient of performance can be enhanced by altering various parameters and specifications. The usage of different fluids, nozzles of different materials, use of thermoelectric modules, and inclusion of heat exchangers can also alter the COP of the device. Change in design can also change the amount of airflow and vortex formation eventually altering the output hence the coefficient of performance. Using compressed air at low temperature and high pressure also enhances the coefficient of performance. This review throws light on various dimensions to improve the performance of vortex tube.

References

1. Aljuwayhel NF, Nellis GF, Klein SA (2005) A parametric and internal study of the vortex tube using a CFD model. Int J Refrig 28(3):442–450
2. Liu JY, Gong MQ, Wu JF, Cao Y, Luo EC (2005) An experimental research on a small flow vortex tube at low temperature ranges. In: Proceedings of the twentieth international cryogenic engineering conference (ICEC20). Elsevier Science, Amsterdam, pp 165–168
3. Bazgir A, Heydari A, Nabhani N (2019) Investigation of the thermal separation in a counter-flow Ranque-Hilsch vortex tube with regard to different fin geometries located inside the cold-tube length. Int Commun Heat Mass Transfer 108:104273
4. Xue Y, Arjomandi M (2008) The effect of vortex angle on the efficiency of the Ranque-Hilsch vortex tube. Exp Thermal Fluid Sci 33(1):54–57
5. Xue Y, Arjomandi M, Kelso R (2012) Experimental study of the flow structure in a counter flow Ranque-Hilsch vortex tube. Int J Heat Mass Transf 55(21–22):5853–5860
6. Guo X, Zhang B, Liu B, Xu X (2019) A critical review on the flow structure studies of Ranque-Hilsch vortex tubes. Int J Refrig
7. Attalla M, Ahmed H, Ahmed MS, El-Wafa AA (2017) Experimental investigation for thermal performance of series and parallel Ranque-Hilsch vortex tube systems. Appl Therm Eng 123:327–339
8. Valipour MS, Niazi N (2011) Experimental modeling of a curved Ranque-Hilsch vortex tube refrigerator. Int J Refrig 34(4):1109–1116
9. Bej N, Sinhamahapatra KP (2016) Numerical analysis on the heat and work transfer due to shear in a hot cascade Ranque-Hilsch vortex tube. Int J Refrig 68:161–176
10. Xue Y, Arjomandi M, Kelso R (2013) The working principle of a vortex tube. Int J Refrig 36(6):1730–1740
11. Ahlborn B, Groves S (1997) Secondary flow in a vortex tube. Fluid Dyn Res 21(2):73
12. Matveev KI, Leachman J (2019) Numerical investigation of vortex tubes with extended vortex chambers. Int J Refrig 108:145–153
13. Moraveji A, Toghraie D (2017) Computational fluid dynamics simulation of heat transfer and fluid flow characteristics in a vortex tube by considering the various parameters. Int J Heat Mass Transf 113:432–443
14. Sadi M, Farzaneh-Gord M (2014) Introduction of annular vortex tube and experimental comparison with Ranque-Hilsch vortex tube. Int J Refrig 46:142–151
15. Farzaneh-Gord M, Sadi M (2014) Improving vortex tube performance based on vortex generator design. Energy 72:492–500
16. Avcı M (2013) The effects of nozzle aspect ratio and nozzle number on the performance of the Ranque-Hilsch vortex tube. Appl Therm Eng 50(1):302–308
17. Kaya H, Uluer O, Kocaoğlu E, Kirmaci V (2019) Experimental analysis of cooling and heating performance of serial and parallel connected counter-flow Ranquee-Hilsch vortex tube systems using carbon dioxide as a working fluid. Int J Refrig 106:297–307
18. Samuels DC (1998) A finite-length instability of vortex tubes. Eur J Mech B/Fluids 17(4):587–594
19. Kirmaci V, Kaya H, Cebeci I (2018) An experimental and exergy analysis of a thermal performance of a counter flow Ranque-Hilsch vortex tube with different nozzle materials. Int J Refrig 85:240–254
20. Kandil HA, Abdelghany ST (2015) Computational investigation of different effects on the performance of the Ranque-Hilsch vortex tube. Energy 84:207–218
21. Devade K, Pise A (2014) Effect of cold orifice diameter and geometry of hot end valves on performance of converging type Ranque Hilsch vortex tube. Energy Procedia 54:642–653
22. Hamdan MO, Al-Omari SA, Oweimer AS (2018) Experimental study of vortex tube energy separation under different tube design. Exp Thermal Fluid Sci 91:306–311
23. Bazgir A, Nabhani N, Eiamsa-Ard S (2018) Numerical analysis of flow and thermal patterns in a double-pipe Ranque-Hilsch vortex tube: influence of cooling a hot-tube. Appl Therm Eng 144:181–208

24. Yun J, Kim Y, Yu S (2018) Feasibility study of carbon dioxide separation from gas mixture by vortex tube. Int J Heat Mass Transf 126:353–361
25. Rafiee SE, Sadeghiazad MM (2014) Three-dimensional and experimental investigation on the effect of cone length of throttle valve on thermal performance of a vortex tube using k–ε turbulence model. Appl Therm Eng 66(1–2):65–74
26. Lu G, Zhou G (2016) Numerical simulation on performances of plane and curved winglet type vortex generator pairs with punched holes. Int J Heat Mass Transf 102:679–690
27. Kaya H, Günver F, Kirmaci V (2018) Experimental investigation of thermal performance of parallel connected vortex tubes with various nozzle materials. Appl Therm Eng 136:287–292
28. Liang F, Wang H, Wu X (2019) Study on energy separation characteristics inside the vortex tube at high operating pressure. Therm Sci Eng Prog 14:100432
29. Korkmaz ME, Gümüşel L, Markal B (2012) Using artificial neural network for predicting performance of the Ranque-Hilsch vortex tube. Int J Refrig 35(6):1690–1696
30. Gao CM, Bosschaart KJ, Zeegers JCH, De Waele ATAM (2005) Experimental study on a simple Ranque-Hilsch vortex tube. Cryogenics 45(3):173–183
31. Li N, Zeng ZY, Wang Z, Han XH, Chen GM (2015) Experimental study of the energy separation in a vortex tube. Int J Refrig 55:93–101
32. Kumar A, Subudhi S (2017) Cooling and dehumidification using vortex tube. Appl Therm Eng 122:181–193
33. Agrawal N, Naik SS, Gawale YP (2014) Experimental investigation of vortex tube using natural substances. Int Commun Heat Mass Transfer 52:51–55
34. Shamsoddini R, Abolpour B (2018) A geometric model for a vortex tube based on numerical analysis to reduce the effect of nozzle number. Int J Refrig 94:49–58
35. Syed S, Renganathan M (2019) Numerical investigations on flow characteristics and energy separation in a Ranque Hilsch vortex tube with hydrogen as working medium. Int J Hydrogen Energy 44(51):27825–27842
36. Dincer K (2011) Experimental investigation of the effects of threefold type Ranque-Hilsch vortex tube and six cascade type Ranque-Hilsch vortex tube on the performance of counter flow Ranque-Hilsch vortex tubes. Int J Refrig 34(6):1366–1371
37. Piralishvili SA, Polyaev VM (1996) Flow and thermodynamic characteristics of energy separation in a double-circuit vortex tube—an experimental investigation. Exp Thermal Fluid Sci 12(4):399–410
38. Zheng N, Liu P, Shan F, Liu J, Liu Z, Liu W (2016) Numerical studies on thermo-hydraulic characteristics of laminar flow in a heat exchanger tube fitted with vortex rods. Int J Therm Sci 100:448–456
39. Rattanongphisat W, Thungthong K (2014) Improvement vortex cooling capacity by reducing hot tube surface temperature: experiment. Energy Procedia 52:1–9

Implementation of Six-Sigma Tools in Hospitality Industry: A Case Study

Nishant Bhasin, Harkrit Chhatwal, Aditya Bassi, and Shubham Sharma

Abstract This paper presents a case study in which the Six-Sigma concept was implemented in the improvement of the process of operation of a restaurant to meet the customer expectations and was investigated to study the impact. During this investigation, the problem with large pickup order lead time as per the voice of customer (VOC) was examined through the application of DMAIC (Define, Measure, Analyze, Improve, and Control) concept of Six-Sigma. The study includes the proper analysis of the current system depicting the existing problems within the restaurant. The methodology aims to analyze the root cause of existing problems and then based on the root cause helps to design an improvement plan through proper process mapping. At the end of the investigation, a few solutions were proposed and were implemented practically. The calculations indicated that the sigma level of the process increased from 0 to 2.2 sigma. The increase in the sigma level clearly depicts that the implemented solutions have a positive impact on the process.

Keywords Six-sigma · Hospitality industry · VOC · DMAIC

1 Introduction

Six-Sigma is based on continuous quality improvement in which is based on a philosophy of reducing variation and shifting the mean toward the center. It follows statistical approaches and methodologies which targets to reduce the defects up to 3.4 defects per million opportunities. Apart from the data-driven approaches, it also provides the enterprise level in an organization, a business strategy which will help in controlling and improving business process. It is done by increasing the effectiveness and efficiency of all operations to meet customer demands and expectations. It was first given and adopted by Motorola in 1980s [1], and it was later adopted by many big organizations like General Electric which enhanced its popularity everywhere

N. Bhasin (✉) · H. Chhatwal · A. Bassi · S. Sharma
Amity University, Noida, Uttar Pradesh 201313, India
e-mail: nishant9900@gmail.com

© The Author(s), under exclusive license to Springer Nature Singapore Pte Ltd. 2021

R. M. Singari et al. (eds.), *Advances in Manufacturing and Industrial Engineering*, Lecture Notes in Mechanical Engineering,
https://doi.org/10.1007/978-981-15-8542-5_104

across the globe [2]. In the early stage, it was mostly implemented in the manufacturing industries. But after realizing its true potential, it was used for manufacturing as well as service sector industries [3].

Six-Sigma originated from Frederick Gauss's concept of normal distribution which is a bell-shaped curve. For a service process, the sigma capability (Z-value) is a metric that tells how well a process is performing. Higher is sigma capability, the better is the process performance. As the defects go down, the sigma capability goes up. This methodology is driven by the customer and is implemented from the top in any organization. For the measurement of output variation, 3 sigma concept is used which was introduced by Walter Shewhart in 1922. It is related to a process with a yield of 99.97%. The use of quality systems and its scope was taken to a new level by Motorola by increasing the quality standard to Six-Sigma. It uses statistical tools such as SPC, FMEA, Gage R&R are among the main tools which are used in Six-Sigma.

It provides organization a structure of their processes to streamline which also targets the high-level management [4]. The Six-Sigma concept involves the implementation of DMAIC methodology which is used when an existing process needs any improvement. If there is a need for a new process, then DFSS (Design for Six-Sigma) is used. DFSS has many approaches to product and processes [5].

2 Literature Review

Six-Sigma can have a variety of meanings and definitions based on the perspective of different people [6]. According to Minitab, it is a tool driven by information which is used for reducing waste, variation, and making the existing processes more efficient [7]. It consists of DMAIC which helps the organizations to do projects based on Define, Measure, Analyze, Improve, and Control. It also has the belt-based training in which people are trained and provided designations such as Champion, Master BB (Black Belt), BB, Green Belt [8]. It has a large amount of applications in different companies with their respective issues [9].

Six-Sigma has some limitations when it is implemented in service industries which act as barriers [10–12]. It is because the process of gathering data is very difficult sometimes in service sector as compared to the manufacturing sector. It is sometimes difficult to measure the customer satisfaction in the service industries because there are only two outcomes either the customer is satisfied or dissatisfied [13]. Apart from the limitations, there are some successes also when it comes to the implementation of Six-Sigma in the service sector. It has helped some companies to reduce their work order completion time and material gathering time [14]. Some companies have implemented Six-Sigma tools to reduce variation caused by defects and eliminate the non-value adding activities to increase their profit maximization [15]. Several studies showcased successful cases of Six-Sigma application in the service organization such as hospitals, food industry, hotels, and airlines [16]. There are various service organizations that have seized the product improvement methodology [17].

Kivela and Kagi stated that "Six-Sigma is a product enhancing approach that is well renowned in the service-oriented business" [18].

However, Six-Sigma can only be a success in the service industry if the management stops thinking about results and start to think about streamline the processes so that a business can get the most benefit out of it [19, 20]. For this, applying the appropriate methodology is very important to improve a business performance is very essential. The main thing is that one must prove to the people working in an organization the benefits of Six-Sigma so that they can accept the change enthusiastically before applying any Six-Sigma tool. To apply these tools in the service industries, a few things important to know are that all work is a process, all processes possess variability, and all processes have data that explains variability [21, 22].

3 Implementation and Results

The implementation study of lean Six-Sigma tools being reported here was carried out in a local Burger Restaurant in which there was a problem reported by the customers with the large pickup order lead time. As mentioned earlier, during the conduct of the implementation study being reported here, the time taken with the packaging of the food was chosen as the scope for studying where some Six-Sigma tools were applied and their results were duly examined. The activities carried out in these phases are reported in the following subheadings.

3.1 Define Phase

The first step was the define phase. This phase begins with identifying the voice of customer (VOC) and converting it into critical to quality (CTQ). The VOC to CTQ translation matrix is shown in Table 1. The translation matrix clearly depicts that the customer does not want to wait for picking up the order and the order must be made available for pickup within 16 min. Now, the lead time is in minutes which are continuous data.

Subsequently, the information was gathered about the operation of the restaurant and SIPOC (supplier, input, process, output, and customer) chart was developed as shown in Fig. 1.

Table 1 VOC to CTQ translation matrix

V.O.C	CTQ	CTQ measure	CTQ specification
Does want to wait much for picking up the order	Lead time	Time in minutes	<16 min

S (Supplier)	I (Input)	P (Process)	O (Output)	C (Customer)
Customer for Pick-up	Order for Pick-up	Pick-up order process	Complete Order	Customer for Pick-up
Restaurant	Food for order		Payment	Cashier

Fig. 1 SIPOC chart

Process: Receive order → Order sent to kitchen → Make order → Prepare order for pickup → Customer picks-up the order and pays.

After that project charter was developed which consists of Business case, Problem statement, Goal statement, Project scope, Definition of defect, CTQs, Benefits. The project charter is shown in Fig. 2 in which the problem with the large time taken for pickup orders (20 min) is listed in the problem statement and the goal statement is also shown in which it is expected that the lead time will be reduced by 20%.

Project Charter

Business Case

It takes an average of 20 minutes for pick-up orders at the restaurant. If the total lead time for pick-up orders is not reduced then it may harm the restaurant's reputation which will eventually lead to less customers in the restaurant. Improving the lead time will increase the revenue generation and will result in higher customer satisfaction.

Problem Statement

Customer complaints are there regarding the time taken for their pick-up orders at the restaurant itself. As of now, the total lead time of pick-up orders is an average of 20 minutes.

Goal Statement

The goal is to reduce the total lead time for pick-up orders from an average of 20 minutes by 20 % by end of July, 2019.

Scope

In Scope: The dine in food and only burgers.

Definition of a Defect

Any case where the total cycle time from the moment the customer places their order to the moment they receive their food at the restaurant is greater than 16 minutes.

CTQs

- Lead time for pick-up orders.

Benefits

- Enhanced customer satisfaction.
- Increased revenue generation.

Roles

Team Members	Nishant Bhasin, Harkrit Chhatwal, Aditya Bassi, Shubham Sharma

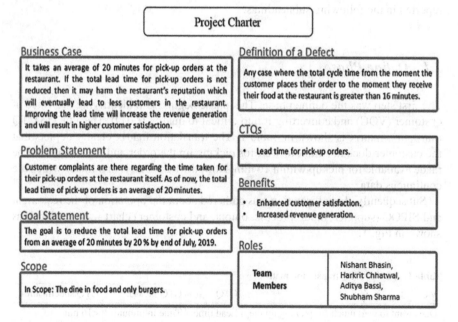

Fig. 2 Project charter

3.2 Measure Phase

For the measure phase, the operation definition was stated which says that pickup order lead time is the total cycle time in minutes from the moment the customer places their order to the moment they receive their food at the restaurant. As per the V.O.C, this cycle time should not exceed 16 min. So, the current process is defective as the average time for pickup order is around 20 min. So, for the measure phase, data collection was done and 60 pickup order lead time readings were taken and their average was 20.34 min. Now, sigma level was calculated for the baseline data using *Z*-Test and keeping the upper specification limit (USL) as 16. So, the DPMO (Defects per million opportunities) was 999,997.7856 and the sigma level came out to be 0. Now, before proceeding further the collected data was tested for normality test and the *P*-value is greater than 0.05, which clearly depicts that the data follows normal distribution. The normality test using Minitab is shown in Fig. 3.

The next step in measuring phase is determining the process capability which is shown in Fig. 4. The results from process capability analysis clearly show that the average mean of the process is more than the USL. The process needs centering as the graph lies outside the specifications as per the V.O.C.

After this, the process flowchart was made as shown in Fig. 5. After examining the process, the time taken to pack the food is a potential cause for the large lead time of the pickup orders. The time taken to prepare the condiments can also be one of the reasons.

After the process flowchart, the cause and effect diagram was developed as shown in Fig. 6. This diagram helps to identify the root cause of the problem of pickup order lead time which will be further examined in the analyze phase.

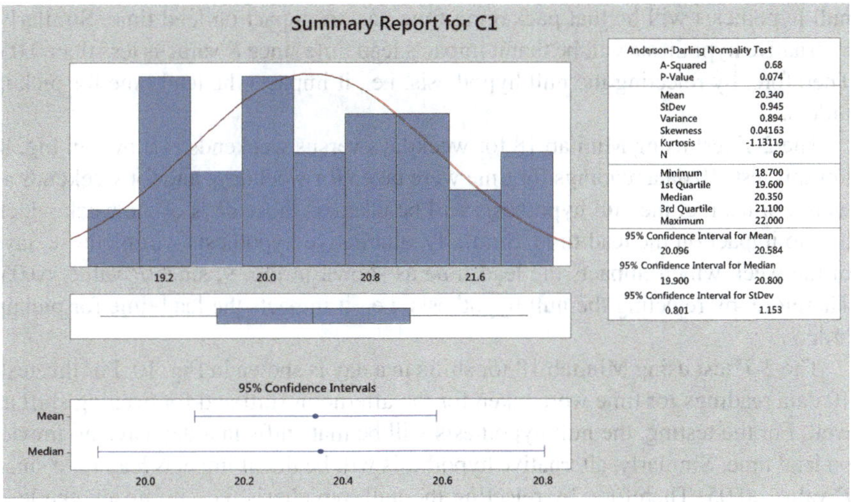

Fig. 3 Normality test for the data collected

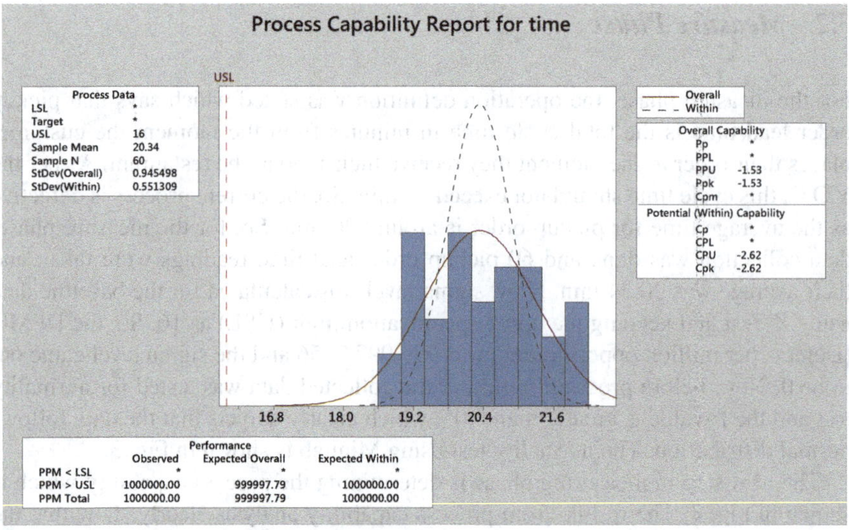

Fig. 4 Process capability analysis for the collected data

3.3 *Analyze Phase*

In this phase, the root causes identified from the cause and effect diagram where subjected to hypothesis testing. These include regression analysis of the packaging time versus lead time, 2-*T* test for weekdays versus weekends and 2-*T* test for shifts in a day. These tests were run using Minitab 18. The regression analysis is shown in Fig. 7. For this test, 10 data readings for time were taken. For the testing, the null hypothesis will be that packaging time has no impact on lead time. Similarly, alternative hypothesis will be that it impacts lead time since *P* value is less than 0.05. Therefore, by rejecting the null hypothesis, i.e., it impacts the lead time for pickup orders.

The 2-*T* test using Minitab 18 for weekdays versus weekends is shown in Fig. 8. For this test, 10 data readings for time were taken for weekdays and for weekends as well. For testing, the null hypothesis will be taken as those days of the week which has no impact on the lead time. Similarly, alternative hypothesis will be those days of the week which impacts the lead time as shown in Fig. 9, since *P* value <0.05. Therefore, by rejecting the null hypothesis, i.e., it impacts the lead time for pickup orders.

The 2-*T* test using Minitab 18 for shifts in a day is shown in Fig. 10. For this test, 10 data readings for time were taken for the afternoon shift and for evening shift as well. For the testing, the null hypothesis will be that shifts in a day have no impact on lead time. Similarly, alternative hypothesis will be that it impacts lead time since *P* value <0.05. Therefore, by rejecting the null hypothesis, i.e., it impacts the lead time for pickup orders. As per analysis phase, the main identified root cause is the

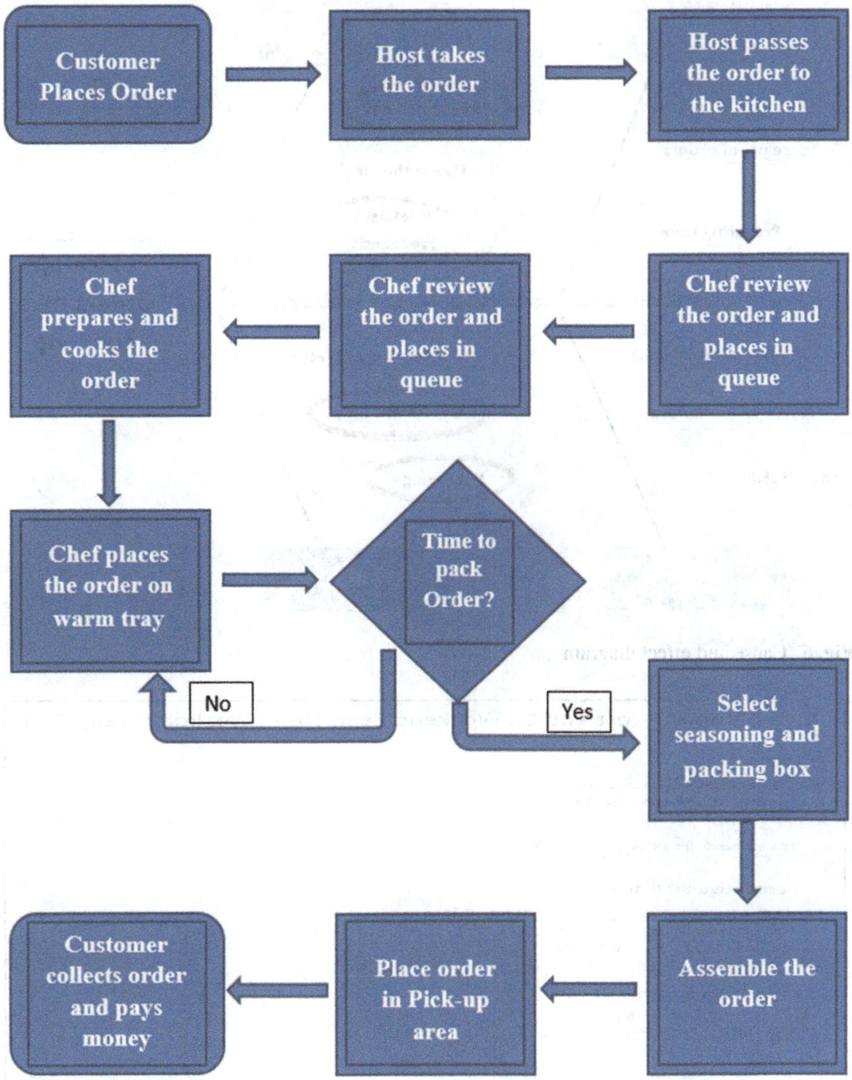

Fig. 5 Process flowchart

packaging of food which is the main cause for the large lead time for the pickup orders which also affects the lead time in case of weekdays versus weekends and shifts in a day.

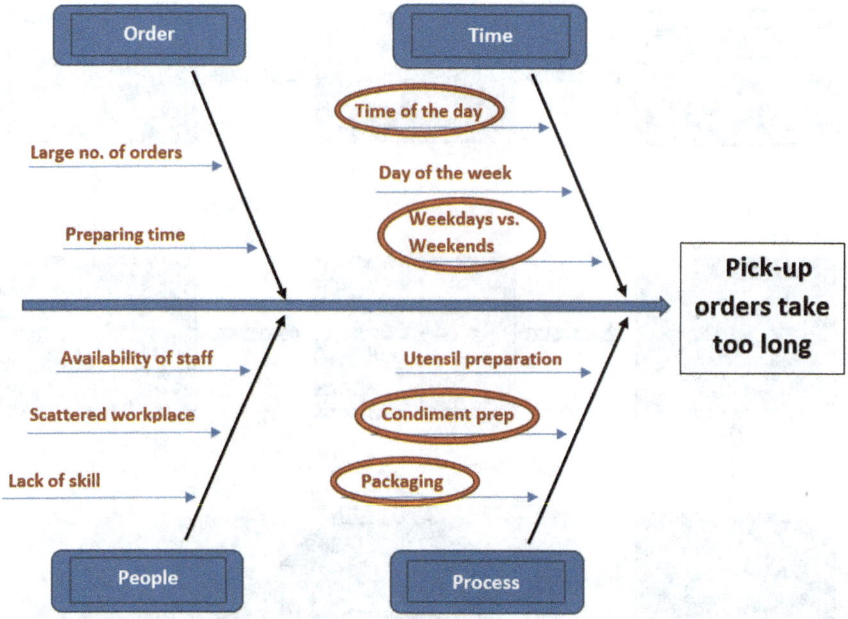

Fig. 6 Cause and effect diagram

Two-Sample T-Test and CI: Weakend Lead Time, Weakdays Lead Time

Method

μ_1: mean of Weakend Lead Time
μ_2: mean of Weakdays Lead Time
Difference: $\mu_1 - \mu_2$

Equal variances are not assumed for this analysis.

Descriptive Statistics

Sample	N	Mean	StDev	SE Mean
Weakend Lead Time	10	20.460	0.763	0.24
Weakdays Lead Time	10	21.160	0.425	0.13

Estimation for Difference

Difference	95% CI for Difference
-0.700	(-1.292, -0.108)

Test

Null hypothesis H_0: $\mu_1 - \mu_2 = 0$
Alternative hypothesis H_1: $\mu_1 - \mu_2 \neq 0$

T-Value	DF	P-Value
-2.53	14	0.024

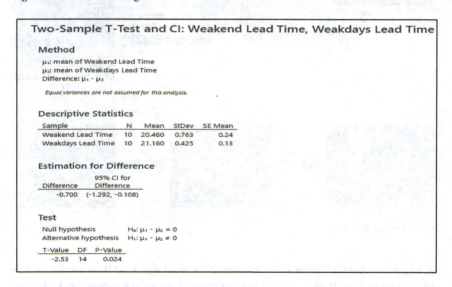

Fig. 7 Regression analysis

Two-Sample T-Test and CI: Weakend Lead Time, Weakdays Lead Time

Method

μ_1: mean of Weakend Lead Time
μ_2: mean of Weakdays Lead Time
Difference: $\mu_1 - \mu_2$

Equal variances are not assumed for this analysis.

Descriptive Statistics

Sample	N	Mean	StDev	SE Mean
Weakend Lead Time	10	20.460	0.763	0.24
Weakdays Lead Time	10	21.160	0.425	0.13

Estimation for Difference

Difference	95% CI for Difference
-0.700	(-1.292, -0.108)

Test

Null hypothesis $H_0: \mu_1 - \mu_2 = 0$
Alternative hypothesis $H_1: \mu_1 - \mu_2 \neq 0$

T-Value	DF	P-Value
-2.53	14	0.024

Fig. 8 2-*T* test for days of the week

Two-Sample T-Test and CI: Afternoon Shift, Evening Shift

Method

μ_1: mean of Afternoon Shift
μ_2: mean of Evening Shift
Difference: $\mu_1 - \mu_2$

Equal variances are not assumed for this analysis.

Descriptive Statistics

Sample	N	Mean	StDev	SE Mean
Afternoon Shift	10	20.510	0.654	0.21
Evening Shift	10	19.930	0.411	0.13

Estimation for Difference

Difference	95% CI for Difference
0.580	(0.059, 1.101)

Test

Null hypothesis $H_0: \mu_1 - \mu_2 = 0$
Alternative hypothesis $H_1: \mu_1 - \mu_2 \neq 0$

T-Value	DF	P-Value
2.37	15	0.031

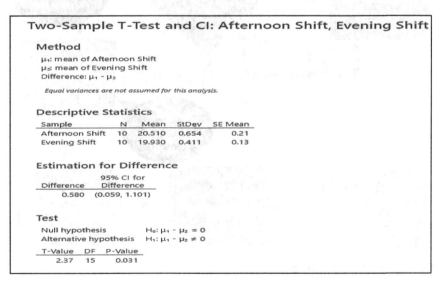

Fig. 9 2-*T* test for shifts in a day

3.4 Improve Phase

During the conduct of this phase, the theory was studied. Large lead time of pickup order was caused as a result of large time taken to pack the food and selecting the condiments of the dishes. Based on the problems, a corrective process map was made

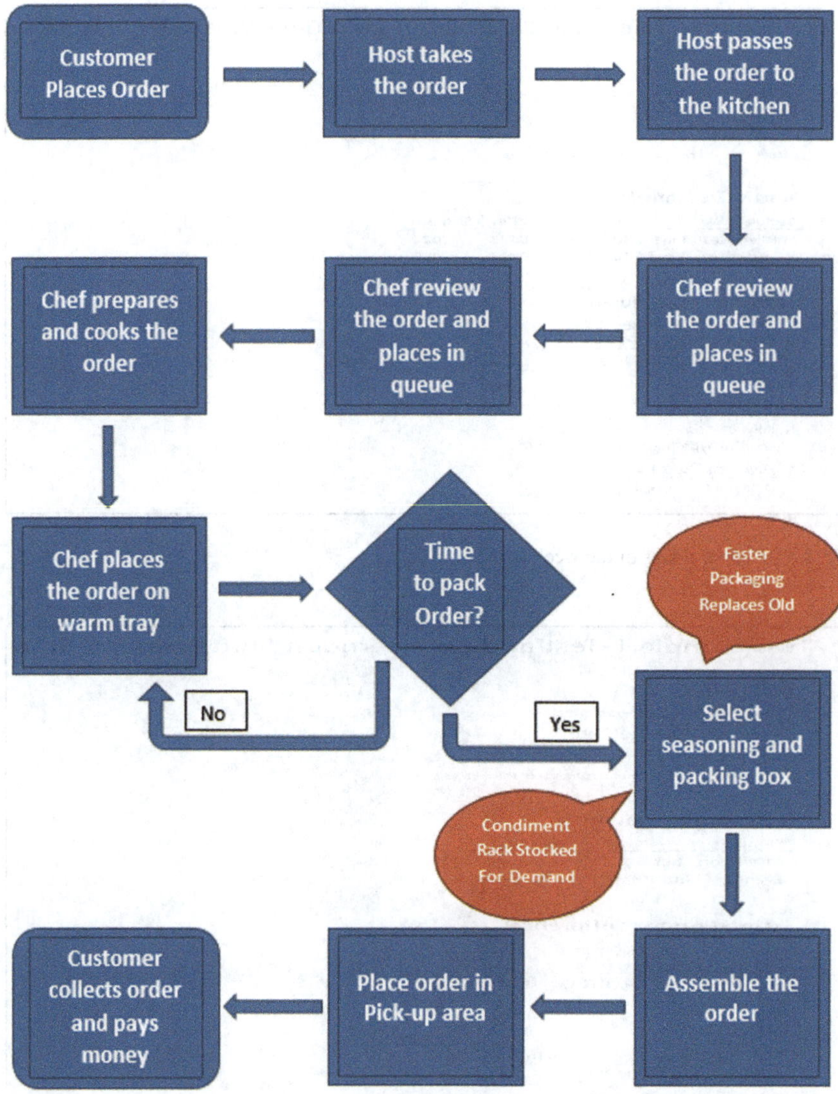

Fig. 10 Corrective process map

as shown in Fig. 10. The old packaging which was very complex and time-consuming was replaced with the new one which was easy to handle and less time-consuming. The condiments were properly stocked in a rack as per the demand.

Now, 60 new lead time readings were taken when the old packaging was replaced with the new one and the lead time of pickup order came out to be an average of 15.32 min. After that, sigma level was calculated for the data after implementing the improved phase using *Z*-test and keeping the upper specification limit (USL) as 16.

Paired T-Test and CI: PREVIOUS LEAD TIME, IMPROVED LEAD TIME

Descriptive Statistics

Sample	N	Mean	StDev	SE Mean
PREVIOUS LEAD TIME	60	20.340	0.945	0.122
IMPROVED LEAD TIME	60	15.323	0.887	0.115

Estimation for Paired Difference

Mean	StDev	SE Mean	95% CI for μ_difference
5.0167	0.4043	0.0522	(4.9122, 5.1211)

μ_difference: mean of (PREVIOUS LEAD TIME - IMPROVED LEAD TIME)

Test

Null hypothesis H₀: μ_difference = 0
Alternative hypothesis H₁: μ_difference ≠ 0

T-Value	P-Value
96.12	0.000

Fig. 11 Paired *T*-test

So, the DPMO (Defects per million opportunities) was 222,838.6336 and the sigma level came out to be 2.2.

The paired *T*-test using Minitab 18 for the baseline data versus the improve phase data is shown in Fig. 11. For this test, 60 data readings for time were taken. For the testing, the null hypothesis will be that there is no difference in the data collected for lead time both the cases. Similarly, alternative hypothesis will be that there is a difference between the lead times collected in both cases. Since, *P* value >0.05. Therefore, by rejecting the null hypothesis, i.e., the improved phase data for the lead time is less than that of the baseline data for the lead time.

3.5 *Control Phase*

While implementing the control phase, it is to be ensured that, the suggestions are implemented and sustained. In order to check whether the solutions are implemented, the owner of the restaurant must ensure that the chef must follow the solution implemented and the kitchen items does not get scattered. Moreover, there must be the minimum number of staff available so that there is no idle time during the working hours due to which the customer should wait for long.

4 Conclusion

Most of the tools used in Six-Sigma can be applied to the hospitality industry. The DMAIC method and tools such as VOC translation matrix, cause and effect diagram and hypothesis testing proved to be very helpful for the conduct of this study. However, different types of customers evaluate the quality of a service in different ways which can prove as a challenge when Six-Sigma tools are applied. The goal in the project charter showed in this paper is achieved as the pickup order lead time average is reduced from 20.34 to 15.32 min, i.e., the lead time average is reduced by more than 20%.

References

1. Black K, Revere L (2006) Six Sigma arises from the ashes of TQM with a twist. Int J Health Care Qual Assur 19(2–3):259–266
2. Antony J, Bhuller AS, Kumar M, Mendibil K, Montgomery DC (2012) Application of six sigma DMAIC methodology in a transactional environment. Int J Qual Reliab Manag 29(1):31–53
3. Laureani A, Antony J (2012) Standards for lean six sigma certification. Int J Prod Perform Manag 61(1):110–120
4. Catherwood P (2002) What's different about six sigma? IEE Manuf Eng August:186–189
5. El-Haik B, Roy DM (2005) Service design for six sigma: a roadmap for excellence. Wiley, Hoboken, NJ
6. Henderson KH, Evans JR (2000) Successful implementation of six sigma: benchmarking general electric company. Benchmarking Int J 7(4):260–281
7. Goh TN (2002) A strategic assessment of six sigma. Qual Reliab Eng Int 18:403–410
8. Zu X, Fredendall LD, Douglas TJ (2008) The evolving theory of quality management: the role of six sigma. J Oper Manag 26(5):630–650
9. Lin C, Chen FF, Wan H, Chen YM, Kuriger G (2013) Continuous improvement of knowledge management systems using six sigma methodology. Robot Comput Integr Manuf 29(3):95–103
10. Antony J (2004) Six sigma in the UK service organizations: results from a pilot survey. Manag Auditing J 19(8):1006–1013
11. Benedetto AR (2003) Adapting manufacturing-based six sigma methodology to the service environment of a radiology film library. J Healthc Manag 48(4):263–280
12. Sehwall L, DeYong C (2003) Six sigma in health care. Int J Health Care Qual Assur 16(6):1–5
13. Hensley RL, Dobie K (2005) Assessing readiness for six sigma in a service setting. Managing Serv Qual 15(1):82–101
14. Holtz R, Campbell P (2004) Six sigma: its implementation in Ford's facility management and maintenance functions. J Facil Manag 2(4):320–329
15. Nourse L, Hays P (2004) Fidelity wide processing wins team excellence award competition. J Qual Participation 27(2):42–48
16. Abou Kamar MS (2014) Six sigma application in the hotel industry: is it effective for performance improvement? Res J Manag Sci 3(12):1–14
17. Pearlman MD, Chacko H (2012) The quest for quality improvement using six sigma at Starwood hotels and resorts. Int J Hosp Tourism Adm 13:48–66
18. Kivela J, Kagi J (2008) Applying six sigma in foodservice organization. Tourism 56:319–337
19. Lahap J, Bahri KA, Noorsa R, Johari MS, Noraslinda NM (20017) Six sigma and the Malaysian hotel industry. J Tourism Hosp 6. https://doi.org/10.4172/2167-0269.1000329
20. Hung H-C, Sung M-H (2011) Applying six sigma to manufacturing processes in the food industry to reduce quality cost. Sci Res Essays 6
21. Manurung S (2011) Improvement material inventory tracking for maintenance and project through lean six sigma methodology
22. Shing CK, Nadarajan S, Chandren S (2014) Scraps management with lean six sigma

Impact of Integrating Artificial Intelligence with IoT-Enabled Supply Chain—A Systematic Literature Review

Ranjan Arora, Abid Haleem, P. K. Arora, and Harish Kumar

Abstract Supply chain plays an important role in differentiating competing organizations. The success of efficient supply chain is dependent on how effectively an organization uses real-time data. Internet of things (IoT)-enabled supply chain helps in getting real-time data which if integrated well with artificial intelligence (AI) helps in analyzing data collected via IoT and thus processing it to get the desired results. It helps in providing real-time visibility and making supply chain proactive, and helping organization in enhancing customer service and increasing revenue along with profitability. This paper evaluates the existing literature in assessing the impact of integrating AI with IoT-enabled supply chain.

Keywords Artificial intelligence (AI) · Internet of things (IoT) · Supply chain management · IoT-enabled supply chain · Literature review

1 Introduction

Survival of the fittest is the mantra that has been followed since ages and still hold true. Thus, organizations who can innovate and use latest technology are the ones that will be able to survive. One of the critical elements that help an organization remains competitive is its supply chain. As per *Council of Supply Chain Management Professionals (CSCMP), 2019*, supply chain consists of new product development, procurement and sourcing, manufacturing, and logistics. An efficient supply chain

R. Arora (✉)
Aligarh, Uttar Pradesh, India
e-mail: ranjan.arora@gmail.com

A. Haleem
Jamia Millia Islamia, New Delhi 110025, India

P. K. Arora
Galgotias College of Engineering and Technology, Greater Noida, Uttar Pradesh, India

H. Kumar
National Institute of Technology Delhi, Delhi 110040, India

© The Author(s), under exclusive license to Springer Nature Singapore Pte Ltd. 2021
R. M. Singari et al. (eds.), *Advances in Manufacturing and Industrial Engineering*,
Lecture Notes in Mechanical Engineering,
https://doi.org/10.1007/978-981-15-8542-5_105

helps in achieving competitive advantage along with maximizing customer value. Supply chain has been transforming over years, with one of the critical elements being the usage of latest technologies. The use of technology has helped in collaborating, collecting data, analyzing, and then making it proactive, agile, and responsive.

It is important to have a proactive supply chain, which gets the right data at the right time. IoT is a medium of getting real-time information. IoT is a grid of objects linked together so that can interact and communicate with each other [1]. Thus, IoT-enabled supply chain helps in providing real-time data by linking all supply chain actors. It is then essential to use this data effectively so that right decisions can be taken. AI plays a vital role as it builds rules/models based on previous learning and then solving problems. Thus, integrating AI with IoT-enabled supply chain helps in taking faster and right decisions at the right time. The aim of AI is not to replace human beings but to augment them and then solve tasks by applying a wide range of intellectual, industrial, and social applications [2]. AI enables a machine to learn from previous experience, fine-tune to new input data, and perform an intelligent task automatically. This helps an organizations' supply chain to be proactive and also competitive. This agile and responsive supply chain enables an organization in increasing revenue, improving customer service, and resource optimization at supply chain level.

The benefits provided by integrating AI with IoT-enabled supply chain has brought the attention of authors to study the same. The objective of this paper is to study the impact of integrating AI with IoT-enabled supply chain. It highlights both the positive impacts along with constraints it brought.

2 Literature Review

There is a lot of research that has been done in the area of supply chain, IoT, and AI. Researchers have also done research in IoT-enabled supply chain with the role AI plays in this. Thus, it is essential to capture the right research done earlier in this area so that it can be analyzed effectively. There are various ways of selecting the previous research and then analyzing it but the authors have taken the following steps for this paper

- Studying existing research done in integrating AI with IoT-enabled supply chain
- Relevant search criteria defined in selected
- Defining inclusions and exclusions
- Relevant literature review collected, collated, and analysed.

By taking this approach, the authors have been able to get the right data points and then they have been analysed to come up with results, discussion, and conclusion. The below sections highlight the key points that have been collected from previous research in three road areas

- IoT-enabled supply chain
- Integrating AI with Internet of things (IoT)-enabled supply chain
- Impact of integrating AI with IoT-enabled supply chain.

The data collected from previous research first describes key points on each of these areas and then the important characteristics of each of these. This has then helped the authors in right analysis, which brings the key results. Also, it is important to bring out areas for future research, so that this review can be taken a step further and adds value.

2.1 Internet of Things (IoT)-Enabled Supply Chain

IoT-enabled supply chain connects all resources via Internet and provides real-time data. If used effectively by all supply chain partners including suppliers and distributors, the effect is many folds. IoT-enabled supply chain interconnects all devices/resources and then makes them communicate with each other. IoT-enabled supply chain tracks and traces products in the supply chain along with making it proactive. It enables real-time visibility which in return increases revenue by enhancing customer service. Its success is dependent on deployment of WSN wireless sensor networks, middleware, radio frequency identification (RFID), software for IoT applications and cloud computing [3]. Thus, it has the potential to transform the business.

Some of the key characteristics highlighted by the previous research are

- IoT-enabled supply chain helps in improving performance and operational efficiency by becoming core component of supply chain strategy implementation [4]
- Helps in overcoming supply chain vulnerability, ambiguity, and complexity [5]
- Enables real-time record updation which will benefit supply chain actors suppliers, distributors, etc. [1]
- Provides real-time information which enables right product at the right place and at the right time [6]
- Harmonization between supply chain partners enabling increase in efficiency [1].

2.2 Integrating AI with Internet of Things (IoT)-Enabled Supply Chain

AI is defined as perceiving, learning, reasoning, and solving the problem automatically [7]. AI is now an integral component and key strategic element of many organizations across sectors [2]. IoT-enabled supply chain has transformed organizations across industries by making it proactive. But huge amount of data collected through this, needs to be processed, analyzed so that right decision can be taken at the right time. It has been identified in earlier studies that applying AI technology can help in significantly improving predictive capability and analytic insights, if applied to big data [8]. Some of the key characteristics of integrating AI with IoT-enabled supply chain are

- AI is used to automate human task by taking knowledge of an intelligent agent and then performing decision-making and giving maximum outputs for a specific purpose [9]. But its purpose is to enhance human capability and not replace it [10]
- It is important to see how AI and humans be complementary in decision making and work cohesively to augment capabilities of each other [2]
- AI can be divided into various sub-components—artificial neural networks (ANN), machine learning, fuzzy logic, and agent-based systems [2]
- AI is a scalable strong analytic tool and plays a significant role by analyzing large amount of data real time by being scalable [7]
- AI-centric technologies provide real-time monitoring of processes which offers significant efficiencies as compared to manual processes [11]
- Organizations have seen enhanced benefits of integrating AI technologies in the development of smart factory of the future and intelligent manufacturing [12]
- AI has been used to resolve following supply chain issues—purchasing, inventory management, freight consolidation, location planning, and scheduling problems [2].

2.3 *Impact of Integrating AI with IoT-Enabled Supply Chain*

Integrating AI with IoT-enabled supply chain helps in making supply chain agile and responsive. It in turn provides real-time visibility along with faster decision making which increases customer satisfaction and revenue for an organization. AI has made a significant impact across industries ranging from manufacturing, healthcare, finance, retail, utilities, logistics, and supply chain. Some of the important points from previous research are highlighted below

- Disruptive technologies have impacted supply chain across sectors. AI has enabled improvement in productivity and yield, conserving water in agriculture supply chain [9]
- Some of the constraints of AI as useful analytics tool are lack of enough training data, centralized architecture, resource constraint, and privacy [7]
- Sales performance improves using AI-driven dashboard along with predictive forecasting capability [2]
- AI techniques if used in inventory control and planning problems are capable of intelligently capturing inventory patterns across the value chain and thus enabling desirable inventory at each stocking point and avoiding bullwhip effect [2]
- AI techniques for predicting future demand have improved forecasting accuracy [13]

- AI may not work in scenarios involving uncertainty and risk due to lack of environmental knowledge [2].

3 Result and Discussion

There is a lot of work done in supply chain, IoT, IoT-enabled supply chain and AI. The previous research has highlighted the impact of IoT-enabled supply chain an organizations business along with how it makes supply chain agile, responsive, and proactive. It is clear from previous researches that usage of right technology makes supply chain agile and responsive. One of the key technologies has been IoT, and further, the use of AI has transformed the supply chain. This paper addresses the impact of integrating AI with IoT-enabled supply chain. It also brings forth the challenges it poses, as it is important to understand under what circumstances it can be used. There has been work done in various industries—manufacturing, healthcare, finance, retail, utilities, logistics, and supply chain.

One of the key factors that most organizations are apprehensive of is that it replaces human beings. But from the analysis of previous studies, it is clear that AI augments human beings and not replaces them.

4 Conclusion

The objective of the paper is to study the impact of integrating AI with IoT-enabled supply chain by highlighting how an organization benefits from the same along with the constraints faced by its usage. Success of an organization in today's environment depends on an agile, responsive, and proactive supply chain. IoT-enabled supply chain helps in interconnecting all available resources of the network. This interconnection helps communication between resources along with collection of data. AI technologies help in processing this data by using previous learning. Thus, AI technologies are unable to process something which is new and coming for the first time. So, AI is not replacement of human but augmentation of human. From the review, it has been clear that one of the efficient ways of making supply chain agile is by enabling it with IoT and integrating with AI. This provides competitive advantage to an organization and increasing revenue by preventing stock-outs and enhancing customer service. But as suggested there are certain limitations involved with using AI, which may be due to lack of previous data and thus should be kept in mind while taking decisions. Also, AI has to complement humans and not remove them. This helps in increase in customer satisfaction, fewer stock-outs, low inventory holding costs, and increase in revenue for an organization. This comes with few drawbacks also like it is ineffective if there is lack of enough training data along with security and privacy constraints.

Like every research, this also has few limitations. The few limitations of this paper are that it has considered few researches done in the past, and secondly, it is not industry specific. Thus, future research can be done by reviewing more studies done in the past. This will help in comprehensively analyzing the previous studies and come up with important recommendations. Also, research can be done by taking one industry sector at a time. Also, statistical tools can be applied once data has been collected by collecting data using surveys.

Acknowledgements The authors have sought continuous feedback from practitioners in this field. This has helped the authors in improving the paper quality along with bringing key dimensions which are essential.

References

1. Dhumale RB, Thombare ND, Bangare PM (2017) Supply chain management using internet of things. Int Res J Eng Technol (IRJET) 4(06):787–791
2. Dwivedi YK, Hughesa L (2019) Artificial intelligence (AI): multidisciplinary perspectives on emerging challenges, opportunities, and agenda for research, practice and policy. Int J Inf Manag
3. Lee I, Lee K (2015) The internet of things (IoT): applications, investments, and challenges for enterprises. Bus Horiz 58(4):431–440
4. Arora R, Haleem A, Arora PK (2020) Impact of integrating artificial intelligence with IoT enabled supply chain—a systematic literature review. Smart Innov Syst Technol 174 (in press)
5. Abdel-Basset M, Manogaran G, Mohamed M (2018) Internet of things (IoT) and its impact on supply chain: a framework for building smart, secure and efficient systems. Future Gener Comput Syst 86:614–628
6. Sivamani S, Kwak K, Cho Y (2014) A study on intelligent user-centric logistics service model using ontology. J Appl Math 1:1–10
7. Singh SK, Rathore S, Park JH (2019) BlockIoTIntelligence: a blockchain-enabled intelligent IoT architecture with artificial intelligence. Future Gener Comput Syst
8. Rubik B, Jabs H (2018) Artificial intelligence and the human biofield: new opportunities and challenges. Cosmos Hist 14(1):153–162
9. Dogo EM, Salami AF, Aigbavboa CO, Nkonyana T (2019) Taking cloud computing to the extreme edge: a review of mist computing for smart cities and industry 4.0 in Africa. In: Edge computing. Springer, Berlin, pp 107–132
10. Katz Y (2017) Manufacturing an artificial intelligence revolution. Available at SSRN 3078224
11. Zhong RY, Xu X, Klotz E, Newman ST (2017) Intelligent manufacturing in the context of industry 4.0: a review. Engineering 3(5):616–630
12. Li BH, Hou BC, Yu WT, Lu XB, Yang CW (2017) Applications of artificial intelligence in intelligent manufacturing: a review. Front Inf Technol Electron Eng 18(1):86–96
13. Jeong B, Jung H, Park N (2002) A computerized forecasting system using genetic algorithm in supply chain management. J Syst Softw 60(3):223–237

Experimental Study for the Health Monitoring of Milling Tool Using Statistical Features

Akanksha Chaudhari, Pavan K. Kankar, and Girish C. Verma

Abstract In this manuscript, the technique for health monitoring of the milling tool has been proposed using statistical features extraction from the raw time domain signal and their trend analysis. The extracted features like mean, kurtosis, skewness are a good indicator of the tool health. An experimental study has been performed in order to obtain the vibration signal using an accelerometer. The surface roughness parameters have been measured using mobile surface measuring instrument "Handy surf". The surface topography has also been performed for the milled surface with the three different conditions of the tool, i.e. healthy, tending to failure, and the blunt tool. The obtained results show good agreement with the statistical trend analysis.

Keywords Slot milling · Feature extraction · Condition monitoring · Surface topography

1 Introduction

Milling is a metal cutting operation in which the excess material from the workpiece is removed by employing a multipoint cutting tool called milling cutter. It is majorly used in manufacturing industries due to its high production rate capability and good dimensional accuracy [1]. Milling vibration can generate noise and degrade the quality of a milled surface [2]. To achieve the better surface finish, speed of the spindle is increased which results in the decrease in feed per tooth, whereas for acquiring high martial removal rate, the depth of cut is increased that ultimately enhances the surface roughness [3]. Phenomena like surface roughness, tool wear and tear, chatter and noise are the results of the unwanted vibration at the cutting tip of the milling process [4]. The amplitude of vibration depends on the severity of tool wear as well as on the tool signatures like helix angle and diameter of the multipoint

A. Chaudhari · P. K. Kankar (✉) · G. C. Verma
System Dynamics Lab, Discipline of Mechanical Engineering, Indian Institute of Technology Indore, Indore 453552, India
e-mail: pkankar@iiti.ac.in

© The Author(s), under exclusive license to Springer Nature Singapore Pte Ltd. 2021

R. M. Singari et al. (eds.), *Advances in Manufacturing and Industrial Engineering*, Lecture Notes in Mechanical Engineering,
https://doi.org/10.1007/978-981-15-8542-5_106

milling tool [5]. The monitoring of responses like cutting force and vibration amplitude has been successfully used to predict the change in the spindle speed of the milling machine [6]. At the same time, its dependency on the machining parameters like spindle speed, feed rate, axial and radial depth of cut has been already reported in several literature [7]. In actual practice, cutting force sensors are problematic to place as their dimensional constraints like the physical size of a workpiece make it not suitable for the milling of medium and large workpiece [8]. Thus, there is a need for real-time monitoring of the tool to diagnose the faults occurring in the tool during operation. Condition monitoring techniques like artificial neural network (ANN) and support vector machine (SVM) have been used to monitor the real-time condition of the rotating mechanical components [9]. Statistical features used as an input feature vector in machine learning models are the good indicative of the health behaviour of a mechanical system [10–12]. Most of the authors have involved the various artificial intelligence techniques to classify the faults occurring in the machining operations [8]. Although it is a well-established technique, the statistical features can also be a good indicator of the real-time condition of the machinery or its components. Very few of them have focused on the capability of statistical features to indicate the deviation in the health of the components [13].

In this manuscript, an attempt has been made to classify the health of the cutting tool using statistical features. Various features from captured data have been extracted and used as the condition indicators of the tool health. An experimental study has been performed to acquire the vibration amplitude during the milling operation. The contours of the surface textures have also been obtained using optical surface profiler, LD-130. The 3D surface plots show the strong agreement of the surface profile with the observed vibration signals.

2 Experimental Setup

The experimental setup used in the present study is shown in Fig. 1. It comprises of CNC milling machine, a work-holding bench, a triaxial accelerometer of sensitivity 10.52 mV/g along *x*-direction, 10.57 mV/g along *y*-direction and 10.51 mV/g along *z*-direction, respectively, mounted on the workpiece in the *x*–*z* plane and OROS data acquisition system. The slot milling experiments were performed on the SS316 workpiece of dimension 170 × 100 × 20 mm using an end-milling cutter of diameter 12 mm. As the operation is performed, the amplitude of the vibration in the three mutually perpendicular directions *x*-, *y*- and *z*-directions has been acquired using OROS data acquisition system. Sampling frequency of gathering data is kept at 25 kHz [14]. The signals acquired in the time domain for three different directions for three different health condition of the tool.

The sample signal for healthy conditions along *x*-, *y*- and *z*-directions is shown in Fig. 2. Figures 3 and 4 mention the signal captured for tending to failure condition and failure condition, respectively.

Fig. 1 Experimental setup of CNC milling machine

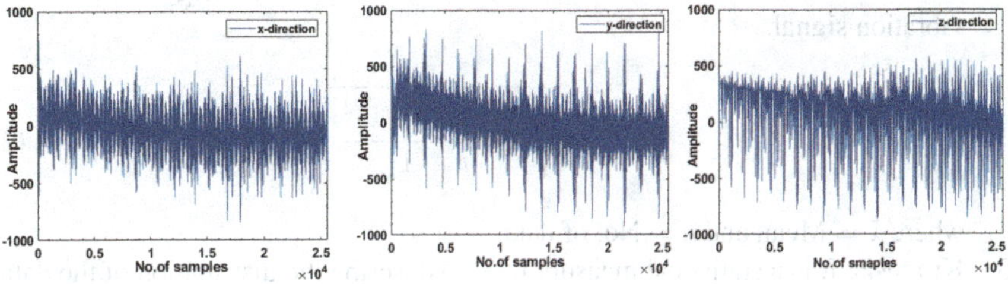

Fig. 2 Time domain signals in healthy condition

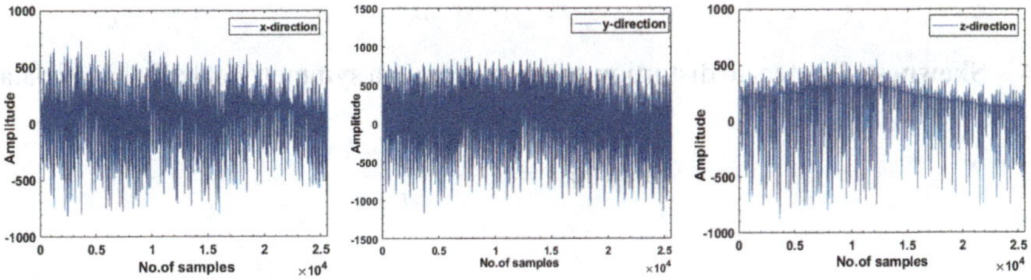

Fig. 3 Time domain signals for tending to failure condition

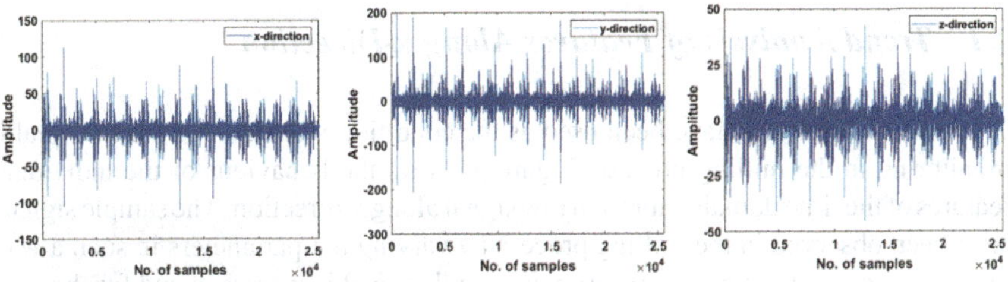

Fig. 4 Time domain signal for failure condition

3 Features Extraction

The signal has been analysed in the time domain and further processed to obtain various statistical features. Statistical features are the good indicators of the condition of the mechanical equipment [15]. It can indicate the deviation in the health of the system by analysing the change in the magnitude of the numerical values of the feature parameters. The four statistical features extracted for the signal are below [16]:

1. **Maximum Amplitude**: It is the value of the maximum deviation of the signal from its mean position. For a given signal "y", the maximum amplitude is given as

$$\text{Max.amplitude}(y\_\text{max}) = \max(y) \tag{1}$$

2. **Standard Deviation**: Standard deviation is the measure of energy content in the vibration signal.

$$\sigma = \sqrt{\frac{\sum_{i=1}^{n}(x_i - \bar{x})^2}{n-1}} \tag{2}$$

where $\bar{x}$ = Mean and n = No. of data.

3. **Kurtosis**: It is a statistical measure used to describe the distribution of the data.

$$\text{Kurtosis} = \frac{\sum_{i=1}^{n}\frac{(x_i - \bar{x})}{n}}{\sigma^4} \tag{3}$$

4. **Skewness**: It refers to distortion or asymmetry in a symmetrical bell-shaped data.

$$\text{Skewness} = \frac{\sum_{i=1}^{n}(x_i - \overline{x})^3}{n} \tag{4}$$

4 Results and Discussions

4.1 Trend Analysis of Features Along x-Direction

The extracted features have been used as the condition indicator for the tool health monitoring in the milling process. Figure 5 shows the behaviour of the individual features of the time domain vibration raw signal along x-direction. The sample signals have been observed in the milling process by varying the parameters in such a way that as compared to previous experiment spindle speed is increased and for the next run, rest parameters are kept constant while the depth of cut is increased. Figure 5a

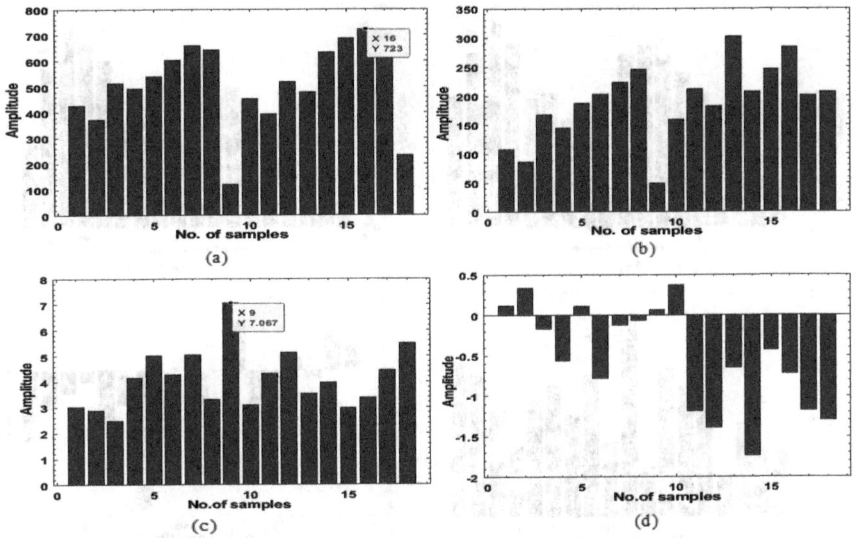

Fig. 5 **a** Maximum amplitude of the vibration signal. **b** Standard deviation of the raw signal. **c** Kurtosis of the vibration signal. **d** Skewness of the vibration signal

shows that initially, and the tool is healthy and as the spindle speed is increased the amplitude of the maximum vibration decreases. The amplitude of the vibration is measured in mm/s^2. For the next pass, rest parameters are constant, and the depth of cut is increased; from this, we can be seen that abrupt increment in the maximum amplitude of the vibration. The healthy tool shows this trend. Further, it can be noticed that after few passes, the amplitude increases continuously, and the tool tends to show the failure criteria. After the transition, the tool fails (blunts) in the 9th pass and 18th and the amplitude of the vibration decreases to the minimum value [14]. A similar trend is observed in the standard deviation of the signal. Kurtosis is the third moment of the statistical distribution. If the data is normally distributed about its mean position, the kurtosis is found to be 3.

As seen in Fig. 5c, for healthy conditions, i.e. for experiment number 1, 10, etc., the kurtosis lies around the value of 3. The deviations in the value are observed with the deterioration in tool health. Figure 5d shows the variation in the skewness of the signals.

4.2 *Trend Analysis of Features Along y-Direction and z-Direction*

Figures 6 and 7 show the trend of the features for the raw signals in *y*-direction and *z*-direction, respectively. It also infers similar trends as shown by the signals captured

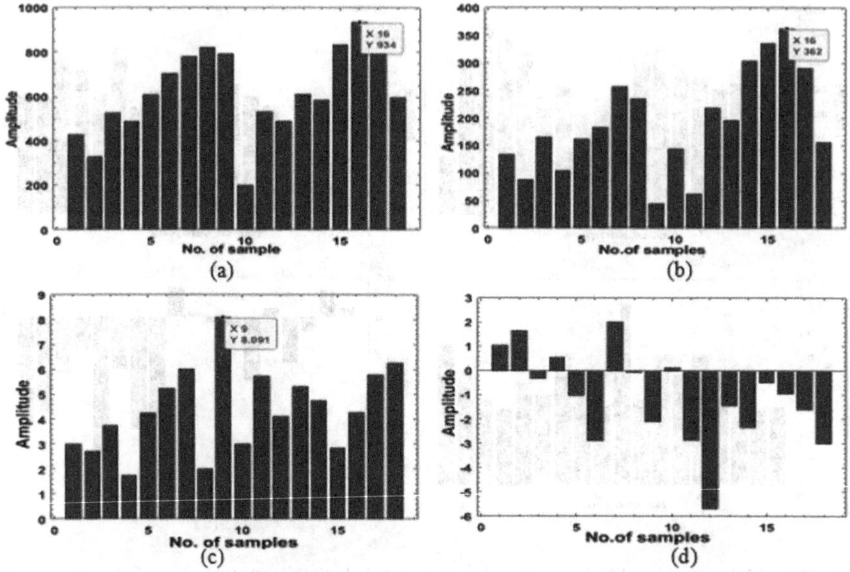

Fig. 6 **a** Maximum amplitude of the vibration signal in *y*-direction. **b** Standard deviation of the raw signal in *y*-direction. **c** Kurtosis of the vibration signal in *y*-direction. **d** Skewness of the vibration signal in *y*-direction

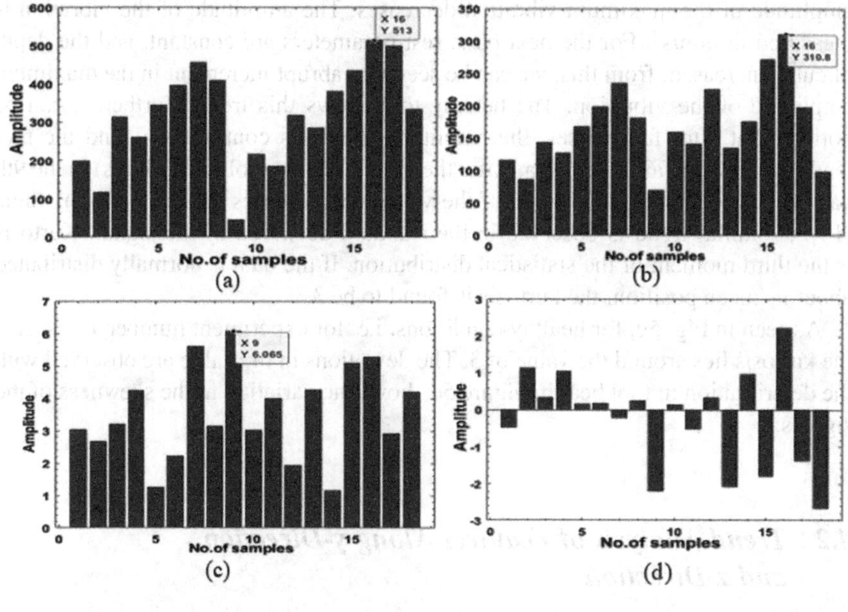

Fig. 7 **a** Maximum amplitude of the vibration signal in *z*-direction. **b** Standard deviation of the raw signal in *z*-direction. **c** Kurtosis of the vibration signal in *z*-direction. **d** Skewness of the vibration signal in *z*-direction

in x-direction. Thus, these features are proved to be a good indicator of the health of the milling tool.

5 Surface Topography

Surface topography shows both the profile shape and surface roughness. It includes the waviness and the surface finish. It comprises the small local deviation of a surface from the ideal flat plane. Surfaces have never an ideal geometrical shape but instead include different deviations, which is characterized by the statistical parameters such as the variance of the height, the slope and the curvature. However, it has been observed that surface topography is a non-stationary random process [17]. It means the variance of the height distribution is related to the sampling length and hence is not unique for a surface; the texture of the surface is an important factor that influences friction, noise reduction, wear resistance, corrosion and lubrication. In this study, experimental procedures have been adopted to obtain surface topography results. Figure 8 shows the setup of LD 130.

Surface topography results have been obtained for the three different samples belonging to all three different categories. The local surface area of 2×2 (mm^2) has been considered for surface topography evaluation. The machine is set up in such a way that it considers the no-measurement zone for 40% of the total length, equally divided at the beginning and the termination. The obtained results are shown in Fig. 9. As seen in Fig. 9a–c, the "R_a" value increases as the tool condition deteriorates.

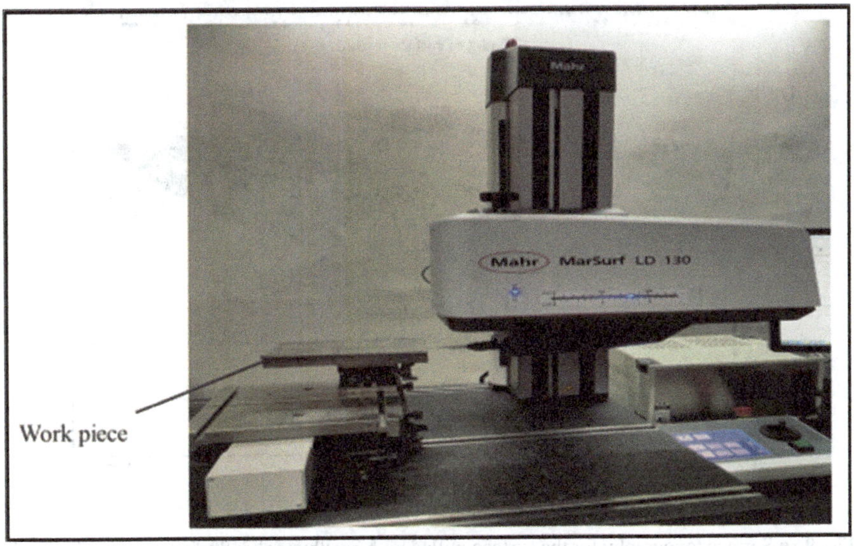

Fig. 8 Experimental setup for surface topography measurement

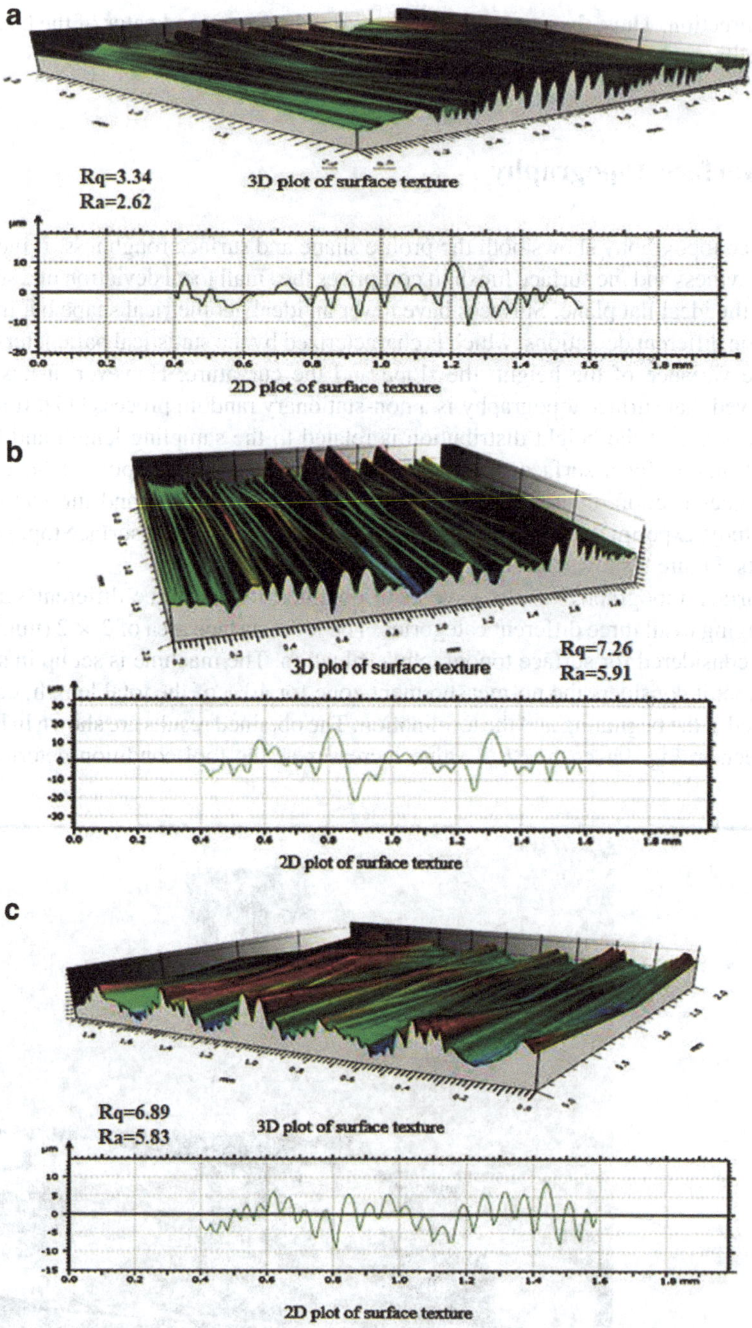

Fig. 9 **a** Surface topography for the surface milled with healthy tool. **b** Surface topography for the surface milled with tool tending to failure. **c** Surface topography for the surface milled with failed tool

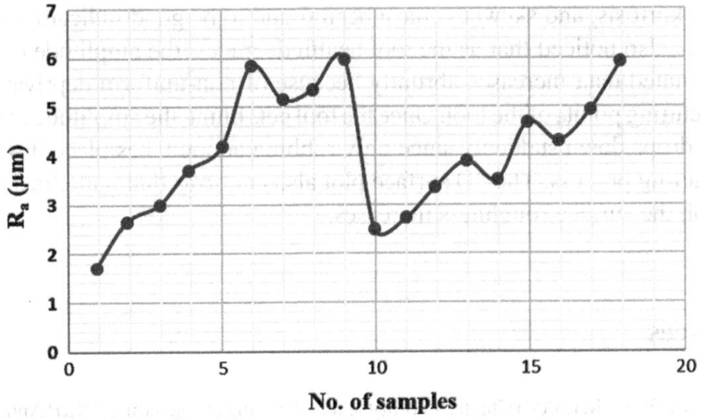

Fig. 10 Variation of average surface roughness

Figure 10 shows the variation in the average surface roughness for all the samples. The value is obtained using a mobile surface measuring instrument "Handy surf".

5.1 Important Parameter for Surface Roughness

1. **Arithmetic mean value (R_a)**: The average surface (R_a) is the arithmetic average of the absolute values of the roughness profile.

$$R_a = \frac{1}{n}\sum_{i=1}^{n} |h_i| \tag{5}$$

2. **Root mean square (R_q)**: The root mean square, RMS is the root mean square average of the profile height deviations from the mean line, recorded within the evaluation length.

$$R_q = \sqrt{\frac{\sum_{i=1}^{n} (h_i)^2}{n}} \tag{5}$$

6 Conclusion

In this study, the performance of the end mill cutter in terms of surface roughness has been observed on the workpiece of SS316 (170 mm × 100 mm × 20 mm). The condition of the tool health and surface roughness by altering parameters like spindle speed and depth of cut has been noted. Statistical features like mean, standard deviation, kurtosis, and skewness have been found to be good indicators of the tool health. It is also noticed that as the tool health degrades, the amplitude of vibration of accumulated data increases abruptly because of non-uniform deprivation in the different cutting points of the tool. Once the tool gets blunt, the amplitude of measured vibration drops down suddenly since only rubbing action takes place instead of the desired cutting process. The 3D surface plot also confirms that with the degradation of the tool, the surface roughness increases.

References

1. Altintas Y, Weck M (2004) Chatter stability of metal cutting and grinding. CIRP Ann 53(2):619–642
2. Shimana K, Kondo E, Chifu T, Nakao M, Nishimura Y (2016) Real-time estimation of machining error caused by vibrations of end mill. Procedia CIRP 46:246–249
3. Prasad BS, Babu MP (2017) Correlation between vibration amplitude and tool wear in turning: numerical and experimental analysis. Eng Sci Technol Int J 20(1):197–211
4. Babu MP, Prasad BS (2016) Experimental investigation to predict tool wear and vibration displacement in turning—a base for tool condition monitoring. J Manuf Sci Prod 16(2):103–114
5. Amin AN, Patwari AU, Sharulhazrin MS, Hafizuddin I (2010) Investigation of effect of chatter amplitude on surface roughness during end milling of medium carbon steel. In: International conference on industrial engineering and operations management, Dhaka, Bangladesh
6. Yue C, Gao H, Liu X, Sy L, Wang L (2019) A review of chatter vibration research in milling. Chin J Aeronaut 32(2):215–242
7. Singh KK, Kartik V, Singh R (2017) Modeling of dynamic instability via segmented cutting coefficients and chatter onset detection in high-speed micromilling of Ti6Al4V. J Manuf Sci Eng 139(5):051005
8. Zhou Y, Xue W (2018) Review of tool condition monitoring methods in milling processes. Int J Adv Manuf Technol 96(5–8):2509–2523
9. Kankar PK, Sharma SC, Harsha SP (2011) Fault diagnosis of ball bearings using machine learning methods. Expert Syst Appl 38(3):1876–1886
10. Sharma A, Amarnath M, Kankar PK (2016) Feature extraction and fault severity classification in ball bearings. J Vib Control 22(1):176–192
11. Honarvar F, Martin HR (1997) New statistical moments for diagnostics of rolling element bearings. J Manuf Sci Eng 119(3):425–432
12. Upadhyay R, Manglick A, Reddy DK, Padhy PK, Kankar PK (2015) Channel optimization and nonlinear feature extraction for Electroencephalogram signals classification. Comput Electr Eng 45:222–234
13. Kankar PK, Sharma SC, Harsha SP (2012) Vibration-based fault diagnosis of a rotor bearing system using artificial neural network and support vector machine. Int J Model Ident Control 15(3):185–198
14. Gangadhar N, Kumar H, Narendranath S, Sugumaran V (2014) Fault diagnosis of single point cutting tool through vibration signal using decision tree algorithm. Procedia Mater Sci 5:1434–1441
15. Madhusudana CK, Budati S, Gangadhar N, Kumar H, Narendranath S (2016) Fault diagnosis studies of face milling cutter using machine learning approach. J Low Freq Noise Vibr Act Control 35(2):128–138
16. Prakash J, Kankar PK (2020) Health prediction of hydraulic cooling circuit using deep neural network with ensemble feature ranking technique. Measurement 151:107225
17. Sahoo P (2011) Surface topography. In: Tribology for engineers. Woodhead Publishing, Cambridge, pp 1–32

PVT Aware Analysis of ISCAS C17 Benchmark Circuit

Suruchi Sharma, Santosh Kumar, Alok Kumar Mishra, D. Vaithiyanathan, and Baljit Kaur

Abstract Due to rapid technology scaling in the nanometer regime accompanied by supply voltage reduction, high leakage currents such as subthreshold leakage, junction leakage, and gate leakage have become prominent sources of power consumption in CMOS VLSI circuits. Consequently, it is imperative to estimate and decrease leakage power at the nanometer regime. However, this unabated scaling makes the CMOS circuits more prone to process (P), voltage (V), and temperature (T) variations at nanometer technologies. This work discusses a comprehensive investigation of different circuit-level leakage power reduction techniques such as power gating (PG), drain gating (DG), LECTOR, and GALEOR and analyzes the impact of PVT variations on leakage power dissipation and delay using ISCAS C17 benchmark circuit.

Keywords ISCAS C17 benchmark circuit · Leakage power reduction techniques · Propagation delay · PVT corner

1 Introduction

Scaling of CMOS devices leads to performance enhancement and higher chip packing density for the last few decades. Contrarily, there is a significant increment in subthreshold leakage current, causing leakage power dissipation [1]. The International Technology Roadmap for Semiconductors (ITRS) has reported that leakage power dissipation as the substantial contributor attributed to total power dissipation in the nanometer regime. Additionally, leakage current in CMOS circuits constitutes the current that is flowing in the circuit during standby and inactive mode [2]. It is anticipated that leakage power will increase to 32 times for each device by 2020 [3]. To further explain the significant components of leakage current, an NMOS (n-type MOSFET) transistor [4] is shown in Fig. 1. It demonstrates six leakage

S. Sharma (✉) · S. Kumar · A. K. Mishra · D. Vaithiyanathan · B. Kaur
Department of Electronics and Communication Engineering, National Institute of Technology Delhi, Delhi 110040, India
e-mail: suruchisharma@nitdelhi.ac.in

© The Author(s), under exclusive license to Springer Nature Singapore Pte Ltd. 2021
R. M. Singari et al. (eds.), *Advances in Manufacturing and Industrial Engineering*, Lecture Notes in Mechanical Engineering,
https://doi.org/10.1007/978-981-15-8542-5_107

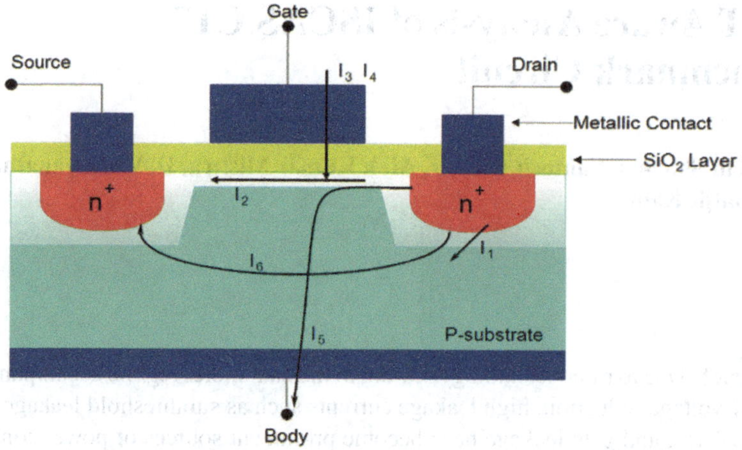

Fig. 1 The cross-sectional view of MOSFET illustrating different leakage current mechanism at the nanometer regime

currents flowing through a single NMOS transistor where I_1 represents reverse-biased leakage current that occurs between drain and source when these are reverse biased to well junctions, I_2 is the subthreshold current which occurs in CMOS transistor amidst source and drain terminals when the voltage at gate terminal is less than the threshold voltage, I_3 is gate-oxide tunneling current which occurs when electron tunneling takes place due to low oxide thickness concomitant toward the high electric field, I_4 is hot-carrier injection current which occurs when due to high electric field across the interface, carriers get injected into the oxide layer, I_5 is gate-induced drain leakage (GIDL) current which occurs between drain and source because of high electric field, and I_6 is punch-through current occurs due to merging of depletion region of both sides.

These leakage currents depend on device's operating conditions; for instance, subthreshold current occurs when the device is in inactive mode, gate-oxide tunneling current and hot-carrier injection current occur when the device is in standby mode, reverse-bias leakage current, GIDL current, and punch-through current occur when the device is in standby or inactive mode. These different leakage current mechanisms adversely affect the performance of the circuit at nanometer technology nodes. Therefore, it is indispensable to estimate and minimize leakage power dissipation at the early stages of circuit design. This work discusses a variety of leakage power reduction techniques at gate level such as PG, DG, LECTOR, and GALEOR [5]. Till now, the leakage power reduction techniques, as mentioned above, are used in academic research only. The adoption of these techniques in industrial benchmark circuits is limited in number because these circuits have not been included in the libraries of computer-aided design (CAD) tools yet.

Furthermore, the scaling at the nanometer regime makes the device more susceptible to physical and environmental parametric variations [6]. Therefore, the effect of these variations on the circuit must be examined by considering the sources of variations being process (P), voltage (V), and temperature (T). These parameters are taken into account by independently varying P, V, and T over their allowable ranges and analyzing the subsequent combinations for best- and worst-case performances. Here, the leakage mentioned above, power reduction techniques are implemented on a combinational benchmark circuit of the ISCAS85 family, i.e., C17 at 45 nm technology node using Cadence Virtuoso. Additionally, PVT analysis is also essential to make this a variation tolerant design.

So, C17 is a combinational benchmark circuit that belongs to ISCAS85 family with five inputs (i.e., *A*, *B*, *C*, *D*, and *E*) and two outputs (i.e., V_{01}, V_{02}) [7]. All benchmark circuits of the ISCAS85 family are used for logic testing. C17 is chosen for evaluation as it is the smallest circuit amongst the ISCAS85 family. This benchmark circuit consists of six NAND gates, as shown in Fig. 2, where N_i refers to NAND gates, and *i* ranges from 1 to 6. The CMOS equivalent of the conventional C17 benchmark circuit consists of twelve NMOS and PMOS transistors, as shown in Fig. 3, where

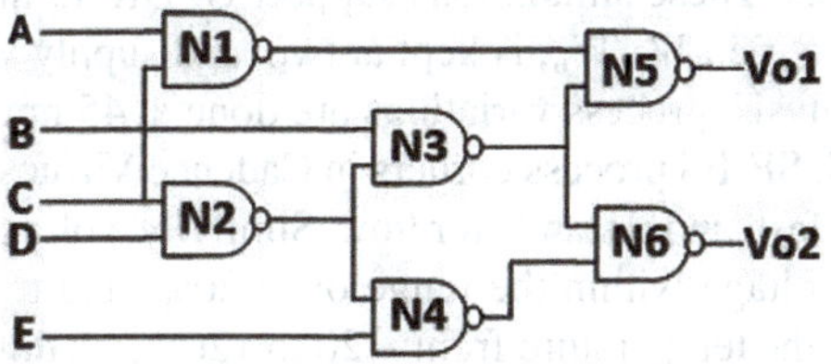

Fig. 2 Gate level circuit of ISCAS C17

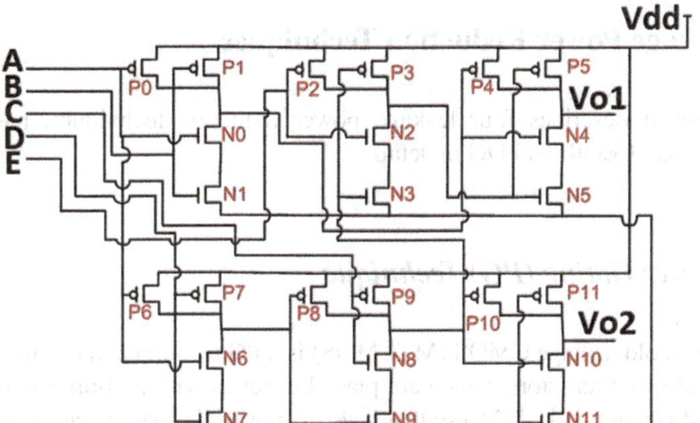

Fig. 3 CMOS equivalent circuit of ISCAS C17

P_i refers to PMOS transistor, N_i refers to the NMOS transistor, and i ranges from 0 to 11.

The research paper is arranged as follows. Section 1 has introduced the need for leakage power reduction and its PVT analysis in CMOS VLSI circuits. Additionally, this section also provides a brief introduction of the C17 benchmark circuit and its selection for this paper. In Sect. 2, we describe our simulation setup. Section 3 discusses the various leakage power reduction techniques that have been implemented using C17.

Furthermore, Sect. 4 discusses the need for PVT analysis for this circuit. Sect. 5 presents the simulation results of several leakage power reduction techniques applied on the C17 circuit at 45 nm technology node and its PVT analysis. Lastly, Sect. 6 concludes this paper and presents its future perspectives.

2 Simulation Setup

Cadence Virtuoso simulations have been at a 45 nm technology node to implement C17 for PVT analysis. These simulations support GPDK 45 nm model library file. Here, the aspect ratio, i.e., W_p/W_n, is kept at two, and supply voltage (V_{DD}) is kept at 1 V. For PVT analysis, process variations are done at 45 nm technology node by selecting TT, FF, SS, SF, FS process corners in Cadence Virtuoso. Here, T stands for typical, F stands for fast, and S stands for flow. Similarly, voltage variations are done by varying supply voltage within the range of $\pm$ 20%, and temperature variations are done by altering the temperature from −20 to 120 °C as this range is considered as the commercial temperature range.

3 Leakage Power Reduction Techniques

This segment describes four leakage power reduction techniques, namely PG, LECTOR, GALEOR, and DG in detail.

3.1 Power Gating (PG) Technique

Multi-threshold-voltage CMOS (MTCMOS) is a PG technique that consists of very high V_{th} sleepy transistors, which are placed in series with pull-up network (PUN) and pull-down network (PDN) so that leakage power dissipation can be minimized. Figure 4 shows the diagram of the PG technique for leakage power reduction. This technique uses two sleep transistors of very high threshold voltage, which have been placed for leakage power reduction [8]. These sleepy transistors are managed by sleep signals, as shown in Fig. 4. For the operation in active mode, the sleep signal

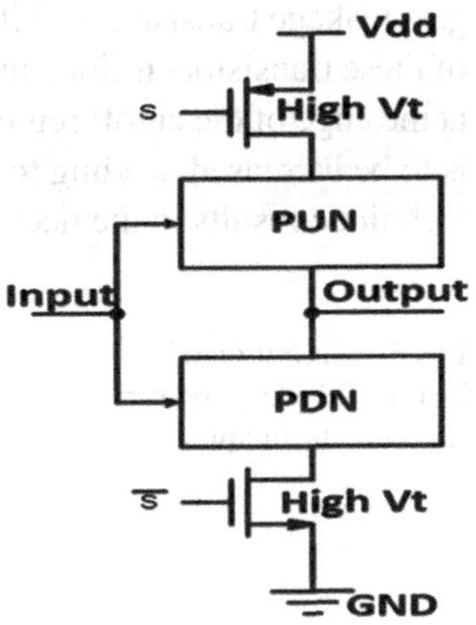

Fig. 4 Illustration of the power gating technique for leakage power reduction

(i.e., *S*) is set logic '0', and $\overline{S}$ is set logic '1' so that both the sleep transistors are in their on-state and circuit executes according to all logic inputs provided. Although in sleepy or standby mode, the *S* is set logic '1', and $\overline{S}$ is set logic '0'. In this case, both the sleep transistors are operating in their off-state so that no current flows from supply to ground. Consequently, data loss occurs in this technique. Therefore, it can be deduced that this technique renders better leakage power reduction if data confinement is not required in standby mode [9].

3.2 LECTOR Technique

This technique involves the insertion of two leakage control transistors (LCT_1 and LCT_2) amongst PUN and PDN, as illustrated in Fig. 5. Here, LCT_1 is a PMOS transistor, and LCT_2 is an NMOS transistor. The wired configuration of both these leakage transistors is done to ensure that each one of the LCT is always at the edge of the cutoff region unaffected of any applied input combination. Therefore, path resistance gets incremented from supply to the ground, leading to a considerable reduction in leakage power. This technique is appropriate for both standbys as well as inactive mode. However, the outcome of this technique leads to the degradation of signal quality when the scaling of technology is done [10].

3.3 GALEOR Technique

In this technique, two GLTs transistors (i.e., GLT1 and GLT2) are placed between the PUN and PDN of the CMOS circuit, as illustrated in Fig. 6. Here, GLT1 is an NMOS transistor which is arranged between PUN and output node. Moreover, GLT2 is a PMOS transistor that is located between the output node and the PDN. These gate leakage transistors (GLTs) have high threshold voltage, and wired configuration of these transistors is done in such a way so that one of the GLT transistors is always at the edge of the cutoff region. Therefore, path resistance from supply to the ground is to be increased, leading to a substantial reduction in leakage power. However, this technique results in the decrement of output [11].

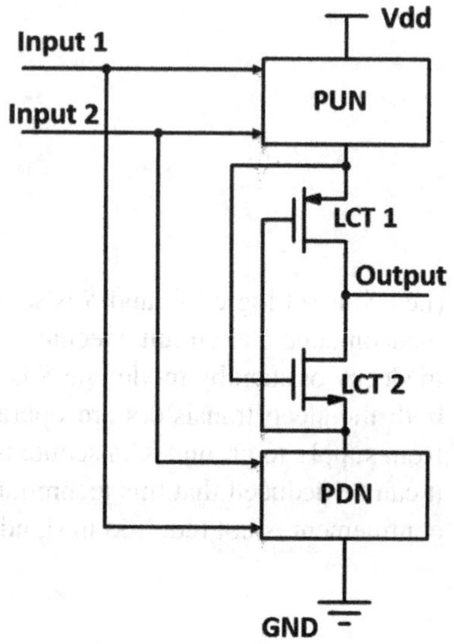

Fig. 5 Illustration of LECTOR leakage power reduction technique

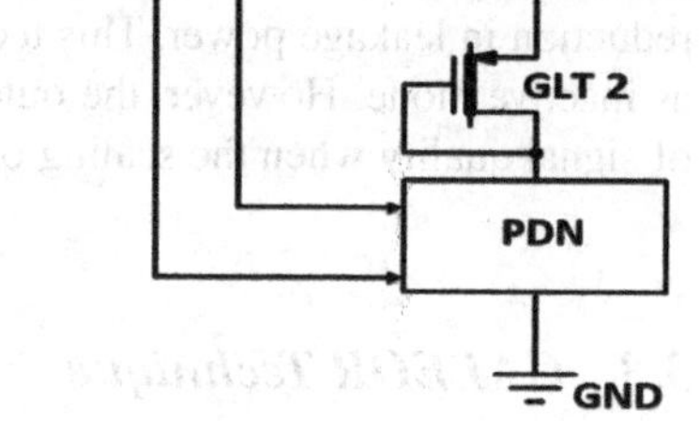

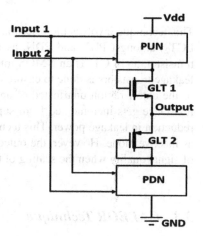

Fig. 6 Illustration of GALEOR leakage power reduction technique

3.4 Drain Gating (DG) Technique

This technique involves the insertion of two sleep transistors between PUN and PDN, as shown in Fig. 7. For the active mode operation of the technique, the sleep signal (i.e., *S*) is kept at logic '0', and $\overline{S}$ is kept at logic '1' so that both the sleepy transistors are kept in the on-state, and the circuit works typically according to the given logic input. However, in standby mode, *S* is kept at logic '1', and $\overline{S}$ is kept at logic '0' so that both the transistors remain in the off-state and supply to the ground path is broken. This technique reduces gate leakage current and subthreshold leakage current but requires two power supplies for its operation. In this technique, the propagation delay between input and output can be managed along with leakage power reduction [12].

4 PVT Analysis

PVT analysis is done by independently varying P, V, and T over their allowable ranges so that best- and worst-case performances can be evaluated. The main objective of PVT analysis is to make circuit variability intolerant so that circuit performance remains unaffected when variations in terms of PVT occur in the structure. PVT variations lead to a mismatch in characteristics from nominal, producing yield loss [13].

Primarily, process variations occur because of process imperfections such as line edge roughness (LER), lithography and well proximity effects (LPE and WPE), random dopant, and oxide thickness fluctuations (RDF and OTF). LPE and WPE are

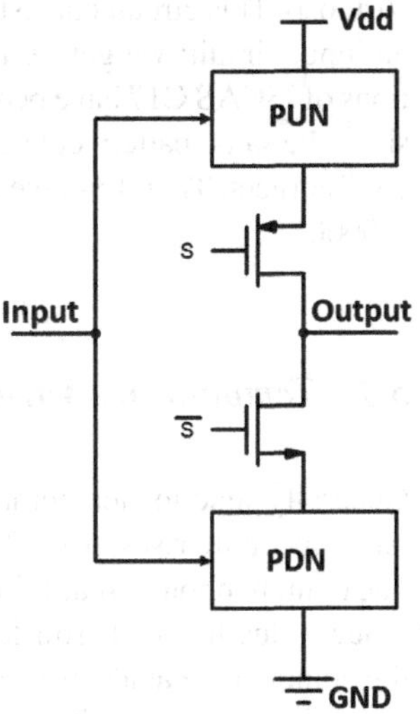

Fig. 7 Illustration of drain gating leakage power reduction technique

mainly due to photolithography limitations and occur as a result of the correlation between adjacent structures. RDF is induced by a mismatch in the number of dopants carriers across the channel. LER basically accounts for deviations in the shape of the gate and a perfect level edge, and this parameter has not scaled down with technology and thus has become a significant source of variation at nanometer technology nodes. Furthermore, OTF accounts for atom-level interface roughness between the gate dielectric and silicon. Process corners illustrate the worst-case combination of processing and environment for the delay, power consumption, and functionality [14].

Voltage variations are dynamic environmental variations and are caused by the voltage drop across from the parasitic resistance of an IC. The duration of these voltage fluctuations ranges from nanoseconds to microseconds. However, supply voltage returns to its supposed value just before a new voltage drop occurs. With continuous scaling of CMOS technology and supply voltage is followed by threshold voltage scaling, resulting in degradation of circuit performance and increment of leakage current. Therefore, an optimized value of supply voltage is a requisite [15].

Temperature variations are also dynamic environmental variations and are caused by the power dissipated by transistors and ambient temperature fluctuations. A circuit has regions of high activity, generally called hotspots, and these hotspots dissipate more power contributing to temperature variations [16].

5 Results and Discussions

In this section, a comparative analysis is performed to estimate leakage power dissipation and propagation delay for different leakage power reduction techniques while considering PVT variations for ISCAS C17 combinational circuit using Cadence Virtuoso. This circuit consists of five inputs resulting in 32 input combinations as for an input circuit; we get $2n$ input combinations. Initially, all the 32 input combinations of ISCAS C17 have been analyzed. It is observed that all 32 input combinations show the same pattern concerning leakage power and propagation delay concerning PVT corners. Therefore, we have shown results for only one input combination, i.e., '00000'.

5.1 Temperature Variations

Generally, junction temperature gets raised to 85 °C during regular circuit operation. At temperature rises, two effects dominate the drain current, i.e., (i) mobility lessens due to more phonon scattering, resulting in delay enhancement (ii) threshold voltage reduces due to low Fermi-level, increasing leakage current. Figures 8 and 9 show the comparative analysis of various leakage power reduction techniques concerning the conventional circuit concerning leakage power and delay in terms of variations

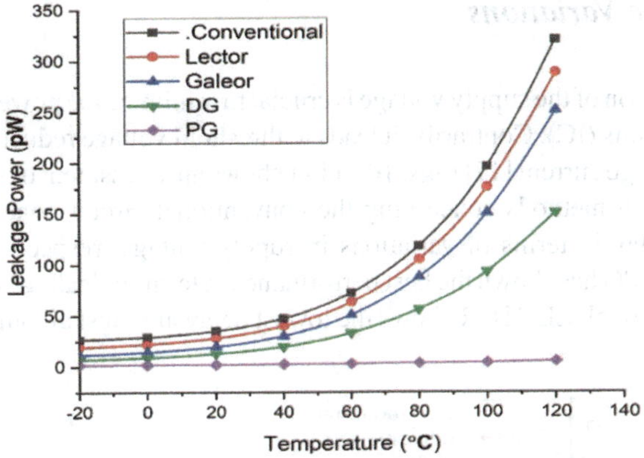

Fig. 8 Sensitivity of leakage power to temperature variations for conventional as well as different leakage power reduction techniques

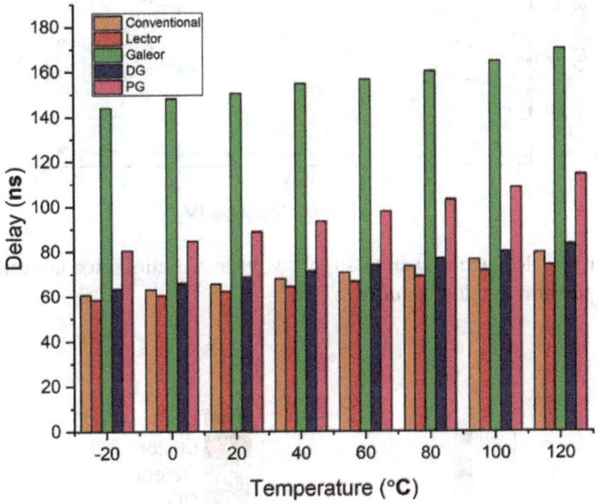

Fig. 9 Sensitivity of delay to temperature variations for conventional as well as a different technique

in temperature, respectively. It can be recognized in Fig. 8 that PG shows the best performance by drawing the lowest leakage power at raised temperatures amongst all others. However, LECTOR shows the lowest delay amongst all techniques as transistors are always working at the edge of the cutoff region, resulting in delay reduction (refer to Fig. 9) due to a smaller number of hotspots formation.

5.2 Voltage Variations

The minimization of the supply voltage is crucial to minimize the power dissipation of integrated circuits (IC). Contrarily, it leads to threshold voltage reduction resulting in increased leakage current [17] Figs. 10 and 11 shows an assessment of various leakage power reduction methods concerning the conventional circuit concerning leakage power and delay in terms of variations in supply voltage, respectively. Figure 10 represents that PG has shown the best performance in terms of leakage power amongst all others. Although LECTOR shows the lowest delay amongst all other, as shown in

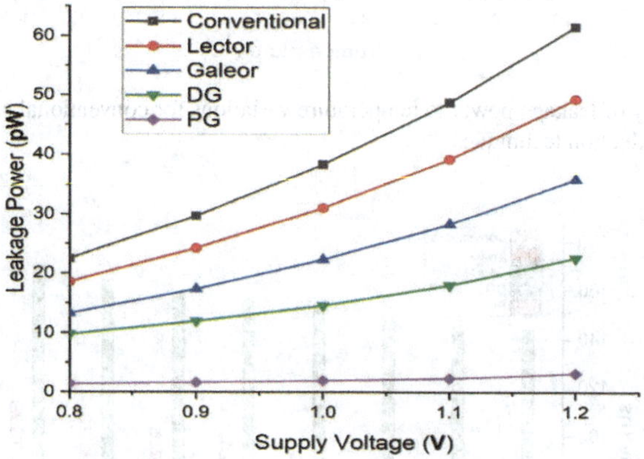

Fig. 10 Sensitivity of leakage power to supply voltage variations for conventional as well as different leakage power reduction techniques

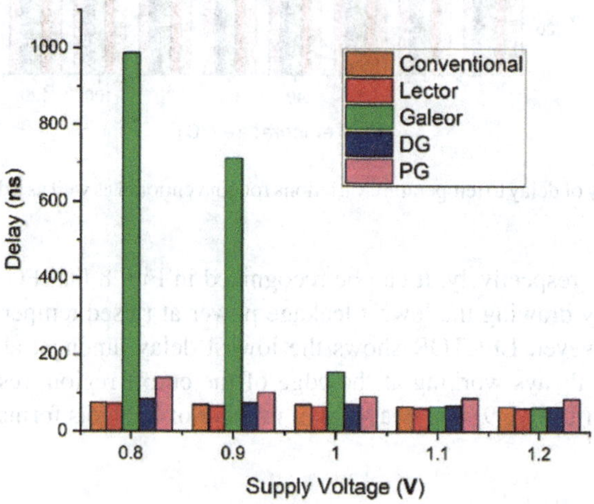

Fig. 11 Sensitivity of delay to supply voltage variations for conventional as well as different leakage power reduction techniques

Fig. 11, increment in a delay regarding PG can be traded off by decrement in leakage power.

5.3 *Process Corner Variations*

The effect of environmental and process variations on transistors can be categorized as typical (also called nominal), fast, or slow. Process corners are defined when processing variations are combined with environmental variations. The two complementary types of transistors with independent characteristics are present in CMOS technology, so the speed of each transistor can be distinguished [18]. A conventional two-letter designation has been used to describe process corners, where the first letter represents an NMOS device, and the second letter represents a PMOS device. Hence, five classic corners are defined as typical–typical (TT), fast–fast (FF), slow–slow (SS), fast–slow (FS), and slow–fast (SF). The distribution of process variations is provided by the manufacturing company. The first three corners (i.e., TT, FF, and SS) are recognized as smooth corners because, at these corners, both devices have been affected evenly. The last two corners (i.e., FS and SF) are known as 'skewed' corners because at these corners, one device will switch much faster than the other, resulting in imbalanced switching of the devices. Hence, it causes one edge of the output to have much less skew than the other [19].

Figures 12 and 13 show the comparative analysis of conventional as well as different leakage power reduction techniques concerning leakage power and delay, respectively, in terms of variations in process parameters. It can be observed from Fig. 12 that PG has shown the best performance in terms of leakage power amongst all others. Figure 13 shows that the best performance can be obtained from the SS

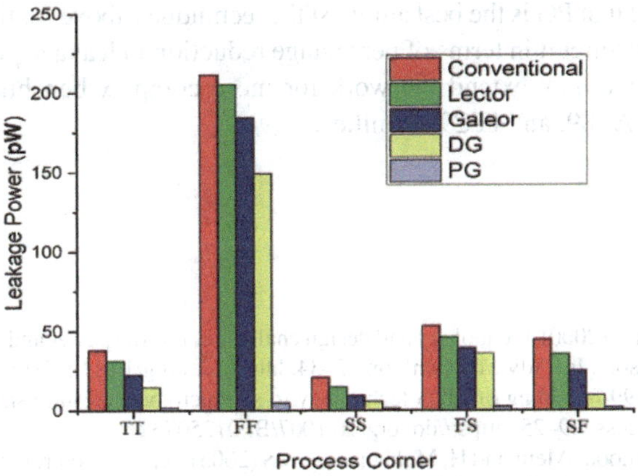

Fig. 12 Sensitivity of leakage power to process corner variations for conventional as well as different leakage power reduction techniques

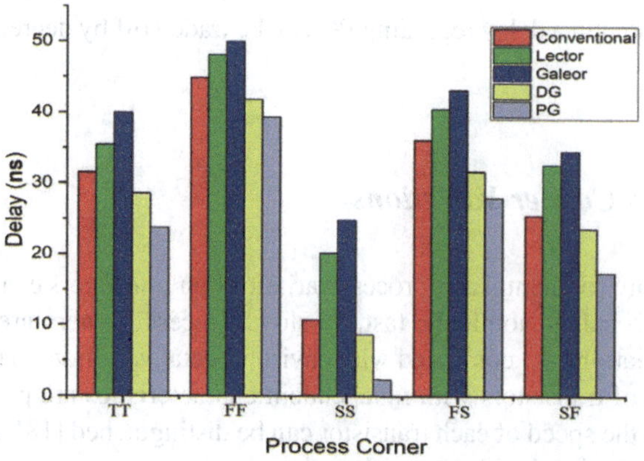

Fig. 13 Sensitivity of delay to process corner variations for conventional as well as different techniques

process corner, viz. smallest delay and worst-case performance is provided by the FF process corner, viz. highest delay.

6 Conclusion and Future Scope

In this article, we have analyzed different leakage power reduction techniques for CMOS VLSI circuits at the circuit level using Cadence Virtuoso 45 nm technology node. It is observed that PG provides the highest percentage decrease in the leakage power and propagation delay concerning variations in temperature, supply voltage, and process corners amongst all leakage power reduction techniques. Therefore, it can be deduced that PG is the best amongst the techniques above for the conventional C17 benchmark circuit in terms of percentage reduction in leakage power and delay. In the future, we may extend the work for more complex benchmark circuits of ISCAS85, ISCAS89, and ITC99 families.

References

1. De V, Borkar S (2000) Technology and design challenges for low power and high-performance microprocessors. In: Adv Met Conf, pp 25–34. https://doi.org/10.1145/313817.313908
2. Mead CA (1994) Scaling of MOS technology to submicrometer feature sizes. Analog Integr Circ Sig Process 6:9–25. https://doi.org/10.1007/BF01250732
3. Roy K, Mahmoodi-Meimand H, Mukhopadhyay S (2003) Leakage control for deep-submicron circuits. In: Lopez JF, Montiel-Nelson JA, Pavlidis D (eds) VLSI circuits and systems. SPIE, p 135

4. Roy K, Mukhopadhyay S, Mahmoodi-Meimand H (2003) Leakage current mechanisms and leakage reduction techniques in deep-submicrometer CMOS circuits. Proc IEEE 91:305–327. https://doi.org/10.1109/JPROC.2002.808156
5. Bendre V, Kureshi AK (2015) An overview of various leakage power reduction techniques in deep submicron technologies. In: 2015 International conference on computing communication control and automation. IEEE, pp 992–998
6. Kim KK, Bin KY, Choi M, Park N (2007) Leakage minimization technique for nanoscale CMOS VLSI. IEEE Des Test Comput 24:322–330. https://doi.org/10.1109/MDT.2007.111
7. Bryan D (1985) The ISCAS'85 benchmark circuits and netlist format. North Carolina State University, pp 695–698
8. Kao JT, Chandrakasan AP (2000) Dual-threshold voltage techniques for low-power digital circuits. IEEE J Solid-State Circ 35:1009–1018. https://doi.org/10.1109/4.848210
9. Mutoh S, Aoki T, Matsuya Y et al (1995) 1-V power supply high-speed digital circuit technology with multithreshold-voltage CMOS. IEEE J Solid-State Circ 30:847–854. https://doi.org/10.1109/4.400426
10. Hanchate N, Ranganathan N (2004) LECTOR: a technique for leakage reduction in CMOS circuits. IEEE Trans Very Large Scale Integr Syst 12:196–205. https://doi.org/10.1109/TVLSI.2003.821547
11. Katrue S, Kudithipudi D (2008) GALEOR: leakage reduction for CMOS circuits. In: Proceedings of the 15th IEEE international conference on electronics, circuits and systems, ICECS 2008, pp 574–577. https://doi.org/10.1109/ICECS.2008.4674918
12. Park JC, Mooney VJ, Pfeiffenberger P (2004) Sleepy stack reduction of leakage power. Lecture notes in computer science (including subseries Lecture notes in artificial intelligence and Lecture notes in bioinformatics), vol 3254, pp 148–158. https://doi.org/10.1007/978-3-540-30205-6_17
13. Shah AP, Neema V, Daulatabad S (2016) Effect of process, voltage, and temperature (PVT) variations in LECTOR-B (leakage reduction technique) at 70 nm technology node. In: International conference on computing, communication and security, ICCCS 2015. https://doi.org/10.1109/CCCS.2015.7374173
14. Yang S, Khursheed S, Al-Hashimi BM et al (2013) Improved state integrity of flip-flops for voltage scaled retention under PVT variation. IEEE Trans Circ Syst I Regul Pap 60:2953–2961. https://doi.org/10.1109/TCSI.2013.2252640
15. McConaghy T, Breen K, Dyck J, Gupta A (2013) Variation-aware design of custom integrated circuits: a hands-on field guide. In: Variation-aware design of custom integrated circuits: a hands-on field guide. Springer Science+Business Media, New York
16. Windels J, Van Praet C, De Pauw H, Doutreloigne J (2009) Comparative study on the effects of PVT variations between a novel all-MOS current reference and alternative CMOS solutions. In: Midwest symposium circuits and systems, pp 49–53. https://doi.org/10.1109/MWSCAS.2009.5236154
17. Haghdad K, Anis M (2008) Design-specific optimization considering supply and threshold voltage variations. IEEE Trans Comput Des Integr Circ Syst 27:1891–1901. https://doi.org/10.1109/TCAD.2008.2003288
18. Weste NEH, Harris DM (2011) CMOS VLSI design: a circuits and systems perspective, 4th edn. Pearson, Cambridge
19. Alioto M, Consoli E, Palumbo G (2015) Variations in nanometer CMOS flip-flops: Part I—Impact of process variations on timing. IEEE Trans Circ Syst I Regul Pap 62:2035–2043. https://doi.org/10.1109/TCSI.2014.2366811

Correction To: Advances in Manufacturing and Industrial Engineering

Ranganath M. Singari, Kaliyan Mathiyazhagan, and Harish Kumar

Correction to:
R. M. Singari et al. (eds.), *Advances in Manufacturing and Industrial Engineering*, Lecture Notes in Mechanical Engineering,
https://doi.org/10.1007/978-981-15-8542-5

In the original version of chapters "Designing of Fractional Order Controller Using SQP Algorithm for Industrial Scale Polymerization Reactor" and "Design of Delay Compensator for a Selected Process Model", the following belated correction has been incorporated:

The author name "Vivek Joshi" has been changed to "V. Joshi".

The chapters and book have been updated with this change.

The updated version of these chapters can be found at
https://doi.org/10.1007/978-981-15-8542-5_38
https://doi.org/10.1007/978-981-15-8542-5_83

© The Author(s), under exclusive license to Springer Nature Singapore Pte Ltd. 2021

R. M. Singari et al. (eds.), *Advances in Manufacturing and Industrial Engineering*, Lecture Notes in Mechanical Engineering,
https://doi.org/10.1007/978-981-15-8542-5_108

9789811585418VOL02